M Greenwood
The Rockefeller University
N. Y City NY
360 - 1319
area code (212)

Cyclic AMP

Cyclic AMP

G. ALAN ROBISON
Department of Pharmacology
Vanderbilt University
School of Medicine
Nashville, Tennessee

REGINALD W. BUTCHER
Department of Biochemistry
University of Massachusetts
Medical School
Worcester, Massachusetts

EARL W. SUTHERLAND
Department of Physiology,
Vanderbilt University
School of Medicine
Nashville, Tennessee

With contributions by
Th. Posternak and Joel G. Hardman

ACADEMIC PRESS
New York and London **1971**

ACADEMIC PRESS, INC.
111 Fifth Avenue, New York, New York 10003

United Kingdom Edition published by
ACADEMIC PRESS, INC. (LONDON) LTD.
Berkeley Square House, London W1X 6BA

LIBRARY OF CONGRESS CATALOG CARD NUMBER: 70-137598

PRINTED IN THE UNITED STATES OF AMERICA

For Jill, Joan, and Claudia

Contents

Preface

This was a difficult book to write because the subject refused to sit still. It was not simply that it behaved like a naughty child at the photographer's, for that would be expected in any viable field of scientific research. Rather, it has seemed to us more like an imaginary child who, in the course of having his picture taken, suddenly grew to adult proportions and then left the studio badly in need of a shave.

One of the purposes we had in mind when we started writing was to stimulate interest in the subject, especially among comparative and clinical endocrinologists, but this no longer seems necessary. The justification now for a monograph on cyclic AMP is that a large amount of information has accumulated about the compound, and it would seem useful to have this information summarized all in one place. We have attempted to provide such a summary, hoping that it will be helpful to those who would like to become more familiar with the subject. We have tried to keep in mind the needs and interests of graduate students and investigators in most of the biological and biomedical sciences, and hope the book will also be of interest and value to clinical investigators. Insofar as it will be used in graduate teaching programs, we have assumed that most students will have had basic courses in modern biochemistry and physiology.

Although the subject grew more rapidly than we had anticipated, we did realize, at a fairly early stage, the need for help. Accordingly, Chapter 3, on the chemistry of cyclic nucleosides, was written by Theo Posternak, and Chapter 11, summarizing current information about cyclic GMP (the only 3′,5′-mononucleotide other than cyclic AMP known to occur in nature), was written by Joel Hardman. The only other noncollaborative chapter was the first one (by E. W. S.), representing a retrospective account of how cyclic AMP came to be discovered.

Of course we are indebted to numerous other colleagues, at Vanderbilt and elsewhere, who read various sections of the manuscript at different stages of development and made many valuable suggestions. We are also grateful to the many investigators who were thoughtful enough to send us manuscripts of their papers in advance of publication and to those who gave us permission to use some of their figures to help illustrate the text.

We should also like to express our gratitude to Dr. Eugene Garfield and his colleagues at the Institute for Scientific Information. Without the help of the modern information services provided by this organization, a monograph of this type, attempting complete coverage of a rapidly expanding field of research, would be almost impossible.

Of course we have no illusions that our goal of complete coverage has been attained. By adding addenda to most of the earlier chapters at the galley proof stage, we have attempted to make the book as up-to-date as possible. It is an absolute certainty, however, that numerous important discoveries concerning cyclic AMP have yet to be made, not to mention reported in the open literature. It is even possible that the material summarized in this monograph will ultimately come to be seen as only a small fraction of the complete story. Our hope is that the reader will at least be prepared for these future advances and may even anticipate some of them. We also hope that by setting down what has been learned in the past, this will in and of itself contribute to future progress.

G. Alan Robison
Reginald W. Butcher
Earl W. Sutherland

March, 1971

CHAPTER 1

An Introduction*

A heat-stable factor mediating the action of epinephrine and glucagon on the activation of liver phosphorylase was found in 1956. This heat-stable factor, cyclic AMP, appears throughout the animal kingdom and is present in microorganisms. The biological role of cyclic AMP has been extended from a relatively simple role in liver glycogenolysis to the point where it appears to be a key regulatory agent in most mammalian tissues. The purpose of this introductory chapter is to review the developments which led to the discovery of the heat-stable factor, with a perspective gained from these years of continuing investigations with cyclic nucleotides.

As a student I was intrigued and puzzled by the actions of hormones. I'm still fascinated by the complex manifestations which occur when a small amount of one of these chemicals is omitted or injected. And I wonder how many years will pass before we understand rather completely the actions of epinephrine, insulin, and cortisol, to name just three of the number of hormones known to exist and function. Our progress is limited in part by our general lack of knowledge of the fundamental biochemistry and biophysics of various processes such as contractility, secretion, rhythmicity, and transport or permeability phenomena. The

*By Earl W. Sutherland.

1

hormones appear to regulate existing functions—to accelerate or inhibit intrinsic activities—and often we know little of the basic physiology or biochemistry at the molecular level.

My earliest publications beginning in 1942 dealt with enzymes. Those early years in the Cori laboratory were interesting ones with opportunity for contact with a number of outstanding investigators. In addition to Dr. Carl Cori, I had the opportunity to collaborate with Sidney Colowick, Christian de Duve, Theo Posternak, Mildred Cohn, and others. While enzymology was my only productive activity at this time, I frequently added various hormones to various enzymes, usually with negative results. Occasionally I would note some activity of insulin or a glucocorticoid but when pursued in greater depth these effects on soluble enzymes invariably turned out to be nonspecific. In any event, some time was spent during this period exploring ways to study the action of one or more hormones. I concluded that it was difficult to produce or evaluate the action of hormones in broken cell preparations, and at a minimum such studies should be correlated with studies on intact cells. In 1955, in Brussels, I summarized some of my ideas in the introductory section of a lecture entitled "Hormonal Regulatory Mechanisms." This portion of the lecture is reproduced below.

> Although hormonal effects and actions have been studied for more than half a century, the precise mechanism of action of any hormone has not yet been clearly established. There have been numerous reports describing the action of certain hormones on enzymes, cell extracts or on cellular particles, but in no case has the described effect been generally accepted as a complete explanation for the mechanism of action of that hormone. In some cases the effect described has not been clearly localized as to exact site of action, while in other cases it has not been clear that the observed hormonal effect was necessarily related to the physiological effects noted in intact animals. These remarks are not intended to reflect a feeling of pessimism, but to indicate the general difficulties which have been encountered in the analysis of hormone action. Two outstanding difficulties which will be considered briefly are (a) establishment of reproducible hormonal effects in a definite soluble system, and (b) extrapolation of *in vitro* effects to the physiological action of the hormone in the intact animal.
>
> If we examine the first problem more carefully, we are tempted to feel that some hormones act at a level of organization more complex than that existing in a soluble homogeneous system. At present it seems that hormones act on specialized colonies of cells and the action is generally lost if the cell structure is disrupted. Basic enzymatic reactions proceed in mammals without the presence of the hormones and the mammalian hormones do not influence the metabolism of unicellular organisms. Moreover, the hormones act in tremendous dilutions and very small amounts suffice to produce marked changes in intact animals. The possibility exists that some of the hormones exert their action on highly organized groups of molecules.
>
> On the other hand, there is reason to feel that the hormones may act at the molec-

ular level. For example, recalling the elucidation of the mechanism of the bio-synthesis of urea, we find a period when urea synthesis could be studied only in the whole animal; later its formation was studied in perfused livers, then in liver slices, and was finally obtained in extracts of cells. By analogy, it seems that a systematic analysis of hormone action might also proceed from studies of the intact animal to isolated tissues and finally to soluble systems. Regarding the inability of mammalian hormones to act on unicellular organisms, we now have numerous examples where seemingly identical metabolic reactions occurring in vertebrate and invertebrate cells actually differ in fine or gross details, e.g., in nature of the enzyme catalyst, in cofactor requirement, or even in the actual chemical reaction. Moreover, in some cases the amounts of hormone necessary for action can be calculated to be in the range where interaction with a single enzyme or other cell constituent would be possible. In such calculations consideration must be given to the amounts of hormone required for maximal or half-maximal effect, possibility of fixation of hormones and the possibility that when steady states exist, very small changes in an enzyme or other cell component may result in drastic overall changes. This point will not be pursued here except to mention Clark's outstanding contributions in this area (Clark, 1937), and to add that if we compare his calculations for the molecules of neurohormones necessary for half-maximal activity, to the number of molecules of an enzyme per unit weight of tissue, the figures are of the same order of magnitude. It seems likely that a useful working hypothesis will include the possibility of hormonal action at a (soluble) molecular level.

The second major problem is that of correlating results of *in vitro* experiments with the classically described and reproducible actions of the hormone *in vivo*. In certain cases interesting effects of hormones on broken cell preparations have been reported; however, it often is not apparent what relation these observations have to the known action of the hormone *in vivo*. Therefore, two large gaps exist in the analysis of hormonal action (gaps which experiments to date have not closed), i.e., what specific metabolic events are some hormones influencing in the intact animal and what is the physiological significance of the observations made with broken cell preparations.

Despite the difficulties which have been encountered in analysis of hormonal action, considerable progress has been made toward understanding how these agents may regulate metabolic processes. We have now come to realize that almost all known metabolic reactions are catalyzed by enzymes, and we know that these enzymatic reactions are influenced not only by the enzymes but by various cofactors which may accelerate the reactions, and by other factors which may either accelerate or depress the rate of the reactions. The biochemical analysis of metabolic reactions has proceeded rapidly in recent years and may now have reached a stage where in selected cases a more exact analysis of hormonal action may be anticipated. For example, recent reports by Villee *et al.* (1955) of a stimulatory effect of estradiol on a soluble DPN-linked isocitric dehydrogenase are of great interest.

At this time it is intended to discuss some approaches and methods which have contributed to our knowledge of hormonal action with emphasis upon approaches which may lead to understanding of the intimate nature of hormonal action. Certain reports dealing with hormonal effects on enzymes or on broken cell preparations will not be considered because the physiological significance of the results is not clear. Future analysis of these effects and correlation of these findings with *in vivo* findings may be very fruitful, however.

PREPARATIONS AND APPROACHES FREQUENTLY EMPLOYED IN ANALYSIS OF HORMONE ACTION

In Vivo **Preparations**

Much of the knowledge that we possess regarding hormonal action has been obtained from studies of relatively intact animals and, in some cases, the knowledge is essentially limited to such studies.

The limitations of studies with intact animals have been evident and it seems unlikely that definitive analysis of hormone action will be gained from these studies alone. The difficulties in analysis arise in part because the hormonal control mechanisms are highly integrated, because compensatory mechanisms of various types are brought into action when hormonal imbalance is present or created, and because the complex reactions within a given organ may be influenced by the metabolism or metabolites of other organs. These three difficulties may not allow a decision as to whether a given hormonal effect is a primary effect or a secondary effect.

Nevertheless, it is a logical necessity to have results observed in the intact animal correlated with *in vitro* results in order to establish the physiological significance of *in vitro* findings.

In Vitro **Experiments**

Intact Cell Preparations

Perfused Organs. Perfused organ preparations are intermediate in nature between intact animal preparations and other *in vitro* preparations. In some cases it is possible to maintain the organ cells in relatively well preserved conditions for some time, and at the same time the organ cells may be separated from the influence of other organs. Perfused organs have been used to study the action of several hormones with varying degrees of success. Although technically some perfusions are not simple, such preparations may be of considerable value in selected cases, and perhaps organ perfusion should be used more widely. If slicing or more drastic treatment of cells results in a loss of response to the cells to the hormone, and if a latent period before hormonal action is anticipated, use of perfused organ may yield valuable information which could not be obtained readily with slices or homogenates.

Slices. Tissue slices have been widely used and are valuable preparations for the study of relatively intact cells. A number of slices can be obtained from a single organ, thus increasing the number of experiments which can be carried out with a single animal and consequently the effect of individual animal variation can be minimized. Tissue slice techniques have been reviewed recently and will not be discussed in detail here (Elliot, 1955). The slice technique has been applied especially to liver, brain, and kidney, but slices from numerous other organs have been studied.

Others. Preparations containing relatively intact muscle cells have posed special difficulties. Cells present in muscle slices are rather extensively damaged especially when skeletal muscle has been used. The rat diaphragm has been used widely and some factors in the use of diaphragms have been reviewed by Krahl (1951) and by others. Fiber bundle preparations of skeletal muscle have been obtained (Richardson *et al.,* 1930) and small single muscle preparations have been used, e.g., a single frog muscle. Occasionally muscle minces have been used (Krebs and Eggleston, 1938).

Broken Cell Preparations

In vitro experiments utilizing intact cells are simpler in many respects than experiments involving intact animals. Nevertheless, the activities of the intact cell are numerous and complex; therefore, the natural tendency has been to break the cells and search for hormonal effects in broken cell preparations. Permeability problems are less obvious in these preparations although the desirability of removing cell membrane permeability relations remains to be proved. Even though cell membranes may be broken many preparations contain particles where membrane phenomena are operative and in these cases it is apparent that complex relations exist even though some simplification of preparation has been achieved. Preparation of homogenates, particulate fractions from homogenates, and extraction of enzymes from tissue have been reviewed (Potter, 1955; Hogeboom, 1955; Morton, 1955). The potential value of studies with broken cell preparations is obvious; precise analysis of hormonal action in such systems should be simpler and more conclusive. Extrapolation of hormone effects in these preparations to the physiological action of the hormone in the intact animals must be done with extreme care, however (Sutherland, 1949). A very simple example of the difficulty of such extrapolation was encountered when insulin was added to a reaction mixture containing purified phosphoglucomutase. Zinc-free insulin stimulated the activity of this enzyme, but it soon became apparent that the enzyme was very sensitive to traces of heavy metals and that zinc-free insulin was a moderately good metal binding agent and, in fact, was somewhat more effective than crystalline serum albumin.

During an exploratory period a number of intact cell preparations were studied, including rat diaphragms, various sliced preparations (especially liver slices), minces, and so forth. To these were added crude pituitary extracts, insulin, steroids, and later epinephrine. When insulin was tested on liver slices it was possible to rediscover glucagon. For some while we did not know that the hyperglycemic–glycogenolytic factor in insulin preparations had been studied years before by Burger and his colleagues (Burger and Brandt, 1953) in Germany. Anyway, for 6 months Christian de Duve and I had a very pleasant time studying the distribution of glucagon (Sutherland and de Duve, 1948). Later, after completing some work with two mutase enzymes, I concentrated my efforts on the mechanism of action of epinephrine and glucagon.

I. GLYCOGENOLYSIS

Studies on the glycogenolytic action of epinephrine and glucagon were attractive for several reasons. The effects on glycogen breakdown and glucose production in liver were rapid, large, and reproducible. Liver slices could be used and a number of slices could be prepared from the liver of one rabbit. In addition, the basic biochemistry of glycogen breakdown had been established through the classical work of the Cori's and others, with phosphorylase, phosphoglucomutase, and glucose-6-phosphatase being the basic enzymes involved. This sequence is illustrated in Fig. 1.1. However, other enzymes were known to participate in the synthesis or degradation of glycogen, and at this stage even the possible hydrolysis of glycogen was thought to deserve consideration. In general, however, the chief problem in logical analysis at this stage was the accepted theory or assumption that glycogen phosphorylase catalyzed the *in vivo* synthesis of glycogen as well as its degradation. Certainly *in vitro* this enzyme catalyzed the synthesis as well as the degradation of glycogen, and under the usual *in vitro* conditions synthesis was favored. At the same time it was well known that the injection of epinephrine caused a rapid breakdown of glycogen in both skeletal muscle and liver, and did not cause an increased synthesis of glycogen. It seemed unlikely that phosphorylase activation would lead to glycogenolysis only, so some other factor was suspected.

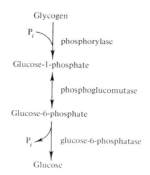

Fig. 1.1. Enzymes involved in the conversion of liver glycogen to glucose.

Several rather simple questions were asked at this early stage. For example, were the relative potencies of epinephrine and its analogs similar to those seen in intact animals? Were we, in other words, studying a phenomenon that was significant at higher levels of organization? Also, was the increased breakdown of glycogen caused by epinephrine or glucagon really a result of phosphorolysis rather than hydrolysis? All our evidence indicated that phosphorolysis was involved (McChesney *et al.,* 1949). Or could the increased glucose release be a more primary event than glycogenolysis? It seemed possible, for example, that glucose might be pumped out of the cell in some sort of extrusion process rather than overflowing from a high level of endogenous glucose. Measurement of glucose levels (Sutherland, 1950) convinced us that the increased glucose output resulted from an overflow phenomenon and not from an extrusion process in which glucose was actively transported from inside the cell.

II. ACTIVATION OF PHOSPHORYLASE

The activation of phosphorylase was deduced from studies of intermediates in the slices. It was found that epinephrine or glucagon caused a lowering of glycogen associated with increased levels of glucose-1-phosphate, glucose-6-phosphate, and free glucose in the slices (Sutherland, 1950). The slices were incubated in high concentrations of inorganic phosphate so that any effect on phosphate concentration would be minimal. Actually the slices were forming large amounts of inorganic phosphate from organic phosphate sources during the incubation so that the inorganic phosphate was present in much higher concentrations than the hexose phosphates. We now know that this is one of the important reasons for the tendency of phosphorylase to always catalyze the breakdown rather than the synthesis of glycogen (Larner *et al.,* 1959). A real understanding of how glycogen synthesis was effected came later, following the discovery of the UDPG transglucosylase by Leloir and his colleagues (see Section III, C in Chapter 5), but during the period under discussion the existence of a separate enzyme for glycogen synthesis was only suspected. A considerable effort was expended one summer with a medical student, Robert Haynes, searching for some increase in a different or smaller type of glycogen. This was not found, and I do not recall publishing this negative project. In any event, these and other experiments carried out during this period led to the conclusion that phosphorylase, rather than phosphoglucomutase or glucose-6-phosphatase, was rate-limiting in the conversion of glycogen to glucose in the liver. The evidence suggested that epinephrine and glucagon were acting to somehow increase the activity of this enzyme.

Once this conclusion had been reached from studies of intermediates in slices, it was simple to demonstrate an activation of phosphorylase in liver or muscle, and an example is shown in Fig. 1.2. Phosphorylase activity in extracts from slices which had been incubated in the presence of the hormones was much greater than in extracts from control slices.

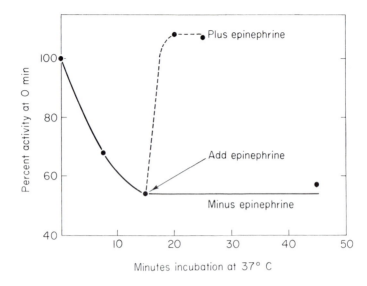

Fig. 1.2. Effect of epinephrine on phosphorylase activity in dog liver slices. After incubating in a glycylglycine-phosphate buffer (pH 7.4) for the indicated times, the slices were homogenized and phosphorylase activity determined. From Rall, Sutherland, and Wosilait (1956a).

It could have been possible, of course, for the activity of phosphorylase to have been increased as the result of a change in the concentration of some regulatory metabolite, i.e., by an allosteric mechanism, in which case it might not have been detected by this type of experiment. As it was, however, it seemed reasonable to conclude that the greater activity of phosphorylase in the extracts implied a greater activity in the intact cells.

III. CHEMICAL NATURE OF PHOSPHORYLASE ACTIVATION

Although phosphorylase activation by epinephrine or glucagon could be shown easily if intact cell preparations were studied, it was soon found that all or virtually all response to the hormones was lost if the cells were broken. Many varied and repeated attempts to demonstrate hormone effects in broken cell preparations were negative. We considered the possibility, previously expressed by others, that hormones might require intact cells structure or even colonies of cells in order to express their activity. Nevertheless, I decided to study the activation and inactivation of the enzyme phosphorylase, hoping that with further insight we might proceed more reasonably in our studies with broken cell preparations.

On arrival at Western Reserve University in 1953, I was joined by Dr. Walter Wosilait. Over a 2-year period we purified liver phosphorylase to homogeneity and studied the inactivation process. Before this time the Cori's had shown that muscle phosphorylase *a* with a molecular weight of about 480,000 was converted to phosphorylase *b* (requiring adenosine 5′-phosphate for activity) with a molecular weight one-half that of *a*. The mechanism of action of this phosphorylase-rupturing enzyme was unknown. We found that active phosphorylase from dog liver had a molecular weight of about 240,000, i.e., the molecular weight of the inactive form from rabbit muscle.* We purified from liver a soluble enzyme which inactivated liver phosphorylase (Wosilait and Sutherland, 1956). On inactivation the molecular weight was apparently unchanged. The inactive form differed from that of muscle phosphorylase *b* by having very little increase of activity when incubated in the presence of adenosine 5′-phosphate (5′-AMP). It seemed likely to us that at least with liver phosphorylase the inactivation might well be accomplished by a minor change, such as the loss of one or two amino acids, loss of an amide group or two, or possibly loss of a group containing phosphate or sulfate. We found that inorganic phosphate was lost during inactivation, one mole being released for each 120,000 gm of enzyme. This effect, which we later established was due to a relatively specific phosphatase, is illustrated in Fig. 1.3.

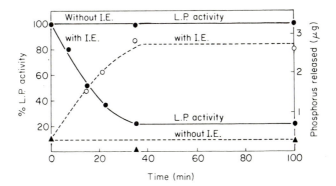

Fig. 1.3. Effect of incubating with "inactivating enzyme" (I. E.) on liver phosphorylase (L. P.) activity and on the release of inorganic phosphate. Phosphate was measured in the supernatant fluid after TCA precipitation. From Wosilait and Sutherland (1956).

---

*The molecular weight of muscle phosphorylase has just recently been revised, to 370,000 and 185,000 for the *a* and *b* forms, respectively (Seery *et al.,* 1967; De Vincenzi and Hedrick, 1967).

At about this time Ted Rall joined me and remained as an active productive collaborator for about 7 years, until sabbatical time in 1961. Jacques Berthet from Louvain joined us for a while a year or so later. We examined the reactivation of phosphorylase in slices, extracts, and homogenates. As might be expected, the reactivation process was associated with the incorporation of phosphate into the molecule. Radioactive phosphate was rapidly incorporated into phosphorylase in liver slices, and epinephrine and glucagon greatly increased this incorporation. An experiment demonstrating this effect is illustrated in Fig. 1.4. By this time Krebs and Fischer (1956) had found that ATP was essential for the conversion of phosphorylase b to a in rabbit muscle extracts. It became clear from our own experiments that in liver the concentration of active phosphorylase represented a balance between inactivation of phosphorylase (by the phosphatase) and a reactivation of the enzyme by a process in which phosphate was donated to the protein.

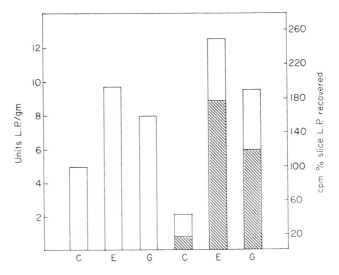

Fig. 1.4. Effect of epinephrine (E) and glucagon (G) on ^{32}P incorporation into liver phosphorylase (L. P.). Dog liver slices were incubated for 20 min in glycylglycine-phosphate buffer containing ^{32}P. Hormones present during last 5 min of incubation, except for control slices (C). Cross-hatched portions of bars at right indicate radioactivity made TCA-soluble after enzymatic inactivation of L. P. From Rall, Sutherland, and Wosilait (1956a).

This knowledge was important in our search for the action of hormones in broken cell preparations. Knowing that ATP and magnesium ions were necessary in the activation process, we began to add ATP and Mg^{2+} to our homogenates as we studied hormone effects. Again, however, our scientific wisdom delayed our progress, although in this case for just a matter of weeks rather than the months or years of delay caused by our "knowledge" of phosphorylase equilibrium. Our problem here was that we knew that fluoride ions prevented or antagonized the conversion of active phosphorylase to inactive phosphorylase. By this time we were adding purified dephosphophosphorylase (inactive phosphorylase) to our homogenates, and it seemed logical to add fluoride to help preserve any active phosphorylase which might be formed. Fortunately, we frequently incubated homogenates with ATP and magnesium ions in the absence of fluoride, and in these experiments we began to see pronounced effects of epinephrine or glucagon (Rall *et al.,* 1957). We know now that fluoride ions stimulate adenyl cyclase in broken cell preparations to a level where hormone stimulation can usually not be seen. This stimulatory effect of fluoride is still unexplained and indeed has not been studied in any detail. By analogy with other effects of fluoride in systems containing nucleotides and magnesium ions we might expect the basic action to be an inhibitory one, e.g., to preserve some active intermediate in the cyclase reaction. However, it is known that fluoride can accept a phosphate group, so perhaps it can also accept a pyrophosphate group and thus stimulate the cyclization. Some experiments regarding the effect of fluoride ions are sadly lacking even today.

In any event, the ability to demonstrate an effect of hormones in broken cell systems was another very important landmark in our studies with hormones. The discovery of phosphorylase activation and the discovery of the chemical nature of the change when phosphorylase activity changed were probably equally important landmarks — but not as exciting as finally establishing hormone effects in broken cell systems. These effects appeared to be physiologically significant. At this point I would like to quote from the introduction and summary of the January 1957 paper by Rall, Berthet, and myself.

> The concentration of active phosphorylase in liver represents a balance between inactivation by liver phosphorylase phosphatase (inactivating enzyme) and reactivation by dephosphophosphorylase kinase. The enzymatic inactivation of phosphorylase proceeds with the release of inorganic phosphate, while the reactivation of dephosphophosphorylase requires magnesium ions and ATP and proceeds with the transfer of phosphate to the enzyme protein.
>
> It has been shown in liver slices that epinephrine and glucagon displace this balance in favor of the active phosphorylase. This report is concerned with the demonstration of a similar effect in cell-free liver homogenates; i.e., an increased formation of active phosphorylase occurred in cell-free homogenates in the presence of sympathomimetic amines and glucagon. The relative activities of the sympathomimetic amines in homogenates were found to be similar to the relative activities determined by liver slice technique or by injection into intact animals.
>
> It has been possible to show that the response of the homogenates to the hormones

occurred in two stages. In the first stage, a particulate fraction of homogenates produced a heat-stable factor in the presence of the hormones; in the second stage, this factor stimulated the formation of liver phosphorylase in supernatant fractions of homogenates in which the hormones themselves were inactive.

1. The formation of liver phosphorylase from dephosphophosphorylase in cell-free homogenates of dog and cat liver was increased markedly in the presence of either epinephrine or glucagon in low concentration.

2. The relative activities of sympathomimetic amines in homogenates were similar to those observed in liver slices and in the intact animals.

3. The response to the hormones in liver homogenates was separated into two phases: first, the formation of an active factor in particulate fractions in the presence of the hormones and, second, the stimulation by the factor of liver phosphorylase formation in supernatant fractions of homogenates in which the hormones themselves had no effect.

4. The active factor was heat-stable, dialyzable, and was purified considerably by chromatography on anion and cation exchange resins.

The loss of hormonal activity when the particulate fraction was removed, and its restoration when this fraction was added back, are illustrated in Fig. 1.5. The enzymatic activity responsible for the production of the heat-stable factor was later referred to as adenyl cyclase, after the factor itself had been identified as cyclic AMP. The properties of adenyl cyclase will be described in more detail later (Chapter 4). I

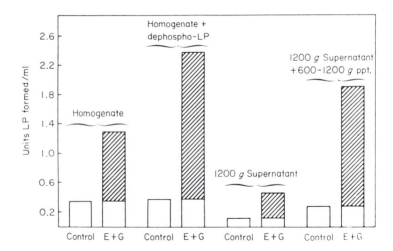

Fig. 1.5. Effect of epinephrine and glucagon (E + G) on phosphorylase activation in whole and fractionated cat liver homogenates. Homogenates or fractions thereof were incubated with ATP and Mg^{2+} in the presence and absence of the hormones. Incubation mixtures were supplemented with inactive liver phosphorylase (dephospho-LP) where indicated. Phosphorylase activity measured before and after 10 min incubation at 30°C. Each bar represents the amount of LP formed during incubation. The effect of the hormones is indicated by the cross-hatched areas. From Rall, Sutherland, and Berthet (1957).

suppose, in general, I was disappointed that adenyl cyclase in animals was not soluble. Instead, the cyclase system is built into particulate material which in some cases at least appears to be derived from plasma membranes. In retrospect, this seems to be an appropriate place for a hormone-sensitive enzyme to be, and its presence in these membranous fractions will probably be more interesting than if it had been soluble. In many cases the system is very labile, although in some cases, such as brain, the preparations can be washed repeatedly in hypotonic and hypertonic aqueous solutions, can be lyophilized, and the dry powder can be extracted with dry ether for hours. These powders respond nicely to epinephrine in the presence of added ATP and magnesium ions. However, if the powder was washed with slightly wet ether, activity was lost, and usually the epinephrine effects disappear before much of the activity noted in the presence of fluoride disappears. We have encountered this loss of hormone response (with retention of considerable activity when tested with fluoride) in a number of tissues. Davoren disintegrated avian erythrocyte membranes to tiny amorphous particles and these particles retained considerable cyclase activity when tested with fluoride ion. Later Øye found that these preparations had lost their ability to respond to epinephrine (Øye and Sutherland, 1966). It seems likely that many years from now work will be continuing on adenyl cyclase, which is probably a lipid–protein complex with a distinct orientation within the membranes.

IV. THE IDENTIFICATION OF CYCLIC AMP

Characterization of the heat-stable compound progressed rapidly, although the very small amounts of the chemical present in tissues or in our preparations created some problems. Within a matter of months, however, the chemical had been isolated and later crystallized. It contained adenine, ribose, and phosphate in ratios of 1 : 1 : 1 and was not inactivated by a number of phosphatases or diesterases. An enzyme present in animal tissue did inactivate it rapidly, however, and formed adenosine 5′-phosphate as the product of hydrolysis.

At about this time we suddenly found ourselves cooperating with Dr. David Lipkin at Washington University in St. Louis. The story of this cooperation was recounted in a Harvey Lecture in 1962, and I include this section below:

> Four years ago we asked Dr. Leon Heppel for an enzyme that might help us characterize a substance that we had isolated from animal tissue preparations. He kindly supplied us with this enzyme; the enzyme did not hydrolyze our compound, a result we had not expected. We thanked Dr. Heppel, and in our letter stated: 'However, it did not attack our nucleotide (or polynucleotide) and, therefore, did not

support our tentative idea that the compound may be a double diester, i.e., a di-nucleotide. . . We hope we can accumulate enough in the next several months to do more on the structure.'

Some weeks later Dr. Heppel called us stating that Dr. David Lipkin had written from St. Louis to ask for the same enzyme to help characterize an unknown nucleo-tide that his group had isolated from a barium hydroxide digest of adenosine tri-phosphate. Dr. Heppel informed us and Dr. Lipkin that the tentative structures proposed by both groups were identical. Very soon the groups exchanged samples and found that they were indeed identical chemicals.

I enjoy recalling this period of time because of the free and friendly exchange of information and ideas. Also because we knew that a really good chemist – Dr. Lip-kin – was active in the field completing the determination of structure and molecular weight (Lipkin *et al.*, 1959); thus two very amateur chemists – Dr. Rall and I – were freed so that we could continue our studies of the formation of this chemical agent in animal tissues.

Before proceeding with these studies, we might glance at the structure of this agent (Fig. 1.6). At first glance, this four-ringed structure seems rather complex, and to most of us amateur chemists it is.* Most of us, however, had learned that many metabolic changes involve adenosine triphosphate and we have accepted ATP as a common, everyday chemical; this chemical is similar to ATP, perhaps simpler. The two rings above represent the familiar purine, adenine, which is attached to the ribose ring. This ring is almost flat and is attached through its 5'-position to a phosphate group. ATP has two more phosphate groups attached to this phosphate, whereas adenosine-3',5'-phosphate (cyclic adenylate) simply has the single phosphate joined again to the ribose in the 3'-position.

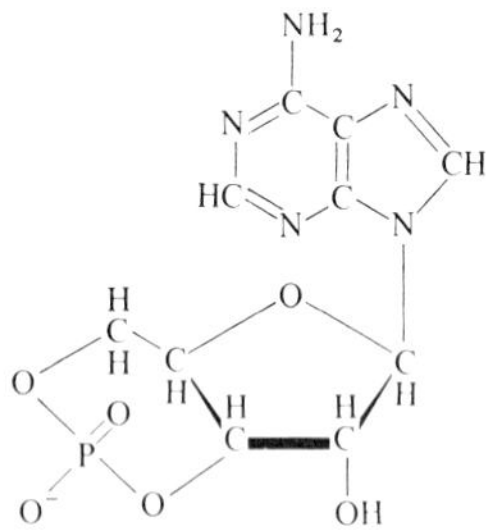

Fig. 1.6. Structural formula of cyclic AMP (adenosine 3',5'-monophosphate).

We have been unable to make this compound with our models. Our problem has been a technical one – we just have not used our models enough to wear away bits of paint on the corners. Dr. Lipkin, however, was able to join all the atoms together and sent us a picture of the model – which shows it can be done. Actually, we understand that the planar ring of the ribose is not absolutely flat, and if our models could repro-duce nature in this respect, we should be able to use our unchipped atom to create this four ring. In any event, when this fourth ring is formed, a very stable molecule results and it can be boiled at a variety of pH's without destroying even the glycosidic

---

*The chemistry of cyclic AMP is discussed in more detail in Chapter 3.

bond that is ordinarily fairly sensitive in related compounds. The reason for this stability is not clear, but this stability was most helpful to the biologists when they isolated this compound.

I mentioned that an inactivating enzyme was found which converted cyclic AMP to 5′-AMP. This phosphodiesterase was later studied, and will be described in more detail in Chapter 4. It became apparent that the level of cyclic AMP, like the level of active liver phosphorylase, represented a balance between the activities of two opposing enzymes (Fig. 1.7). Adenyl cyclase catalyzes the formation of cyclic AMP from ATP, while the phosphodiesterase catalyzes its hydrolysis to 5′-AMP.

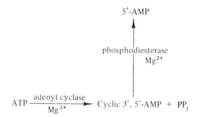

Fig. 1.7. The reactions catalyzed by adenyl cyclase and phosphodiesterase.

As mentioned before, I have considered that three landmarks were most prominent to this stage: first, the discovery of the activation of phosphorylase; second, with Dr. Wosilait, the discovery of the nature of the change in the phosphorylase molecule; and third, with Dr. Rall, the ability to study hormone effects in broken cell preparations. Then a new area began to develop, i.e., the elucidation of the biological role of cyclic AMP (in addition to the role found in our early studies with phosphorylase). Here Dr. R. W. Butcher and also Dr. Alan Robison have been most active, although others have been active in certain areas. For example, the work of Peter Davoren has been mentioned, and he also found some interesting phenomena in intact avian erythrocytes. Richard Makman studied the dramatic changes of cyclic AMP in *E. coli*. Ivar Øye and others have collaborated for shorter periods of time. A collaborative effort with Dr. Theo Posternak was very productive, and some of the derivatives he produced have been most helpful in evaluating the biological role in certain tissues. During recent years a great number of investigators outside my group have made contributions and it is our intent to point these out in subsequent sections. It is especially pleasant to acknowledge the fine cooperation and collaboration found here at Vanderbilt, especially with Dr. C. R. Park and his group, and also the fine collaboration with groups outside this University. The biological role of cyclic AMP already appears extensive, but nevertheless studies are still in the early stages and will undoubtedly be proceeding for many years.

Before closing I will mention that much of my own effort during the past several years has been engaged with studies of cyclic guanylic acid. Here Dr. Joel Hardman has been most active in developing this field. Mr. James Davis, my invaluable right-hand man for years, is also applying most of his effort in this area. Methodology has been difficult but enough progress has been made so that Dr. Hardman will summarize the current status in Chapter 11. I hope that by the time this is published we will have made some additional progress. Methodology continues to limit the rate of progress in this area, as has also been the case, I might add, with studies of cyclic AMP.

The foregoing words were written during the summer of 1967. Obviously there have been opportunities for revision, but, since I was dealing for the most part with history, I did not feel the need to take advantage of these opportunities. In retrospect, however, now that the book is about to appear, it is obvious that the later paragraphs do suffer from a certain lack of perspective. It seemed appropriate then, for example, to single out my own colleagues for special mention, whereas this seems much less appropriate today. On the other hand, it would be extremely difficult in a discussion of this sort to give proper credit for the numerous important advances that have been made by others during the past few years. All of this reflects the fact that progress in cyclic nucleotide research occurred much more rapidly than any of us had anticipated. We have attempted to summarize this progress in the pages to follow. In the meantime, this introductory chapter should be regarded as a personal account of how the subject appeared to me several years ago, before most of this progress had been made.

CHAPTER 2

Cyclic AMP and Hormone Action

I. INTRODUCTION

Cyclic AMP is now recognized as a versatile regulatory agent which acts to control the rate of a number of cellular processes. It occurs in all animal species investigated, including bacteria and other unicellular organisms, although it has so far not been detected in higher plants. In those cells in which it occurs, cyclic AMP seems to play primarily a regulatory role, and thus differs from many biochemical agents which may at times play a regulatory role but which have other functions besides. Cyclic AMP does not seem to be essential, for example, in the same sense that ATP and the calcium ion are essential. Many basic cellular processes come to a complete halt in the absence of one or the other of these agents, whereas cyclic AMP seems to act in most cases to either increase or decrease the rates of cellular processes which would occur at one rate or another even in its complete absence.

Even this generalization may be invalid, however. Information about cyclic AMP is accumulating at such a rapid rate that we are probably not in a very good position to generalize about it at all. It seems possible, for example, that if some cells which normally contain cyclic AMP were completely depleted of it, they might well be incapable of anything approaching normal function. Furthermore, it is now very clear that if *all* of the cells of a multicellular organism were devoid of cyclic AMP, or lacked mechanisms for regulating its formation, then that organism could not long survive. Thus the distinction we have tried to draw between agents which are essential for certain basic processes to occur, on the one hand, and agents which regulate these processes, on the other, is a fuzzy one in some respects. Still, there does seem to be some basis for a distinction between compounds which play a regulatory role primarily and those which may at times play a regulatory role but only in addition to some other function.

One other generalization we might attempt, again with the understanding that it might be invalidated even before these words are in print, is that the intracellular level of cyclic AMP always seems to change in response to some external demand, and never in response to a change in the function of the cell in which it occurs. Cellular metabolism is, of course, regulated in response to either type of change. Perhaps the most familiar example of a metabolic change in response to a change in cell function is the rapid mobilization of glycogen which occurs in muscle in response to muscular work. There are numerous other examples of cells changing their rate of doing work, and consequently the rate at which they utilize energy, but in not a single instance is such a change known to lead to a change in the intracellular level of cyclic AMP. By contrast, as we will attempt to make clear in the pages to follow, there are numerous examples of changes in the external environment leading to large and physiologically important changes in the level of cyclic AMP. It is almost as if cyclic AMP had been placed in cells to serve as a "Director of Foreign Affairs," acting therein to regulate cell function in response to changes in the external environment, sometimes for the good of the cell in which it occurs, but perhaps more often for the good of the cell population as a whole. Other regulatory mechanisms seem to have evolved to keep the metabolic machinery of cells in tune with changes in their normal functional activity.

Although many extracellular influences may regulate the level of cyclic AMP—and we should emphasize that the study of bacteria and other unicellular organisms from this point of view is only just beginning— certainly one of the most important in multicellular animals is the hormonal influence. Even allowing for the possibly distorted perspective that might arise from the fact that cyclic AMP was discovered in the course of endocrinological research, it still seems difficult to overestimate the importance of this influence. Since most of the research on cyclic AMP to date has been concerned directly or indirectly with hormones, this aspect of the subject will be the major focus of attention throughout most of this book.

II. HORMONES

Hormones are often defined as chemical agents which are released from one group of cells and travel via the bloodstream to affect one or more different groups of cells. This is the classic definition first proposed by Starling and Bayliss. Another definition, proposed by Huxley (1935), would place less emphasis on the mode of travel of these agents and more emphasis on their biological function. According to this definition, hormones are regarded primarily as information-transferring molecules, the essential function of which is to transfer information from one set of cells to another, for the good of the cell population as a whole.* This definition has several useful features. First, since it includes neurotransmitter substances released from nerve endings, it emphasizes the similarities instead of the differences between the endocrine and nervous systems. A second potentially useful feature of this definition is that it is broad enough to include agents released from unicellular organisms, which agents may play a role similar to that played by hormones in multicellular organisms (see, for example, Tomasz, 1965) and which may also act similarly at the biochemical level.

This definition is also useful because it helps (or should help) to dis-

*The adjective "good" in this context refers to anything which has survival value.

tinguish hormones from other classes of biologically active compounds. Vitamins, for example, are concerned primarily with energy metabolism and not with the transfer of information. Since the concepts of information and entropy are closely related in cybernetic theory, we might say, on the basis of Huxley's definition, that hormones are concerned more with producing changes in entropy than they are with producing changes in energy. It is of course clear that the amount or direction of energy flow within cells may be altered as one of the consequences of hormone action.

Does the definition proposed by Huxley include all of the chemical agents included in the classic definition of hormones? It is not entirely clear that this is the case. We are reminded at this point of certain Eskimo societies whose language contains just one word for things which fly (so that there is no convenient way to distinguish conceptually between a bird and an airplane, for example) but more than twenty words to describe different types of snow. Such a language would not be useful in a modern industralized society, but does seem useful and was perhaps essential under the conditions which led to its development. The point to be made is that our concepts of reality are heavily influenced by the language we use, even though the things or phenomena described are the same regardless of the words used. We implied previously that the classical definition of hormones may have been too exclusive, but it may also, in another sense, be too inclusive, in the same sense that the Eskimo word for things that fly would be too inclusive under certain conditions. In terms of the classical definition, there seem to be at least two distinct types of hormones. These two types have not always been distinguished from each other, perhaps, at least in part, because they were subsumed under a common definition.

One type includes epinephrine, glucagon, insulin, gastrin, secretin, parathyroid hormone, calcitonin, and many of the hormones released from the anterior and posterior lobes of the pituitary gland. These and certain other hormones, together with the neurotransmitter substances released from nerve endings, seem clearly to play the kind of information-transferring role ascribed to them by Huxley's definition. Cells respond to these hormones more or less rapidly, and the response is very often of short duration, or at least the magnitude of the response at any given instant is closely related to the amount of the hormone present. These chemical messengers are of special interest from the standpoint of this monograph, because so many of their effects result from their ability to produce immediate changes in the intracellular level of cyclic AMP.

The other type or class of hormones would include the steroid hormones, thyroid hormone, and at least one of the hormones produced by the adenohypophysis, namely growth hormone. These hormones play an information-transferring role in a sense, but differ in many ways from the hormones mentioned previously. Cells respond to these hormones much less rapidly than to the others, and the changes which are produced by them may persist for long periods, regardless of the present concentration of the hormone. Indeed these differences have often been used, implicitly if not always explicitly, to place these hormones in a separate category. They play an important *maintenance* role, and in their prolonged absence cells may become incapable of responding to other hormones. Consequently they have been referred to as *permissive* hormones. Since cells and tissues do not develop properly in their absence, they have also been called *developmental* hormones. It is sometimes difficult to distinguish these hormones, in terms of their biological activity, from certain other biologically active compounds. Vitamin D, for example, which resembles the steroid hormones chemically, also resembles these hormones in the kinds of effects it produces (and indeed resembles them more in this regard than the other substances classified as vitamins). These effects in all cases seem to involve an alteration in protein synthesis, and in no case has this been shown to involve an immediate change in the level of cyclic AMP. If we regard the hormones of the first group as primarily messengers, then we might say that the hormones of this second class function more as *maintenance engineers.*

Having drawn this distinction between the two types of hormones in bold relief, we have now to put in some shading. This tells us that the distinction is not always as clear as the human urge to classify things might like it to be. For example, we referred to glucagon as an example of the first (primarily messenger) type of hormone, and yet we know that glucagon may at times, as during postnatal development, play an important role as a developmental hormone. Conversely, the glucocorticoids, which function primarily as maintenance engineers, may at times produce rapid responses which may or may not involve protein synthesis. Thus, even though a case could probably be made for reserving the word "hormone" for one or the other of these two groups of substances, and devising a new name for the other group, it might not be useful to do so. It would perhaps be better to realize that the substances which we call hormones have at least two different types of functions. Some hormones function primarily as messengers but may at times serve, on a part-time basis, as maintenance engineers. The other group of hormones serve as maintenance engineers most of the time, but may occasionally act as messengers. Although we have only a few words for describing different types of snow, this is not harmful so long as we understand that different types of snow exist, and that these differences may be important under some circumstances. The point of this discussion has been that it may also not be harmful to define a group of divergent substances as hormones, so long as we understand that there may be some important differences between them.

III. RECEPTORS AND SECOND MESSENGERS

The receptor concept was introduced by pharmacologists in the last century to explain the selective toxicity of plant alkaloids and certain synthetic chemotherapeutic agents. The concept was later applied to the actions of epinephrine and other hormones, and was systematically developed in a classic monograph by A. J. Clark (1937). Since that time, most concepts of hormone action have included the idea of a receptor. This can be loosely defined as that part of the cell with which the hormone interacts to produce a response. According to the more detailed definition provided by Schueler (1960), the receptor is regarded as a pattern of forces forming a part of some biological system in or on the cell and having roughly the same dimensions as a certain pattern of forces presented by the hormone molecule, such that between the two patterns an interaction may occur. The initial result of this interaction, in terms of classical receptor theory (Clark, 1937; Ariens, 1966), is the production of a stimulus, or signal, which under appropriate conditions may lead to the production of one or more cellular responses.

It is now becoming clear that the receptors with which many hormones combine are closely related to—and may even be part of—an adenyl cyclase system. In these cases, depending upon whether the cyclase is stimulated or inhibited, the result of the hormone–receptor interaction (i.e., the stimulus) will be an increase or a decrease in the intracellular level of cyclic AMP. When the stimulus is an increase in the level of cyclic AMP, then we have suggested that the hormone might appropriately be regarded as a first messenger, with cyclic AMP as the second messenger. This basic concept is illustrated in Fig. 2.1. According to this concept, the first messenger carries the required information to the cell, and the second messenger transfers this information to the cell's internal machinery. The end result of the increased level of cyclic AMP will depend entirely on the nature of the cell in which it occurs and on the prevailing conditions. How cyclic AMP acts to carry out this function is not known in all cases, but some of its known effects are discussed in Chapter 5.

Although cyclic AMP is the only second messenger recognized to date, it seems possible or even likely that other second messengers will be

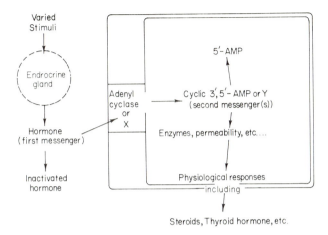

Fig. 2.1. Schematic representation of the second messenger concept.

discovered. This seems especially likely in the case of certain hormones which are not known to be capable of stimulating adenyl cyclase. Possibly one or more of the other cyclic nucleotides discussed in Chapter 11 will be found to function in this capacity.

The hormones which are known at the time of this writing to utilize cyclic AMP as a second messenger for the production of one or more of their effects are listed in Table 2.I Also indicated in this table are some of the tissues in which an effect on cyclic AMP has been demonstrated,

TABLE 2.I. Hormones Which Utilize Cyclic AMP as a Second Messenger

Hormone	Tissue	Chapter
Catecholamines	Various tissues	6 and 8
Glucagon	Liver, pancreatic islets, and adipose tissue	7 and 8
ACTH	Adrenal cortex and adipose tissue	8 and 9
LH or ICSH	Ovarian and testicular tissue	9
Vasopressin	Various epithelial tissues	10
Parathyroid hormone	Kidney and bone	10
TRH	Anterior pituitary	10
TSH	Thyroid tissue	10
MSH	Frog skin	10
Prostaglandins	Various tissues	8 and 10
Histamine	Brain	10
Serotonin	*Fasicola hepatica*	10 and 12

either *in vitro* or *in vivo* or both. The effects which are either known or thought to be mediated by cyclic AMP are discussed in the chapters indicated in the third column. It should be emphasized that many other hormones not listed in Table 2.1 may act by this same mechanism. Among the more likely candidates at the time of this writing are gastrin and angiotensin. These hormones are discussed briefly in Chapter 10 and elsewhere, but are not listed in the table because an effect by either of them on the formation or accumulation of cyclic AMP has not been experimentally demonstrated.

All of these hormones are either known or thought to act by stimulating adenyl cyclase, and in the case of some of them (e.g., glucagon and ACTH) there is evidence to suggest that this may be their only important biological action. Most of them are highly selective in that they stimulate adenyl cyclase in only one or a few tissues, just as most tissues respond to only one or a few hormones. There are, of course, many exceptions to these rules of specificity. The catecholamines are an outstanding exception among hormones, since they stimulate adenyl cyclase in a great variety of tissues. Rat adipose tissue is perhaps the most striking exception among tissues, since it responds to a large variety of hormones. Whether a given hormone stimulates adenyl cyclase in a given tissue presumably depends upon whether or not the tissue contains receptors for that hormone.

A question we might consider at this point is how the receptors for these hormones are related to adenyl cyclase. Are the receptors part of the enzyme system itself, or are they part of some other system which secondarily stimulates adenyl cyclase as only one of several functions? It should be made clear at the outset that the answer to this question is unknown. Progress in attempting to answer it has been hampered by the unavailability of a highly purified hormone-sensitive adenyl cyclase preparation, which is in turn related to the particulate and highly labile nature of the enzyme (assuming that it is one enzyme). What information we have about adenyl cyclase is summarized in Chapter 4. Here we note only that the mammalian enzyme seems invariably to be located either in the cell membrane or else in certain membranous constituents of cells, and that it has the characteristics of a lipoprotein. The protein component of the enzyme is probably intimately associated with the lipid matrix of the cell membrane. Possibly the pattern of forces which we define as the receptor for a given hormone is contributed by both of these components.

We have tended to favor the view that hormone receptors constitute an integral part of the adenyl cyclase system because it seems to be the simplest hypothesis capable of explaining the data already at hand. In support of this hypothesis, it may be noted that in those hormonal responses which are mediated by cyclic AMP, the stimulation of adenyl cyclase is the earliest event known to occur, and attempts to measure or detect an earlier event have been uniformly unsuccessful. Particulate preparations of adenyl cyclase from a variety of tissues can be washed repeatedly with hypotonic media without losing their capacity to respond to hormones, and some have been carried through several additional purification procedures, including lyophilization and extraction with dry ether (Klainer *et al.,* 1962). These preparations moreover retain all of the characteristics exhibited by the receptors in intact tissue. For example, preparations from tissues which respond to the catecholamines with an increase in cyclic AMP have the characteristics of an adrenergic β-receptor, while those from adrenal cortical tissue have the characteristics of the ACTH receptor. Studies with the catecholamines, to be discussed in more detail in Chapter 6, have been especially useful because these hormones affect a large number of tissues to elicit a corresponding variety of effects. Every β-receptor effect which has been studied in detail seems to be mediated by cyclic AMP, and there are no known exceptions to this generalization. There is, in other words, no example of an effect thought to be mediated by β-receptors for which there is evidence that it is initiated by any system other than adenyl cyclase. These and other considerations, including the cellular location of adenyl cyclase and also the great rapidity with which hormones affect it, seem to provide a reasonable basis for the hypothesis that the enzyme and the receptors are part and parcel of the same system.

One possible model of the adenyl cyclase system, which may be useful as a working hypothesis, is illustrated in Fig. 2.2. In this model the protein

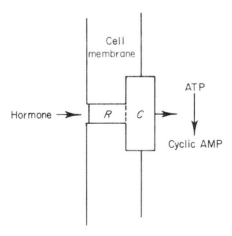

Fig. 2.2. Possible model of the protein component of the membrane adenyl cyclase system. From Robison *et al.* (1967a).

component of the system is pictured as consisting of at least two types of subunits, a regulatory subunit (R), facing the extracellular fluid, and a catalytic subunit (C), with its active center directed toward the interior of the cell. The receptor is here regarded as a part of the regulatory subunit. Interaction with a hormone leads to a conformational perturbation which is extended through the regulatory subunit to the catalytic subunit, thereby altering the activity of the latter. One regulatory enzyme which is known to be composed of subunits such as this is bacterial aspartate transcarbamylase, the first enzyme in the pathway leading to pyrimidine biosynthesis. Gerhart and Schachman (1965) studied the enzyme from *E. coli,* and found that one type of subunit possessed all of the catalytic activity, while the other type of subunit possessed all of the binding sites for CTP, the normal allosteric inhibitor in this species. In the present scheme, the catalytic subunit might be expected to be basically similar in all tissues, whereas the structure of the regulatory subunit would vary, thus accounting for hormonal specificity. This would be similar to aspartate transcarbamylase from bacteria, where the catalytic subunit seems to be similar in all species, but where the regulatory subunits differ in their sensitivity to allosteric effectors, depending on their source.

Whether adenyl cyclase is in fact composed of distinct subunits is of course unknown. The enzyme has been obtained in a soluble form from several bacterial species (Hirata and Hayaishi, 1967; Brana, 1969; Tao and Lipmann, 1969), but its chemical constitution has not been studied. These bacterial enzymes do not respond to mammalian hormones, but adenyl cyclase from *Brevibacterium liquifaciens* is stimulated by pyruvate and several other α-keto acids (Hirata and Hayaishi, 1967). It is tempting to suppose that the mechanism by which pyruvate stimulates this enzyme is basically similar to that by which hormones act in mammalian systems, but whether this is valid or not remains to be seen.

In cells in which two or more structurally different hormones are known to be capable of stimulating adenyl cyclase, the question arises as to whether all of the hormones stimulate the same enzyme, or whether there are separate enzymes for each hormone. The available evidence suggests that in rat adipocytes all of the hormones stimulate the same enzyme, since the effects of maximally effective concentrations of hormones are not additive (Butcher *et al.,* 1968; Birnbaumer *et al.,* 1969). Instead, the activity of adenyl cyclase in the presence of two hormones only reaches the activity produced by the more effective of the two. In terms of the model depicted in Fig. 2.2, this result could be explained by postulating a different regulatory subunit for each hormone, all attached to the same catalytic subunit, or else different receptors on the same regulatory subunit. In those cases where an effect is produced by two or more hormones of markedly different size, such as epinephrine and glucagon in the liver, it is conceivable that the pattern of forces with which the smaller hormone interacts may be a part of the presumably larger pattern of forces with which the larger hormone interacts. This would be in line with an early study by Cornblath (1955), showing that even submaximal doses of glucagon and epinephrine were not additive in stimulating phosphorylase activation in liver slices. On the other hand, Bitensky et al. (1968) have presented preliminary data to suggest that glucagon and epinephrine may interact with separate adenyl cyclase systems. More work will be required to establish this point.

Some investigators (for example, Hechter et al., 1966) have suggested that the receptors for hormones which stimulate adenyl cyclase are probably part of some separate system, which may in turn affect other systems besides adenyl cyclase. In support of this hypothesis, the point has been made that the hormonal sensitivity of adenyl cyclase preparations can often be destroyed under conditions where the sensitivity to stimulation by fluoride remains, the implication being that the receptor system must have been closely related to adenyl cyclase at one time, but became detached during the procedure which destroyed the hormonal sensitivity. However, this evidence seems quite equivocal, and does not support one hypothesis more than the other. In terms of the hypothesis represented by Fig. 2.2, the loss of hormonal sensitivity would be equivalent to the loss of sensitivity to an allosteric effector. This is very commonly seen with allosteric enzymes, including purified aspartate transcarbamylase. In the case of the enzyme from E. coli, for example, the sensitivity to CTP can be destroyed by relatively gentle means while the catalytic activity remains largely unaffected (Gerhart and Pardee, 1963).

Furthermore, if the loss of hormonal sensitivity were due simply to a physical detachment of the receptor system from the adenyl cyclase system, it might be thought that at least some restoration of sensitivity could be achieved by means of recombination experiments. Many attempts of this type have been made, and all of them have been unsuccessful. It is recognized, of course, that if the two systems were held in close juxtaposition by means of other elements within the cell membrane, then their separation might well be irreversible. It can only be concluded that the loss of hormonal sensitivity tells us very little about the relation of receptors to adenyl cyclase. The other arguments which have been raised against the unitarian hypothesis seem equally equivocal.

Although major emphasis in this section has been placed on those hormones which increase intracellular cyclic AMP levels by stimulating adenyl cyclase, it is clear that a similar result could in theory be brought about by inhibiting phosphodiesterase. However, the only hormone which has so far been shown to be capable of inhibiting phosphodiesterase directly is triiodothyronine (Mandel and Kuehl, 1967), and even in this case the physiological significance of the action is dubious. The phosphodiesterase therefore seems to be a less prominent control point for hormone action than adenyl cyclase.

We have now to mention that several hormones produce some of their effects by causing a fall in the intracellular level of cyclic AMP. It is clear that if a second messenger is acting in the presence or absence of hormonal stimulation to maintain the rate of a given process, then a response will result from a lowering of its concentration. The catecholamines, the prostaglandins, insulin, and melatonin have now been shown to produce one or more of their effects by this means. The mechanism or mechanisms by which these hormones produce this effect on the level of cyclic AMP are unknown. It will be noted that the catecholamines and the prostaglandins were also listed in Table 2.1, and indeed they are the only hormones known (again at the time of this writing) which are capable of causing both an increase and a decrease in the intracellular level of cyclic AMP. Insulin and melatonin are discussed in Chapters 7 and 10, respectively.

A possibility which can be mentioned here is that some hormones may act by altering the level of some second messenger other than cyclic AMP, which could then go on to influence the activity of adenyl cyclase or phosphodiesterase. Other second messengers could influence not only the activity but the rate of synthesis of these enzymes, and indeed it is possible that some permissive hormones may act by just such a mechanism.

In summary, the question of whether the receptors with which hormones interact to stimulate adenyl cyclase are part of the cyclase system itself, or whether they are part of some separate system which in turn influences adenyl cyclase, cannot at the present time be answered, since there is no conclusive evidence one way or the other. We continue to favor the hypothesis that the receptor is a part of the adenyl cyclase system itself, if only because this seems to be the simplest hypothesis which is consistent with the available facts. Simplicity is not necessarily a virtue in and of itself, but the assumption of a separate receptor system, capable of influencing not only adenyl cyclase but many other systems as well, leads to theoretical complications of considerable magnitude. The possibility has already been mentioned that in certain cases one second messenger could influence the production or accumulation of another second messenger, but the recognition of this as a theoretical possibility is not the same as assuming that this is *always* the mechanism by which adenyl cyclase is affected by hormones. In fact it should be obvious from this very example that eventually a point must be reached beyond which this multiplication of receptors will have to stop. One of the main points of this section has been that for many hormones the best place to stop may be the adenyl cyclase system itself.

IV. THE INTRACELLULAR LEVEL OF CYCLIC AMP

A. Normal levels

The widespread distribution of cyclic AMP among animal species was mentioned previously, and will be further discussed in Chapter 12. In multicellular organisms, cyclic AMP has been found in most tissues studied, including the mammalian body fluids listed in Table 2.II. It

TABLE 2.II. Mammalian Body Fluids in Which the Level of Cyclic AMP Has Been Measured

Fluid	Level (approx.)	Reference
Plasma	$2 \times 10^{-8}\ M$	Broadus *et al.* (1970a)
Cerebrospinal fluid	$2 \times 10^{-8}\ M$	W. F. Henion and E. W. Sutherland (unpublished)
Gastric juice	$2 \times 10^{-8}\ M$	J. A. Oates and G. A. Robison (unpublished)
Milk	$10^{-6}\ M$	Kobata *et al.* (1961)
Urine	$10^{-6}\ M$	Butcher and Sutherland (1962)

apparently does not occur in the red blood cells of some mammals (Sutherland *et al.*, 1962) but does occur in the nucleated erythrocytes of other species (Davoren and Sutherland, 1963a) and also in mammalian leukocytes and platelets (Robison *et al.*, 1969). It has been reported to be absent in the electric organ of the eel (Williamson *et al.*, 1967), so this may constitute another exception.

The concentration of cyclic AMP in plasma, CSF, and gastric juice is normally quite low, on the order of $10^{-8}\ M$, but its concentration in milk and urine is relatively high, in the $10^{-6}\ M$ range. The high urinary level is the more extraordinary when it is realized that cyclic nucleotides are the only organic phosphate compounds which have been detected in urine (Price *et al.*, 1967). The original cellular source of cyclic AMP in urine and other body fluids is uncertain. This problem is discussed in more detail in Chapter 11, in connection with the excretion of cyclic nucleotides in general.

Some cells, such as avian erythrocytes (Davoren and Sutherland, 1963a), resemble many unicellular organisms in that they are capable of pumping cyclic AMP out into the extracellular fluid against a concentration gradient. Some mammalian cells (e.g., renal cortical cells) may also be capable of this, although most seem to release only a small fraction of their intracellular level. Although small compared to the level within cells, this release may be quite substantial at times. Glucagon, for example, stimulates cyclic AMP formation in the liver, and the injection of large doses of this hormone may lead to transient increases in the blood level which are as much as 40 times higher than normal (Broadus et al., 1969, 1970b). The relationship between intra- and extracellular cyclic AMP levels in general needs further systematic study.

In most cases, the intracellular concentration of cyclic AMP at any given instant will depend primarily on the relative activities of adenyl cyclase and phosphodiesterase. Current knowledge about both of these enzymes is summarized in Chapter 4. Their relative activities are undoubtedly influenced by a number of factors in addition to hormones, and not all of these factors are understood. The concentration of cyclic AMP might therefore be expected to vary somewhat from one tissue to another, but in general, in the absence of stimulation by exogenous hormones, the concentration has been found in a variety of tissues to be similar, on the order of 0.1 to 0.5 nanomoles/gm of tissue (wet weight), or about 0.5 to 2.6 picomoles/mg of protein. Assuming an even distribution throughout the intracellular water, the concentration would thus be somewhere in the general range of 10^{-7} M. This is much less than the concentrations of other adenine nucleotides. The concentration of ATP is usually on the order of about 1 to 5 μmoles/gm of tissue (wet weight), or 5×10^{-3} M. The concentration of 5'-AMP is usually about one-tenth of this, with the concentration of ADP most often falling somewhere in between. These nucleotides therefore exist in cells in concentrations which are some 1000 to 10,000 times greater than that of cyclic AMP.

The actual concentration of free cyclic AMP may of course be much less than 10^{-7} M, due to protein binding and the like, and it is possible that in certain parts of some cells it may be much higher, as a result of compartmentalization. The subcellular distribution of cyclic AMP, like the relation between intra- and extracellular cyclic AMP, is in need of further study.

B. Effects of Hormones

The concentrations of many cell constituents remain fairly stable under normal physiological conditions. Among small molecules, the hydrogen ion and ATP could be cited as prime examples. The concentration of cyclic AMP differs in that it is subject to large fluctuations, and of the several factors which may be responsible for these fluctuations, none is more dramatic than the effect of hormones. Glucagon, for example, is capable of increasing the concentration of cyclic AMP in the isolated perfused rat liver to a level which is more than 80 times higher than normal (Robison *et al.*, 1967b), and the effect of ACTH in the adrenal cortex is even more striking. In this case the concentration of cyclic AMP may increase to such a level that it actually approaches the concentration of ATP (Grahame-Smith *et al.*, 1967). These extraordinary effects are not likely to be of physiological significance, since the hormone concentrations which are necessary to produce them are much higher than any which could occur under normal conditions. They are mentioned here primarily to illustrate the enormous capacity which cells have for effecting changes in their content of cyclic AMP.

We are now beginning to realize that the changes in the level of cyclic AMP which are physiologically important are often small compared to the changes which we know *can* occur under certain experimental conditions. Cannon (1939), considering the mammalian organism from an engineering standpoint, concluded that it was a well-designed structure because each organ seemed to have built into it a large margin of safety. Perhaps the ability to form cyclic AMP can be taken as an illustration of this at the cellular level. For example, although high concentrations of glucagon are capable of causing an 80-fold increase in the hepatic level of cyclic AMP, physiological concentrations may produce only a threefold increase. This is nevertheless sufficient to produce a maximal stimulation of glucose output, and increasing the concentration of glucagon (and hence of cyclic AMP) beyond this point seems to have no further effect. Similarly in the adrenal cortex the concentration of cyclic AMP required to produce a maximal increase in the rate of steroidogenesis is small relative to the levels which can be reached in response to supramaximal concentrations of ACTH (Grahame-Smith *et al.*, 1967). In classical receptor theory, when a drug is capable of producing a stimulus greater than necessary for the production of a maximal response, or conversely when a maximal response occurs in response to a stimulus which is less than maximal, then that tissue is said to contain "spare receptors" (Ariens, 1966). This concept is illustrated especially well by some of the aforementioned studies with cyclic AMP. It is, of course, recognized that different cellular processes may be stimulated by different levels of cyclic AMP, and also that in some cells the response may be related to the level of cyclic AMP over a wider range.

In many tissues the intracellular level of cyclic AMP is one of the most important and also one of the most sensitive factors regulating cell function. Small changes in the level of cyclic AMP may lead to effects which are very large relative to the functional capacity of the affected cells. These changes may occur in both directions, an increase above the baseline leading to one effect, a decrease below the baseline leading to the opposite effect. Sometimes these changes in the level of cyclic AMP occur in response to different hormones. In the liver, for example, glucagon tends to increase the level of cyclic AMP while insulin tends to lower it, and many hepatic functions reflect a subtle balance between the opposing actions of these two hormones (Chapter 7). In certain other cases opposite changes in the level of cyclic AMP may occur in response to the same hormone acting under different conditions. Epinephrine, for example, is capable of interacting in some tissues with two different types of receptors. Interaction with one type leads to an increase in the level of cyclic AMP while interaction with the other leads to a decrease, and the net effect of epinephrine in these tissues will depend on whichever type of receptor predominates. This is discussed in more detail in Chapter 6.

The specificity of hormones which stimulate adenyl cyclase was mentioned previously in Section III. In many tissues this specificity seems to be absolute. For example, glucagon is not known to be capable of stimulating adenyl cyclase in adrenal cortical preparations at any concentration, just as ACTH is not known to be capable of stimulating hepatic adenyl cyclase. In certain other tissues, however, the specificity tends to decrease as hormone concentrations are increased beyond their normal physiological levels. Physiological concentrations of glucagon do not affect the heart, for example, but higher concentrations are capable of stimulating cardiac adenyl cyclase, and at these concentrations glucagon becomes a positive inotropic agent. Rat adipose tissue is even less specific in that it contains receptors not only for glucagon but for ACTH and several other hormones as well. Whether these hormones ever reach adipose tissue in the rat in concentrations high enough to be effective lipolytic agents is problematical, but it is at least clear that they have this potential. Further comparative study of what receptors are present in which tissues may provide clues to some of nature's more interesting experiments. Many of these receptors do not seem to serve any useful purpose, but they do not seem to have been harmful either.

C. Effects of Drugs

Many drugs may act by stimulating or inhibiting adenyl cyclase, although only a few examples are known at present. The most thoroughly studied example is isoproterenol, a synthetic catecholamine which stimulates adenyl cyclase in a variety of tissues. The effects of this drug are discussed in some detail in Chapter 6.

In contrast to adenyl cyclase, phosphodiesterase is known to be a site of action for several drugs. Among the drugs which are known to be capable of inhibiting this enzyme in some tissues are the methylxanthines (Butcher and Sutherland, 1962), puromycin (Appleman and Kemp, 1966), and some of the benzothiadiazine derivatives (Schultz *et al.,* 1966; Moore, 1968), and there is no doubt that at least some of the effects of these drugs can be related to this action. The ability of high concentrations of triiodothyronine to inhibit phosphodiesterase was mentioned previously. Whether this effect is physiologically significant is still unclear, but it could well account for some of the pharmacological activity of this hormone. In addition, imidazole stimulates phosphodiesterase activity under some conditions (Butcher and Sutherland, 1962), and it is possible that this is the basis for some of its pharmacological activity. These drugs are discussed again in Chapter 4 and also in Section V,C below.

Diazoxide, a benzothiadiazine derivative, may be an example of a drug which has the capacity to inhibit phosphodiesterase in some tissues and adenyl cyclase in others. This possibility is discussed in Chapter 6, Section III,C.

D. Effects of Ions

Changes in the ionic environment surrounding intact cells may have a profound effect on the intracellular level of cyclic AMP. The mechanisms responsible for the effects of ions on the level of cyclic AMP have not been the subject of any kind of systematic study, and it is possible that a variety of mechanisms may be operative. For example, increasing the ratio of K^+ to Na^+ causes an increase in the level of cyclic AMP in brain slices (Sattin and Rall, 1967) and rat diaphragm muscle (Lundholm *et al.,* 1967), but tends to produce the opposite effect in the isolated rat heart (Namm *et al.,* 1968). Since relatively large changes in either the absolute or relative concentrations of Na^+ and K^+ are not known to have a large effect on either adenyl cyclase or phosphodiesterase activity in broken cell preparations, the reasons for these effects in intact cells seem quite obscure.

It has been suggested from time to time that this or that hormone may stimulate adenyl cyclase secondarily to a change in the ionic environment. Although some agents may affect the accumulation of cyclic AMP by such a mechanism, it should be noted that most hormones which stimulate adenyl cyclase are capable of doing so in highly washed particulate preparations in the absence of any added inorganic ion (except Mg^{2+}, which is necessary for the adenyl cyclase reaction). There is evidence that K^+ facilitates the ability of hormones to stimulate adenyl cyclase in adipose tissue (Chapter 8) and possibly elsewhere, but K^+ does not seem to be absolutely required. Calcium, on the other hand, does seem to be required in order for ACTH to stimulate adenyl cyclase (Bär and Hechter, 1969d; Birnbaumer *et al.*, 1969), although it is not required for most hormones and may even prevent the action of some, such as that of vasopressin in the toad bladder (Orloff and Handler, 1967; see also Chapter 10). Higher concentrations of Ca^{2+} tend to inhibit adenyl cyclase from most sources studied.

The fluoride ion is of special interest because it has a remarkable stimulatory effect on the activity of adenyl cyclase in broken cell preparations from most mammalian tissues studied, although it apparently does not stimulate bacterial adenyl cyclase (Tao and Lipmann, 1969). Studies of adenyl cyclase in the developing rat brain (Schmidt *et al.*, 1970) suggest that the stimulatory effect of fluoride in this tissue is in fact the reversal of an inhibitory influence which does not become apparent until about 9 days after birth. Up to that time, brain adenyl cyclase resembles bacterial adenyl cyclase in its lack of sensitivity to fluoride. On the other hand, the factor or factors responsible for fluoride sensitivity are present in sea urchin eggs even before fertilization (Castaneda and Tyler, 1968), and the magnitude of the fluoride effect has been reported to be reduced as a result of fertilization. The effect of fluoride on adenyl cyclase is discussed again in Chapter 4, and the principle point to be made here is that fluoride has never been observed to stimulate the formation of cyclic AMP in an intact cell. The reasons for this are as obscure as the mechanism of the fluoride effect itself.

In summary, the observations which have been made to date on the effects of inorganic ions on the formation or accumulation of cyclic AMP are such that generalizations do not seem possible. This is an area in which a great deal of additional research is required.

E. Third Messengers

One of the functions of cyclic AMP as a second messenger is to stimulate the production or release of several hormones, including insulin, thyroid hormone, some steroid hormones, and also some of the anterior pituitary hormones. We suggested several years ago that these hormones might be thought of as "third messengers." Upon reconsideration, however, it can be seen that this terminology might lead to some semantic difficulties, because we now know that these "third messengers" may at times act as first messengers to stimulate the formation of cyclic AMP in *their* target cells.

A schematic illustration of this is provided in Fig. 2.3. As discussed in more detail in Chapter 10, it now seems reasonably clear that the thyrotropin-releasing hormone (TRH) produced in the hypothalamus stimulates adenyl cyclase in the thyrotropic cells of the anterior pituitary gland. The resulting rise in the level of cyclic AMP in these cells stimulates the release of thyroid-stimulating hormone (TSH), which in turn travels

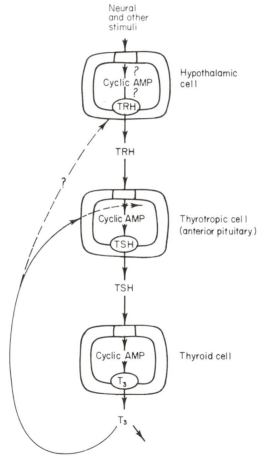

Fig. 2.3. Illustrating the participation of cyclic AMP in the regulation of thyroid function. Details in text.

to the thyroid gland to stimulate adenyl cyclase there. This results in an increase in the level of cyclic AMP which in turn leads to the release of thyroid hormone, principally triiodothyronine (T_3). In many cells T_3 may act to facilitate the action of hormones which stimulate adenyl cyclase, but in thyrotropic cells it acts to inhibit the action of TRH. We can thus see that cyclic AMP may regulate endocrine function at a great variety of levels. In the case of thyroid function, we can note that it might be involved at more levels than we have so far mentioned. For example, it is known that the hypothalamus contains cells which respond to norepinephrine with an increase in the level of cyclic AMP. Although there is no evidence at the present time that cyclic AMP has anything to do with the production or release of TRH, this is obviously a possibility which will have to be considered sooner or later. At present very little is known about the factors which regulate the release of the various hypothalamic releasing hormones.

V. EXPERIMENTAL APPROACHES USED IN THE STUDY OF HORMONES WHICH STIMULATE ADENYL CYCLASE

Most of the hormones listed in Table 2.I are known to be capable of stimulating adenyl cyclase. In order to justify a claim that a given hormone produces a given effect as a result of this action, the following criteria should be satisfied.

(1) The hormone should be capable of stimulating adenyl cyclase in broken cell preparations from the appropriate cells, while hormones which do not produce the response should not stimulate adenyl cyclase.

(2) The hormone should be capable of increasing the intracellular level of cyclic AMP in intact cells, while inactive hormones should not increase cyclic AMP levels. It should be demonstrated that the effect on the level of cyclic AMP occurs at dose levels of the hormone which are at least as small as the smallest levels which are capable of producing a physiological response. The increase in the level of cyclic AMP should precede or at least not follow the physiological response.

(3) It should be possible to potentiate the hormone (i.e., increase the magnitude of the physiological response) by administering the hormone together with theophylline or other phosphodiesterase inhibitors. The hormone and the phosphodiesterase inhibitor should act synergistically.

(4) It should be possible to mimic the physiological effect of the hormone by the addition of exogenous cyclic AMP.

Each of these four criteria or approaches is subject to certain qualifications, as discussed in the following paragraphs.

A. Studies with Broken Cell Preparations

These studies are useful for several reasons, one of which is that the environment in the vicinity of the adenyl cyclase can be greatly simplified and therefore better controlled than in more organized systems. In most mammalian tissues which have been studied, using conventional homogenization techniques, adenyl cyclase has been located in the low speed or "nuclear" fraction, which contains fragments of the cell membranes. Often these preparations can be washed repeatedly, thus eliminating many metabolites and soluble enzymes, and in some cases can be taken through additional purification steps with the retention of hormonal sensitivity. The only additions which have to be made under these conditions are ATP and Mg^{2+}. The ultimate goal here would of course be to obtain a preparation containing *only* adenyl cyclase, but to date it has not been possible to purify adenyl cyclase from mammalian sources beyond a certain point without destroying its sensitivity to hormonal stimulation. The ability to be stimulated by fluoride sometimes outlasts its ability to respond to hormones, but even this is very labile. The loss of adenyl cyclase activity in a skeletal muscle homogenate maintained at 4°C is illustrated in Fig. 2.4. The addition of sulfhydryl reagents has been found to be helpful in preserving activity in preparations of avian and amphibian erythrocytes (Øye and Sutherland, 1966; Rosen and Rosen, 1969) but these agents have not been found to be helpful in mammalian systems.

The great lability of adenyl cyclase is perhaps the chief limitation to be kept in mind during studies with broken cell preparations. However, once a satisfactory preparation has been obtained, it becomes possible to carry out a number of useful experiments. Most preparations can be quick-frozen and stored at −70°C for long periods (and can be thawed at least once) without altering the properties of adenyl cyclase in any obvious way. Aliquots of a given preparation treated and stored in this manner yield results which are highly reproducible. Often the data cannot be related quantitatively to data obtained from studies with more highly organized systems, not only because of the altered circumstances of the adenyl cyclase (dilution effects, absence of endogenous inhibitors or activators, etc.) but also because of possible differences in the way the hormone is handled (e.g., metabolism, tissue uptake, and the like). However, in those cases where structural analogs and/or competitive inhibitors are

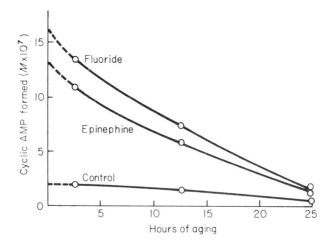

Fig. 2.4. Loss of adenyl cyclase activity in a rabbit skeletal muscle homogenate at 4°C. Zero time was taken as the time of muscle excision. Aliquots of the homogenate (equivalent to 40 mg tissue, wet wt.) were taken at intervals and incubated for 15 min at 30°C in the presence of 2 mM ATP, 3 mM Mg^{2+}, and 6.7 mM caffeine. Concentration of epinephrine was 10^{-4} M and of fluoride 10^{-2} M.

available, then the relative potencies of these agents on adenyl cyclase should be the same as their relative potencies in producing or inhibiting the physiological response in the intact tissue. And in all cases it should be possible to compare the active hormone with a series of inactive hormones.

One other point to be considered in this section is that most tissues contain several types of cells in which the adenyl cyclases may differ with respect to the hormones which stimulate them. In some cases techniques are available for separating the various cell types prior to homogenization. Where such techniques are not available, the analysis of the experimental data may be difficult.

B. Studies with Intact Tissues

Studies with broken cell preparations may tell us something about what the intact cells are capable of, and often provide information which is unobtainable by any other means, but the results are not always directly applicable to events occurring in more highly organized systems. Studies with intact tissues may be especially important from this point of view because they can often be carried out under conditions where the physiological response and the change in cyclic AMP can be measured simultaneously. The information obtained is therefore more directly relevant to physiological and clinical situations. Many of the points which can be tested with broken cell preparations (e.g., relative potencies of analogs, inhibition by competitive antagonists) can also be studied in intact tissues. In addition, the change in cyclic AMP and the physiological response can be compared as functions of the dose of the hormone needed to elicit the response and the time required for the response to be manifested. The rate or magnitude of some cellular processes may be directly related to the level of cyclic AMP, while in other cases the relation may be more complex, as when a change in cyclic AMP acts as a trigger to initiate a series of reactions. Obviously the more steps there are between the initial action of cyclic AMP and the response chosen for measurement, the more complex will the relation between the two events appear. It is equally clear that if a response occurs in the absence of a change in the level of cyclic AMP, then the response cannot be mediated by cyclic AMP.

As with broken cell preparations, possible complications arising from different cell types need always to be kept in mind. Adipose tissue, for example, contains cells which respond to prostaglandin E_1 with an increase in the level of cyclic AMP, as well as true adipocytes which respond with a fall in the level of cyclic AMP (Chapter 8). As a result, the relation of the antilipolytic effect of prostaglandin E_1 to its effect on the level of cyclic AMP in adipose tissue could not be understood until studies were carried out with isolated fat cells. Many and probably most organs contain cells whose adenyl cyclases differ in their sensitivities to hormones. The kidney, for example, contains cells which are sensitive to vasopressin, parathyroid hormone, and epinephrine. Chase and Aurbach (1968) showed that vasopressin stimulates adenyl cyclase primarily in cells from the renal medulla, while the parathyroid-sensitive cyclase was located primarily in the cortex. The degree of overlap may nevertheless be such that attempts to relate any one of the functions of these hormones to the level of cyclic AMP in the kidney may be difficult.

Another danger in these studies stems from the fact that cyclic AMP levels in most tissues are subject to extremely rapid fluctuations, so that in most cases the speed of tissue fixation becomes a highly critical factor. A good illustration of this was provided by Namm and Mayer (1968). They injected epinephrine into thoracotomized rats and froze the hearts 15 sec later by clamping between two blocks of metal at the temperature of liquid nitrogen. The frozen hearts were later pulverized and extracted, and cyclic AMP measured. By this technique it was possible to show that epinephrine had caused a highly significant threefold increase in the cardiac level of cyclic AMP. When the hearts were instead removed by scissors and then frozen by immersion in dichlorodifluoromethane at the temperature of liquid nitrogen, no increase in the level of cyclic AMP in response to epinephrine could be demonstrated. This requirement for rapid fixation has been especially frustrating to investigators interested in relating the level of cyclic AMP in various areas of the brain to behavioral effects. Many psychopharmacological agents are thought to produce their effects by altering the rate of release or reuptake of brain catecholamines, and it is known that the catecholamines may profoundly affect the level of cyclic AMP when incubated with brain slices (Kakiuchi and Rall, 1968a). However, until methods are developed which will allow the rapid fixation of discrete parts of the brain while the brain is in a functional state, the relation between cyclic AMP and behavior cannot be directly studied.

An interesting fact mentioned previously is that although fluoride stimulates adenyl cyclase in most broken cell preparations which have been studied, an effect of this ion on the level of cyclic AMP in an intact tissue has never been demonstrated. Fluoride had no effect, for example, when it was added to a suspension of washed avian erythrocytes, under conditions where epinephrine stimulated cyclic AMP formation dramatically (Øye and Sutherland, 1966). After hemolysis or cell breakage, however, and in the presence of ATP and Mg^{2+}, fluoride produced its characteristic stimulatory effect. Similar observations have been made with brain and many other tissues, and an example from adipocytes is provided by the data in Table 2.III. The reasons for the lack of effect of fluoride in intact tissues, in the face of its pronounced activity in broken cell preparations, are presently unknown.

TABLE 2.III. Effect of Epinephrine and Fluoride on Cyclic AMP Formation by Adipocytes Before and After Homogenization

Experiment[a]	Conditions	Cyclic AMP (mmoles $\times 10^{-7}$/gm)	
		Whole cells	Broken cells
I	Control	2.4	7.8
	Epinephrine (1 μg/ml)	20.4	8.7
	NaF (0.01 M)	3.2	33.3
II	Control	3.8	13.0
	Epinephrine (1 μg/ml)	10.8	96.1
	NaF (0.01 M)	4.9	145.2

[a]Experiments I and II were performed at different times using different preparations of adipose tissue. For the experiments involving whole cells, adipocytes were prepared by the method of Rodbell and incubated as described previously (Butcher et al., 1968) in the absence of added ATP or Mg^{2+}. For the experiments with broken cells, the adipocytes were homogenized in an isotonic tris buffer (experiment I) or in a hypotonic glycylglycine-$MgSO_4$ buffer (experiment II) and incubated in the presence of 2 mM ATP and 3 mM $MgSO_4$. Caffeine (1 mM) was present in all cases.

C. Phosphodiesterase Inhibitors

The potentiation of a hormone by theophylline or other phosphodiesterase inhibitor can in general be taken as good presumptive evidence that the hormone acts by stimulating adenyl cyclase. Although this evidence is relatively weak by itself, for reasons to be considered below, it nevertheless represents an important criterion which should be established. Table 2.IV contains a list of some of the responses for which

TABLE 2.IV. Some Hormonal Effects Which Are Enhanced
in the Presence of a Methylxanthine

Response	Hormone	Tissue	Reference
Phosphorylase activation	Catecholamines	Liver	Northrop and Parks, 1964a
	Catecholamines	Muscle	Hess *et al.,* 1963
Lipolysis	Catecholamines	Adipose	Vaughan and Steinberg, 1963
Positive inotropism	Catecholamines	Cardiac	Rall and West, 1963
	Histamine	Cardiac	Dean, 1968
Relaxation	Catecholamines	Smooth muscle	Lundholm *et al.,* 1966
Amylase release	Catecholamines	Parotid gland	Babad *et al.,* 1967
Insulin release	Catecholamines	Pancreas	Turtle *et al.,* 1967
	Glucagon	Pancreas	Sussman and Vaughan, 1967
	ACTH	Pancreas	Lebovitz and Pooler, 1967
Steroidogenesis	ACTH	Adrenal	Halkerston *et al.,* 1966
	LH (ICSH)	Ovary	Dorrington and Kilpatrick, 1967
	Angiotensin	Adrenal	Kaplan, 1965
Increased permeability	Vasopressin	Toad bladder	Orloff and Handler, 1967
	Vasopressin	Ventral frog skin	Baba *et al.,* 1967
Melanocyte dispersion	MSH	Dorsal frog skin	Abe *et al.,* 1969a
Thyroid hormone release	TSH	Thyroid	Bastomsky and McKenzie, 1967; Ensor and Munro, 1969
HCl production	Histamine	Gastric mucosa	Robertson *et al.,* 1950; Harris and Alonzo, 1965
Amylase release	Acetylcholine	Exocrine pancreas	Kulka and Sternlicht, 1968
TSH release	TRH	Anterior pituitary	Bowers *et al.,* 1968b
Facilitation of neuromuscular transmission	Catecholamines	Nervous	Breckenridge *et al.,* 1967; Goldberg and Singer, 1969
Reduced discharge frequency	Catecholamines	Cerebellar	Siggins *et al.,* 1969
Enzyme synthesis	Catecholamines	Liver	Wicks, 1968
Inhibition of aggregation	Prostaglandins	Platelets	Cole *et al.,* 1970
Calcium mobilization	Parathyroid	Bone	Wells and Lloyd, 1968a

this type of potentiation has been demonstrated. In many of these instances the participation of cyclic AMP has been established, whereas in others this has only been surmised. Caffeine and theophylline have been the drugs most widely studied from this point of view, but several other agents, including puromycin (Appleman and Kemp, 1966) triiodothyronine (Mandel and Kuehl, 1967), and some of the benzothiadiazine derivatives (Schultz *et al.,* 1966; Moore, 1968) have also been shown to be capable of inhibiting phosphodiesterase in at least some tissues.

The chief drawback with all of these agents, not only as tools for basic research, but also from the point of view of their use in clinical medicine, is that they are not very specific. In all cases relatively high concentrations are required in order to inhibit phosphodiesterase, and at these concentrations the drugs may have many other actions. These other actions could tend to promote the action of the hormone being studied (leading to a false positive, insofar as this approach is regarded as a criterion) or they might tend to inhibit it (leading to a false negative). Order of potency studies may be helpful in distinguishing the effects of phosphodiesterase inhibition from effects resulting from some other action. Of the methylxanthines, for example, theophylline has been found in all mammalian tissues studied to be a more potent inhibitor of phosphodiesterase than either caffeine or theobromine. This order of potency may not and indeed probably does not hold for all of the actions of these drugs.

In the case of a false negative, the fact that it is false must of course be established. Many of the actions of these drugs may tend to oppose those of cyclic AMP, or of agents which act by way of cyclic AMP. Some which have been demonstrated for the methylxanthines, at concentrations equal to or less than those which inhibit phosphodiesterase, include stimulation of phosphorylase phosphatase (Sutherland, 1951a) and glycogen synthetase phosphatase (De Wulf and Hers, 1968a), inhibition of the release of cyclic AMP from avian erythrocytes (Davoren and Sutherland, 1963a), and inhibition of the rise in cyclic AMP caused by adenosine in guinea pig brain slices (Kakiuchi *et al.,* 1969; Sattin and Rall, 1970).

It should also be established that an active phosphodiesterase is present in the preparation being studied, and that the drug to be used is capable of inhibiting it. It is becoming clear that the properties of phosphodiesterase, like those of many other enzymes, may differ from one tissue or species to another. Although all mammalian tissues which have been studied to date contain a phosphodiesterase which is susceptible to inhibition by the methylxanthines, for example, some unicellular organisms contain an enzyme which is apparently not affected by these compounds (Brana and Chytil, 1966; Chang, 1968).

Still another important point to be considered in studies with phosphodiesterase inhibitors is the dose of the hormone. If this is large enough to produce a maximal response, then clearly no amount of phosphodiesterase inhibition can increase it further, even though the level of cyclic AMP may be increased substantially. As a general rule, the most appropriate dose to use is the smallest dose which will yet give a measurable and reproducible response. Under these conditions the effects of hormones which stimulate adenyl cyclase will usually be increased in a clear-cut and dramatic fashion by drugs which inhibit phosphodiesterase activity.

A point related to the preceding one is that in studies with isolated tissue or organ systems *in vitro,* phosphodiesterase inhibitors may be incapable of producing a response in the absence of added hormones. In general, the effect of phosphodiesterase inhibition in any given tissue will depend upon the ongoing rate of adenyl cyclase activity, and this may vary with the temperature and other factors. In the isolated toad bladder at 22°C, for example, the addition of theophylline by itself causes a substantial increase in permeability, but at 2°C the presence of vasopressin or oxytocin is necessary in order for theophylline to be effective (Bourguet, 1968). A good illustration of this point is provided by the action of diazoxide in the liver. Sokal (1968) observed that diazoxide had no apparent effect on carbohydrate metabolism in the isolated perfused rat liver, and concluded from this that no part of the hyperglycemia which this drug produces *in vivo* (Tabachnick *et al.,* 1965) could be attributed to a direct effect on the liver. Other investigators, however, had reported that diazoxide was capable of inhibiting liver phosphodiesterase (Schultz *et al.,* 1966; Moore, 1968). We therefore tested diazoxide in the isolated perfused liver in the presence of a small concentration of glucagon which itself had a barely detectable effect on glucose output, and found that under these conditions diazoxide had a very striking effect (J. H. Exton and G. A. Robison, 1969, unpublished observations).

Finally, in this section, we can note that imidazole and possibly many other agents are capable of stimulating phosphodiesterase activity *in vitro.* To the extent that this also occurs in intact cells, these agents ought to antagonize hormones which act by stimulating adenyl cyclase. Although the effect of imidazole on the level of cyclic AMP in intact cells has not to our knowledge been measured, imidazole has now been tested in several systems as a possible antagonist of hormones which are either known or thought to act by stimulating adenyl cyclase. In each of these cases (listed in Table 2.V) imidazole was found to inhibit the response to the hormone. In several cases it was found that the inhibitory effect of imidazole could be overcome by higher concentrations of the hormone, as would be expected if imidazole was acting by stimulating

TABLE 2.V. Some Hormonal Effects Which Are Antagonized by Imidazole

Response	Hormone	Tissue	Reference
Positive inotropism	Catecholamines	Cardiac	Kukovetz and Pöch, 1967c
Relaxation	Catecholamines	Smooth muscle	Bueding et al., 1967; Wilkenfeld, and Levy, 1969
Lipolysis	Catecholamines	Adipose	Goodman, 1969a
	ACTH	Adipose	Nakano et al., 1970
TSH release	TRH	Pituitary	Wilber et al., 1969
Ca^{2+} mobilization	Parathyroid	Bone	Wells and Lloyd, 1968b
H^+ transport	Gastrin[a]	Gastric mucosa	Alonso et al., 1965
Inhibition of aggregation	Prostaglandins[a]	Platelets	Rossi, 1968
Steroidogenesis	ACTH	Adrenal	Linarelli and Farese, 1970

[a]The effect of imidazole was opposite to that of the hormone, but was not studied in the presence of the hormone.

phosphodiesterase. If it can be established that this occurs in intact cells, imidazole may provide an additional useful tool for the study of hormone action.

D. Exogenous Cyclic AMP

Although it may at times be possible to mimic the effect or effects of a hormone by the application of exogenous cyclic AMP, this may not be possible in all cases because of the permeability problem. Most cells are relatively impermeable to phosphorylated compounds in general, and cyclic AMP is additionally subject to rapid hydrolysis by phosphodiesterase. Nevertheless, by the application of large concentrations of the nucleotide, it has in several cases been possible to reproduce hormone effects in intact tissues. In some of these cases it has been necessary to include a phosphodiesterase inhibitor in order to see an effect. In certain other cases, where enough is known about the steps between the increase in cyclic AMP and the physiological response being investigated, this fourth criterion can often be established by the use of broken cell preparations. In certain cases where cyclic AMP itself is inactive when applied to intact cells, it may be possible to mimic the effect of the hormone with one of the acyl derivatives of cyclic AMP. The synthesis of these derivatives is described in detail in the next chapter. Their effects, and some of the problems pertaining to their use, are discussed in more detail in Chapter 5 (Section II).

E. Hormones which May or May Not Affect Adenyl Cyclase

The foregoing four criteria were set up for the purpose of identifying hormones which act by stimulating adenyl cyclase. Although no one of the four criteria is conclusive when considered by itself, the establishment of all four criteria would appear to constitute strong evidence that the hormone in question does act by this mechanism. Although it may be possible to design additional fruitful experiments as new information becomes available, this conclusion seems almost inescapable when the response has been studied in detail from all four points of view and where no discrepancies have been detected. In these cases we might say that a response mediated by cyclic AMP has been well established.

Other hormones might affect the level of cyclic AMP, either directly or indirectly, by stimulating or inhibiting the activity of either adenyl cyclase or the diesterase. For these hormones the aforementioned four criteria will not be applicable, at least in the form in which we have stated them here. Whereas it is possible to establish that a given hormone response is caused by an increase in the level of cyclic AMP even when the mechanism by which cyclic AMP acts in unknown (assuming the response can be elicited by exogenous cyclic AMP), it may be more difficult to prove the converse, i.e., that a particular response is mediated by a *decrease* in the level of cyclic AMP. It might seem reasonable, if the hormone can be shown to produce a decrease in the level of cyclic AMP, and if the effect in question is the opposite of that produced by an increase in the level of cyclic AMP, to conclude that the observed effect is the result of the decreased level of cyclic AMP. This conclusion is far from inescapable, however, as is well illustrated by the stimulatory effect of insulin on liver glycogen synthetase. This effect would be expected because cyclic AMP inhibits this enzyme (by stimulating its conversion to a less active form) and because insulin lowers the level of cyclic AMP in the liver. The effect of insulin is opposite to that of glucagon and epinephrine, both of which increase the level of cyclic AMP in the liver. On the basis of these data the probability seems high that insulin might stimulate glycogen synthetase in the liver, in part if not entirely, by virtue of its ability to lower the level of cyclic AMP in this organ. The hooker is that insulin also increases the activity of glycogen synthetase in muscle, a tissue in which it apparently does *not* lower the level of cyclic AMP. Now the possibility has to be considered that the mechanism by which insulin acts in muscle may also apply to its action in the liver, and this hypothesis cannot really be tested until more is known about its action in muscle. This is a good example of a situation where all of the evidence from one tissue is compatible with a mechanism involving a decrease in the level of cyclic AMP, but where evidence from another tissue suggests that caution is called for.

VI. TWO MAJOR PROBLEMS

In this chapter we have considered, in general terms, some aspects of hormone action related to cyclic AMP. Once it is established that a given response to a given hormone is mediated by cyclic AMP, two major problems remain to be solved before it can be said that the mechanism of action of the hormone is well understood. The first problem, which we considered briefly in Section III of this chapter, is to understand how the hormone acts to alter the level of cyclic AMP. Substantial progress here is not likely to come until methods are available for isolating in pure form adenyl cyclase and other constituents of the cell membrane.

The second problem, no less important than the first, is to understand how cyclic AMP acts to produce the observed response. Information relating to this problem varies considerably from one response to another, being meagre in some cases yet rather extensive in others. Here we wish only to note that no general principles have emerged at the present time which would enable us to understand the various actions of cyclic AMP within any kind of unitarian framework, and it would appear that for the time being each response will have to be considered on an individual basis. This is not to say that general principles will always be lacking in this field, only that our present knowledge is insufficient for them to be formulated. We have reserved Chapter 5 for a general discussion of this topic, while many of the actions of cyclic AMP will be dealt with in the appropriate sections of later chapters.

But first we are going to review what is known about cyclic AMP as a chemical compound, and also what we know about the enzymes which are involved in its synthesis and metabolic breakdown. Readers who are primarily interested in gaining a quick overview of the subject may wish to move directly to Chapter 5, reserving the intervening chapters for later reference.

VII. ADDENDUM

With reference to the first of the two major problems mentioned in Section VI above, we should like to call attention to the fine discussion of proteins by Dickerson and Geis (1969). We continue to believe that when the mechanism by which CTP inhibits the catalytic activity of aspartate transcarbamylase is understood, then a substantial step will have been taken toward understanding the mechanism by which hormones stimulate adenyl cyclase. It is becoming increasingly obvious, however, that there is a great discrepancy between what we know about enzymes that function in an aqueous environment and those that function in the lipophilic environment of the cell membrane.

Attention can also be called to an interesting paper by Lefkowitz *et al.* (1970). These investigators used high pressure disintegration in the presence of phosphatidylethanolamine to obtain a "solubilized" preparation of adenyl cyclase from an adrenal tumor. This preparation could be fractionated with the retention of sensitivity to stimulation by ACTH. The relation between ACTH-binding and the stimulation of adenyl cyclase activity was very striking in these experiments, emphasizing further the close relation that exists between hormone receptors and adenyl cyclase.

Bär and Hechter (1969a) have objected to our designation of cyclic AMP as a second messenger on the grounds that other factors may be generated to couple the initial step of hormone reception to the final step of cyclic AMP formation, the implication being that cyclic AMP would then be a messenger of some higher order. We feel that when these other factors are discovered, an adequate terminology can probably be devised to deal with the situation. Perhaps these factors will be analogous or related to those which allow the properties of common table salt to emerge from the properties inherent in metallic sodium and gaseous chlorine, in which event we would not want to refer to them in the same terms we reserve for discrete molecular entities. We commonly think of sodium chloride as composed of two components rather than three or more, and it was in a similar sense that we thought of cyclic AMP as a second messenger in hormone action rather than as a third or fourth.

Turning finally to the second major problem, understanding the mechanism of action of cyclic AMP, considerable progress was made in this area since this chapter was written. These newer findings are summarized in Chapter 5 and especially in the addendum to that chapter.

CHAPTER 3

Chemistry of Cyclic Nucleoside Phosphates and Synthesis of Analogs[*]

I. CHEMISTRY OF CYCLIC NUCLEOSIDE PHOSPHATES

A. Introduction

The properties of 3',5'-adenylic acid and its analogs are primarily of biological and biochemical interest, and in consequence the greater part of this monograph is devoted to these aspects. This book will therefore

[*]By Dr. Th. Posternak, Department of Biological Chemistry, University of Geneva.

be read largely by those who are not organic chemists. As a result, the chemical introduction provided by this chapter is elementary in scope: its object is to provide a background for readers unfamiliar with the basic principles of the chemistry of nucleic acids, nucleotides, and nucleosides, and, in particular, with some special chemical properties of the cyclic nucleoside phosphates, mainly the 3′,5′-phosphates. Cyclic 3′,5′-adenylic acid is in fact a simple derivative of a common nucleotide found throughout nature, 5′-adenylic acid or adenosine 5′-monophosphoric acid (AMP).

B. Nucleic Acids, Nucleosides, and Nucleotides

It will be recalled that nucleic acids give on hydrolysis the following substances: (1) phosphoric acid; (2) a sugar: ribose (I) from ribonucleic acid (RNA) and deoxyribose (II) from deoxyribonucleic acid (DNA); (3) the purine bases adenine (III) and guanine (IV), and the pyrimidine bases cytosine (VI) and thymine (V) from DNA, and cytosine (VI) and uracil (VII) from RNA.

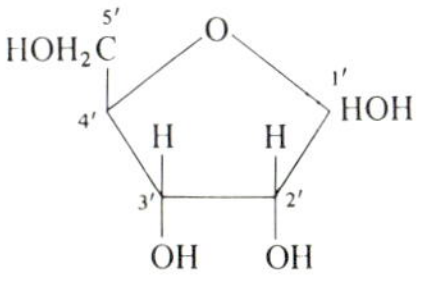

D - Ribose

(I)

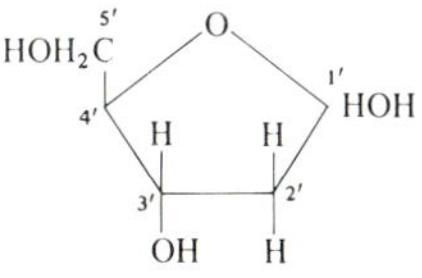

D - Deoxyribose

(II)

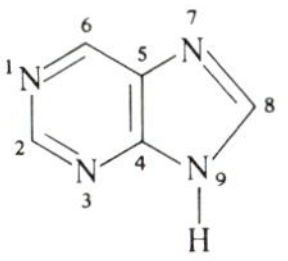

Purine

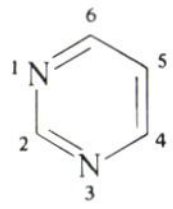

Pyrimidine

Numbering system of purine and pyrimidine derivatives

A nucleoside consists of a sugar and a base joined by a glycosidic bond. In the naturally occurring nucleosides, the sugar is linked to N-9 in the purine bases and N-3 in the pyrimidine bases (Gulland and Story, 1938). It has been shown that ribose or deoxyribose are found in the furan form (I and II) (Levene and Tipson, 1937) and that the glycosidic bond configuration is β (Lythgoe et al., 1947). The ribonucleosides have the

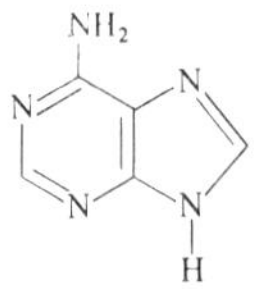

Adenine
(III)

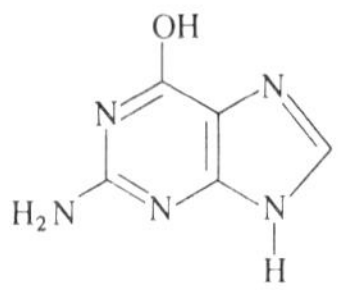

Guanine
(IV)

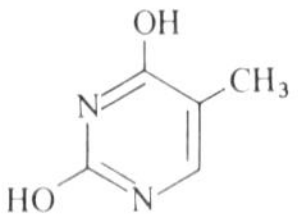

Thymine
(V)

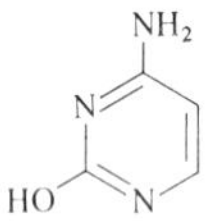

Cytosine
(VI)

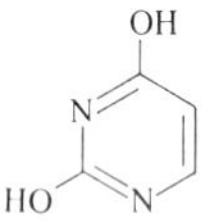

Uracil
(VII)

following names according to the kind of base they contain: adenosine, guanosine, uridine, and cytidine.

The most general method of synthesizing the ribonucleosides is by treating suitably protected base derivatives with an acetohalogeno-ribofuranoside; these derivatives must have an unsubstituted N-9 (purine derivative) or N-3 (pyrimidine derivative) position.

Nucleic acids are known to consist of long chains in which the nucleo-sides are joined by $3',5'$-phosphodiester linkages.

The chain subunits known as nucleotides are in fact nucleoside phos-phates. The ribonucleotides have the following names: adenylic acid (AMP); guanylic acid (GMP); cytidylic acid (CMP), and uridylic acid (UMP).* The name of the nucleotide is often preceded by the number of the sugar carbon to which the phosphate group is joined. In this way, $5'$-adenylic acid, for example, may be described as adenosine $5'$-phosphate or $5'$-AMP.

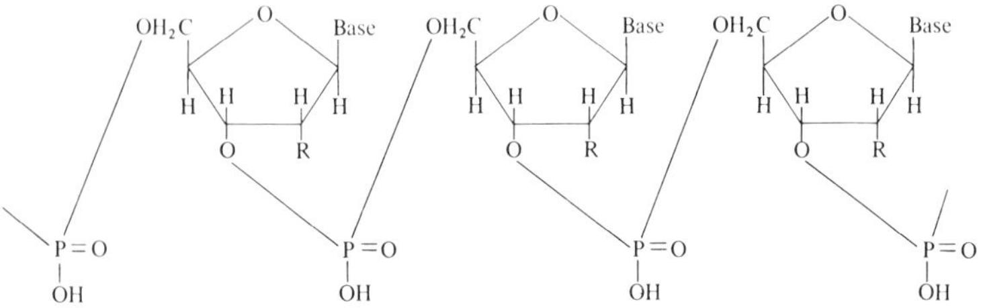

RNA   R = OH
DNA   R = H

*Thymidylic acid (TMP) is a deoxyribonucleotide.

In general, nucleoside 5'-phosphates are most conveniently prepared by synthesis. The nucleoside is first converted to a 2',3'-isopropylidene derivative (VIII) by condensation with acetone. This is then subjected to phosphorylation at the 5'-position, for instance using dibenzylphosphochloridate, to yield another intermediate (IX). The protecting groups are then removed: the benzyl residues by catalytic hydrogenation to give toluene, the isopropylidene residue by controlled acid hydrolysis. A nucleoside 5'-phosphate (X) is thus obtained (Michelson and Todd,

(VIII)   (IX)   (X)

1949). The nucleoside 2'- and 3'-phosphates are most easily obtained by chemical or enzymatic degradation of nucleic acids.

C. Cyclic Phosphodiesters

1. Some Reactions of Phosphate Esters

Before proceeding further, it may be useful to consider some general properties of the phosphate group. The presence of the oxygen atom in this group results in an appreciable positive charge on the phosphorus. The latter is therefore susceptible to nucleophilic attack by either neutral or anionic reagents with unshared electron pairs. This results in the displacement of a group X, as in the following two examples:

$$\begin{array}{c} R_1 \\ \diagdown \\ P \\ \diagup \diagdown \\ R_2 \quad X \end{array} \; \overset{O}{} \; + \; ROH \; \longrightarrow \; \begin{array}{c} R_1 \\ \diagdown \\ P \\ \diagup \diagdown \\ R_2 \quad OR \end{array} \; \overset{O}{} \; + \; H^+ \; + \; X^- \qquad (1)$$

$$\begin{array}{c} R_1 \\ \diagdown \\ P \\ \diagup \diagdown \\ R_2 \quad X \end{array} \overset{O}{} + \begin{array}{c} R_3 \\ \diagdown \\ P \\ \diagup \diagdown \\ R_4 \quad O^- \end{array} \overset{O}{} \longrightarrow \begin{array}{c} R_1 \\ \diagdown \\ P \\ \diagup \diagdown \\ R_2 \quad O \end{array} \overset{O}{} \begin{array}{c} O \quad R_3 \\ \diagdown \diagup \\ P \\ \diagdown \\ R_4 \end{array} + X^- \qquad (2)$$

2. Formation of Cyclic Phosphodiesters

The term cyclic phosphodiester describes a compound in which one phosphoric acid residue is esterified by condensation with two hydroxyl groups found in the same molecule. According to the respective positions of these two OH groups, it is evident that cyclic esters of variable ring size could exist. Only those having a 5-membered ring (derived from a 1,2-glycol) or a 6-membered ring (derived from a 1,3-glycol) concern us here. The formation of a phosphodiester with a 5-membered ring can be readily effected in a number of ways.

(i) A dialkyl phosphate becomes particularly alkali-labile when the carbon adjacent to the one carrying the phosphoryl group is bound to a hydroxyl group (XI). By nucleophilic attack, the latter forms a bond with the phosphorus with elimination of RO^- and H^+ which combine to form ROH.

The cyclic phosphate (XII) is sensitive to hydrolysis and under more vigorous conditions the original monophosphate (XIV) is formed together with a second monophosphate ester (XIII) which results from migration of the phosphate group. It can be seen that the new derivative is an isomer of the original compound.

The formation of the cyclic phosphodiester is also governed by steric factors. If the original compound derives from a cyclic dialcohol possessing a 5'- or 6'-membered ring, the reaction is greatly facilitated if this dialcohol has the *cis* configuration, because of the proximity of the hydroxyl groups. These structural features are found in the ribonucleic acids, where the furanose ring possesses two *cis* hydroxyl groups. RNA is, in fact, known to be attacked by alkali under relatively mild conditions to give, as intermediates, nucleoside 2',3'-phosphodiesters (XV). These are converted by subsequent hydrolysis to a mixture of nucleoside 2'- and 3'-phosphates. No nucleoside 5'-phosphates are formed under these conditions.

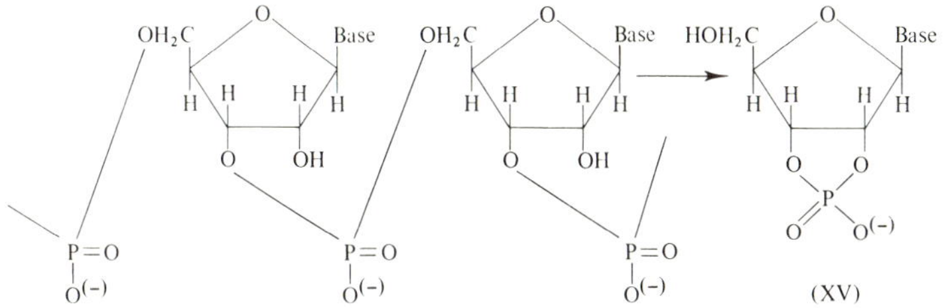

(XV)

The resistance of DNA to the action of alkalis is explained by the absence of an OH group in the 2'-position. In this case the formation of cyclic phosphodiesters is impossible.

It is worthy of mention that ribonucleases (enzymes which hydrolyze RNA to nucleotides) act in an analogous way, i.e., by causing the intermediate formation of cyclic phosphodiesters which, by further hydrolysis, are converted to a mixture of 2'- and 3'-nucleotides.

A similar mechanism pertains to the rearrangement undergone by phosphomonoesters of 1,2-glycols when treated with acid. (These compounds are, interestingly enough, quite stable in the presence of alkali.) Once again, the intermediate is a phosphodiester (XVII) which then undergoes hydrolysis.

(ii) Another way of forming a cyclic phosphodiester from a phosphomonoester involves the use of a now standard reagent, dicyclohexyl-

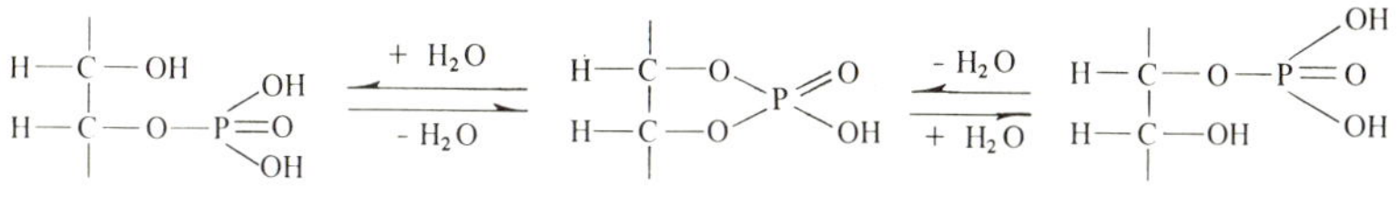

(XVI)   (XVII)   (XVIII)

carbodiimide (XIX), where R_1 = cyclohexyl, commonly referred to as DCC (Khorana et al., 1957; Smith et al., 1961). The probable course of the reaction is as follows:

The addition product (XXI) is the site of an intramolecular nucleophilic attack on the phosphorus by the neighboring OH group, thus forming a cyclic phosphodiester (XXIII) in addition to dicyclohexylurea (XXII). In a subsequent reaction, 5-membered cyclic phosphodiesters may react with dicyclohexylurea in the presence of DCC with the formation of a dicyclohexylurea phosphoryl derivative. This reaction is of interest as it leads to the preparation of phosphodiesters containing not only 5-membered rings, but also 6-membered rings (derivatives of 1,3-glycols) and 7-membered rings (derivatives of 1,4-glycols) (Khorana et al., 1957; Smith et al., 1961).

The formation of a phosphodiester with a 5-membered phosphate ring necessitates a pentagonal cyclic glycol with *cis* 1,2-OH groups. In the case of a hexagonal cyclic glycol in the chair configuration, the two neighboring OH groups must be one axial and the other equatorial (*cis* configuration) (Fig. 3.1A) or both equatorial (*trans* configuration) (Fig.

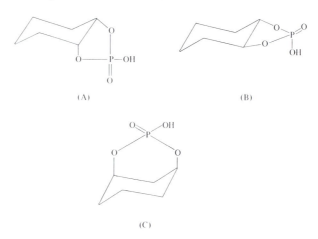

(A)   (B)

(C)

Fig. 3.1. Fusion of a chair form with a cyclic phosphate ring. A, Derivative of an axial-equatorial cyclic 1,2-glycol; B, derivative of a diequatorial cyclic 1,2-glycol; C, Derivative of diaxial cyclic 1,3-glycol.

3.1B). The *trans* configuration involving two 1,2 axial hydroxyls is excluded.

It should be evident that in order to form a 6-membered cyclic phosphodiester from a cyclic 5-membered glycol, the 1,3-OH groups must be *cis*. If the 6-membered cyclic phosphodiester derives from a 6-membered cyclic 1,3-glycol, as in C in Fig. 3.1, the most favorable orientation of the two OH groups would be diaxial. When the phosphate group has neighboring hydroxyls in both 1,2- and 1,3-positions in favorable orientation, it is usually the former which react, giving rise to a pentagonal ring.

Another reaction which always takes place when a monophosphate is treated with DCC is an attack by a second molecule of phosphate ester on the intermediate addition product (XXIV). The product of this reaction is a pyrophosphate (XXV) which may be an obligatory intermediate. This pyrophosphate (for instance XXVI) can be converted to a cyclic phosphate (XXVII) under certain conditions (e.g., treatment with DCC in boiling pyridine) with the regeneration of the original monophosphate ester (X).

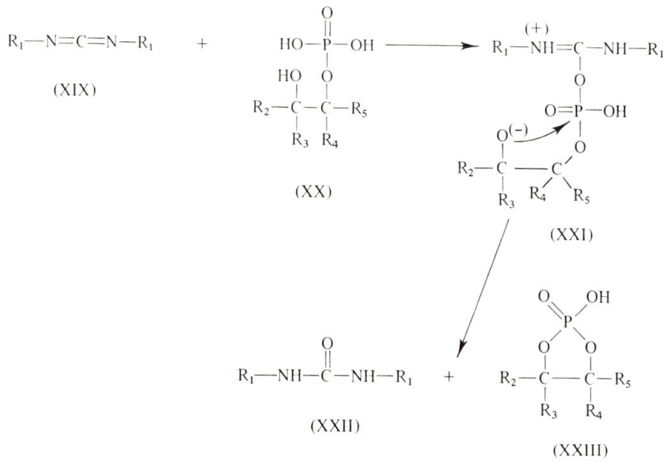

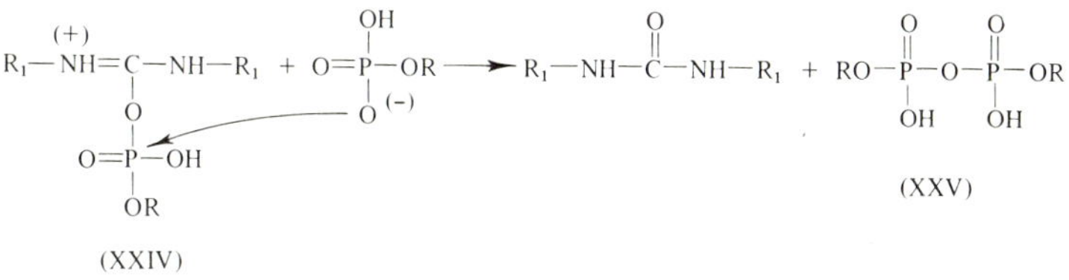

R_1—N=C=N—R_1 $\quad$ + $\qquad$ HO—P(=O)—OH $\quad\longrightarrow\quad$ R_1—NH=C(+)—NH—R_1

(XIX)

HO O
R_2—C—C—R_5
R_3 R_4

(XX)

O=P(—OH)
$O^{(-)}$ O
R_2—C——C(—R_4,—R_5)
R_3

(XXI)

R_1—NH—C(=O)—NH—R_1 $\quad$ + $\qquad$ R_2—C——C—R_5 with O—P(=O)—OH bridge, R_3 R_4

(XXII) $\qquad\qquad$ (XXIII)

R_1—NH=C(+)—NH—R_1 $\quad$ + $\quad$ O=P(—OH)(—OR) $\quad\longrightarrow\quad$ R_1—NH—C(=O)—NH—R_1 $\quad$ + $\quad$ RO—P(=O)(OH)—O—P(=O)(OH)—OR

O=P(—OH)(—OR) $O^{(-)}$

(XXIV) $\qquad\qquad$ (XXV)

R_1 = cyclohexyl

Base—(sugar)—CH_2—O—P(=O)(OH)—O—P(=O)(OH)—O—CH_2—(sugar)—Base

H H   H H $\qquad$ H H   H H
HO   OH $\qquad\qquad$ HO   OH

(XXVI)

$\downarrow$ DCC

CH_2—O—(sugar)—Base with O=P(—HO)—O cyclic $\qquad$ + $\qquad$ $H_2O_3POH_2C$—(sugar)—Base

H H   H H $\qquad\qquad\qquad\qquad\qquad\qquad$ H H   H H
HO   OH $\qquad\qquad\qquad\qquad\qquad\qquad\qquad\qquad$ HO   OH

(XXVII) $\qquad\qquad\qquad\qquad\qquad\qquad\qquad$ (X)

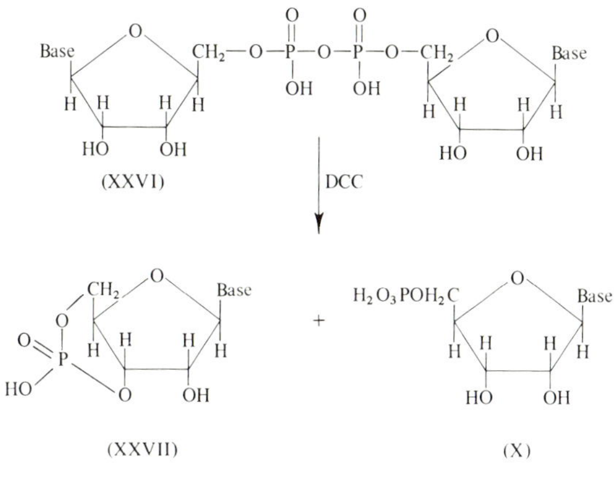

(iii) More recently, Borden and Smith (1966a,b) have synthesized nucleoside 3′,5′-cyclic phosphates by a cyclization reaction in which the starting material is a phosphodiester of the type (XXIV), in which X is a 4-nitrophenyl or 2,4-dinitrophenyl residue. (Nitrophenols are used instead of phenol because they are more acidic, thus facilitating the elimination of X in the form of a nitrophenoxide.) The cyclization step is accomplished in anhydrous dimethylsulfoxide. The sodium salt of the starting material is treated in the presence of potassium-t-butoxide in t-butyl alcohol. Under these conditions the OH groups are ionized and their anions can make a nucleophilic attack on the phosphorus, thereby displacing the nitrophenoxide residue with the formation of the cyclic diester (XXVII). The advantage of this reaction is that it takes place under mild conditions, at 20°C.

Several procedures have been described for the preparation of the starting compounds (XXXI). In one, a nucleoside 5′-phosphate (XXVIII) is allowed to react with 4-nitrophenol or 2,4-dinitrophenol in the presence of DCC. This is now a standard method of esterification. In another procedure a suitably protected nucleoside, such as the 2′,3′-isopropylidene derivative (XXIX) of guanosine, is allowed to react with nitrophenyl phosphate (XXX) and DCC. The protecting group is then removed by mild acid treatment, leaving the 4-nitrophenyl ester of guanosine 5′-phosphate.

(iv) Nucleoside 3′,5′-cyclic phosphates can also be prepared by treating the corresponding nucleoside triphosphate with barium hydroxide. In fact, cyclic AMP (XXXIII) was first synthesized by degrading ATP (XXXII) with hot barium hydroxide (Cook *et al.*, 1957; Lipkin *et al.*, 1959a). The mechanism of this reaction can be represented by Eq. (1) in

(XXXIII)

(XXXII)

Section I, C, 1, except that in this case the group displaced by nucleophilic attack on the phosphorus is the pyrophosphate anion. This reaction is also of interest because it is analogous to the reaction catalyzed by adenyl cyclase, as discussed in the next chapter.

3. Conformation of Cyclic Nucleotides

As ribose in the furanose form has its 2'- and 3'-OH groups in the *cis* configuration, the formation of a 2',3'-phosphodiester is favored from a steric point of view. The conformation of a 3',5'-phosphodiester leads to an appreciable strain, as we are dealing with a bicyclic system consisting of a 6-membered ring considered to be in the chair conformation (this having the least energy) and a 5-membered (furanose) ring. These two rings are fused in a *trans* relationship to one another. On considering a model of such a system (Fig. 3.2), it is evident that the furanose ring must adopt the half-chair form in which two neighboring carbon atoms take

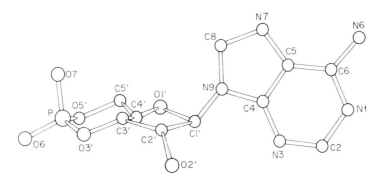

Fig. 3.2. Conformation of a nucleoside 3',5'-cyclic phosphate. This is one of two conformations in which cyclic AMP exists in crystals, with the conformation of the ribose phosphate ring system being the same in both. From Watenpaugh *et al.* (1968).

up their positions on either side of the plane determined by the other atoms. C-3' in this case is pointing on the same side of the plane defined by C-1', C-2', and O as the C-4'–C-5' bond, while C-4' is directed on the opposite side. This conformation has been confirmed by NMR spectroscopy (Jardetzky, 1962).

The considerable strain existing in the nucleoside 3',5'-phosphodiesters may explain why their synthesis requires fairly energetic conditions. For example, their preparation by the action of DCC on a 5'-nucleotide required heating in boiling pyridine for several hours.

4. Chemical Properties of Cyclic Phosphodiesters

The ring of a cyclic phosphodiester can be opened by acid or alkaline hydrolysis to form a phosphomonoester. The rate of this process differs according to the number of atoms in the ring. Five-membered cyclic esters of phosphoric acid as well as phosphonic and sulfuric acids show extraordinarily high rates of hydrolysis compared to their acyclic or 6-membered analogs. If the rates of hydrolysis in 1 N HCl at the same temperature are compared, for example, it can be said that ethylene phosphate (XXXIV) is hydrolyzed approximately a million times more rapidly than dimethyl phosphate (XXXVI). However, a 6-membered phosphodiester such as trimethylene phosphate (XXXV) is hydrolyzed only about ten times faster than dimethyl phosphate. The high rate of hydrolysis exhibited by 5-membered cyclic phosphodiesters has been

(XXXIV)   (XXXV)   (XXXVI)

explained by the appreciable strain found in the ring. X-ray studies (Usher *et al.*, 1965; Steitz and Lipscomb, 1965) of methylethylene phosphate have shown the existence of deformations, particularly in the angles between the phosphorus and the two ring oxygen atoms. It has been shown by the use of ^{18}O in alkaline medium that dimethyl phosphate is hydrolyzed principally by rupture of the C—O bond, whereas in the case of ethylene phosphate it is the P—O bond which breaks. The considerable strain that would be found in the intermediate addition product of the cyclic phosphate with a molecule of water has also been implicated (Dennis and Westheimer, 1966). The nucleoside cyclic 2',3'-phosphodiesters are hydrolyzed with the same facility as the other phosphodiesters with 5-membered rings.

Information concerning the hydrolysis of 3′,5′-phosphodiesters is limited. Although cyclic AMP is somewhat more stable than the corresponding 2′,3′-phosphate, its hydrolysis in alkaline medium is nevertheless much faster than that of trimethylene phosphate. The time required for the disappearance of 50% of the starting material ($t/2$) in 1 N NaOH at 100°C is about 12 min in the case of cyclic AMP, whereas for trimethylene phosphate it is on the order of 20 hr. Presumably the greater lability of the nucleotides stems from the extra strain resulting from the *trans* fusion of the phosphate ring with the sugar furanose ring.

It has been shown, both in acid and in alkaline medium, that the bond which is preferentially broken in a ribonucleoside 3′,5′-phosphodiester is the 5′-bond. This is undoubtedly a result of less steric protection, and it is for this reason that cyclic AMP, when treated with Dowex-50 in the H⁺ form (Sutherland and Rall, 1958) gives mostly ribose-2(3)-phosphate and only 5% ribose-5-phosphate (the riboside bond also being broken under these conditions). When treated for 26 min with 1 N HCl at 100°C, cyclic 3′,5′-CMP gives 2′(3′)-CMP and 5′-CMP in the proportion of 6 : 1. Treated for 8 min under the same conditions, cyclic UMP gives a mixture of 2′(3′)-UMP and 5′-UMP in the proportion of 5 : 1 (Smith *et al.,* 1961).

The results with acid hydrolysis stand in contrast to the action of the specific phosphodiesterase which attacks only the 3′-bond of cyclic AMP with quantitative formation of 5′-AMP. The action of this enzyme is discussed in the next chapter.

5. Hydrolysis of the Glycoside Bond

It is known that the glycoside bond of purine bases is more acid-labile than that of the pyrimidines, and it is also known that the nucleosides derived from deoxyribose are easier to hydrolyze than those derived from ribose. There is an empirical rule which states that the ease of hydrolysis of an N-glycoside increases with the basicity of the nitrogen. A basic nitrogen carries an unshared electron pair and its basicity is due to the tendency of this pair to fix a proton. During the acid hydrolysis of an N-riboside (XXXVII) the first event is considered to be precisely this fixation of a proton (XXXVIII) followed by the formation of the free base and of the carbonium ion (XXXIX). The latter gives finally the hydrolysis product (XL) by addition of OH⁻. The presence of an electron-attracting hydroxyl group slows the reaction XXXVIII ⟶ XXXIX, which is why the hydrolysis of deoxyribosides, in which this group is absent, is usually more rapid than that of ribosides.

According to another representation (Micheel and Heesing, 1961;

Garrett *et al.,* 1966), the primary product (**XXXVIII**) would undergo a transposition consisting of the migration of a proton to the ring oxygen accompanied by opening of the ring (**XLI**). By hydrolysis, the free base and compound (**XLII**) are then formed and the latter undergoes a ring closure to (**XL**) by loss of a proton.

The nitrogen of the pyrimidine bases is adjacent to a $C = O$ group and thus has amide characteristics. Its basicity is therefore diminished as a result of the possibility of mesomerism:

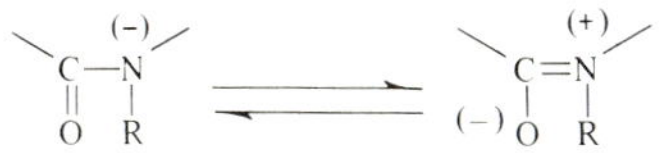

According to the empirical rule mentioned above and to the mechanism just described, the hydrolysis of a pyrimidine nucleoside would be more difficult than that of a purine nucleoside, and this has in fact been shown by experiment.

In the case of the nucleoside 2',3'-phosphodiesters, the sensitivity of the phosphate ring to acid hydrolysis is such that its rupture is accomplished much more rapidly than hydrolysis of the glycoside bond. On the other hand, the phosphodiester ring of 3',5'-phosphodiesters is much more resistant to hydrolysis, and consequently it has been possible to study its influence on the rate of glycosidic hydrolysis. The data in Table 3.I was obtained by Smith et al. (1961). It can be seen from this table that

TABLE 3.I. Hydrolysis of Cyclic 3',5'-Nucleotides in 1 N HCl at 100°C

	$t/2$ in min required for:	
Cyclic 3',5'-nucleotide	Disappearance of the starting material	Hydrolysis of the glycosidic bond of the corresponding 2' (3') or 5'-nucleotide[a]
3',5'-AMP	30[b]	2–4
3',5'-GMP	28[b]	1.5
3',5'-UMP	8[c,d]	
3',5'-TMP	5[e]	[f]
3',5'-CMP	26[g,h]	[h]

[a] The speed of nucleotide glycoside hydrolysis is only slightly affected by the position of the phosphate residue.

[b] In the case of 3',5'-GMP and -AMP, $t/2$ corresponds to half the time required for liberation of the purine bases. This liberation may be preceded by opening of the phosphodiester ring, in which case the latter reaction would be the rate-determining step.

[c] Uracil represents 67% of the decomposition products, the rest being 6% 5'-UMP and 24% 2'(3')-UMP.

[d] Uridine 5'-phosphate is unaffected under these conditions.

[e] In 0.1 N HCl at 100°C.

[f] Thymidine 5'-phosphate is unaffected under these conditions.

[g] The products are 5'-CMP (14%) and 2'(3')- CMP (86%).

[h] Cytidylic acid is unaffected under these conditions.

the presence of the cyclic phosphate group produces a marked stabilization of the glycoside bond of the purine bases. On the other hand, the labilization of this bond is evident in cyclic 3′,5′-UMP and 3′,5′-TMP, although this effect is not seen in the case of 3′,5′-CMP. At the present time it is difficult to suggest a mechanism to explain these interesting effects.

It may be added that the rates of acid hydrolysis of the glycoside bonds of various nucleosides and corresponding nucleotides have also been compared. Whereas in the case of 5′-AMP a labilization due to the phosphate residue has been shown, for 5′-deoxycytidylic acid the reverse effect has been found, i.e., a greater stability to hydrolysis than deoxycytidine (Venner, 1966).

6. Spectrophotometric Constants of Cyclic Nucleotides

The ultraviolet absorption spectra of cyclic AMP at various pH's are essentially the same as those of adenosine and other adenosine nucleotides. The extinction coefficients for several cyclic nucleotides at the wavelength of maximum absorbancy are listed in Table 3.II.

D. Preparation of Cyclic GMP*

The synthesis of cyclic GMP, the possible biological significance of which is discussed in Chapter 11, presents certain difficulties. The preparation of guanosine-5′-4-nitrophenyl phosphate, which could be an intermediate (see Section I, C, 2), gives poor yields. The direct cyclization in pyridine solution in the presence of DCC is accomplished with difficulty because of the poor solubility of 5′-GMP (Strauss and Fresco, 1965). It is better to use a benzoylguanosine 5′-phosphate as the starting material. The procedure described below represents an improvement on the method of Smith *et al.* (1961).

> Two-hundred milligrams of 5′-GMP (Pabst, sodium salt, 0.47 mmole) were dissolved in 20 ml H_2O. The solution was cooled in ice, and 4 gm Amberlite IR-120 (H^+) were added. After shaking for 7.5 min, the suspension was filtered with suction on a Buchner filter and the resin washed with ice cold water. Total volume: 100 ml. The total phosphate was estimated, and 320 mg (1.05 mmole) of 4′-morpholine-*N*, *N*′-dicyclohexylcarboxamidine were added. The solution was evaporated to dryness at low temperature and left for 4–5 hr under high vacuum.
> **Benzoylation.** The dry residue (0.38 mmole) was taken up in a mixture of 7.2 ml of anhydrous pyridine and 1.2 ml of benzoyl chloride (10.5 mmoles). It was allowed

*Section I.D written in collaboration with J. G. Falbriard.

TABLE 3.II. Composition and λ_{max} of Cyclic Nucleotides[a]

Cyclic Nucleotide	Composition	λ_{max} (mμ)	(ϵ_{max})
3',5'-AMP	$C_{10}H_{12}N_5O_6P \cdot H_2O$ (crystalline free acid)	258 256	(ϵ 14650) at pH 7.0 (ϵ 14500) at pH 2.0
3',5'-GMP	$C_{10}H_{12}N_5O_7PCa/2 \cdot 5\ H_2O$ (Ca salt)	262 254 236	(ϵ 12400) at pH 12.0 (ϵ 12950) at pH 7.0 (ϵ 11350) at pH 1.0
3',5'-CMP	$C_9H_{12}N_3O_7P$ (Crystalline free acid)	272 279	(ϵ 9340) at pH 7.0 (ϵ 12430) at pH 2.0
3',5'-UMP	$C_9H_{11}N_2O_3P \cdot C_6H_{15}N$ (Crystalline triethyl-ammonium salt)	260 261	(ϵ 7740) at pH 12.0 (ϵ 9940) at pH 7.0
3',5'-TMP		264.5 264.5	at pH 12.0 at pH 1.0
2',3'-AMP	$C_{10}H_{11}O_6N_5Ba/2$ (Ba salt)	260	(ϵ 13150) Neutral salt
2',3'-GMP	$C_{10}H_{11}O_7N_5PBa/2$ (Ba salt)	252	(ϵ 12150) Neutral salt
2',3'-CMP	$C_9H_{11}O_7N_3PBa/2$ (Ba salt)	268 232	(ϵ 8400) and (ϵ 8150) Neutral salt
2',3'-UMP	$C_9H_{10}O_8N_2PBa/2$ (Ba salt)	258.5	(ϵ 9570) Neutral salt

[a] From Brown et al. (1952) and Smith et al. (1961).

to react in the dark, shaking from time to time. A slight precipitation was formed together with a slight brown coloration. After 1 hr it was cooled in ice and 20 ml water were added in small portions; after 5 min it was extracted 4 times with 30 ml volumes of chloroform. The chloroform extracts were dried over Na_2SO_4. A single spot was obtained when an aliquot was subjected to chromatography in a system containing isopropanol and 1% ammonium sulfate (2:1 v/v). After filtration, the Na_2SO_4 was washed with chloroform and the solutions were evaporated to dryness in vacuo and left under high vacuum for 4–5 hr.

Partial Saponification. The residue was taken up with 12 ml of a mixture of 2 parts pyridine to 1 part water. Saponification was effected by addition of 12 ml of 2 N NaOH. The mixture was allowed to sit at room temperature for 4½ min. The pH was adjusted to 7.5 with 20 gm Amberlite Ir-120 (pyridinium form). Twenty-five milliliters of water were then added; the exchanger was removed by filtration and washed with 150 ml H_2O. The filtrate was evaporated to dryness in vacuo and left for several hours under high vacuum. The residue was suspended in 20 ml water. Twelve grams of Dowex-50 (H^+) were added to lower the pH to 2. The solution was then diluted with 20 ml water. After filtration and washing with 150 ml water, the filtrate was extracted 4 times with 75 ml ether. Yield: 8.95 mg P (65.5% of the phosphate of the original GMP). The aqueous solution was neutralized with pyridine and evaporated to dryness at low temperature. (R_f of N-benzoyl-5'-GMP = 0.50 in the system described above.)

Cyclization and Saponification. The residue was taken up in 70 ml anhydrous pyridine; 100 mg of 4'-morpholine-N,N'-dicyclohexylcarboxamidine and 206 mg of DCC were added. After boiling under reflux for 3 hr a further 412 mg DCC were added and after 5 hr another 412 mg DCC. After the last addition, the solution was left for 5 hr under reflux. After removal of the solvent under reduced pressure, the residue was extracted with 150 ml H_2O and the extract evaporated to dryness. The residue from this step was taken up with a mixture of 5 ml 95% alcohol and 10 ml concentrated ammonium hydroxide. The solution was heated for 2 hr at 105 °C in a sealed tube. After evaporation to dryness, it was taken up in 80 ml H_2O and extracted 3 times with 40 ml ether. The aqueous solution was concentrated to a small volume and applied to a DEAE-cellulose (HCO_3^-) column of dimensions 35 × 2.8 cm. A linear gradient elution was carried out with 7 liters of triethylammonium bicarbonate (pH 7.5), 0.002–0.1 M. Fractions of 20 ml (8.5/hr) were collected. The substance was detected spectrophotometrically at 252 mμ, with the appropriate peak usually being found in the fractions 70–110. These fractions were evaporated to dryness at low temperature and the triethylamine removed by several coevaporations with methanol. The triethylammonium salt of cyclic GMP (composition $C_{13}H_{20}N_7O_7P$) was thus obtained. Yield: 17% of the starting GMP. Extinction coefficient: $\lambda_{max} = 252–253$ mμ; $\epsilon_{max} = 12650$. $R_f = 0.225$ in the system isopropanol-concentrated ammonia-water (7 : 1 : 2).

II. DERIVATIVES OF CYCLIC AMP

A. Introduction

As discussed again in Chapter 5 (Section II), cyclic AMP is a relatively inert chemical when applied to intact cells or when injected into intact animals. Among the possible reasons for this are poor penetration through cell membranes (which seems to be characteristic of phosphorylated compounds in general) and rapid destruction by the phosphodiesterase. The preparation of analogs and derivatives was undertaken in order to obtain substances favoring either a better penetration to the interior of the cells or a greater resistance to the action of phosphodiesterase or both (Posternak *et al.,* 1962; Falbriard *et al.,* 1967). In addition, it was felt that the study of these derivatives might: (1) help to establish correlations between structure and activity at the biochemical level; and (2) lead to the preparation of compounds that might act as antagonists of the original substance.

Considering the structure of cyclic AMP, it is evident that there are three reactive functions capable of being transformed: (a) the N^6-amino group; (b) the 2'-hydroxyl group; and (c) the phosphoryl group. Moreover, introduction of substituents in the purine skeleton is possible. These possibilities are represented in Fig. 3.3.

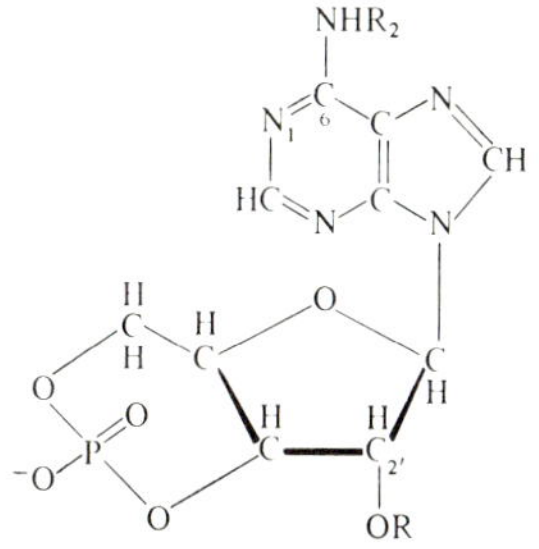

$$
\begin{array}{ll}
\text{I} & R_1 = R_2 = H \\
\text{II} & R_2 = H \\
\text{III} & R_1 = H \\
\text{IV} & R_1 = R_2
\end{array}
$$

Fig. 3.3. Basic structure of adenosine $3',5'$-monophosphate and current derivatives. R_1 and R_2 are acyl groups unless otherwise indicated.

B. Some Esters and An Oxide

As an example of the conversion of the phosphoryl group, ethyl and methyl esters of cyclic AMP have been prepared by the action of the corresponding alkyl iodides on the silver salt of the cyclic nucleotide. The possible biological activity of these derivatives has yet to be evaluated.

A nitrogen oxide has been obtained by the action of perphthalic acid on cyclic AMP. By analogy with the results of similar reactions with other adenine derivatives, it is almost certain that we are dealing here with the N^1-oxide. As in the case of the alkyl esters, this derivative also has not been tested for biological activity.

C. Acyl Derivatives

The derivatives which have so far shown the most interesting biological properties, as discussed again in Chapter 5, Section II, are those carrying an acyl group either in position N^6 or $2'-O$ or simultaneously in both positions. The basic idea here was to introduce into the molecule lipophilic fatty acid residues of variable length, in hopes that this might facilitate passage across cell membranes. As noted previously, these derivatives did prove to be generally more potent than cyclic AMP when applied to intact cells, but the reason for this is still not entirely clear.

The general preparative method is as follows. Cyclic AMP is dissolved in anhydrous pyridine in the form of a salt with either 4'-morpholine-N,N'-dicyclohexylcarboxamidine (Smith *et al.*, 1961) or triethylamine and treated with an acid anhydride. A considerable difference was found in the speeds of acylation at room temperature: the reaction on the oxygen took place much more rapidly than that on the nitrogen. Brief treatment by the acylating reagent gave a 2'-O-monoacyl derivative of cyclic AMP. Using the 4'-morpholine-N,N'-dicyclohexylcarboxamidine salt the nearly quantitative formation of the 2'-O-acyl derivatives was observed according to the following times (as shown by paper chromatography): acetyl derivative, 2 hr; butyryl derivative, 8 hr; octanoyl derivative, 12 hr. More prolonged treatment by the acylating reagent (from 5 to 8 days) produced a N^6-2'-O' diacyl derivative.

Another method for preparing these diacyl derivatives consists of treating a salt of cyclic AMP in pyridine solution with an acid chloride. The products obtained are in general more difficult to purify, and this alternate method possesses no advantages over the one described above.

The two acyl residues have very different susceptibilities to the action of alkali. The 2'-O-acyl group can therefore be selectively eliminated by brief treatment at room temperature (about 5 min in 0.1 N NaOH) with the formation of a N^6-monoacyl derivative of cyclic AMP.

The products have been characterized by paper chromatography (Whatman No. 1), the preferred solvent system being ethanol and 0.5 M ammonium acetate (5:2). The R_f values, which are listed in Table 3.III, increase quite regularly with the length and number of acyl residues. All the 2'-O-acyl derivatives have practically the same λ_{max} as cyclic AMP itself (258–260 mμ) with values of ϵ ranging from 13,000 to 15,000. When there is substitution at N^6, the λ_{max} is usually displaced toward 270–273 mμ. A benzoyl group at this position produces a λ_{max} of 280 mμ.

Of all the acyl derivatives prepared, the butyric acid derivatives have been employed most frequently in biological experiments. The dibutyryl and N^6-monobutyryl derivatives have generally been found to be the most potent, and have in several instances produced an action on intact cells where cyclic AMP itself was inactive. The preparation of these derivatives is here described in detail.

N^6-2'-O-Dibutyryl Cyclic AMP (Fig. 3.3, $R_1 = R_2 = C_4H_7O$)

Two-hundred-fifty milligrams of cyclic AMP were dissolved in the minimum quantity of cold aqueous 0.4 M triethylamine. The solution was evaporated to dryness *in vacuo*, taken up in anhydrous pyridine, reevaporated to dryness, and the whole operation repeated once more. Finally, it was dried for 5 hr under high vacuum. The residue was taken up in 7.5 ml of hot anhydrous pyridine; after cooling, 3.75 gm butyric anhydride were added and the mixture boiled for 3–4 min during which opera-

TABLE 3.III. R, λ_{max}, and ϵ_{max} of Derivatives of Cyclic AMP

Compound	$R_F{}^a$	λ_{max} (mμ)	ϵ_{max}
Cyclic AMP	0.34	258	14100
2'-O-Monoacetyl	0.47	260	14050
N^6-Monoacetyl	0.48	273	14100
N^6-2'-O-Diacetyl	0.61	273	14030
2'-O-Monobutyryl	0.61	258	13400
N^6-Monobutyryl	0.63	272	15100
N^6-2'-O-Dibutyryl	0.81	270	14840
N^6-Monohexanoyl	0.63	274	16900
N^6-2'-O-Dihexanoyl	0.80	274	–
2'-O-Monooctanoyl	0.75	259	14600
N^6-Monooctanoyl	0.78	273	18460
N^6-2'-O-Dioctanoyl	0.88	272	14800
N^6-Monolauryl	0.82	273	17200
N^6-Monostearyl	0.84	273	24000
N^6-Monobenzoyl	0.60	280	19450
N^6-Monoadamantane carboxylate	0.55	272	14200
N^1-Oxide	0.23	231.5	40100
		260	9540
Ethyl ester	0.66	260	–

^{a} Solvent system: ethanol-0.5 M ammonium acetate (5:2); Whatman No. 1.

tion the product dissolved almost completely. The solution was immediately cooled and left at room temperature protected from light and moisture. The reaction was followed by paper chromatography. From time to time, 0.02 ml samples were removed and diluted with 0.03 ml water. After 4 hr, 0.01 ml of this mixture was subjected to paper chromatography in the system ethanol-0.5 M ammonium acetate (5:2) on Whatman No. 1 paper. After drying, spots were located by their fluorescence in ultraviolet light. After 5 to 8 days, the formation of the diacyl derivatives was practically quantitative. Attempts to accelerate the reaction by using higher temperatures led to complications; after boiling for 3 hr in the presence of pyridine and butyric anhydride, the major product formed had an R_f of 0.56 and differed from the expected dibutyrate.

The reaction mixture was cooled in ice, 3.5 ml water were added in small quantities with stirring, and the mixture allowed to stand for 5 hr at 4°C. During this period any excess anhydride reagent and any mixed anhydride formed from the carboxylic acid and the organic phosphate derivative were destroyed. The mixture was evaporated (30°–40°C), first with a water pump and then dried for 24 hr under high vacuum. Butyric acid was thus removed. The residue was taken up in 1 ml of a solution containing 150 mg of $BaI_2 \cdot 2H_2O$. When solution was complete, 10 ml absolute alcohol and 40 ml freshly distilled anhydrous ether were added. After several hours at 4°C, the precipitate was recovered by centrifugation and washed with an alcohol–ether (1:4) mixture. It was purified by dissolving in 1.5–2 parts of water followed by the addition of alcohol and ether. The combined supernatants from the first precipitation were evaporated to dryness *in vacuo* and taken up by a solution of 75 mg $BaI_2 \cdot 2 H_2O$ in 0.5 ml H_2O. By adding 6 ml alcohol and 25 ml dry ether, a second precipitation of crude barium salt was obtained which was purified by dissolving in water and re-precipitating with alcohol and ether. Found after drying *in vacuo* over H_2SO_4 a total yield of 210–230 mg N^6-$2'$-O-dibutyrate (as barium salt). Composition, after drying at 110°C *in vacuo* over P_2O_5, $C_{18}H_{23}N_5O_8PBa/2$, 3 H_2O.

N^6-Monobutyryl Cyclic AMP (Fig. 3.3, $R_1 = H$, $R_2 = C_4H_7O$)

Two-hundred-fifty milligrams of cyclic AMP as the triethylamine salt were con-verted to the N^6-$2'$-O-dibutyryl derivative as described above. After hydrolysis of the excess anhydride and evaporation of the pyridine and butyric acid (first with a water pump and then under high vacuum), the residue was taken up in 20 ml alcohol. It was cooled in ice and neutralized to phenolphthalein with 2 N NaOH. Then excess 2 N NaOH was added to give a final concentration of 0.1 N. The solution was kept for 4–5 min out of the ice bath, then returned to the ice bath and rapidly adjusted to pH 2.0 with 2 N NaOH, was evaporated to dryness at low temperature. The well-dried residue was thoroughly extracted with boiling absolute alcohol. Sodium chloride was removed by filtration on a sintered glass filter. The alcoholic extracts containing from 60 to 80% of the original quantity of P were evaporated to dryness. The residue was dissolved in 1 ml water (warming if necessary) and the theoretical quantity of solid barium acetate added. The precipitation was completed by addition of alcohol. Yield: 200 mg of N^6-monobutyryl cyclic $3',5'$-AMP as the barium salt. Composition, after drying at 110°C *in vacuo* over P_2O_5, $C_{14}H_{17}N_5O_7PBa/2$, 2 H_2O.

$2'$-O-Monobutyryl Cyclic AMP (Fig. 3.3, $R_1 = C_4H_7O$, $R_2 = H$)

Equimolar quantities of cyclic AMP (250 mg) and $4'$-morpholine-N,N'-dicyclo-hexylcarboxamidine (220 mg) were dissolved in hot anhydrous pyridine (7.5 ml). After cooling, 3.75 g of butyric anhydride were added and the resulting mixture left at room temperature with moisture excluded. Paper chromatography showed the nearly quantitative formation of the $2'$-O-acyl derivative after 8 hr. The reaction mixture was cooled in ice. After decomposition of excess anhydride reagent, the procedure of isolating the $2'$-O-monobutyrate as the barium salt was exactly as described for the isolation of the barium salt of the N^6-$2'$-O-dibutyryl derivative. Found, after drying *in vacuo* over H_2SO_4, 0.296 mg $2'$-O-butyrate (as barium salt). Composition, after drying at 110°C *in vacuo* over P_2O_5, $C_{14}H_{17}N_5O_7PBa/2$, 2 H_2O.

III. RECENT PROGRESS

Professor Posternak wrote and submitted the manuscript for this chapter by the promised date. Therefore, since the results of a number of pertinent studies have been published since that time, we will briefly mention some of them in this section.

Watenpaugh *et al.* (1968) determined the structure of cyclic AMP by single crystal X-ray diffraction, and obtained data suggesting that there were two molecules of cyclic AMP per asymetric unit, with different conformations about the glycosidic bonds. Since the conformation of the ribose phosphate part of either molecule was as described by Professor Posternak in Section I, C, 3, we substituted his original line drawing in Fig. 3.2 with a model based on the calculations of Watenpaugh and his colleagues. The other conformation is illustrated in Fig. 3.4. Although the ribose rings are identical in both molecules, the conformation about the glycosidic bonds is quite different.

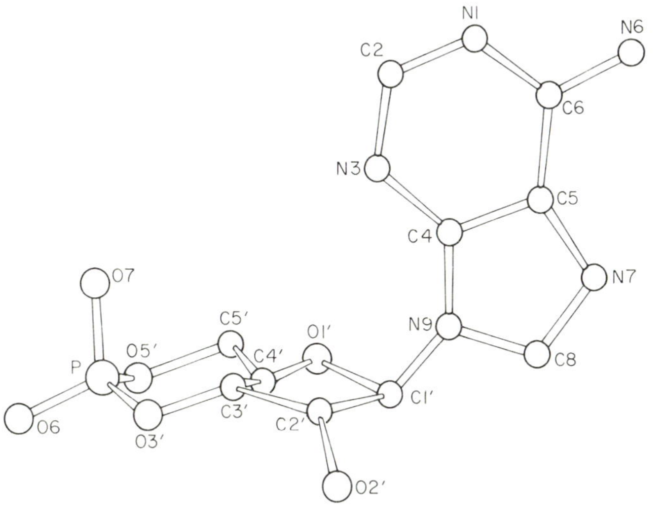

Fig. 3.4. Molecular configuration of cyclic AMP, differing from that shown in Fig. 3.2 by having a different torsion angle between the planar adenine moiety and the furanose ring. Cyclic AMP exists in crystals in both conformations. From Watenpaugh *et al.* (1968).

Coulter (1968) studied the crystal structure of cyclic UMP, and concluded that it could also exist in two conformations. If we regard the bases in these molecules as being planar, which is apparently the case, then it would appear that one of the two possible molecules of cyclic UMP resembles the cyclic AMP molecule illustrated in Fig. 3.2, in that the torsion angles between the base and the ribose are similar. However, the other cyclic UMP molecule does not seem to resemble the other cyclic AMP molecule illustrated in Fig. 3.4. We do not pretend to understand the reasons for these differences, or how they might be expected to influence the chemical and biological properties of the molecules. For further details, interested readers are referred to the original papers.

Greengard *et al.* (1969b) measured the enthalpy of hydrolysis of the 3'-bond of cyclic AMP, and obtained a mean value of $-14,100 \pm 200$

cal/mole. It had been concluded previously, based on studies of the equilibrium of the adenyl cyclase reaction (Greengard *et al.*, 1969a), that the free energy of hydrolysis of the 3'-bond was −11, 900 cal/mole. It would thus appear that the large exergonic change associated with the hydrolysis of cyclic AMP represents primarily a change in enthalpy, with only a small change in entropy. It might have been expected that the difference in entropy between cyclic AMP and 5'-AMP would be large. However, as pointed out by Watenpaugh *et al.* (1968), 5'-AMP exists in only one conformation whereas cyclic AMP exists in two. This might partially account for the relatively small entropic change calculated by Greengard and his colleagues. These investigators also measured the enthalpy of hydrolysis of the 3'-bond of cyclic GMP. The value obtained, −10,500 cal/mole, was considerably less than the value obtained for cyclic AMP. On the assumption that the entropy change during the hydrolysis of cyclic GMP was similar to that occurring during the hydrolysis of cyclic AMP, which might not be valid, Greengard and his colleagues calculated the free energy of the 3'-bond in cyclic GMP to be much less than it appears to be in cyclic AMP. In any event, the reasons for the observed differences in the enthalpy of hydrolysis are poorly understood.

Mention can be made of several potentially interesting analogs of cyclic AMP. Thomas and Montgomery (1968) described the synthesis of the 3',5'-cyclic phosphate of 6-mercaptopurine ribonucleoside, while Hanze (1968) synthesized the cyclophosphate of tubercidin (7-deaza-adenosine). The latter compound was roughly as active as cyclic AMP itself in stimulating the activation of liver phosphorylase, indicating that the nitrogen in the 7-position is not an important feature of cyclic AMP. Professor Posternak and his colleagues have recently reported the synthesis of an isomer of cyclic AMP, in which the ribose was attached to the nitrogen in the 3- instead of the 9-position (Posternak *et al.*, 1969). This compound was found to be at least as active as cyclic AMP in stimulating the release of TSH from the anterior pituitary *in vitro* (Cehovic, 1969). By contrast, it did not share the ability of cyclic AMP to stimulate the release of growth hormone (Cehovic *et al.*, 1970a,b).

Drummond and Powell (1970) tested several analogs of cyclic AMP as activators of phosphorylase *b* kinase and also as substrates for brain phosphodiesterase. Compounds involving structural alteration of the phosphate moiety (phosphothioate and phosphonate analogs) or the sugar moiety (the D-xylofuranosyl analog) were found to be relatively ineffective as activators of the kinase and were also poor substrates for the phosphodiesterase. By contrast, the tubercidin analog was an even better substrate than cyclic AMP itself.

Swislocki (1970b) has pointed out that the dibutyryl derivative is converted fairly rapidly to the N^6-monobutyryl derivative when incubated in bicarbonate buffers, emphasizing the importance of using butyric acid controls when studying the biological activity of these compounds.

Hubert-Habart and Goodman (1969) described a method for converting nucleosides to cyclic 3′,5′-nucleotides with *cis* fused rings. A useful procedure for the preparation of ^{32}P-labeled cyclic AMP, starting with inorganic phosphate, has been described by Symons (1970).

CHAPTER 4
Formation and Metabolism of Cyclic AMP

Developments leading to the discovery of adenyl cyclase and phosphodiesterase as the enzymes responsible for the formation and metabolism of cyclic AMP were outlined in Chapter 1. Our purpose in this chapter will be to summarize currently available information about the properties and preparation of these enzymes.

I. ADENYL CYCLASE

The reaction catalyzed by adenyl cyclase can be written as shown in Fig. 4.1.

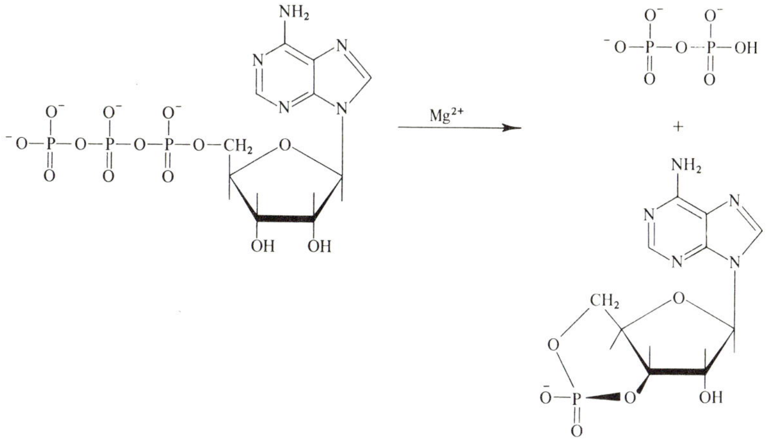

Fig. 4.1. The adenyl cyclase reaction.

The mechanism of this reaction has not been established, although it is known that a divalent cation (Mg^{2+} or Mn^{2+}) is required and that pyrophosphate is formed stoichiometrically with cyclic AMP (Rall and Sutherland, 1962; Hirata and Hayaishi, 1967). The reaction can be reversed under some conditions, as discussed in Section I,E below.

A. Distribution

Table 4.I contains a list of the mammalian cells and tissues in which the existence of adenyl cyclase has been demonstrated, together with the hormones which have been found to stimulate it, where this has been studied. In the dog, the only cells studied which did not contain measurable adenyl cyclase activity were red blood cells (Sutherland *et al.,* 1962), although the erythrocytes of certain other species may differ in this regard, and certainly the nucleated erythrocytes in birds and frogs do contain adenyl cyclase (Davoren and Sutherland, 1963a; Rosen and Rosen, 1969). The possible relation of hormone receptors to adenyl cyclase was discussed in Chapter 2, and the effects of these hormones will be discussed in more detail in Chapters 6 through 10.

TABLE 4.I.   Distribution and Hormonal Sensitivity of Mammalian Adenyl Cyclase

Tissue	Hormone	Reference
Liver	Glucagon and epinephrine	Sutherland and Rall, 1960
Skeletal muscle	Epinephrine	Sutherland and Rall, 1960
Cardiac muscle	Catecholamines	Murad *et al.*, 1962
	Glucagon	Murad and Vaughan, 1969; Levey and Epstein, 1969a
	Triiodothyronine	Levey and Epstein, 1969b
Kidney	Vasopressin	Brown *et al.*, 1963
	Parathyroid	Chase and Aurbach, 1968
Bone	Parathyroid	Chase *et al.*, 1969a
	Calcitonin	Murad *et al.*, 1970a
Brain	Catecholamines	Klainer *et al.*, 1962
Adrenal	ACTH	Grahame-Smith *et al.*, 1967
Corpus luteum	LH and prostaglandins	Marsh, 1970 a and b
Ovary	LH	Dorrington and Baggett, 1969
Testes	LH and FSH	Murad *et al.*, 1969a; Kuehl *et al.*, 1970a
Thyroid	TSH	Klainer *et al.*, 1962
	Prostaglandins	Zor *et al.*, 1969b
Parotid	Catecholamines	Malamud, 1969; Schramm and Naim, 1970
Pineal	Catecholamines	Weiss and Costa, 1968a
Lung	Epinephrine	Klainer *et al.*, 1962
Spleen	Epinephrine	Klainer *et al.*, 1962
Adipose	Epinephrine	Klainer *et al.*, 1962
	Several	Rodbell *et al.*, 1968
Brown adipose	Catecholamines	Butcher and Baird, 1969
Platelets	Prostaglandins	Butcher *et al.*, 1967; Wolfe and Shulman, 1969
Leucocytes	Catecholamines and prostaglandins	Scott, 1970
Erythrocytes	None demonstrated	Sheppard and Burghardt, 1969
Uterus	Catecholamines	Dobbs and Robison, 1968
Pancreas	None demonstrated	Cohen and Bitensky, 1969
Anterior pituitary	Several	Fleischer *et al.*, 1969; Zor *et al.*, 1969c
Vascular smooth muscle	None demonstrated	Klainer *et al.*, 1962

Adenyl cyclase has also been detected in many lower forms, as listed in Table 4.II. It should be noted that cyclic AMP itself has been measured in a number of other animal species, in addition to those listed, suggesting the existence in these cells of an enzyme for catalyzing its formation. It would thus appear that adenyl cyclase is very widely distributed throughout the animal kingdom. Higher plants have not been extensively studied from this point of view, but to date, in those plants which have been examined, neither adenyl cyclase nor cyclic AMP has been detected. It seems possible, therefore, that plant cells do not contain cyclic AMP.

TABLE 4.II. Species Distribution of Adenyl Cyclase

Mammals	Sutherland and Rall, 1960
Birds	Davoren and Sutherland, 1963a
Amphibians	Rosen and Rosen, 1969
Fish	Sutherland et al., 1962
Insects	Sutherland et al., 1962
Segmented worms	Sutherland et al., 1962
Liver flukes	Mansour et al., 1960
Bacteria	Hirata and Hayaishi, 1967; Tao and Lipmann, 1969

B. Cellular Distribution

It was established in the early experiments (Chapter 1) that most of the adenyl cyclase activity in homogenates of most mammalian tissues occurred in the low-speed or "nuclear" fraction. This is the fraction in liver homogenates which sediments at a centrifugal force of 600 g (De-Duve et al., 1955), and which was later shown to contain fragments of the cell membrane in addition to nuclear material (Neville, 1960). Since adenyl cyclase could not be detected in preparations of canine erythrocytes, which lack nuclei, but did occur in the nucleated erythrocytes of pigeons, the possibility had to be considered that nuclei were the principal source of the enzyme.

In experiments with broken cell preparations from pigeon erythrocytes, it was found that the methods of homogenization and fractionation employed had much to do with the sedimentation characteristics of adenyl cyclase (Davoren and Sutherland, 1963b). If the cells were disrupted by water lysis and centrifuged at 600 g and then 78,000 g, adenyl cyclase and DNA were both found primarily in the 600 g precipitate. If the cells were lysed, but the lysate was layered over a 20% glycerol solution, a significant fraction of the cyclizing system was now found in the 78,000 g precipitate. Davoren designed and built a pressure homogenizer which made possible the extensive fragmentation of the cell membranes with little damage to the nuclei. When this procedure was followed by centrifugation over the 20% glycerol system, the adenyl cyclase was now found to be associated almost completely with the 78,000 g fraction, while the DNA remained in the 600 g precipitate. If the 78,00C g precipitate was fractionated in a density gradient system, the adenyl cyclase was associated with the major protein component, the density of which centered at 1.18. Although the earlier liver studies had excluded mitochondria as a cellular location for adenyl cyclase, rat liver was fractionated again since the density figure above was near that of mitochondria. After pressure homogenization, the mitochondria (as judged by cytochrome oxidase content) were localized in the 10,000 g precipitate, while the adenyl cyclase system was in the 78,000 g precipitate (Fig. 4.2). The major particulate protein peaks from density gradient centrifugation of dog and avian erythrocyte and rat liver particles were subjected to electron microscopy, and in the dog erythrocyte and rat liver systems structures appearing to be cell membranes were noted. In the avian erythrocytes, however, the membranes appeared to have been shattered to a fine debris.

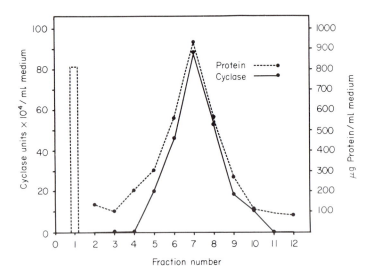

Fig. 4.2. Gradient density fractionation of pigeon erythrocyte "78,000 g particles" derived from 4 ml of packed cells. Fraction 1 consisted primarily of material which sedimented during the density gradient centrifugation, and was devoid of adenyl cyclase activity. From Davoren and Sutherland (1963b).

The data obtained in these studies indicated that the adenyl cyclase system of avian erythrocytes and rat liver were components of the cell membrane. This has more recently been confirmed by other investigators (Pohl *et al.,* 1969; Rosen and Rosen, 1969). In addition, the experiments with the whole nucleated erythrocyte suggested that the active site of the cyclizing enzyme was on the inside of the membrane, for the ineffectiveness of ATP in the medium would strongly militate against an exterior location, at least for the active site.

Data suggesting that the active site of adenyl cyclase is located on the inside of the membrane was obtained by Øye and Sutherland (1966) in experiments with proteolytic enzymes on the adenyl cyclase and ATPase activity of turkey erythrocytes. The ATPase activity was found to follow the adenyl cyclase closely during fractionation, including gradient density studies. Since ATP in the medium is hydrolyzed by whole cells, one might assume that a catalytic site is external, in contrast to the hypothesized inside site of adenyl cyclase. Treatment of whole cells with pepsin, trypsin, and bromelin decreased the ATPase activity somewhat without any significant change in the level of the adenyl cyclase (Table 4.III).

TABLE 4.III. Effect of Treatment with Various Enzymes on Adenyl Cyclase and ATPase Activities of Whole Avian Erythrocytes[a]

	Cyclase (% of control)	ATPase (% of control)
Control (saline)	100	100
Albumin	100	100
Pepsin	100	75
Trypsin	109	67
Phospholipase C	188[b]	92
Neurominidase	113	108
Bromelin	113	67

[a] Cells were suspended in an isotonic buffer and incubated for 15 min at 37°C in the presence or absence of a 0.25% (w/v) suspension of the indicated protein. They were then washed and incubated in the presence of ATP, Mg^{2+}, and epinephrine. Data from Øye and Sutherland (1966).

[b] Associated with rapid hemolysis.

Neuramidase was ineffective on both enzyme systems. In other studies, when hemolyzed cells (the nucleated ghosts) were exposed to the proteolytic enzymes, both activities were abolished. The conclusions from these data might be schematically represented as in Fig. 4.3.

Although the fragments studied by Davoren and Sutherland were insensitive to stimulation by epinephrine, Øye and Sutherland (1966) succeeded in preparing membranes from turkey erythrocytes which did respond to epinephrine. The adenyl cyclase activity had a density of 1.18

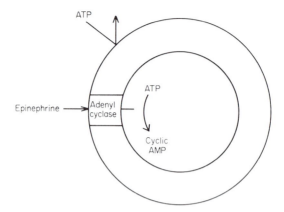

Fig. 4.3. Illustrating the relation of ATP to adenyl cyclase in intact avian erythrocytes. Exogenous ATP does not enter cells and is not converted to cyclic AMP. Endogenous ATP is converted to cyclic AMP in the presence of epinephrine.

as determined by density gradient centrifugation, which was in agreement with the density found by Davoren. In phase contrast microscopy, such preparations appeared to consist of large pieces of the membrane and were essentially free of smaller particles.

In some cells adenyl cyclase may occur in membranous constituents other than the plasma membrane. Rabinowitz *et al.* (1965) found that most of the adenyl cyclase in homogenates of rabbit psoas muscle sedimented with the microsomal fraction, and preparations from other tissues may also contain some activity in this fraction (Sutherland *et al.*, 1962; Entman *et al.*, 1969a). De Robertis *et al.* (1967) found that adenyl cyclase in homogenates of rat cerebral cortex was almost wholly particulate, most being found in the mitochondrial fractions with a slightly smaller amount present in the microsomal fractions. The submitochondrial fractions separated by sucrose gradient showed that the specific activity of the adenyl cyclase was low in the myelin and mitochondrial subfractions and high in the three subfractions containing nerve endings. After hypoosmotic shock of the mitochondrial fraction, almost all adenyl cyclase remained in particulate subfractions. After gradient separation of the main subfraction, it was found that adenyl cyclase concentrated in two fractions which contained mainly synaptic membranes.

In bacteria also adenyl cyclase seems to be associated with particulate fractions (Brana, 1969; Ide, 1969), although the enzyme from these cells can be solubilized relatively easily (Hirata and Hayaishi, 1967; Tao and Lipmann, 1969).

C. Purification

Attempts to obtain adenyl cyclase from the cells of multicellular organisms in a highly purified form have to date been unsuccessful. Preparations containing adenyl cyclase have been washed in hypotonic solutions, frozen, washed again with hypotonic and salt solutions, and then refrozen with good recovery of activity. After a wash with 0.1% Triton solution, much of the activity in several tissues was solubilized or dispersed in a 1.8% Triton solution. Concentrations of adenosine triphosphatase, pyrophosphatase, and adenosine 3',5'-phosphate phosphodiesterase relative to that of adenyl cyclase were lowered by these procedures. However, to date no cyclase preparations have been entirely free from these other enzymes. Increases of specific activity on a protein basis have been relatively small, e.g., only two- or three-fold for brain preparations and fifteenfold or so for liver preparations (Sutherland *et al.*, 1962). Some preparations from heart and liver have been lyophilized, and the dried powders could be extracted with dry diethyl ether in the cold with little or no loss of activity.

As mentioned previously, bacterial adenyl cyclase seems to be relatively easily solubilized, and the enzyme from *Brevibacterium liquefaciens* has been purified extensively (Hirata and Hayaishi, 1967). Preparations from this source have the highest specific activity of any adenyl cyclase studied, and are virtually free of contaminating phosphodiesterase, ATPase, and pyrophosphatase activity. Adenyl cyclase from *B. liquefaciens* is sensitive to stimulation by pyruvate and several other α-keto acids, and it is tempting to suppose that pyruvate may act in this instance by a mechanism analogous to the mechanism by which hormones act in other cells. It is yet unclear, however, that the bacterial enzyme is closely related to the enzyme found in higher forms.

D. Properties

Particulate preparations of adenyl cyclase can be frozen rapidly in hypotonic media and stored at −70°C for long periods without loss of activity. At higher temperatures, however, the activity of adenyl cyclase prepared from the cells of multicellular organisms is lost more or less rapidly (see Fig. 2.4) and has been found to be similarly unstable in all preparations studied to date. The reasons for this are unclear. Sulfhydryl reagents have been helpful in maintaining activity in preparations from avian and amphibian erythrocytes (Øye and Sutherland, 1966; Rosen and Rosen, 1969) but have not been found effective in mammalian preparations.

The ability of the fluoride ion to stimulate adenyl cyclase has been observed in most preparations from multicellular organisms (Rall and Sutherland, 1958). Fluoride does not stimulate bacterial adenyl cyclase (Tao and Lipmann, 1969), and in the developing rat brain, where this has been studied (Schmidt et al., 1970), an effect of fluoride cannot be detected for at least the first 5 days postpartum. The development of adenyl cyclase activity in washed particulate preparations from rat brain, measured in the presence and absence of fluoride, is illustrated in Fig. 4.4. One interpretation of these data is that in the course of development an inhibitory influence which can be reversed by fluoride is superimposed on the adenyl cyclase system. The nature of this influence is presently obscure, and other interpretations are possible. The effect develops at

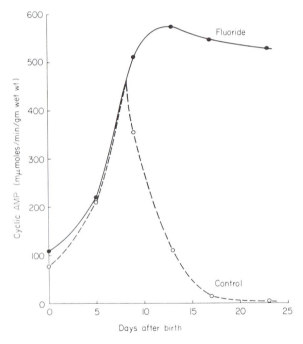

Fig. 4.4. Development of the fluoride effect in rat brain. Brains from rats of different ages were pooled and homogenized in hyptotonic Mg-glycylglycine buffer. Washed particulate (50,000 *g*) preparations were incubated for 12 min at 34°C in Hepes buffer (40 m*M*, pH 7.4) containing 2 m*M* ATP, 3 m*M* MgSO$_4$, and 3.3 m*M* theophylline, in the absence (dashed line) and presence (solid line) of 10 m*M* NaF. Each tube contained particles from 100 mg of tissue in a total volume of 2 ml. The amount of cyclic AMP formed during the incubation was taken as a measure of adenyl cyclase activity. From Schmidt *et al.* (1970).

different times in different organs since, for example, hepatic adenyl cyclase can be stimulated by fluoride even in preparations from newborn rats (Wicks, 1969).

The mechanism of the stimulatory effect of fluoride is poorly understood. This has been studied to some extent in preparations of fat cell ghosts (Bär and Hechter, 1969a; Birnbaumer *et al.,* 1969; Vaughn and Murad, 1969). The maximum velocity of the cyclase reaction was markedly increased by fluoride with little or no change in the apparent affinity for ATP. The optimal concentration of fluoride from one adult preparation to another may vary, but has generally been found to be broad, with peak effects in the range of 5 to 10 mM. Birnbaumer *et al.* (1969) plotted the increment in cyclase activity as a function of the concentration of fluoride and obtained a sigmoidal curve, in contrast to the hyperbolic curve generated by increasing concentrations of ACTH. They found that increasing the temperature from 30° to 37°C increased the effect of small concentrations of fluoride markedly, whereas the ACTH effect was not much affected by this change. The inability of fluoride to stimulate the formation of cyclic AMP in intact cells was mentioned previously (see Table 2.III). The reason for this is unknown.

The effects of ions other than fluoride have also been studied to some extent. Manganese ions can replace Mg^{2+} in at least some preparations (Sutherland *et al.,* 1962), and in at least one case, fat cell ghosts in the presence of fluoride, Mn^{2+} is even more effective than Mg^{2+} (Birnbaumer *et al.,* 1969). Copper and zinc ions are strongly inhibitory, and cobalt is also inhibitory to some extent (Sutherland *et al.,* 1962; Birnbaumer *et al.,* 1969). Calcium has been found to be inhibitory in all preparations studied, although there is at least one hormone (ACTH) which seems to require Ca^{2+} in order to stimulate adenyl cyclase (Bär and Hechter, 1969b). Relatively large changes in the concentrations of Na^+ or K^+ have little effect on the activity of adenyl cyclase in broken cell preparations, although such changes may alter the level of cyclic AMP in intact cells substantially (Chapter 2, Section IV,D). The pH optimum for the formation of cyclic AMP in most preparations is broad, being relatively flat between pH 7.2 and 8.2 (Sutherland *et al.,* 1962).

One of the most striking features of adenyl cyclase, its sensitivity to hormones, will not be discussed in detail here because it is more appropriately discussed elsewhere in connection with the hormones themselves. The relation of hormone receptors to adenyl cyclase was discussed previously in Chapter 2 (Section III), and it is perhaps unnecessary to reemphasize our ignorance in this area. Adenyl cyclase from a unicellular organism has in no case been shown to respond to a hormone from a multicellular animal, which is perhaps not surprising. In metazoa, sensitivity to hormones seems to develop at different times in different tissues. In tadpole erythrocytes, for example, adenyl cyclase can be stimulated by fluoride but not by epinephrine. After metamorphosis, however, adenyl cyclase from frog erythrocytes does respond to epinephrine (Rosen and Erlichman, 1969). The development of hormonal sensitivity by mammalian adenyl cyclase has not been carefully studied in broken cell preparations. It would appear, however, that sensitivity of the hepatic system to glucagon develops a few days before birth (Greengard, 1969a) whereas that of the brain enzyme to norepinephrine does not develop until a few days after birth (Schmidt *et al.*, 1970). To date, these studies have not contributed to our understanding of how these hormones stimulate adenyl cyclase. The analogy between the effects of hormones in higher species and the effect of pyruvate in *B. liquefaciens* was mentioned previously, but whether this analogy is valid or not remains to be seen.

The properties of adenyl cyclase in preparations from mammalian and other metazoan tissues are suggestive of a lipoprotein structure (Sutherland *et al.*, 1962; Øye and Sutherland, 1966). However, even in the case of the soluble systems obtained from bacteria (Hirata and Hayaishi, 1967; Tao and Lipmann, 1969), the chemical structure of the enzyme has not been well defined. We can hope that the detailed study of these apparently simpler bacterial systems will be helpful in leading to a better understanding of the nature of adenyl cyclase as it occurs in higher forms.

E. Reversal of the Reaction

It was mentioned previously that the adenyl cyclase reaction could be reversed under some conditions. This was demonstrated by Greengard *et al.* (1969a) using the highly purified enzyme from *B. liquefaciens*. Magnesium ions were required in order for the reaction to proceed in either direction, and pyruvate was found to stimulate in both directions. Although the conversion of cyclic AMP to ATP probably does not occur naturally to any significant extent (at least under the conditions prevailing within the cells of higher organisms), the finding that it *could* occur was of interest because it led to the conclusion that the free energy of hydrolysis of the 3'-bond of cyclic AMP was even greater than that of the α-β bond of ATP. As discussed in the preceding chapter, this was later confirmed by means of calorimetric experiments (Greengard *et al.*, 1969b).

F. Methods of Assay

Cyclic AMP produced from ATP in broken cell preparations can be measured by any of the methods mentioned in the Appendix for the assay of tissue levels of cyclic AMP. It has become more common, however, to measure the rate of conversion of labeled ATP to cyclic AMP (Krishna *et al.,* 1968a; Bär and Hechter, 1969a). This procedure works well so long as care is taken to avoid contamination of the counting fluid with ATP. The concentration of ATP in the reaction mixture may exceed that of cyclic AMP by several orders of magnitude, so that if even a small fraction of the original ATP is counted it may be enough to invalidate the results. Krishna and his colleagues have recommended passage of the reaction mixture through Dowex 50 followed by precipitation of the remaining ATP by $ZnSO_4$-$Ba(OH)_2$. It is important to note that the order of this procedure cannot be reversed, because incubating ATP with barium hydroxide leads to substantial conversion to cyclic AMP (Cook *et al.,* 1957). This would lead to such a high background that valid estimates of apparent adenyl cyclase activity would be difficult if not impossible to obtain.

A major factor complicating the quantitative interpretation of most adenyl cyclase measurements is contamination by other enzymes, including ATPase, pyrophosphatase, various deaminases, and phosphodiesterase. An ATP-regenerating system is often included in the reaction mixture to reduce the influence of ATPase, although such systems may introduce complications of their own. For most qualitative purposes, such as determining whether adenyl cyclase is or is not stimulated by a given hormone, the use of an ATP-regenerating system does not appear to be necessary (e.g., Marcus and Aurbach, 1969; Makman, 1970).

Some degree of inhibition of phosphodiesterase does seem to be necessary in most preparations, and a methylxanthine is usually included in the reaction mixture for this purpose. As with the ATP-regenerating system, this may introduce complications, e.g., the methylxanthine may inhibit adenyl cyclase (even though this is usually overshadowed by the greater inhibition of phosphodiesterase). However, this seems to be a necessary evil, for in many preparations cyclic AMP does not accumulate at all in the absence of a phosphodiesterase inhibitor. More selective inhibitors of phosphodiesterase may be useful here. When measuring the rate of conversion of labeled ATP to cyclic AMP, a useful technique may be to add an excess of cold cyclic AMP, as described by Weiss (1969a).

Adenine and adenosine, in contrast to phosphorylated compounds, penetrate cells quite readily and are partially converted to ATP. A useful procedure for estimating adenyl cyclase in intact cells has thus been to incubate tissues with labeled adenine or adenosine, and then measure the rate of labeling of cyclic AMP. This procedure has been employed by Kuo and DeRenzo (1969), Humes and Kuehl and their colleagues (Humes *et al.,* 1969; Kuehl *et al.,* 1970b), Shimizu *et al.* (1969), Moskowitz and Fain (1970), and Marquis *et al.* (1970), among others. The question of whether a small pool of ATP provides a disproportionate amount of the substrate for adenyl cyclase has yet to be answered by these experiments.

II. PHOSPHODIESTERASE

A. Discovery and Initial Studies

The discovery and the initial studies of the phosphodiesterase were mentioned in Chapter 1. Briefly summarized, an enzyme capable of destroying the biological activity of cyclic AMP was found in extracts of heart, brain, and liver (Sutherland and Rall, 1958). The enzyme from dog heart was partially purified, and the product of the reaction was shown to be 5'-AMP. Activation of the phosphodiesterase by Mg^{2+} and inhibition by caffeine was also reported.

The problems involved with the assay of cyclic AMP levels in tissues necessitated further work with the phosphodiesterase. The phosphorylase assay system was known to be sensitive to activation or inhibition by substances found in tissues including Ca^{2+}, glucose-1-P, and UDPG, as well as unknown substances (Butcher et al., 1960, Murad, 1965). Thus, the availability of a specific enzyme for the destruction of cyclic AMP was very desirable.

Although the phosphodiesterase activity in extracts of brain was considerably higher than that in heart (Sutherland and Rall, 1958), the heart enzyme appeared to be a better candidate for purification because it was more amenable to fractionation. Approximately 60% of the phosphodiesterase activity in homogenates of beef heart was found to be associated with particulate fractions after low-speed centrifugation. The soluble phosphodiesterase from beef heart was purified by the steps outlined in Section III,D of the Appendix. The final preparations were purified between 80- and 140-fold (as compared to the initial supernatant fractions). The activity associated with the particulate fractions was not elutable under several conditions, including hypertonic and hypotonic washing, rehomogenization, and freezing. However, at least in preliminary experiments, the particulate and soluble forms of the enzyme were very similar enzymatically.

The K_m of the purified enzyme was around $1 \times 10^{-4}\,M$. It was inhibited by the methylxanthines, theophylline being about six times as potent as theobromine or caffeine (Fig. 4.5). The inhibition of the enzyme by theophylline appeared to be competitive. Imidazole was found to stimulate the phosphodiesterase. As shown in Fig. 4.6, the maximal activity of the enzyme in tris buffer appeared to be at approximately pH 7.9. In the

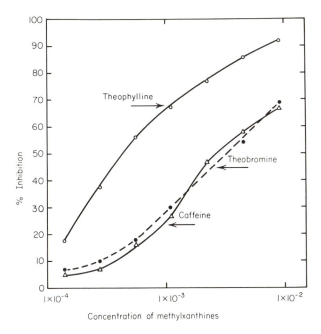

Fig. 4.5. Inhibition of phosphodiesterase by methylxanthines. Cyclic AMP was incubated with the purified beef heart enzyme for 30 min at 30°C, in the presence of the indicated concentrations of the drugs. From Butcher and Sutherland (1962).

presence of imidazole, the pH optimum was shifted downward to approximately 7.4, with a general broadening of the curve. Since the pK of imidazole is around 7.0, it seemed likely that this shift was due at least in part to the protonated form of imidazole being active and the free base inactive. The purified phosphodiesterase preparation was essentially free of interfering enzymes, and this preparation has been used successfully with not only the phosphorylase assay but also with several more direct, chemical systems.

The phosphodiesterase has been identified in a great many tissues (Butcher and Sutherland, 1962). However, to date only a few have been studied in any detail. Drummond and Perrott-Yee (1961) reported the preparation of phosphodiesterase from rabbit brain which was only slightly more active than the whole homogenate. Cheung (1967) studied rat brain phosphodiesterase in greater detail, but reported no substantial purification. Nair (1966) obtained preparations from dog heart which were similar in specific activity to the beef heart enzyme (Butcher and Sutherland, 1962).

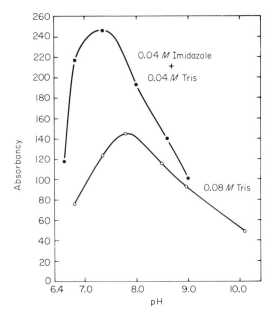

Fig. 4.6. Effect of pH on phosphodiesterase activity in the presence and absence of imidazole. After incubating for 15 min at 30°C, the reaction was stopped by boiling in acid. The samples were then neutralized and incubated for 15 min with *Crotalus atrox* venom. The phosphate released from the 5'-AMP was assayed by the Fiske-SubbaRow method, which includes a final reading on a spectrophotometer. Absorbancy is thus a measure of phosphodiesterase activity. From Butcher and Sutherland (1962).

B. Kinetic Studies

The K_m of the heart enzyme appears to be around $1 \times 10^{-4} M$ (Butcher and Sutherland, 1962; Nair, 1966), and of the brain enzyme somewhat higher (Cheung, 1967). Recently, however, Brooker *et al.* (1968) have reported the presence of a second phosphodiesterase activity in crude brain fractions with a K_m around $1 \times 10^{-6} M$.

C. Substrate Specificity

The purified heart phosphodiesterase appears to be specific for purine 3',5'-mononucleotides. However, initially it appeared that cyclic pyrimidine nucleotides were also well hydrolyzed. Crude preparations of heart phosphodiesterase hydrolyzed cyclic UMP at 66% the rate of cyclic AMP (Butcher and Sutherland, 1959; Sutherland and Rall, 1960). Later, Drummond and Perrott-Yee (1961) reported that cyclic

UMP was hydrolyzed at only 11% of the rate of cyclic AMP by crude preparations of phosphodiesterase from brain, and this was also found by Nair (1962, 1966) using his most purified dog heart preparation. However, this discrepancy was resolved when Hardman and Sutherland (1965) reported the existence of a second phosphodiesterase in crude preparations of heart with a specificity for cyclic UMP. It was found that the second enzyme was lost quite early in the purification and that the purified beef heart phosphodiesterase preparations hydrolyzed cyclic UMP at 17% of the rate of cyclic AMP. The phosphodiesterase is essentially inactive against cyclic 2′,3′-AMP (Butcher and Sutherland, 1962; Nair, 1966) and against polyadenylate and polyuridylic acids, and RNA and DNA (Sutherland and Rall, 1958; Butcher and Sutherland, 1962; Nair, 1966). Interestingly, Breckenridge (1964) obtained evidence that indicated that the nonterminal phosphates of 2′,5′-ADP and 3′,5′-ADP are hydrolyzed by the phosphodiesterase. To date, no other activities of the purified enzyme preparations have been reported. The purified heart phosphodiesterase was essentially without activity against acylated derivatives of cyclic AMP. Blecher *et al.* (1968) reported that dibutyryl cyclic AMP was hydrolyzed by fat homogenates, but did not determine clearly that this was due to hydrolysis of the derivative per se or if it was being metabolized to free cyclic AMP and then broken down.

D. Metal Requirements

Drummond and Perrott-Yee (1961) found that the brain enzyme required Mg^{2+} but that partial activity was obtained in the presence of Co^{2+} and Mn^{2+}. The hydrolysis of cyclic 3′,5′-AMP was differentiated from that of cyclic 2′,3′-AMP by the requirement for Mg^{2+}. These authors also noted that magnesium at concentrations greater than 0.8 mM were inhibitory. Cheung (1967) found that Mn^{2+} could completely replace Mg^{2+}, and that Co^{2+} and Ni^{2+} could partially replace magnesium. About 30% activity was found in the absence of added ions and this was abolished by the addition of 1 mM EDTA. In earlier studies with the heart enzyme, EDTA abolished phosphodiesterase activity, and Mg^{2+} could not completely restore the activity (Butcher and Sutherland, unpublished).

E. Phosphodiesterase Inhibitors

The methylxanthines have remained as the best known inhibitors of the phosphodiesterase. As mentioned above, theophylline acted competitively on the phosphodiesterase from beef heart (Butcher and Sutherland, 1962). Later, caffeine was reported to act noncompetitively on the phosphodiesterase from dog heart (Nair, 1966) and competitively against the brain phosphodiesterase (Cheung, 1967). The methylxanthine inhibition of the phosphodiesterase has been of considerable use in studying the role of cyclic AMP in hormone actions as pointed out previously in Chapter 2.

Cheung (1967) reported that ATP and other triphosphates are potent inhibitors of the brain phosphodiesterase. Nucleoside triphosphates at a final concentration of 3 mM produced inhibitions of from 75% (ATP) to 41% (TTP). Cheung also found that citrate was an inhibitor of the brain phosphodiesterase, producing a 50% inhibition at 12 mM. Isocitrate, malate, oxalate, pyruvate, and tartrate were mildly inhibitory and several other organic acids were without effect. Cheung suggested that the phosphodiesterase in its active form acted as a metalloenzyme complex, and that the nucleotides and citrate were acting as Mg^{2+} chelators. He also found that pyrophosphate inhibited the enzyme.

Other compounds which have been reported to inhibit phosphodiesterase activity in one or more tissues include puromycin (Appleman and Kemp, 1966), triiodothyronine (Mandel and Kuehl, 1967), diazoxide (Senft et al., 1968a; Moore, 1968), and papaverine (O'Dea et al., 1970; Kukovetz and Pöch, 1970b).

F. Phosphodiesterase Activators

Cheung (1967) reported that his diesterase preparation from brain was stimulated by imidazole, but that there was no significant change in the pH optimum. Nair (1966) also found imidazole stimulatory, but only at a very high concentration. In addition, the stimulation was only of the order of 1.5-fold. However, these experiments were apparently at pH 8.0 and therefore one would not expect much of a stimulation to occur (c.f. Fig. 4.6). In addition, Nair found that ammonium salts at concentrations ranging from 10 to 100 mM produced 1.5-fold stimulation of his purified preparations from dog heart. While it is very unlikely that the stimulation of the cyclic nucleotide phosphodiesterase by imidazole is of any physiological significance, it has been used by several workers in studies of the possible participation of cyclic AMP in hormone actions (see Table 2.V). A recent finding which may be of endocrinological significance is that at least some cells contain a protein factor capable of stimulating the phosphodiesterase (Cheung, 1970a).

G. Subcellular Distribution

Very little additional information about the phosphodiesterase activity associated with particulate fractions has been published since 1962. Cheung and Salganicoff (1967) reported that in brain the phosphodiesterase was mostly microsomal and that considerable soluble activity was concentrated inside the nerve endings. Triton X-100 caused an apparent activation of the enzyme, a process which was termed unmasking of latent activity. DeRobertis *et al.* (1967) found that the phosphodiesterase activity in rat brain homogenates was about 60% particulate. The highest specific activity was found in the soluble fraction, and the particulate activity was found primarily in fractions rich in nerve endings. After osmotic shock, about half of the activity associated with 11,500 *g* precipitates became soluble. At least part of the phosphodiesterase activity in this fraction was suppressed as the tonicity of the reaction medium was increased from hypo- to isotonic.

H. Unicellular Organisms

Phosphodiesterase activity has been detected in bacteria (Brana and Chytil, 1966) and in cellular slime molds (Chang, 1968). The slime mold enzyme was found to have a molecular weight of about 300,000 and a K_m similar to that found for the brain enzyme, but differed from mammalian phosphodiesterase in that it was not sensitive to inhibition by the methylxanthines. Brana and Chytil had noted that the *E. coli* enzyme was also insensitive to methylxanthine inhibition.

III. ADDENDUM

A large number of studies dealing with adenyl cyclase or phosphodiesterase have been published since this chapter was written. A few of these are mentioned in the text, because the authors were kind enough to let us see the manuscripts before publication, and others have been added to Table 4.I. Most of the newer papers on adenyl cyclase, however, represent extensions of earlier studies of hormonal stimulation, and are therefore discussed in the appropriate sections of Chapters 6 through 10. We hope the index will be useful here.

The interesting paper by Lefkowitz *et al.* (1970) on adrenal cortical adenyl cyclase was mentioned in the addendum to Chapter 2, and can be mentioned again here because it emphasizes the close relation between hormone binding and activation of adenyl cyclase, in that case by ACTH. Rodbell and his colleagues have more recently studied the effects of glucagon and other agents on the adenyl cyclase activity of liver cell membranes. One of several intriguing suggestions to come from these studies is that guanine nucleotides (GTP or GDP or both) may play an obligatory role in regulating the transfer of information from the glucagon receptor to the catalytic site of adenyl cyclase. A useful practical contribution has been the introduction of a nitrogen analogue of ATP (nitrogen replacing oxygen between the two terminal phosphoryl groups) as a substrate for adenyl cyclase. This compound can be converted to cyclic AMP but is not affected by ATPase. The results of these studies will be published soon in the *Journal of Biological Chemistry,* probably before this monograph becomes available.

By using Lubrol-PX instead of Triton, Levey (1969) was able to "solubilize" most of the adenyl cyclase activity present in particulate fractions of a cat heart homogenate. However, as in the earlier study mentioned in section I.C., hormonal sensitivity was lost as a result of this procedure.

The Km of adenyl cyclase for ATP has been estimated in a variety of preparations to be on the order of 1 to 5×10^{-4} M (Bär and Hechter, 1969a; Birnbaumer *et al.,* 1969; Drummond and Duncan, 1970). This is about an order of magnitude less than the overall concentration of ATP normally present in cells, but may or may not be less than the concentration in contact with adenyl cyclase. The nature of the fluoride effect is still poorly understood, but the results of a study by Schramm and Naim (1970) support the view that it probably represents the reversal of an inhibitory influence not present under all conditions. Although the stimulation of adenyl cyclase by NaF may be an artifact resulting from cell breakage, continued study of this phenomenon may nevertheless lead to an improved understanding of how adenyl cyclase functions in the intact cell.

Some recent studies of the substrate specificities of adenyl (and guanyl) cyclase and phosphodiesterase(s) are summarized later in Chapter 11. It is possible that the two activities mentioned in section II.B of the present chapter represent primarily the activities of separate enzymes, and that the low Km activity may be of major importance in the metabolism of cyclic AMP *in vivo.* The results of some recent studies by W. J. Thompson and M. M. Appleman that bear on this point will probably be published sometime during 1971.

CHAPTER 5

Some Actions of Cyclic AMP

I. INTRODUCTION

The effects of cyclic AMP which are known at the time of this writing are summarized in Table 5.I. This table includes effects which have been observed at all levels of organization from relatively highly purified enzymes up to the organ level, and which can reasonably be regarded as direct effects of the nucleotide. The more complex effects which have been observed in whole animals are discussed in the next section (Section II). One of the questions to be considered in this chapter is *how* cyclic

AMP produces these various effects. By which mechanism or mechanisms does it act?

We should stress at the outset that no simple answer to this question can be given at the present time. Although we are beginning to understand some of the effects of cyclic AMP, almost nothing is known about some of the others, so whether there is some basic action common to all of them is quite unknown. A few general points can nevertheless be made.

It should first of all be clear from an inspection of Table 5.I that not all of the effects of cyclic AMP will occur in all tissues. This is basically why hormones which affect cyclic AMP in different tissues produce different effects. An endocrinological precept of long standing has been that hormones influence the *rates* of basic processes, and do not endow cells with capabilities that they previously did not have.* This principle can be extended to cyclic AMP. No amount of the nucleotide will cause the liver to produce steroid hormones, for example, because the liver does not contain enzymes capable of manufacturing these hormones. Thus glucagon is not a steroidogenic hormone, at least in the species which have been examined to date. It is of course possible to imagine that it may be possible in the future to endow cells with capabilities which they ordinarily do not have (see, for example, Sinsheimer, 1969). There are at least two ways by which genetic engineers might in theory turn glucagon into a corticosteroidogenic hormone. Since adrenal cortical cells already contain a cyclic AMP-sensitive system for manufacturing steroid hormones (see Chapter 9), the simplest way would be to alter adenyl cyclase in these cells in such a way that it became sensitive to glucagon. This might involve changing only a few amino acids. A more complicated way would be to induce the liver to synthesize steroidogenic enzymes, in which event the engineers who attempted such a feat would have to remember to build in sensitivity to cyclic AMP as well. Even taking these hypothetical possibilities into account (and they may not be as farfetched as they sound), a reasonably safe generalization would seem to be that cells do only what they are equipped to do, and in many cases they are stimulated to do so by a change in the level of cyclic AMP.

Included in Table 5.I are several effects on enzymes, and the question can be asked whether all of the effects of cyclic AMP are initially the

---

*An obvious semantic difficulty arises when we consider first and second messengers which act to induce the synthesis of enzymes not previously present within the cell, since it could be argued that as a result the cell can now do something which it could not previously do. It is nevertheless clear that the cell had the capacity for synthesizing the enzyme all along, and that in the presence of the hormone it was only stimulated to utilize this capacity.

TABLE 5.I. Known Effects of Cyclic AMP

Enzyme or process affected	Tissue	Change in activity or rate	Reference
Protein kinase	Several	Increased	See Table 5.IV
Phosphorylase kinase activation	Muscle	Increased	Krebs *et al.*, 1964; Drummond *et al.*, 1965; De Lange *et al.*, 1968
Phosphorylase phosphatase inactivation	Adrenal	Increased	Merlevede and Riley, 1966
Phosphorylase activation	Several	Increased	Rall and Sutherland, 1958; Mohme-Lundholm, 1963; Northrop and Parks, 1964b; and many others. See Haugaard and Hess, 1965, for a review
Glycogen synthetase kinase	Muscle	Increased	Huijing and Larner, 1966a,b
Glycogen synthetase kinase	Liver	Increased	Bishop and Larner, 1969
Glycogen synthetase inactivation	Muscle	Increased	Rosell-Perez and Larner, 1964; Appleman *et al.*, 1966b
Glycogen synthetase inactivation	Liver	Increased	De Wulf and Hers, 1968a,b
Phosphofructokinase activation	*F. hepatica*	Increased	Stone and Mansour, 1967a
Fructose-1,6-diphosphatase inactivation	Kidney	Increased	Mendicino *et al.*, 1966
Tyrosine transaminase induction[a]	Liver	Increased	Wicks, 1968, 1969
PEP carboxykinase induction[a]	Liver	Increased	Yeung and Oliver, 1968a,b; Wicks, 1969
Glucose-6-phosphatase induction[a]	Liver	Increased	O. Greengard, 1969a,b
Serine dehydratase induction[a]	Liver	Increased	Jost *et al.*, 1969
β-Galactosidase induction[a]	*E. coli*	Increased	Perlman and Pastan, 1968; Ullmann and Monod, 1968
Tryptophanase induction[a]	*E. coli*	Increased	Pastan and Perlman, 1969a
Pyruvate kinase	Loach embryo	Increased	Milman and Yurowitzki, 1967
Glycogenolysis	Liver	Increased	Bergen *et al.*, 1966; Levine and Lewis, 1969
Gluconeogenesis	Liver	Increased	Exton and Park, 1968a,b
Urea formation	Liver	Increased	Menahan and Wieland, 1967; Exton and Park, 1968b
Net K[+] efflux	Liver	Increased	Northrop, 1968; Exton and Park, 1968b
[45]Ca efflux	Liver	Increased	Friedmann and Park, 1968
Ketogenesis	Liver	Increased	Bewsher and Ashmore, 1966; Heimberg *et al.*, 1969

Enzyme or process affected	Tissue	Change in activity or rate	Reference
Lipogenesis	Liver	Decreased	Berthet, 1960
Permeability	Toad bladder	Increased	Orloff and Handler, 1967
Ca²⁺ resorption	Bone	Increased	Vaes, 1969a
Permeability	Ventral frog skin	Increased	Bastide and Jard, 1968; Baba *et al.*, 1967
Permeability	Kidney tubules	Increased	Grantham and Burg, 1966
Permeability	Intestine	Increased	Field *et al.*, 1968
Renin production	Kidney	Increased	Michelakis *et al.*, 1969
Amino acid uptake	Liver	Increased	Mallette *et al.*, 1969a; Chambers and Bass, 1970
Amino acid uptake[b]	Adipose	Decreased	von Bruckhausen, 1968
Amino acid → protein	Liver	Decreased	Pryor and Berthet, 1960
Lipolysis[b]	Adipose	Increased	Chapter 8
Steroidogenesis	Several	Increased	Chapter 9
Oxygen consumption	Adipose	Increased	Reed and Fain, 1968a
Clearing factor lipase	Adipose	Decreased	Wing and Robinson, 1968
Glucose oxidation	Several	±	Chapters 6 and 10
6-Phosphogluconate dehydrogenase	Adipose	Decreased	Bray, 1967
Amylase release[b]	Parotid	Increased	Babad *et al.*, 1967
	Exocrine pancreas	Increased	Kulka and Sternlicht, 1968
Insulin release	Pancreatic islets	Increased	Sussman and Vaughan, 1967; Malaisse *et al.*, 1967
ACTH release	Anterior pituitary	Increased	Fleischer *et al.*, 1969
TSH release	Anterior pituitary	Increased	Bowers *et al.*, 1968b
GH release	Anterior pituitary	Increased	Gagliardino and Martin, 1968
Thyroid hormone release	Thyroid	Increased	Ensor and Munro, 1969
Calcitonin release	Thyroid	Increased	Care and Gitelman, 1968; Care *et al.*, 1969
Histamine release	Leukocytes	Decreased	Lichtenstein and Margolis, 1968
HCl secretion	Gastric mucosa	Increased	Harris and Alonso, 1965; Way and Durbin, 1969
Fluid secretion	Insect salivary glands	Increased	Berridge and Patel, 1968
Transmission	Peripheral nervous	Enhanced	A. Goldberg and Singer, 1969

Discharge	Cerebellar Purkinje	Decreased	Siggins *et al.*, 1969
Melanocyte dispersion	Dorsal frog skin	Increased	Bitensky and Burstein, 1965; Novales and Davis, 1969
Relaxation[b]	Smooth muscle	Increased	Dobbs and Robison, 1968; Moore *et al.*, 1968
Force of contraction[b]	Cardiac muscle	Increased	Kukovetz and Pöch, 1970a
Aggregation	Slime Mold	Increased	Bonner *et al.*, 1969
Aggregation[c]	Platelets	Decreased	Marcus and Zucker, 1965
Cell growth	Hela cells	Decreased	Ryan and Heidrick, 1968
Phosphofructokinase deinhibition[c]	Several	Increased	Mansour, 1966; Stone and Mansour, 1967b
DPHN oscillations[c]	Yeast	Increased	Chance and Schoener, 1964; Cheung, 1966
Release of protein from polysomes[c]	Liver	Increased	Khairallah and Pitot, 1967
Amino acid uptake[c]	Uterus	Increased	Griffin and Szego, 1968
Lac mRNA synthesis	*E. coli*	Increased	Varmus *et al.*, 1970; de Crombrugghe *et al.*, 1970
DNA synthesis	Thymocytes	Increased	Whitfield *et al.*, 1970a,b
Flagella formation	*E. coli*	Increased	Yokota and Gots, 1970
Stalk cell formation	Slime mold	Increased	Bonner, 1970
Melanocytogenesis[b]	Fish skin	Increased	Chen and Tchen, 1970
Luteinization	Granulosa cells	Increased	Channing and Seymour, 1970
Ca^{2+} uptake by SR	Cardiac muscle	Increased	Entman *et al.*, 1969b; Shinebourne and White, 1970
Rate of contraction[b]	Cardiac muscle	Increased	Krause *et al.*, 1970
Membrane potential[b]	Smooth muscle	Increased	Somlyo *et al.*, 1970
Ascorbic acid depletion	Adrenal	Increased	Earp *et al.*, 1970

[a] The term "induction" is used to describe the selective stimulation of the rate of enzyme synthesis without suggesting any mechanism. Indeed, as described in more detail in Section IV, it now seems clear that cyclic AMP may act by several mechanisms to "induce" these enzymes.

[b] Effect of cyclic AMP itself absent or very small compared to the effect of one or another of the acyl derivatives (Chapter 3 and Section II of this chapter).

[c] These effects are shared by similar concentrations of one or more other adenine nucleotides, and are therefore of uncertain physiological significance.

result of an alteration in the activity of some enzyme. We might venture the guess that this is probably true, although it might not be. It is at least conceivable, for instance, that the increased permeability to water which cyclic AMP produces in the toad bladder could be a result of some physical change in the cell membrane not involving an enzyme. This example may serve to underline the depth of our ignorance in this area. In a very real sense, our ignorance regarding the mode of action of cyclic AMP reflects our ignorance of the nature of basic cell processes in general.

Not all of the effects listed in Table 5.I were measured at the same level of organization, and hence are not all mutually exclusive. It is known, for example, that some of the effects observed at the enzyme level are responsible for some of the effects observed at higher levels, just as the latter will contribute to effects observable at higher levels still. It might seem desirable to minimize this type of duplication by listing each effect at the lowest level possible in the light of present knowledge, and excluding whatever effects they may have at higher levels. The extent of our ignorance is such, however, that we thought the table might serve a more useful purpose if it included these other effects. It represents a compromise on this point in that we have at least excluded effects observed at the level of the whole animal. In later chapters, each of the hormonal effects known to be mediated by cyclic AMP will be discussed in detail in terms of as many levels of organization as our understanding permits.

Most of the effects listed in Table 5.I are highly specific for cyclic AMP in that they occur at low concentrations and cannot be reproduced by other nucleotides. Some of the listed effects, on the other hand, have been found in at least some tissues to require relatively high concentrations and can be produced by similar concentrations of other adenine nucleotides. These effects may not be physiologically significant because, as pointed out in Chapter 2, the level of cyclic AMP is generally much lower in most cells than the levels of other nucleotides. Thus a large change in the concentration of cyclic AMP, which would have a substantial effect on a system sensitive to cyclic AMP only, would represent a negligible change in the overall nucleotide concentration and would have a correspondingly small effect on systems sensitive to one or more of these other nucleotides. These effects are included in Table 5.I on the grounds that they may be physiologically significant in certain tissues. Some of the reported effects of cyclic AMP have been deliberately excluded because it is almost certain that they are not important. An example of such an effect would be that on tryptophan pyrrolase. The activity of this enzyme is enhanced by all purine derivatives which have been tested (Gray, 1966), and now there is evidence that cyclic AMP must first be metabolized to 5'-AMP before it can act (Chytil, 1968).

A final point which can be made in this section, even though it was alluded to previously, is that even when the same system occurs in more than one tissue, the response to cyclic AMP may not be identical. An example of this, to be discussed in more detail later in this chapter, is the phosphorylase system, which not only differs in its sensitivity to cyclic

AMP from one tissue to another but may also be differentially affected by other regulatory mechanisms.

II. EFFECTS OF EXOGENOUSLY ADMINISTERED CYCLIC AMP

It was mentioned in Chapter 2 that if an effect of a hormone is mediated by cyclic AMP, then it ought to be possible to mimic that effect by the application of exogenous cyclic AMP. This criterion is subject to some special qualifications which can be discussed here.

Table 5.II contains a list of all the effects of cyclic AMP which have been observed when the nucleotide or one of its derivatives has been administered to whole animals. Other effects will almost certainly be

TABLE 5.II. Effects of Exogenous Cyclic AMP in Intact Animals

Effect	References
Hyperglycemia	Posternak *et al.*, 1962; Northrop and Parks, 1964a; Bergen *et al.*, 1966; R. A. Levine, 1968a; Bieck *et al.*, 1969
Hyperlacticacidemia	Henion *et al.*, 1967
Hypercalcemia and hypophosphatemia	Rasmussen *et al.*, 1968; Wells and Lloyd, 1969
Growth hormone release	Gagliardino and Martin, 1968; Müller *et al.*, 1969
Thyroid hormone release	Bastomsky and McKenzie, 1967; Ahn and Rosenberg, 1968; Ahn *et al.*, 1969
Insulin release	Bressler *et al.*, 1969
Maintenance of adrenal weight	Ney, 1969
Increased blood steroid levels	Imura *et al.*, 1965; Bieck *et al.*, 1969
Hepatic enzyme induction	Yeung and Oliver, 1968b; Wicks *et al.*, 1969
Inhibition of ovulation	Ryan and Coronel, 1969
Fall in blood pressure and increased cardiac output	Levine and Vogel, 1966; Starcich *et al.*, 1967; Levine *et al.*, 1968
Diuresis	Abe *et al.*, 1968
Antidiuresis	R. A. Levine, 1968b
Phosphaturia	Russell *et al.*, 1968; Rasmussen *et al.*, 1968
Drowsiness (peripheral administration)	Henion *et al.*, 1967
Convulsions (intraventricular administration)	S. J. Strada *et al.*, 1968 unpublished; Krishna *et al.*, 1968c
Hyperthermia and hyperphagia (hypothalamic implantation)	Breckenridge and Lisk, 1969
Enzyme induction in reproductive organs	Singhal *et al.*, 1970; Singhal and Lafreniere, 1970

detected as more detailed investigations are carried out. Most of the effects which have been reported to date can indeed by regarded as mimicking the effects of one or another of the hormones which cause an increase in the intracellular level of cyclic AMP, and as such will be discussed in other chapters wherever appropriate. Our purpose here is to discuss the action of exogenous cyclic AMP in its most general aspects.

Cyclic AMP is, in general, a relatively inert compound when injected into animals or applied to intact cells and tissues. Whereas broken cell preparations may respond to concentrations of cyclic AMP in the physiological range, roughly 10^{-8}–10^{-6} M (e.g., Butcher *et al.*, 1965; Robison *et al.*, 1965; Denton and Randle, 1966; Grahame-Smith *et al.*, 1967; Walsh *et al.*, 1968), these concentrations are usually without effect in intact tissues. However, by increasing the concentrations it has in many cases been possible to produce effects in intact tissues, as indicated in Tables 5.I and 5.II. When these effects are not also produced by equimolar concentrations of other nucleotides, and especially when they are the same as those of a hormone known to promote the accumulation of cyclic AMP in the tissue being studied, then it seems reasonable to assume that a physiologically significant amount of cyclic AMP has accumulated within the intracellular space. This accumulation will be limited by the rate at which the nucleotide penetrates the cell membrane and by the rate at which it is destroyed by the phosphodiesterase either during or after its entry into the cell. Several cases have been reported where an ineffective dose of cyclic AMP has been rendered effective by pretreatment of the tissue or organism with theophylline (e.g., Northrop and Parks, 1964a; Weiss *et al.*, 1966; Wicks, 1968; Goodman, 1969b; Novales and Davis, 1969).

It is generally thought that nucleotides and other anionic phosphorylated compounds penetrate cell membranes poorly, if at all, and this generalization does seem to be widely applicable (Leibman and Heidelberger, 1955; Roll *et al.*, 1956; Berne, 1964). Some observations, such as the finding that the administration of ATP tended to raise hepatic levels of ATP in shocked dogs (Sharma and Eiseman, 1966), might seem to suggest otherwise, but can probably be understood in terms of extracellular dephosphorylation followed by uptake and subsequent rephosphorylation (Villar-Trevino *et al.*, 1963). Adenosine and other nucleosides, in contrast to their phosphorylated derivatives, appear to enter most types of cells relatively easily (Jacob and Berne, 1960). Still, the generalization that nucleotides do not cross cell membranes may not apply equally to all nucleotides in all cells. In the case of cyclic AMP, penetration into avian erythrocytes (Davoren and Sutherland, 1963a) and heart muscle (Robison *et al.*, 1965) could not be detected at all by the use of ordinary assay techniques, although when rat hearts were exposed to tritiated cyclic AMP for 30 min, a small amount of tritium was found to have accumulated in the intracellular space. In marked contrast, the toad bladder was found to accumulate tritium at a rapid rate when exposed to tritiated cyclic AMP (Gulyassy and Edelman, 1967), with some of the radioactivity in the form of cyclic AMP even after 1 hr (Gulyassy, 1968). More recently the accumulation of cyclic AMP by the liver has been demonstrated (Levine *et al.*, 1969). Although the permeability of most cell membranes to cyclic AMP may be small, it would appear that the nucleotide can penetrate some membranes to at least some extent.

Although this point has not been systematically studied, to our knowledge, there is some indirect evidence to suggest that cyclic nucleotides do cross cell membranes more readily than their straight chain analogs. Urine contains relatively high concentrations of cyclic nucleotides, but is essentially devoid of other organic phosphates (Price *et al.*, 1967). The urinary level of cyclic AMP increases dramatically in response to parathyroid hormone, presumably as a result of extrusion from renal cortical cells (Chase and Aurbach, 1967, 1968). As discussed in Chapter 2 (Section IV, A), cyclic AMP can also be measured in the blood and other body fluids. Plasma levels rise in response to the injection of glucagon (Kaminsky *et al.*, 1969), presumably in this case as a result of extrusion from the liver. Some cells, such as avian erythrocytes (Davoren and Sutherland, 1963a) and certain bacteria (Makman and Sutherland, 1965), are even capable of pumping cyclic AMP into the external medium against a concentration gradient. Analogous examples involving straight chain nucleotides have not to our knowledge been reported. The release of ADP from platelets and other cells is known to occur (Mustard and Packham, 1970), as is the release of ATP from chromaffin cells (Douglas, 1968), but not necessarily against a concentration gradient. It is of course recognized that studies of efflux do not necessarily reflect the relative rates at which nucleotides penetrate cells from the outside. For example, the release of ATP from chromaffin cells is thought to occur as a result of emiocytosis, and it seems possible that some such mechanism might participate in the release of cyclic nucleotides as well.

In hopes of obtaining compounds which might penetrate cell membranes more readily, a number of derivatives of cyclic AMP were synthesized, as described in detail in Chapter 3. In general, these compounds have been found to be less active than cyclic AMP in tissue extracts but more active when applied to intact tissues (Henion *et al.*, 1967). The relative potencies of a series of these derivatives in activating liver phosphorylase are shown in Table 5.III. For the experiments with

TABLE 5.III. Relative Potencies of Some Derivatives of Cyclic AMP
in Stimulating the Activation of Dog Liver Phosphorylase

	Relative potency	
Derivative	Liver extracts[a]	Liver slices[b]
Cyclic AMP[c]	100	2
N^6-monoacetyl	28	69
N^6-monobutyryl	56	100
N^6-Monooctanoyl	78	48
$2'$-O-monobutyryl	8	35
$N^6,2'$-O-diacetyl	2	27
$N^6,2'$-O-dibutyryl	2	91

[a] Data from Posternak *et al.* (1962).
[b] Data from Henion *et al.* (1967).
[c] Cyclic GMP was essentially inactive in dog liver extracts but was recently found to be as effective as cyclic AMP when added to the isolated perfused rat liver. The effects of cyclic GMP and other cyclic nucleotides are discussed in more detail in Chapter 11.

liver extracts, the compounds were added at a concentration of 10^{-7} M and phosphorylase activity measured by the usual assay procedure (see Appendix). For the intact cell experiments, dog liver slices were incubated for 15 min in the presence of cyclic AMP or one of the derivatives at a concentration of 10^{-5} M. The slices were then homogenized and assayed for phosphorylase activity. It can be seen that under these conditions the dibutyryl derivative was approximately 50 times more effective than cyclic AMP itself, when applied to the intact cells, whereas it was approximately 50 times *less* active when added to liver extracts. The N^6 derivatives were more effective than the $2'$-O derivatives in this system, although this may not be the case for all systems which are affected by cyclic AMP. The liver contains an enzyme or enzymes capable of removing the acyl substituents from the $2'$-O- and probably from the N^6-position as well, and apparently these enzymes are either lost or inactivated during the preparation of the extracts. In addition to these experiments with dog liver slices, the greater effectiveness of the dibutyryl and $2'$-O-monobutyryl derivatives has been demonstrated in the isolated perfused rat liver (Levine and Lewis, 1967, 1969).

In general, the effects of these derivatives have been found to be qualitatively the same as those of cyclic AMP, when applied to tissues where cyclic AMP itself produced some effect. In several cases one of the derivatives has been effective under conditions where cyclic AMP was completely or almost completely inactive. For example, when rat parotid slices were incubated with 2×10^{-3} M cyclic AMP there was no change in the rate of amylase secretion. By contrast, the addition of either the dibutyryl or the N^6-monobutyryl derivative caused a very striking increase in the rate of amylase secretion (Babad *et al.*, 1967). In this instance the derivatives mimicked the effect of the catecholamines, which stimulate the formation of cyclic AMP in the parotid gland. The greater effectiveness of the dibutyryl derivative has also been demonstrated in adipose tissue (Butcher *et al.*, 1965), adrenal glands (Imura *et al.*, 1965; Bieck *et al.*, 1969), thyroid slices (Pastan, 1966; Rodesch *et al.*, 1969), bone (Vaes, 1968a), smooth muscle (Dobbs and Robison, 1968; Moore *et al.*, 1968), and in the isolated perfused guinea pig heart (Kukovetz, 1968).

Some exceptions have been noted. Kim *et al.* (1968) reported that the dibutyryl derivative was less effective than cyclic AMP in causing relaxation of the rabbit ileum, and Ryan and Heidrick (1968) found the dibutyryl derivative to be less effective as an inhibitor of cell growth in tissue culture. These observations might be expected if the responding cells lacked esterase activity necessary for splitting off the butyric acid moieties.

There is no direct evidence that the derivatives penetrate cell membranes more rapidly than cyclic AMP, although this may occur and may in some cases be the explanation for their greater potency. It is also possible that they are more potent because they are more resistant to inactivation by cyclic nucleotide phosphodiesterase. This greater resistance to the action of phosphodiesterase has been established in the case of the N^6-monoacyl derivatives, and most likely applies to the $2'-O$ derivatives as well (Posternak *et al.*, 1962; Henion *et al.*, 1967; Moore *et al.*, 1968). The finding by Blecher *et al.* (1968) that the dibutyryl derivative was hydrolyzed as rapidly as the parent compound by homogenates of adipose tissue may simply indicate that these homogenates contained high esterase activity.

As noted previously, the action of these derivatives may be limited not only by the ability of the compounds to penetrate the cell membrane but also by the ability of the tissues involved to remove one or more of the substituted acyl derivatives. In the liver phosphorylase system, the N^6 derivatives retained some activity but those substituted at the $2'-O$ position were inactive (Henion *et al.*, 1967). Other systems may, of course, have different requirements. Perhaps by virtue of this resistance to phosphodiesterase the derivatives are allowed to penetrate to a site where they can act as a "slow feed" of cyclic AMP, thus building up a concentration of cyclic AMP which could not be attained if the same distance had to be traversed by the unprotected nucleotide. The released cyclic AMP is of course still subject to rapid hydrolysis by the phosphodiesterase, as illustrated in Fig. 5.1. When the dibutyryl derivative is added to the bath of an isolated rat uterus in the absence of a phosphodiesterase inhibitor, the relaxing effect is relatively transient. When the uterus has been pretreated with theophylline, however, then the effect of the dibutyryl derivative is greatly prolonged (Dobbs and Robison, 1968). Potentiation of the dibutyryl derivative by theophylline has also been noted in adipose tissue (Bieck *et al.*, 1969; Goodman, 1969b) and in intact rats (Wells and Lloyd, 1969).

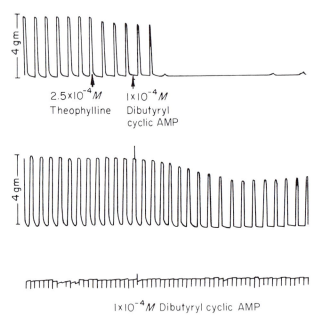

Fig. 5.1. Effect of the dibutyryl derivative of cyclic AMP on uterine motility in the presence (upper panel) and absence (lower panel) of theophylline. Tracings (spontaneous isometric contractions) are from opposite horns of the same uterus from an ovariectomized estrogen-primed rat. From Dobbs and Robison (1968).

Studies with exogenous cyclic AMP and its derivatives may be complicated in many cases by the fact that adenosine and its nucleotides are potent pharmacological agents in their own right (Green and Stoner, 1950). The effects of these agents seem to be due primarily to the presence of the adenosine moiety, and may be exerted at the cell membrane. This seems especially likely in those cases where equimolar concentrations have been found to be approximately equipotent (e.g., Angelakos and Glassman, 1965; Axelsson and Holmberg, 1969). It is also possible that some of the actions of the adenine nucleotides depend upon their conversion to adenosine, in which event an intracellular site of action would seem as likely as an extracellular one.

The powerful vasodilator activity of these compounds is well known (Green and Stoner, 1950; Gordon ahd Hesse, 1961; Rowe *et al.*, 1962; Berne, 1964), and in the case of adenosine this action may play an important role in the regulation of blood flow (Berne, 1964). They may also affect the tone and motility of intestinal smooth muscle, although it is interesting that in this tissue, while 5'-AMP, ADP, and ATP produce different effects in different species, cyclic AMP always produces relaxation (L. Hurwitz *et al.*, 1967, unpublished observations). With the exception of the isolated rabbit ileum (Kim *et al.*, 1968), the dibutyryl derivative has been found to be a more potent relaxant of intestinal smooth muscle than cyclic AMP itself. Adenine nucleotides are vasoconstrictor substances in the kidney (Scott *et al.*, 1965), but the possibility that cyclic AMP might differ in this regard has not to our knowledge been tested.

When injected into anesthetized dogs, adenosine and its nucleotides cause a fall in blood pressure and also tend to depress cardiac contractility (Angelakos and Glassman, 1965), although in this latter respect cyclic AMP is much less potent than any of the other adenine nucleotides (A. M. Lefer and G. A. Robison, 1963, unpublished observations). The intravenous injection of cyclic AMP in unanesthetized dogs (Levine and Vogel, 1966) and human patients (Starcich *et al.*, 1967; Levine *et al.*, 1968) has been reported to cause an increase in cardiac output and a positive chronotropic effect. When studied by direct perfusion of the sinus node, on the other hand, adenosine and its nucleotides were found to produce negative chronotropic effects, although here again cyclic AMP was less potent than the other derivatives (James, 1965, 1967). Guanosine was found to exert a positive chronotropic effect in these experiments.

Adenosine produces a negative inotropic effect in isolated hearts from several species (Buckley *et al.*, 1961; DeGubareff and Sleator, 1965) and may at times induce heart block (Scriabine and Bellet, 1967). These effects of adenosine can be antagonized by the methylxanthines, which are of course well known as positive inotropic agents. Certain other purine and pyrimidine derivatives, including ATP under some conditions (Chen *et al.*, 1946), may also exert a positive inotropic effect, while others may have no effect or variable effects on the perfused heart (Buckley *et al.*, 1961). Cyclic AMP and its derivatives usually fall into this latter category, although Kukovetz (1968) has reported that the dibutyryl derivative causes a positive inotropic as well as a positive chronotropic effect in the isolated perfused guinea pig heart.

Although we have mentioned only the cardiovascular effects in this section, it should be noted that adenosine and its nucleotides may have many other effects throughout the body. The mechanisms by which these various effects are produced are quite unknown, and our principle purpose in mentioning them at all has been to illustrate the nature of the problem involved. The effects of exogenous cyclic AMP should, of course, always be interpreted with caution, but when these effects are the same as those produced by other adenosine derivatives it is clear that the problem of interpretation may be especially difficult.

The dose or concentration of these compounds is another important factor which should be considered. The exposure of intact cells to exogenous nucleotides would in most cases have to be regarded as an unphysiological circumstance at best, and it is not surprising that high concentrations are capable of causing more damage than low. In the case of cyclic AMP and its derivatives, where this point has been studied, the dose response curve has been found to be bell-shaped over a relatively narrow range. This has been demonstrated in the case of glucose release from the liver (Menahan and Wieland, 1967), lipolysis by adipose tissue (Bieck *et al.*, 1968), incorporation of phosphate into phospholipid in thyroid tissue (Burke, 1968), and calcium mobilization from bone (Raisz and Klein, 1969).

An interesting further complication in considering the effects of exogenous application of these agents stems from the evidence that in some tissues adenosine and the adenine nucleotides are capable of stimulating the formation of cyclic AMP. This was actually demonstrated by Sattin and Rall (1970) in guinea pig brain slices. Whether this represents a hormone-like effect exerted at the cell membrane, or whether it depends upon conversion to adenosine, which could act intracellularly, is unclear at the present time. To what extent this type of effect can be extended to other tissues and species is likewise unknown, although certainly many possibilities exist for future investigation. One possibility which is raised by these data is that in at least some cases the effects of exogenous cyclic AMP may be related not to penetration into the cell but to a stimulatory effect on adenyl cyclase. Support for the idea that this may at times occur comes from the old observation that acrasin (which has now been shown to be cyclic AMP in at least some species of cellular slime molds) is capable of stimulating the production of more acrasin (Bonner, 1967, 1969). This is discussed in more detail in Chapter 12.

As for the effects of cyclic AMP and its derivatives when administered to whole animals, these are essentially what would be expected on the basis of studies with isolated tissue preparations. As summarized in Table 5.II, these include predictable effects on some parameters but highly variable effects on others. Such studies may be complicated not only by the nonspecific toxic effects mentioned previously, but also by an effect in one organ opposing the effect in another. The hyperlacticaci-demia, which could arise from skeletal muscle or blood cells or possibly some other source, is followed by a more prolonged fall in blood lactate levels. This is illustrated in Fig. 5.2, and is most likely the result of stimulation of gluconeogenesis (Exton and Park, 1968b). The failure of the nucleotide to affect free fatty acids in any consistent manner (R. A. Levine, 1968a; Bieck *et al.*, 1969) may be partly related to poor penetra-tion into adipose tissue and partly to enhanced utilization by other tissues (Masters and Glaviano, 1969), but perhaps most importantly to stimula-tion of the release of insulin, which tends to antagonize the effect in adipose tissue (Goodman, 1969b). Similarly, the lack of effect on blood calcium levels in intact animals (Wells and Lloyd, 1969) could be ac-counted for by the release of thyrocalcitonin (Care and Gitelman, 1968), which opposes the direct effect in bone. All of these effects will be dis-

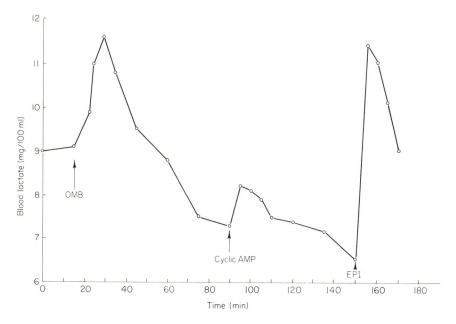

Fig. 5.2 Effect of cyclic AMP, the 2′-O-monobutyryl derivative of cyclic AMP (OMB), and epinephrine (EPI) on the level of blood lactate in an anesthetized dog. Cyclic AMP and the derivative were injected intravenously in doses of 13 μmoles/kg each, while the dose of epinephrine was 4.5 μg/kg. From Henion, Sutherland, and Posternak (1967).

cussed in more detail later, in connection with the hormones which produce them.

The behavioral effects of exogenous cyclic AMP depend upon the route of administration. When the dibutyryl derivative was administered intraperitoneally to mice in a dose of 500 mg/kg, it seemed relatively nontoxic, with drowsiness being the most obvious symptom (Henion *et al.,* 1967). By contrast, when 0.2 μmoles of this derivative were injected into the laterial ventricle in rats (corresponding to a dose of approximately 0.4 mg/kg), the result was an extraordinary increase in respiratory rate followed by tonic convulsions, which recurred at intervals for many hours. This dose was lethal in rats pretreated with theophylline. When 4 mg/kg were administered by this route it rapidly caused lethal convulsions even in rats not pretreated with theophylline (S. J. Strada *et al.,* 1968, unpublished observations). These observations indicate, among other things, that the dibutyryl derivative is not capable of passing the blood-brain barrier. Similar observations on the behavioral effects of cyclic AMP have been made and filmed by Krishna *et al.* (1968c).

III. EFFECTS ON ENZYMES

Of the effects listed in Table 5.I, several occur in a variety of tissues and do not seem to be closely associated with any one hormone or group of hormones. For this reason we have chosen to discuss some of these effects in this chapter, with emphasis on those which have been studied at the enzyme level. The activation of phosphorylase, or at least the acceleration of this process, was the first effect of cyclic AMP to be discovered, and is still the one which has been most carefully studied. We shall therefore begin with it.

A. Phosphorylase Activation

1. Skeletal Muscle

Phosphorylase activation in muscle has been extensively studied by E. G. Krebs and his associates. The sequence of events by which an increase in the level of cyclic AMP leads to an increase in the activity of phosphorylase in this tissue is illustrated diagrammatically in Fig. 5.3. This sequence, according to which the activation of one enzyme leads to

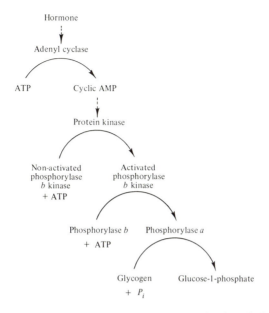

Fig. 5.3. Sequence of events involved in the hormonal activation of phosphorylase in skeletal muscle, leading to glycogenolysis. Modified from Krebs *et al.* (1966).

the activation of another, has aptly been likened to the cascade of reactions involved in blood clotting (Krebs and Fischer, 1960, 1962; Krebs *et al.*, 1966, 1968). Although some amplification may take place at an earlier stage (e.g., one molecule of a hormone may stimulate the formation of more than one molecule of cyclic AMP), most of the amplication involved in hormonal responses mediated by cyclic AMP probably occurs after the formation of cyclic AMP.

The first enzyme in this sequence, and the one which seems to be affected by cyclic AMP, is a protein kinase which was initially referred to as "kinase II" (Krebs *et al.*, 1966). It has now been purified some 300-fold from rabbit skeletal muscle (Walsh *et al.*, 1968), and is perhaps more appropriately referred to as phosphorylase *b* kinase kinase, since it catalyzes the transfer of the terminal phosphate of ATP to phosphorylase *b* kinase, thereby activating the latter enzyme (DeLange *et al.*, 1968). As the most recently discovered of the enzymes in this sequence, relatively little is known about the chemical and physical properties of kinase II, or about the mechanism by which cyclic AMP affects it. The presence of cyclic AMP seems to be required for its activity, and a significant binding of cyclic AMP to the partially purified protein has been demonstrated at nucleotide concentrations as low as 5×10^{-9} *M*. This would be compatible with the allosteric mechanism previously postulated on the basis of other data (DeLange *et al.*, 1968).

Kinase II has been a difficult enzyme to study for several reasons. One is that it has been difficult to separate it from its substrate, phosphorylase *b* kinase, and another is that phosphorylase *b* kinase is autocatalytic, i.e., activated phosphorylase *b* kinase catalyzes the phosphorylation of non-activated phosphorylase *b* kinase. It therefore cannot be unequivocally stated that cyclic AMP affects only kinase II, since the possibility that it also affects phosphorylase *b* kinase directly has not been ruled out. A reasonable guess, however, based on presently available information, would be that cyclic AMP affects only kinase II.

Kinase II is also capable of catalyzing the phosphorylation of several other proteins, including casein and protamine (Walsh *et al.*, 1968). Figure 5.4 illustrates the effect of cyclic AMP on casein phosphorylation. It is evident that this process is completely dependent on cyclic AMP, with an apparent K_m of about 1×10^{-7} *M*. In these experiments, cyclic AMP could not be replaced by any of a variety of other nucleotides tested, including 3'-AMP and 5'-AMP, in concentrations up to 10^{-4} *M*. Walsh *et al.* (1968) also showed that the activation and concomitant phosphorylation of phosphorylase *b* kinase could be greatly enhanced by the addition of kinase II, providing that cyclic AMP was also present. They calculated that the previously demonstrated effect of cyclic AMP

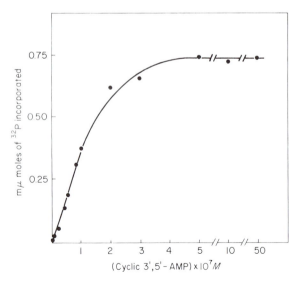

Fig. 5.4. Phosphorylation of casein catalyzed by protein kinase in the presence of increasing concentrations of cyclic AMP. From Walsh, Perkins and Krebs (1968).

on purified phosphorylase *b* kinase could be fully accounted for by contaminating traces of kinase II.

An inhibitory protein which antagonizes the effect of cyclic AMP on kinase II has been partially purified from muscle (Corbin and Krebs, 1969). The physiological significance of this protein is unknown at present.

Phosphorylase *b* kinase has been known for a longer period of time than kinase II, and correspondingly more is known about its properties. It is a large protein — estimates based on sedimentation studies suggest a molecular weight greater than 1,000,000 — which constitutes about 1% of the soluble protein of rabbit skeletal muscle (Fischer and Krebs, 1966). It exists in resting skeletal muscle in a form which has relatively little activity at or below the physiological intracellular pH of 6.9 and exhibits less than maximal activity even at higher pH values. In the presence of ATP and Mg^{2+} it can be converted to a form (activated phosphorylase *b* kinase) having 25 to 50 times as much activity at pH 6.8 and about twice as much activity at pH 8.2. The ratio of kinase activity at pH 6.8 to that at pH 8.2 has been widely used in physiological and biochemical experiments as an index of kinase activation (e.g., Posner *et al.*, 1965; Drummond *et al.*, 1966; Namm and Mayer, 1968; Wollenberger *et al.*, 1969). The ability of cyclic AMP to accelerate this process provided the basis for a useful method of assaying for cyclic AMP (Posner *et al.*, 1964).

As mentioned previously, it would appear that two mechanisms exist for accelerating the phosphorylation and activation of phosphorylase *b* kinase, viz., catalysis by kinase II and autocatalysis by phosphorylase kinase itself. It should be noted that phosphorylase *b* kinase is relatively specific for its normal substrate, phosphorylase *b*. Although it catalyzes its own phosphorylation, and also affects casein at about the same slow rate, other potential substrates tested by DeLange *et al.* (1968) were not affected. Since its autocatalytic activity is low relative to its effect on phosphorylase *b*, and since its activation *in vivo* is exceedingly rapid in response to an increase in the level of cyclic AMP (Drummond *et al.*, 1966), it seems likely that autocatalysis is relatively unimportant physiologically.

Studies in Krebs' laboratory have established that during the process of kinase activation the phosphate from ATP is incorporated at at least two or three and possibly more separate sites, probably serine (and/or threonine) residues. That not all of these sites are associated with enzyme activation was suggested by the findings of DeLange *et al.* (1968) that enzyme activity reached a plateau before phosphorylation was complete, and that cyclic AMP stimulated the rate of kinase activation more than it stimulated the rate of phosphorylation. Chemical evidence for the existence of nonspecific phosphorylation sites unrelated to activation was obtained by Riley *et al.* (1968).

Phosphorylase *b* kinase which has been activated in this manner can be inactivated by a phosphatase (Riley *et al.*, 1968). Although the specificity of this enzyme has not been completely worked out, it is at least known that the enzyme is not identical with the principal muscle phosphatase(s) acting on *p*-nitrophenyl phosphate, serine phosphate, or casein (although it is capable of acting on these substrates at a slow rate). Like most phosphatases, it is activated by metal ions and can be inhibited by fluoride. It can also be inhibited by glycogen, which may in large part explain the ability of glycogen to stimulate the activation of phosphorylase *b* kinase. The effects of glycogen and of cyclic AMP are in this respect additive (Krebs *et al.*, 1964).

In studying the possible mechanism of action of cyclic AMP, prior to the discovery of kinase II, Krebs and his group initially considered two main hypotheses. The first was that cyclic AMP might participate as a coenzyme in a transphosphorylation reaction between ATP and the kinase, while the second was that cyclic AMP might affect the kinase allosterically.

No evidence for the first hypothesis could be obtained (DeLange *et al.*, 1968). Cyclic AMP was not destroyed during the kinase activation reaction (provided the phosphodiesterase was either absent or inhibited) and in the presence of excess cyclic AMP and labeled ATP, no phosphate exchange into the cyclic nucleotide could be demonstrated. These findings led to a consideration of the second hypothesis, that cyclic AMP must be acting by an allosteric mechanism.

However, the evidence did not really support this hypothesis either. DeLange *et al.* (1968) found that cyclic AMP was bound to phosphorylase *b* kinase to an extremely small extent, especially when compared to its binding to phosphofructokinase (see Section III, C), and seemingly far out of proportion to its strong effect as an activator. This led Krebs and his associates to suspect a contaminating factor, which they later found and identified as kinase II.

Although ATP and Mg^{2+} are required in the activation of phosphorylase *b* kinase, there is no evidence for the participation of calcium ions in this process. This is in contrast to an earlier conclusion based on experiments in which the activation of the kinase was not separated from the activation of phosphorylase (Ozawa *et al.*, 1967). We now know that the latter step does require Ca^{2+}, as discussed later in this section and again in Section III, A, 2.

It may be mentioned that phosphorylase *b* kinase can be activated *in vitro* by still another mechanism. This is by incubating the enzyme for short periods with any of several proteolytic enzymes, including trypsin and chymotrypsin, and also a calcium-dependent protease referred to at one time as "kinase activating factor" (Krebs *et al.*, 1968). This is of some theoretical interest because the enzyme produced by this type of proteolytic attack seems to be catalytically very similar to the phosphorylated enzyme produced in the presence of ATP and Mg^{2+}. However, since this irreversible process is not likely to be of physiological significance, and is in any event not affected by cyclic AMP, it will not be further discussed.

Turning now to muscle glycogen phosphorylase itself, it has been known for many years that this enzyme may exist in two forms. Phosphorylase *b* is known to be dimer with a molecular weight of about 180,000 (Seery *et al.*, 1967; De Vincenzi and Hedrick, 1967). It can be converted to phosphorylase *a*, under the catalytic influence of phosphorylase *b* kinase, by a reaction requiring ATP and both calcium and magnesium ions (Krebs *et al.*, 1966, 1968). Phosphorylase *a* exists at least under some conditions as a tetramer, with a molecular weight of about 360,000, so that the reaction may be written as shown in Fig. 5.5.

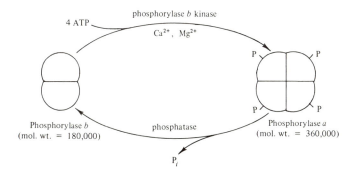

Fig. 5.5 Molecular forms of muscle phosphorylase. Modified from Fischer and Krebs (1966).

Phosphorylase *a* and *b* have often been referred to as the active and inactive forms of the enzyme, respectively, so that the conversion of the latter to the former is often referred to as phosphorylase activation. These terms are now recognized as being somewhat misleading, and not applicable under all circumstances. They were derived from the belief that the activity of phosphorylase *b* was dependent on a concentration of 5'-AMP higher than existed intracellularly, whereas the activity of phosphorylase *a* was independent of this nucleotide. We now know that both forms of the enzyme require 5'-AMP, although the K_m for phosphorylase *a* is so low (Lowry *et al.*, 1964) that for physiological purposes the assumption of independence is probably correct. In other words, the intracellular concentration of 5′-AMP is always high enough to ensure maximal or close to maximal activity of phosphorylase *a*. The other assumption, however, that phosphorylase *b* is always inactive, is almost certainly not correct. We now know, from the work of Morgan and Parmeggiani (1964b), that the intracellular concentration of 5′-AMP may at times greatly exceed the K_m of phosphorylase *b,* so that under these conditions (e.g., severe anoxia) phosphorylase *b* is fully as active as phosphorylase *a*. The effect of 5′-AMP on phosphorylase *b* is to increase the affinity of the enzyme for its substrates, glycogen and inorganic phosphate (Helmreich and Cori, 1964), and this is now recognized as one of the first examples of allosteric activation to have been studied. The conversion of phosphorylase *b* to phosphorylase *a* can properly be spoken of as phosphorylase activation only when the concentrations of 5′-AMP and other allosteric effectors remain at such a level that phosphorylase *b* would in fact be inactive. It is quite possible that the *b* to *a* conversion is physiologically the most important mechanism by which the rate of glycogen breakdown is increased in muscle, but it is surely not, as was at one time supposed, the only such mechanism.

Terminology to deal unambiguously with this type of situation has not been developed, or at least has not come into widespread use. In this monograph we shall continue to refer to the conversion of one molecular form of an enzyme to another more active form as activation of that enzyme, since this is strictly correct under at least some conditions, and may even be correct under most physiological conditions. Effects such as that of 5′-AMP on phosphorylase *b,* where the activator must be present continuously in order for activity to be maintained, and where the formation of a new molecular species does not seem to be involved, will be referred to as *allosteric stimulation,* or, where less is known about the mechanism (i.e., where a specific allosteric site has not been identified), simply as *stimulation* of the enzyme. One other possibility could be considered at this point, and that is that the activity of an enzyme may appear to be greater because there is more of it, due to an increase in the rate of synthesis of the enzyme protein. Experimentally this may be difficult to distinguish from what we have referred to as activation, but, where it has been clearly established that the change results from an increased synthesis of new protein, and not simply from the conversion of one molecular form of the enzyme to another, then we might refer to this as the *induction* of an enzyme. The antonyms of these terms would be *inactivation* for activation, *inhibition* for stimulation, and *repression* for induction. It is recognized that this proposed terminology may have to be modified as new information becomes available.

To return to the phosphorylase b to a conversion, it is known that this involves the addition of the terminal phosphate of ATP onto a specific serine residue, one for each monomer unit (Fischer and Krebs, 1966). The amino acid sequence on either side of the seryl phosphate moiety has been determined. The phosphate is removed as the enzyme is converted back to phosphorylase b under the catalytic influence of phosphorylase phosphatase (Riley *et al.*, 1968). Both during activation and inactivation, the activity of the enzyme in the presence of low concentrations of 5'-AMP (at least under most conditions) is stoichiometrically related to the degree of phosphorylation. The possibility that the phosphate incorporated into phosphorylase a might be one of the phosphates incorporated into phosphorylase kinase during its activation was tested by DeLange *et al.* (1968), but was ruled out. It would appear that the phosphate is incorporated directly from ATP, and does not enter by way of first being incorporated into the kinase.

There is an aspect of uncertainty concerning the state of aggregation of phosphorylase b and a as they exist within the intact cell. Presumably phosphorylase b functions as a dimer, with each monomer unit containing one binding site for 5'-AMP. As emphasized by Morgan and Parmeggiani, and indeed as seems to be the case with allosteric enzymes in general, the kinetics governing the activity of phosphorylase b are complex. ATP competes with 5'-AMP for the same site and is strongly inhibitory, and several other allosteric inhibitors, including glucose-6-phosphate, have been identified (Morgan and Parmeggiani, 1964b). When studied in extracts in the presence of 5'-AMP and ATP at the concentrations which exist intracellularly under normal aerobic conditions, the K_m for inorganic phosphate is greater than 20 mM, whereas the actual intracellular concentration of P_i under these conditions is on the order of 3 mM. The implication of this is that under aerobic conditions, as mentioned previously, the activity of phosphorylase b must be very low. Under anaerobic conditions, on the other hand, the activity of the enzyme may be greatly increased not only by an increase in the ratio of 5'-AMP to ATP but by an increase in the level of P_i as well.

As for phosphorylase a, it was formerly thought to function as a tetramer, but there is now evidence that the catalytically active form of the enzyme is a phosphorylated dimer (Metzger *et al.*, 1968). Thus the basic difference between phosphorylase b and phosphorylase a seems to be the presence of the seryl phosphate moiety in phosphorylase a. This enables the dimer to function catalytically at concentrations of 5'-AMP well below those which normally exist in the cell at all times, whereas the unphosphorylated form, phosphorylase b, requires a concentration of 5'-AMP in excess of the normal aerobic level.

Phosphorylase also contains pyridoxal phosphate, one mole per monomer unit (Fischer and Krebs, 1966). Although this coenzyme is necessary for the activity of both the *a* and *b* forms of the enzyme, the precise nature of its catalytic role has yet to be established.

A requirement for the calcium ion in the phosphorylase *b* to *a* reaction has recently been established (Krebs *et al.*, 1968). This requirement was overlooked for many years because the Ca^{2+} contamination from the reagents and tissue preparations used was often sufficient to allow the reaction to proceed at a maximal rate. Even with relatively highly purified enzymes it has been difficult to demonstrate this requirement consistently without prior addition of a calcium chelating agent. Under these conditions a stimulatory effect of calcium can be seen with both activated and nonactivated phosphorylase *b* kinase. Ozawa *et al.* (1967) estimated that in a dilute extract at pH 6.8 the activity of phosphorylase kinase was more than half-maximal in the presence of as little as 10^{-7} M free Ca^{2+}. On this basis, phosphorylase *b* kinase seems even more sensitive to Ca^{2+} than most actomyosin preparations which have been studied (Gergely, 1964), although it is recognized that in the intact cell the Ca^{2+} requirements of the two systems may be similar. The concentration of free Ca^{2+} in the cytoplasm of resting skeletal muscle is thought to be very low, with contraction being triggered by the sudden release of Ca^{2+} from elements in the sarcoplasmic reticulum (the so-called "relaxing factor") (Gergely, 1964; Sandow, 1965). The low concentration of free Ca^{2+} in resting skeletal muscle is probably one of the main reasons for the low level of phosphorylase *a* seen under these conditions. Electrical stimulation of the gastrocnemius muscle in pithed frogs causes a large and rapid increase in the level of phosphorylase *a*. Since electrical stimulation does not alter the level of cyclic AMP (Posner *et al.*, 1965) whereas it does cause a marked increase in the rate of Ca^{2+} influx (Sandow, 1965), it seems reasonable to suppose that the effect could be due to an increase in the concentration of free intracellular calcium ions, which would enhance the activity of whatever form of phosphorylase *b* kinase was already present. Early experiments (Posner *et al.*, 1965) suggested that there might be a slight increase in the ratio of kinase activity at pH 6.8 to that at pH 8.2 when crude extracts from electrically stimulated muscles were compared with controls, but it has more recently been established (Drummond *et al.*, 1969) that phosphorylase activation may occur under these conditions in the absence of phosphorylase *b* kinase activation.

The conclusion that phosphorylase activation in muscle in response to hormones and electrical stimulation was the result of an increase in phosphorylase kinase activity rather than a decrease in phosphorylase phosphatase activity was initially based on kinetic data (Danforth *et al.*, 1962). This conclusion is compatible with the *in vitro* and *in vivo* studies summarized to this point, although changes in phosphatase activity could also be important at times. For example, Holmes and Mansour (1968) have presented evidence to suggest that glucose may decrease phosphorylase *a* levels by stimulating phosphatase activity. Very little is known about the properties of muscle phosphorylase phosphatase. The possibilities that it is identical with phosphorylase *kinase* phophatase (Riley *et al.*, 1968) and/or the phosphatase involved in the regulation of glycogen synthetase activity (see Section III, C) have not been ruled out.

At this point we might comment briefly on the relative "importance" of cyclic AMP and calcium, as related to phosphorylase activation. A widely held concept up until recently was that increases in the intracellular levels of cyclic AMP and free Ca^{2+} might constitute two separate and independent mechanisms by which the activity of phosphorylase *b* kinase in muscle could be increased. The former mechanism was thought to be involved in the phosphorylase activating effect of epinephrine, for example, while the second was thought to operate in the case of muscular contraction. It seems possible that the effect of epinephrine is produced *primarily* by an increase in the level of cyclic AMP, and that the effect of electrical stimulation is produced *primarily* by an increase in the level of free Ca^{2+}, but the idea that these mechanisms operate separately and completely independently of one another may be an oversimplification. It now seems clear, for example, that in order for the activation of phosphorylase *b* kinase in response to an increase in the level of cyclic AMP to be expressed, a minimum level of free Ca^{2+} must be present. Conversely, it seems possible that some minimum level of activated phosphorylase *b* kinase (possibly dependent upon the level of cyclic AMP) may have to be present in order for an increase in the level of Ca^{2+} to be effective.

Perhaps the minute-to-minute control of muscle phosphorylase kinase activity under normal physiological conditions depends more on subtle changes in the level of *both* cyclic AMP and calcium than on drastic changes in one or the other, such as can be produced experimentally. Indeed, the more we learn about the details of metabolic regulation in general, the clearer does it become that subtlety and moderation are key characteristics of what Cannon (1939) referred to as "the wisdom of the body." To follow up one of Cannon's thoughts on this point, we might say that tissues behave as if they were regulated by expensive cybernetic instruments of the highest quality, rather than by cheap ones which would require drastic environmental changes to set them in motion. In the case of muscle, subtle changes in calcium would help to coordinate the rate of glycogen breakdown with cellular function (in this case contraction) while equally subtle shifts in the level of cyclic AMP would tend to keep the metabolic activity of muscle in line with the requirements of the rest of the body.

Most of the information about phosphorylase activation summarized in the preceding paragraphs came from studies with rabbit muscle preparations, but the enzymes and processes involved in other species seem to be similar. Human muscle phosphorylase could be distinguished immunologically from the rabbit enzyme (Yunis and Krebs, 1962), but in other respects the two enzymes seemed identical. Hanabusa and Kobayashi (1967) reported that phosphorylase *b* kinase was also similar whether obtained from rabbit or human muscle, although it was suggested that the human enzyme might be more sensitive to ATP and Mg^{2+} than the rabbit enzyme. The result which led to this suggestion could also be explained by a lower phosphorylase phosphatase activity in the sample of human muscle, among other possible reasons. The apparent low sensitivity to stimulation by cyclic AMP in these experiments could have been due to the high baseline activity or inadequate assay conditions or both. True species differences in the sensitivity to regulatory factors may exist, however. Frog muscle seems more sensitive than rat muscle to the effects of electrical stimulation, for example, while rat muscle seems more sensitive to epinephrine (Posner *et al.*, 1965; Helmreich and Cori, 1966). Possible reasons for these apparent differences have not yet been investigated at the cellular level. Although the converting enzymes from different mammalian species seem to be interchangeable with respect to substrate, Cowgill (1961) found that lobster muscle phosphorylase kinase would not act on rabbit muscle phosphorylase *b*.

As summarized by Drummond (1967), mammalian white muscle fibers contain more total phosphorylase than red muscle fibers, and the conversion of the *b* to the *a* form in response to tetanic stimulation is more rapid in white muscle. Presumably white muscle contains more phosphorylase *b* kinase, but this point has not been studied. The effects of cyclic AMP and calcium have likewise not been compared in the two types of fibers. It may be noted that the detailed studies of Krebs and Fischer and their colleagues were carried out using leg and back muscles from rabbits, which would provide material containing mainly white muscle fibers contaminated with a smaller amount of red muscle. The near identity of phosphorylase from this source with that from human muscle (which would undoubtedly contain a larger mass of red muscle) suggests that this enzyme at least does not differ in the two types of muscle, however different the regulatory details may prove to be.

2. Cardiac Muscle

The enzymes involved in phosphorylase activation in cardiac muscle have not been studied in as much detail as the corresponding enzymes from skeletal muscle, but the basic processes involved seem to be very similar. There is presently no reason to believe that phosphorylase activation in one type of muscle differs in any fundamental way from phosphorylase activation in another type of muscle. This is not to say that the process necessarily serves the same purpose or that it is equally important in all types of muscle, but only that the molecular mechanisms involved seem to be similar. Readers are referred to the review by Drummond (1967) for a more detailed discussion of this point, and for a good discussion of muscle metabolism in general.

The existence of phosphorylase *a* and *b* in cardiac muscle was demonstrated some years ago, and the converting enzymes, phosphorylase *b* kinase and phosphorylase phosphatase, were partially purified (Rall *et al.*, 1956b). The kinase has more recently been studied in some detail by Drummond and his associates (Drummond and Duncan, 1966; Drummond *et al.*, 1965, 1966). The pH curve is similar to that of the skeletal muscle enzyme, and the enzyme can be activated by incubating with ATP and Mg^{2+}. The rate of this activation is increased by cyclic AMP in the same range of concentrations which are effective in skeletal muscle preparations. A calcium-requiring "kinase activating factor," now known to be a proteolytic enzyme, can also be obtained from cardiac muscle. The equivalent of "kinase II" has not yet been studied in heart muscle preparations, but it is probably safe to predict its existence.

The ratio of kinase activity at pH 7.0 to that at pH 8.2 seems to be greater in extracts from unstimulated cardiac muscle than in extracts from resting skeletal muscle (Hammermeister *et al.*, 1965; Drummond *et al.*, 1966), suggesting that the proportion of the enzyme in the activated form may normally be greater in cardiac muscle. This is conceivably a preparative artifact, but could be a reflection of the fact that cardiac muscle is always active whereas skeletal muscle is not. Kinase activity in the perfused rat heart is rapidly increased in response to epinephrine, which also increases cyclic AMP in this tissue (see Chapter 6).

Although calcium ions may actually inhibit cardiac phosphorylase *b* kinase activity when added to extracts (Drummond and Duncan, 1966), a requirement for Ca^{2+} can nevertheless be demonstrated. Ellis and Vincent (1966) found that if hearts were perfused with a calcium-free buffer, thereby preventing contraction, phosphorylase activation in response to epinephrine did not occur. It was later shown that removal of calcium from the perfusate did not prevent the increase in cyclic AMP in response to epinephrine (Robison *et al.*, 1967a). Namm *et al.* (1968) studied this in perfused rat hearts which were carefully controlled with respect to the rate of coronary flow. They found that the removal of calcium prevented neither the rise in cyclic AMP nor the activation of phosphorylase *b* kinase, but did prevent the conversion of phosphorylase *b* to phosphorylase *a*. These experiments, illustrated in Fig. 5.6, support the view that Ca^{2+} ions are required for the action but not the activation of phosphorylase b kinase. Analogous experiments with intact skeletal muscle would be more difficult because calcium seems to be more tightly bound in this tissue, hence cannot be so readily removed. Ozawa *et al.* (1967) demonstrated a reversible stimulation of cardiac phosphorylase b kinase by calcium ions *in vitro*. They also showed that the addition of cyclic AMP was ineffective if the free Ca^{2+} concentration was reduced to 10^{-7} M or less by the addition of a calcium chelator. Presumably phosphorylase b kinase is activated as usual under these conditions, but the phosphorylase b to a conversion cannot occur.

Friesen *et al.* (1969) observed that when the isolated rat heart was perfused with a buffer containing a higher than normal concentration of calcium (3.6 mM), there was a transient increase in phosphorylase a activity, lagging by a measurable interval behind the equally transient increase in contractile force. The transient nature of these responses, in

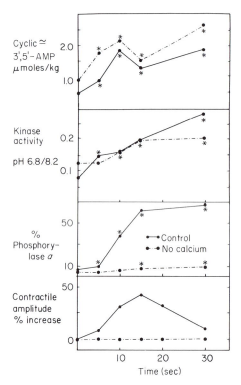

Fig. 5.6. Effect of epinephrine (2 μg added at time zero) on several parameters in the isolated perfused rat heart, in the presence and absence of calcium. Asterisks denote values significantly different from control data ($P < 0.05$). From Namm, Mayer, and Maltbie (1968).

the face of continued perfusion with the higher level of extracellular calcium, is not well understood.

A difference between cardiac and skeletal muscle phosphorylase was noted when the enzymes were put over DEAE-cellulose columns. In contrast to the single peak of activity obtained with the skeletal muscle enzyme, the material from heart yielded three peaks (Davis *et al.*, 1967). All three showed the characteristics of phosphorylase *b*, suggesting the presence of three isozymes. One of the isozymes could be crystallized and was shown to be identical (by immunological and all other criteria studied) to phosphorylase *b* from skeletal muscle. The physiological significance of multiple forms of phosphorylase *b* in cardiac muscle is not known. Similar isozymes have been detected in rat chloroma cells (Yunis and Arimura, 1966b).

3. Smooth Muscle

Insofar as it has been studied, phosphorylase activation in smooth muscle seems to be qualitatively similar to the process in skeletal and cardiac muscle. Mohme-Lundholm (1963) demonstrated the existence of phosphorylase *a* and *b* in bovine tracheal smooth muscle. She found that the interconverting enzymes were also present and seemed to be similar to those of skeletal muscle, i.e., phosphorylase activation occurred in the presence of ATP and Mg^{2+} and could be accelerated by cyclic AMP. Bueding *et al.* (1964) studied phosphorylase *b* from intestinal smooth muscle. The enzyme was similar to phosphorylase *b* from skeletal and cardiac muscle, although immunological as well as kinetic differences were noted.

Diamond and Brody (1966b) made the interesting discovery that in the uterus the ratio of phosphorylase *a* to *b* varied with the state of contraction, being high during contraction and low during the relaxation phase. A variety of agents which stimulated contraction in this muscle, including $BaCl_2$ and oxytocin, were found to increase the level of phosphorylase *a*. Since these agents do not cause an increase in uterine cyclic AMP levels (Dobbs and Robison, 1968), it would appear that in smooth muscle, as in the case of electrical stimulation of skeletal muscle, phosphorylase activation can be affected by more than one mechanism. The effect of calcium ions on the activation of smooth muscle phosphorylase has not yet been studied *in vitro*.

4. Liver

The phosphorylase activating effect of cyclic AMP was initially discovered in liver preparations, as outlined in Chapter 1, and the mechanism of the effect has now been studied in some detail (Riley, 1963; Riley and Wahba, 1969). The results of earlier work had suggested that the activation of phosphorylase kinase, rather than an inhibition of the opposing phosphatase, was involved in the action of cyclic AMP. This has been substantiated by Riley, who has purified phosphorylase kinase from dog liver several hundredfold. The enzyme can be activated by incubation with ATP and Mg^{2+}, and the process can be stimulated markedly by the addition of cyclic AMP. Phosphorylation of the protein was also stimulated by cyclic AMP, and the degree of phosphorylation seems to be more closely related to activity than in the case of the skeletal muscle enzyme. In general, the properties of liver phosphorylase kinase seem to be very similar to those of phosphorylase b kinase from muscle, although to date the low activity at physiological pH relative to that at higher pH values has not been observed in liver preparations.

The important question of whether cyclic AMP affects phosphorylase kinase directly, or whether there is an hepatic equivalent of kinase II which intervenes, cannot at the present time be answered. Riley's preparation was not homogenous, so that the existence of a separate kinase remains a likely possibility. Langan (1968) isolated a protein kinase from liver which seemed similar to the enzyme obtained by Walsh *et al.* (1968) from muscle, but phosphorylase kinase was not tested as a substrate.

Whether there is a requirement for Ca^{2+} in the action of liver phosphorylase kinase is another question which has yet to be answered. Our standard assay system for cyclic AMP, which involves the activation of liver phosphorylase (see Appendix), is not sensitive to the addition of calcium in concentrations up to 10^{-5} M, and higher concentrations are inhibitory. Such observations may not be meaningful, however, because of the possibility that the preparation is already saturated with respect to calcium. High concentrations of Ca^{2+} also inhibit phosphorylase b kinase from rabbit skeletal muscle and heart (Drummond and Duncan, 1968; Krebs *et al.*, 1968).

Phosphorylase phosphatase from liver has been studied in a partially purified form (Wosilait and Sutherland, 1956). The enzyme could be inhibited by heavy metals and also by the addition of fluoride (100% inhibition at a concentration of 0.01 M). Several mononucleotides (including 5′-AMP, but not cyclic AMP) were found to be inhibitory, while the methylxanthines produced a stimulatory effect (Sutherland, 1951a). This effect of the methylxanthines may at times be important, as mentioned previously in Chapter 2.

A difference between inactive (dephosphorylated) liver phosphorylase and phosphorylase b from muscle is that the liver enzyme cannot be stimulated to full activity by the addition of 5′-AMP. This was noted in the early studies (Sutherland and Wosilait, 1956), and has since been confirmed by others (Appleman *et al.*, 1966a; Maddaiah and Madsen, 1966b). An apparent difference between the active forms of the two enzymes was that the molecular weight of active (phosphorylated) liver phosphorylase did not appear to be greater than that of the inactive form, whereas the molecular weight of phosphorylase *a* seemed to be double that of phosphorylase *b*. However, in view of the recent data suggesting that the catalytically active form of phosphorylase *a* is a dimer (Metzger *et al.*, 1968), this apparent difference between the liver and muscle enzymes may not be important physiologically.

Inactive pig liver phosphorylase was recently purified and studied by Appleman *et al.* (1966a). A useful experimental finding was that the activity of this enzyme, like that of inactive phosphorylase from beef adrenals (Riley and Haynes, 1963), could be enhanced by the addition of high concentrations of sodium sulfate. Under these conditions the activity could be further stimulated by the addition of 5'-AMP, but the effect was small compared to its effect on phosphorylase *b*. Many of the properties of inactive liver phosphorylase (including inhibition by sulfhydryl reagents, pH optimum, pyridoxal phosphate content, phosphate incorporation during enzymatic activation, and sedimentation coefficient) were found to be similar to those of phosphorylase *b* from muscle. It had been established previously that liver phosphorylase could serve as a substrate for phosphorylase *b* kinase from heart muscle, and conversely that phosphorylase *b* from heart could be activated by liver phosphorylase kinase (Rall *et al.*, 1956b), and Appleman *et al.* (1966a) showed that phosphorylase *b* kinase from skeletal muscle was also effective on liver phosphorylase.

The kinetic properties of active liver phosphorylase were studied by Maddaiah and Madsen (1966b). The apparent K_m values for glycogen and inorganic phosphate were similar to those reported for muscle phosphorylase *a* (Lowry *et al.*, 1964, 1967). 5'-AMP increased the affinity for both substrates, but, as in the case of phosphorylase *a*, the K_m for this effect was sufficiently low that *in vivo* 5'-AMP would probably always be saturating. All other organic phosphates tested were inhibitory. The effect of UDPG was especially striking, since in the presence of physiological concentrations of substrates and 5'-AMP, the concentration required for 50% inhibition ranged from 0.2 to 0.4 mM, which overlaps with the normal intracellular level of this compound (Mersmann and Segal, 1967). It would thus appear that under normal conditions liver phosphorylase is subject to strong inhibition by UDPG and other organic phosphates present in the liver.

What proportion of the enzyme is in the active form in the so-called "resting" liver, in the intact animal, is not known with certainty. Pre-incubated liver slices from several species were found to contain approximately 50% of the phosphorylase activity measured after the addition of epinephrine or glucagon (Rall *et al.,* 1956a), and purified preparations of phosphorylase could be inactivated still further (to less than 3% of the activity of the original enzyme) by treatment with phosphorylase phosphatase (Wosilait and Sutherland, 1956). Maddaiah and Madsen (1966a) suggested that in rats as much as 60% of the phosphorylase in liver might normally be in the active form, but this is probably an overestimate. Shimazu and Fukuda (1965) showed that after just 30 sec of splanchnic nerve stimulation, in rats, the apparent phosphorylase activity was increased about fourfold over the control level, suggesting that a figure closer to 25% might be a better estimate for the percentage of phosphorylase in the active form in the unstimulated liver. This is close to the figure which may be calculated from the data of Hornbrook and Brody (1963b). The data of Cahill *et al.* (1957b) and Bishop and Larner (1967) suggest that in the dog also about 25% of the liver phosphorylase may be in the active form, under normal steady-state conditions.

Regardless of what percentage of phosphorylase in the liver is in the active form, it should be clear (intuitively, even if we knew nothing about the factors affecting phosphorylase activity) that the actual activity as measured under *in vitro* conditions may be very far from what it is in the environment of the intact liver cell. Failure to appreciate this has led, even within recent years, to some bizarre conclusions. Sokal (1966), for example, measured phosphorylase activity in dilute liver extracts, and found that under the conditions of his experiments glucose-1-phosphate was converted to glycogen at a rate of 8 to 15 μmoles/min/gm of tissue. Since glycogen is not broken down this rapidly in the intact liver — there is evidence that under some conditions the rate of turnover of liver glycogen may approach zero (Stetten and Stetten, 1960) — Sokal concluded that phosphorylase could not be the rate-limiting enzyme in this process. He further suggested (after observing a direct relationship between the degree of phosphorylase activation and the rate of liver glycogenolysis) that the ability of glucagon to stimulate phosphorylase "must be a relatively unimportant side effect of the hormone" (Sokal, 1966). We may recall at this point that the conclusion that phosphorylase *was* the rate-limiting enzyme in liver glycogenolysis was based initially on the measurement of glycolytic intermediates in intact liver cells, as discussed in Chapter 1. The observation which Sokal found puzzling, that phosphorylase may be more active under certain *in vitro* conditions than it is *in vivo,* can be explained in principle by the inhibitory effect of UDPG and other metabolites, although many other factors, some of which were discussed by Maddaiah and Madsen (1966b), may also be involved. Changes in the concentrations of metabolites and possibly other factors could at times play an important role in the control of glycogen metabolism. R. A. Levine (1965), for example, found that in the perfused rat liver the increased rate of glycogenolysis caused by hypoxia was apparently not related to phosphorylase activation. It seems probable that under these conditions the activity of phosphorylase was indeed increased, but by a mechanism not involving the conversion of the inactive to the active form. The mechanism of action of glucagon is discussed in more detail in Chapter 7.

5. Brain

Phosphorylase from brain seems to resemble muscle phosphorylase more than it resembles liver phosphorylase in that the *b* form of the enzyme can be stimulated to the same activity as that of the *a* form by the addition of 5'-AMP (Drummond *et al.*, 1964b; Breckenridge and Norman, 1965). The presence of both phosphorylase kinase and phosphorylase phosphatase in brain has been established, but an effect of cyclic AMP on the interconversion between phosphorylase *a* and *b* in this tissue has not been demonstrated. Rall and Kakiuchi (1966) observed large increases in the level of cyclic AMP in brain slices with no apparent effect on the level of phosphorylase *a,* and attempts to stimulate phosphorylase activation in brain extracts by the addition of cyclic AMP have been unsuccessful.* Since a protein kinase which *is* sensitive to cyclic AMP has been partially purified from brain (Miyamoto *et al.*, 1969), it could tentatively be assumed that brain phosphorylase kinase is not a substrate for this enzyme.

Lowry *et al.* (1967) found that brain phosphorylase *a* had less affinity for glycogen than either muscle phosphorylase *a* or active liver phosphorylase, and found further that the stimulatory effect of 5'-AMP was more pronounced in the case of the brain enzyme. They commented on the apparent paradox that brain, with only 10% as much glycogen as muscle, should contain phosphorylase with a higher K_m for glycogen. Perhaps this is a reflection of the different roles which phosphorylase plays in the two tissues. As pointed out by Lowry and his colleagues, glycogen is definitely an emergency ration for brain, whereas muscle glycogen is regularly called upon during even moderate exercise. Consequently, it may be desirable to have glycogen somewhat less accessible in brain than in muscle.

*Until recently. See Drummond and Bellward, 1970.

The teleonomic significance of the insensitivity of liver phosphorylase to changes in the level of 5'-AMP was also considered by Lowry et al. (1967). Hepatic glycogen, unlike that of brain and muscle, is not primarily intended for local consumption. Therefore, since 5'-AMP levels reflect local rather than systemic needs, control mechanisms based on the concentration of 5'-AMP would be expected to be less prominent than in muscle or brain. We adopted a similar line of reasoning in Chapter 2, where we considered the role of cyclic AMP. It is interesting that just as cyclic AMP seems to change primarily in response to the needs of other tissues, 5'-AMP, to the extent that it functions as a regulatory agent, seems to be concerned primarily with the tissue in which the change occurs. It is almost as if some compounds, such as cyclic AMP and presumably other second messengers, have been elected to specialize in foreign affairs, while certain other metabolites, of which 5'-AMP seems to be a good example, are concerned more with domestic issues. This analogy may, of course, become strained as more systems and species are looked at from this point of view.

6. Adrenal Cortex

Phosphorylase from beef adrenal cortical tissue seems to resemble liver phosphorylase more than muscle or brain phosphorylase in that the inactive form of the enzyme cannot be brought to full activity by the addition of 5'-AMP. It can be stimulated in the presence of high concentrations of Na_2SO_4, however, and under these conditions a slight stimulatory effect of 5'-AMP can be seen (Riley and Haynes, 1963). Kinetic parameters have not been studied in detail, however. Riley and Haynes showed that inactive adrenal phosphorylase could be activated by liver phosphorylase kinase while the active form could be inactivated by liver phosphorylase phosphatase. Conversely, liver phosphorylase was shown to be a substrate for the converting enzymes from the adrenal cortex.

Phosphorylase activation in the adrenal cortex in response to cyclic AMP was first demonstrated by Haynes (1958). Perhaps the most interesting aspect of this response is that it seems to be mediated not by a stimulatory effect on phosphorylase kinase, but rather via an inhibitory effect on the phosphatase (Riley and Haynes, 1963). This effect was studied in some detail by Merlevede and Riley (1966). They found that incubation of a partially purified preparation of adrenal phosphorylase phosphatase with ATP and Mg^{2+} led to an increase in the activity of this enzyme. Since the enzyme could be dialyzed and refractionated without loss of activity, it appeared that the phosphatase could exist in an active and an inactive form. The activation in the presence of ATP and Mg^{2+} could be prevented or rapidly reversed by the addition of physiological concentrations of cyclic AMP. This effect of cyclic AMP was persistent (under the conditions of these experiments), since the enzyme inactivated in its presence could not be reactivated, even after dialysis and refractionation, unless it was first incubated with ATP (or some other nucleotide) in the *absence* of Mg^{2+}.

Although many details relating to this effect remain to be worked out, the very fact that cyclic AMP affects phosphorylase phosphatase in this tissue is of interest. In all other tissues where a stimulatory effect of cyclic AMP on phosphorylase activation has been demonstrated, such as in liver and muscle, the effect seems to be secondary to the stimulation of phosphorylase kinase, and there is no evidence for an effect on phosphorylase phosphatase. By contrast, in the adrenal cortex there is no evidence for an effect of cyclic AMP on phosphorylase kinase, and in fact there is no evidence that kinase activity in this tissue can be altered by any mechanism. The control of phosphorylase activity in the adrenal cortex is thus seen to differ from that in other tissues which have been studied. The teleonomic or evolutionary significance of this unusual arrangement seems quite unclear at the present time. Whether other steroidogenic tissues resemble the adrenal cortex in this respect has yet to be determined.

It may be noted that in certain respects adrenal phosphorylase phosphatase resembles the corresponding enzyme from liver (Wosilait and Sutherland, 1956). The interchangeability with respect to substrate has already been mentioned, and in addition both enzymes can be stimulated by caffeine and inhibited by fluoride. Merlevede and Riley (1966) established that neither of these agents had any effect on the interconversion between the two forms of the adrenal enzyme.

7. Other Tissues

Phosphorylase activity has been detected in a large number of tissues and species in addition to those mentioned in the preceding sections, but the possible effect of cyclic AMP has not been studied in all cases. The example of brain teaches us that sensitivity to cyclic AMP cannot be assumed to be a property of phosphorylase systems in general, and the example of the adrenal cortex shows that even when phosphorylase activation has been demonstrated it might occur by more than one mechanism. Effects of cyclic AMP on phosphorylase activation in adipose tissue (Vaughan and Steinberg, 1965) and toad bladder (Handler and Orloff, 1963) have been observed, but in neither case was the mechanism of the effect investigated. Yunis and Arimura (1966a) studied phosphory-lase activation in rat chloroma (a tumor composed of immature granulo-cytes). Phosphorylase from this source seemed similar in all respects to liver and adrenal cortical phosphorylase. Williams and Field (1961) had previously demonstrated phosphorylase activation in response to glucagon in human leukocytes, and Yunis and Arimura noted a similar though less striking effect when chloroma cells were incubated with glucagon, epinephrine, or piromin (a *Pseudomonas* polysaccharide). The effect of ACTH was apparently not tested. The existence of both phosphorylase kinase and phosphorylase phosphatase in these cells was established, but the effect of cyclic AMP was not studied.

8. Summary

We have discussed the phosphorylase system in some detail because, of all the systems which are now known to be affected by cyclic AMP, it is still the one which has been most carefully studied, and consequently the one about which most is known. It may therefore serve as a useful model when considering other systems affected by cyclic AMP. With this in mind, a few salient points can be summarized.

First, the evidence from skeletal muscle, where the process has been studied in the most detail, suggests that cyclic AMP acts allosterically to increase the activity of a protein kinase (also referred to as kinase II or phosphorylase kinase kinase). This enzyme acts in the presence of ATP and Mg^{2+} to catalyze the phosphorylation and hence activation of phosphorylase b kinase. The activated phosphorylase b kinase then acts, in the presence of ATP and Mg^{2+} and by a reaction which requires Ca^{2+}, to catalyze the phosphorylation of phosphorylase b. The product of this reaction is phosphorylase a, which, under the conditions usually pre-vailing within the cells, is the more active form of the enzyme.

Second, although this basic sequence of events probably applies to a variety of tissues in which phosphorylase activation occurs, it has been established only in the case of skeletal muscle. In at least one tissue, namely adrenal cortical tissue, cyclic AMP seems to act by a different mechanism. In this tissue, instead of stimulating the activation of phos-phorylase kinase, cyclic AMP seems to act by inhibiting the activation of phosphorylase phosphatase. In certain other tissues, such as brain, the phosphorylase system may be completely insensitive to cyclic AMP.

A third point to be kept in mind is that cyclic AMP is only one of several regulatory influences which impinge on the phosphorylase sys-tem. The sensitivity of the system to these other factors also seems to vary from one tissue to another.

Other systems which are affected by cyclic AMP may be equally complex or even more complex, but will not (at least in the pages to follow) be discussed in as much detail.

B. Other Protein Kinases

Following the discovery of the muscle protein kinase by Walsh *et al.* (1968), similar enzymes were soon found in other cells. Those which have been studied at the time of this writing are listed in Table 5.IV. Each of these enzymes can be stimulated by physiological concentrations of cyclic AMP, and seem to be basically similar in most other respects as well.

TABLE 5.IV. Protein Kinases Sensitive to Stimulation by Cyclic AMP

Source of enzyme	Reference
Muscle	Walsh *et al.*, 1968
Liver	Langan, 1968
Brain	Miyamoto *et al.*, 1969
Adipose tissue	Corbin and Krebs, 1969
Adrenal cortex	Gill and Garren, 1970
Toad bladder	Jard and Bastide, 1970
Trout testes	Jergil and Dixon, 1970
Bacteria	Kuo and Greengard, 1969a
Various other phyla	Kuo and Greengard, 1969b

Muscle phosphorylase *b* kinase was shown to be the most effective substrate studied for the enzymes from muscle and adipose tissue, but histones and protamine were also effective. Casein was phosphorylated to some extent, but at a slower rate than the other proteins. The kinases from the other sources listed in Table 5.IV have been studied only with histones or protamine as substrate, and other proteins may be more or less effective. The bacterial kinase differed from the other known kinases in that manganese was a more effective cofactor than magnesium.

Langan (1968) has suggested that the phosphorylation of liver histones might account to some extent for the ability of cyclic AMP to stimulate the induction of several hepatic enzymes (see Table 5.I and Section IV below). The rationale behind this suggestion was that phosphorylation might render these proteins less effective as repressors of DNA template activity. Greengard and his colleagues (Kuo and Greengard, 1969a; Greengard *et al.*, 1969b) have suggested that *all* of the varied effects of cyclic AMP may result from the phosphorylation of a protein, secondary to the stimulation of a protein kinase. Although the evidence to support this hypothesis is not overwhelming, especially when we consider the many effects of cyclic AMP about which little or nothing is known, there is at the same time no evidence which is not compatible with it, and certainly it is true that all of the effects of cyclic AMP which have been studied at the subcellular level are known to require ATP. Research in this area is in a very active stage, and future developments will be awaited with interest.

The mechanism by which cyclic AMP acts to stimulate the activity of these protein kinases is unknown. Most other nucleotides tested, including all of the straight chain adenine nucleotides, have no effect on these enzymes, although cyclic IMP was found to be 80% as effective as cyclic AMP in stimulating the kinase from adipose tissue, when the nucleotides were compared at a concentration of 2×10^{-6} M (Corbin and Krebs, 1969). This is the only nucleotide other than cyclic AMP which has been found to have any significant effect on these enzymes. Since there is no evidence that cyclic AMP participates in the kinase reaction as a cofactor (DeLange *et al.*, 1968), and since cyclic AMP binds very tightly to the enzyme (Walsh *et al.*, 1968), an allosteric mechanism seems likely. Greengard *et al.* (1969a,b) have established that the 3'-phosphate linkage is a "high-energy" bond, and have suggested that this may be important in allowing the nucleotide to interact with the protein, possibly forming a covalent bond. The mechanism by which cyclic AMP binds to these protein kinases, and how this leads to an alteration in their catalytic properties, will undoubtedly be the subject of much intensive research in the years to come.

Gill and Garren (1969) have partially purified a protein from adrenal cortical preparations which has a very high affinity for cyclic AMP. Whether this protein is a protein kinase has not been established at the time of this writing. Section VI can be consulted for recent progress in this area.

C. Glycogen Synthetase

Glycogen synthetase (UDPG:α-1, 4-glucan α-4-glucosyltransferase), also referred to as UDPG-glycogen transglucosylase, was discovered by Leloir and Cardini (1957), and is now recognized as the principle rate-limiting enzyme in the pathway leading from glucose to glycogen. It is subject to control by a mechanism similar to that which controls phosphorylase activity, and is affected in a similar way by cyclic AMP.

Like phosphorylase, glycogen synthetase has been found to exist in two forms. Based largely on their studies of the enzyme from skeletal muscle, Larner and his associates (Rosell-Perez *et al.*, 1962; Friedman and Larner, 1963) concluded that one form was dependent on glucose-6-phosphate for activity, while the other seemed almost independent of this cofactor. They therefore referred to these as the "D" (for dependent) and "I" (for independent) forms of the enzyme, respectively. The physiological implication of this was that under normal conditions the D form would be almost inactive. Glycogen synthetase activity could be increased either by increasing the intracellular level of glucose 6-phosphate (analogous to the effect of 5'-AMP on phosphorylase *b*) or by the conversion of the D to the I form (analogous to the phosphorylase *b* → *a* conversion).

Mersmann and Segal (1967) have more recently suggested that this terminology may be misleading. Working with the liver enzyme, they showed that the conversion of the less active to the more active form of the enzyme was associated not with the elimination of a requirement for glucose-6-phosphate but rather involved the formation of a form which was more sensitive to glucose-6-phosphate. Both forms of the enzyme were shown to require glucose-6-phosphate, but the K_m for the active form was so low that at physiological concentrations this form of the enzyme was almost fully active. By contrast, the K_m for the less active form was so high that under similar conditions this form of the enzyme was almost inactive. Mersmann and Segal also showed that the active form of the enzyme was more sensitive than the inactive form to stimulation by inorganic phosphate, and also that the reason for the greater activity of the active form was a greater affinity for UDPG. Under simulated physiological conditions (0.25–0.5 mM UDPG, 0.05–0.25 mM G-6-P, and 3 mM P_i), the activity of the active form of the enzyme was some 15 to 20 times greater than that of the inactive form. Considering the sensitivity of the inactive form to inhibition by ATP and other metabolites, Mersmann and Segal suggested that the system could be compared to an on-off switch, with activation of the enzyme being tantamount to turning the switch on. They further suggested that the two forms of the enzyme be referred to in the future as synthetase *a* and *b* (the active and inactive forms, respectively), in rough analogy to the muscle phosphorylase system (Krebs *et al.*, 1966).

It is by no means certain, however, that the properties of liver and muscle glycogen synthetase are identical, and indeed, considering the differences which have been noted in the kinetic properties of phosphorylase in different tissues, such a similarity would be surprising. Furthermore, consideration of the data presented by Mersmann and Segal suggests that even in liver the original terminology may not have been so misleading as the authors contend. The form of the enzyme which Larner and his associates refer to as the I form really is independent of glucose-6-phosphate, in a sense, since changes in the level of this metabolite would not be expected to have much effect on its activity. As for the D form, it is at least conceivable, in muscle if not in liver, that an increase in the level of glucose-6-phosphate could lead to an increase in its activity. Pending a resolution of this controversy, however, it may be best to simply refer to the two forms of glycogen synthetase as the active and inactive forms. This would be analogous to the terminology applied to the liver phosphorylase system, and the rationale for the use of the terms "active" and "inactive" is the same as that mentioned earlier in connection with phosphorylase activation (Section III, A, 1).

The interconversion of the two forms of glycogen synthetase from muscle, together with the influence of cyclic AMP thereon, has been studied in some detail by Larner and his colleagues (see Larner, 1966, and Larner *et al.*, 1968, for more detailed reviews) as well as the Buenos Aires group (Belocopitow, 1961; Appleman *et al.*, 1966b; Belocopitow *et al.*, 1967). A number of interesting analogies with the phosphorylase system have emerged from these studies. The inactivation reaction (referred to by these investigators as the conversion of the I to the D form) is catalyzed by a specific kinase, which we may refer to here as synthetase kinase (Friedman and Larner, 1963). This kinase resembles phosphorylase *b* kinase in a number of respects, including stimulation by cyclic AMP, but the two kinases are apparently not identical (Friedman and Larner, 1965; Huijing *et al.*, 1965). The activation of glycogen synthetase (the conversion of the D to the I form) involves a dephosphorylation catalyzed by a relatively specific phosphatase (Friedman and Larner, 1963; Larner *et al.*, 1968). Thus the activity of glycogen synthetase, like that of phosphorylase, appears to depend on a balance between the activities of a kinase and a phosphatase. A major difference is that when the respective kinases are activated, as by an increase in the concentration of cyclic AMP, phosphorylase activity is increased whereas glycogen synthetase activity is decreased, provided all other factors remain constant. The reactions known to be involved in the metabolism of glycogen and the hexose phosphates are represented diagrammatically in Fig. 5.7. Clearly one result of an increase in the level of cyclic AMP, in at least some cells, will be a strong tendency for glycogen level to fall, partly because of an increase in the rate of breakdown, and partly because of a decrease in the rate at which it is synthesized.

The D form of glycogen synthetase from muscle is subject to activation and inhibition by so many metabolites that it is not easy to predict, on the basis of *in vitro* studies, what the activity of the enzyme *in vivo* should be. That it is essentially inactive, however, was suggested by the studies of Danforth (1965). He showed that when 80% or more of the enzyme

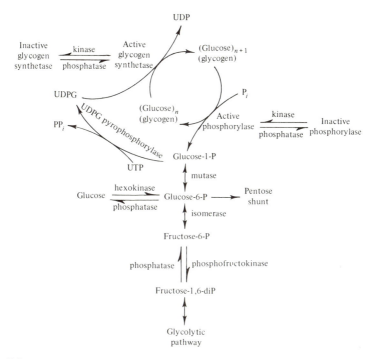

Fig. 5.7. Pathways of glycogen and hexose monophosphate metabolism. Each metabolite is depicted as being contained in only one pool, which may or may not be the case. Note also that not all of these enzymes occur in all cells. See standard textbooks for many details which may have been omitted.

was in the D form, in rat diaphragm muscle, the incorporation of labeled glucose into glycogen was very low. When more of the enzyme was in the I form, however, there was a reasonably good correlation between the percentage of the total enzyme in the I form and the amount of glucose incorporated into glycogen. Of the various activators studied by Rosell-Perez and Larner (1964), glucose-6-phosphate was the most effective, and, at the concentrations known to exist in muscle, the only one which might be expected to have an effect *in vivo*. The concentration of glucose-6-phosphate in mammalian skeletal muscle is on the order of 0.3 mM, which, in the absence of other effectors, might be expected to have a considerable activating effect. The enzyme is inhibited by a variety of agents, including free glucose, inorganic phosphate, fructose-6-phosphate, and several other sugar phosphates, and by all nucleotides tested. Rosell-Perez and Larner did not test UDP, the product of the reaction, but Steiner *et al.* (1965) showed that this was an effective inhibitor of the enzyme from liver.

The inactivation of glycogen synthetase in cardiac and skeletal muscle, catalyzed by synthetase kinase, is remarkably similar to the phosphorylase $b \rightarrow a$ reaction, and this similarity extends even to the amino acid sequence in the area of the serine moiety which becomes phosphorylated (Larner and Sanger, 1965). The requirements for ATP and Mg^{2+} for the two kinases are similar, as is the concentration of cyclic AMP required to stimulate the reaction (Huijing and Larner, 1966a,b; Appleman et al., 1966b). However, whether cyclic AMP affects glycogen synthetase kinase directly or whether it acts by way of a separate protein kinase has not been established. As mentioned previously, glycogen synthetase kinase can be separated from phosphorylase b kinase, but the possibility that both enzymes may be affected by the same protein kinase ("kinase II") has not been ruled out. It is interesting that other procedures known to increase the activity of phosphorylase b kinase, such as incubation with high concentrations of calcium and certain proteolytic enzymes, are likewise effective in causing an irreversible inactivation of glycogen synthetase (Belocopitow et al., 1967).

De Wulf and Hers and their colleagues (De Wulf et al., 1968; De Wulf and Hers, 1968a,b) have studied the control of glycogen synthetase in liver. They found that in homogenates from livers of fed mice, the

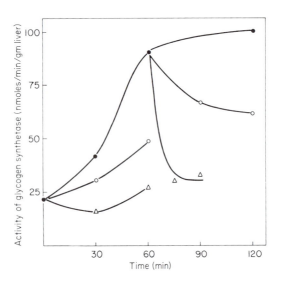

Fig. 5.8. The influence of ATP and cyclic AMP on the activation and inactivation of liver glycogen synthetase. A concentrated mouse liver homogenate, after 0 or 60 min preincubation at $20°C$, was incubated alone (closed circles) or in the presence of ATP and Mg^{2+} (open circles) or ATP, Mg^{2+}, and cyclic AMP (triangles). From De Wulf and Hers (1968a).

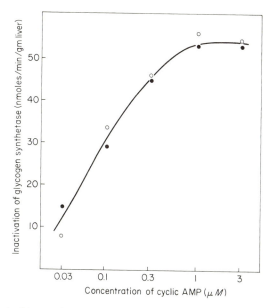

Fig. 5.9. Effect of increasing concentrations of cyclic AMP on the inactivation of liver glycogen synthetase. Extracts were prepared from the livers of mice which had been pretreated with glucose (open circles) or prednisolone (closed circles). Extracts were incubated for 5 min at 20°C in the presence of ATP, Mg^{2+}, fluoride, and caffeine, plus the indicated concentration of cyclic AMP. The data plotted represent the measured *decrease* in synthetase activity. From De Wulf and Hers (1968a).

enzyme was present predominantly in the inactive form. Incubation of the homogenates led to activation of the enzyme, and, since this could be accelerated by caffeine and inhibited by fluoride, it was taken to reflect the action of a phosphatase. As shown in Fig. 5.8, the addition of ATP and Mg^{2+} led to a rapid inactivation which could be accelerated by cyclic AMP. The effect of increasing concentrations of cyclic AMP is illustrated in Fig. 5.9, from which an apparent K_m for cyclic AMP of about 2×10^{-7} M can be calculated. This is similar to values which have been calculated for the activation of dog liver phosphorylase, and also for the effects of cyclic AMP on phosphorylase and glycogen synthetase from muscle. Bishop and Larner (1969) obtained a slightly lower value, about 0.4×10^{-7} M, using a more highly purified system. As discussed previously for muscle, it seems possible that in liver also there is a kinase kinase which mediates the effect of cyclic AMP on both glycogen synthetase and phosphorylase. Perhaps the protein kinase studied by Langan (1968) will be found to serve this function.

Although the two systems seem to be equally sensitive to cyclic AMP in cell extracts *in vitro,* there is evidence that in intact liver cells the synthetase may be somewhat more sensitive than phosphorylase. For example, De Wulf and Hers (1968b) found that glucagon, which increases the level of cyclic AMP in liver (see Chapter 7), produced a near maximal effect on the synthetase at a dose which had a barely detectable effect on phosphorylase. Similarly, Glinsmann and Mortimore (1968) noted that a submaximal concentration of exogenous cyclic AMP was capable of causing a substantial increase in glucose output with little or no effect on phosphorylase activity. It is possible, of course, that these differences could be more apparent than real, owing to the difficulty of keeping phosphorylase in the inactive form in the course of homogenizing the liver. It would nevertheless not be surprising, considering the numerous cell constituents known to affect phosphorylase and glycogen synthetase, if true differences in the sensitivity to cyclic AMP did occur. Such differences could manifest themselves at the level of kinase activation or at the level of the action of the kinase on the inactive forms of phosphorylase and glycogen synthetase. Since even the existence of a kinase kinase in liver is uncertain, there is obviously a good deal of work still to be done in this area.

Other factors in addition to cyclic AMP are important in the control of glycogen synthetase activity. A factor which seems to be especially important is the tissue glycogen concentration. This was first demonstrated by Danforth (1965), who showed that in mouse skeletal muscle *in vivo* and also in incubated rat diaphragms there was a good correlation between the glycogen concentration and the percentage of glycogen synthetase in the I form. The lower the concentration of glycogen, the greater was the percentage of the enzyme in the I form. A similar relationship appears to exist in cardiac muscle (Huijing, 1966; Williams and Mayer, 1966) and liver (De Wulf and Hers, 1968a). There is now evidence that this effect of glycogen is exerted by an inhibitory effect on synthetase phosphatase (Larner *et al.,* 1968), which may or may not be identical with one of the phosphatases known to function in the phosphorylase system. Thus when the level of glycogen falls, the phosphatase is released from its restraining influence, and the balance is shifted toward an increase in the active form of the synthetase. Higher concentrations of glycogen were required in order to demonstrate its inhibitory effect in liver preparations (De Wulf and Hers, 1968a). This makes good sense from the teleonomic standpoint, in view of the higher concentrations of glycogen which normally occur in the liver.

Another factor which seems to be important in the control of glycogen synthetase in the liver, and possibly in muscle as well (Williamson, 1966), is the level of blood glucose. De Wulf and Hers found that glucose infusion in mice led to a very striking increase in hepatic glycogen synthetase activity, and this was later shown to be the result of an increase in the active form of the enzyme (De Wulf and Hers, 1968b). However, the mechanism by which this effect of glucose is brought about is unknown. Since it can also be demonstrated in the isolated perfused rat liver, it cannot be explained entirely in terms of an alteration in the rate of release of a pancreatic hormone. A direct effect on the level of cyclic AMP was ruled out by the demonstration that in the perfused rat liver the intracellular level of cyclic AMP is largely independent of the perfusate glucose concentration (Buschiazzo *et al.,* 1970). Possibly glucose stimulates glycogen synthetase phosphatase. A stimulatory effect of glucose on muscle phosphorylase phosphatase was previously suggested by the work of Holmes and Mansour (1968).

The administration of glucocorticoids also leads to an increase in hepatic glycogen synthetase activity, and De Wulf and Hers (1968b) showed that the steroid effect and the effect of glucose were additive. These effects are thus in the opposite direction to that of cyclic AMP, but seem to be independent of any change in the level of cyclic AMP. At the same time they do not seem to involve any kind of long-lasting change in the sensitivity to cyclic AMP, since the effect of the nucleotide was the same whether the mice from which the livers were obtained had been pretreated with glucose or a steroid or neither or both of these factors. This is partially illustrated in Fig. 5.9, where the open circles represent data from mice injected with glucose, and the closed circles from mice treated with a steroid. It is clear that more research is required before the effects of these agents on glycogen synthetase can be understood.

Søvik *et al.* (1966) have suggested that another important factor in the control of this enzyme may be the level of ATP, since ATP is required for the conversion of the I to the D form. These investigators showed that when ATP levels were reduced in rat diaphragm muscles (by anoxia or treatment with several metabolic inhibitors) there was an increase in the percentage of the enzyme in the I form. There was no consistent correlation under these conditions between the glycogen levels and synthetase I activity. Although a reduction in the availability of ATP might well have an effect on glycogen synthetase activity under certain pathological conditions, the intracellular concentration of ATP under less extreme conditions is a remarkably stable parameter. It seems unlikely, therefore, that the mechanism suggested by Søvik and his colleagues could play an important regulatory role under normal conditions.

The glycogen synthetase system has also been studied in some detail in brain (Goldberg and O'Toole, 1969). Although the system itself seems to be basically similar to the systems in liver and muscle, including sensitivity of the kinase to stimulation by cyclic AMP, it may operate in a somewhat different manner. As discussed by Goldberg and O'Toole, conditions in brain tissue may be such that the D form of the enzyme is actually more active than the I form. Considering the critical dependence of the brain on glycolytic metabolism, it is perhaps not surprising that the control mechanisms regulating the synthesis and breakdown of glycogen in this tissue differ from those operating in liver and muscle. The role of cyclic AMP in the regulation of brain carbohydrate metabolism, if it has any role in this at all, is quite obscure. As mentioned previously (Section III, A, 5), the brain phosphorylase system seems quite insensitive to changes in the level of cyclic AMP, and the effect which Goldberg and O'Toole noted on brain glycogen synthetase activity was much smaller than that observed with the muscle and liver systems.

Glycogen synthetase has been found in many other tissues, and presumably exists in all tissues which contain glycogen. There is suggestive evidence from endocrinological studies (Jungas, 1966) that the system in adipose tissue responds to cyclic AMP in the same fashion as in liver and muscle, but this has not been demonstrated in a cell-free system. Glycogen synthetase in other tissues has not been studied from this point of view.

D. Phosphofructokinase

Phosphofructokinase is another enzyme which may exist in two forms, at least in some species. The most detailed studies of the conversion of inactive to active phosphofructokinase have been carried out using the liver fluke, *Fasciola hepatica*. Stone and Mansour (1967a) showed that in this organism an inactive form of the enzyme could be obtained and that it could be converted to the active form by incubation with ATP, Mg^{2+}, a polyvalent anion (such as inorganic phosphate or a hexose phosphate), and cyclic AMP. The requirement for cyclic AMP was absolute, since no other nucleotide could replace it. Sedimentation studies were consistent with the view that activation was associated with a polymerization process, the active enzyme consisting of three or four subunits. The requirement for ATP and Mg^{2+} suggests the participation of a kinase, but to date such an enzyme has not been isolated. Experiments with yeast (Vinuela *et al.*, 1964) also suggested the existence of two forms of phosphofructokinase, and that the inactive could be converted to the active form by the action of a cyclic AMP-sensitive kinase. Experiments with yeast (Vinuela *et al.*, 1964) and mammalian adipose tissue (Denton and Randle, 1966), where incubation leads to a loss of phosphofructokinase activity which can be prevented by fluoride, were suggestive of the existence of a phosphatase. As in the case of the hypothetical kinase, however, such an enzyme has not been isolated. The evidence for the existence of two forms of phosphofructokinase in any higher form is at present highly equivocal (Mansour, 1966; Denton and Randle, 1966).

In addition to stimulating the conversion of the inactive to the active form, cyclic AMP may also affect phosphofructokinase by acting as an allosteric activator of the active form of the enzyme. This property of cyclic AMP is shared by 5'-AMP and ADP. All three nucleotides compete for an allosteric site with ATP, which is inhibitory (Mansour, 1966; Stone and Mansour, 1967b). This action of cyclic AMP (and of 5'-AMP and ADP) can thus be regarded as a case of deinhibition. A variety of other factors, including inorganic phosphate (stimulatory) and citrate (inhibitory), are known to affect phosphofructokinase, apparently by binding at different allosteric sites, and the complex kinetics involved have been reviewed (Lowry and Passonneau, 1964, 1966). Cyclic AMP is thus only one of a large number of agents capable of affecting the active form of phosphofructokinase allosterically. The concentration required for this effect has generally been found to be of the same order of magnitude as the other nucleotides which are effective, so that the physiological significance of the effect is open to question. Kemp and Krebs (1967) found that cyclic AMP was bound to the enzyme from mammalian skeletal muscle three to four times more tightly than 5'-AMP, in good agreement with kinetic data.

In the case of adipose tissue phosphofructokinase, cyclic AMP is much more effective as an activator than either 5'-AMP or ADP, especially in the presence of citrate (Denton and Randle, 1966). Increases in the level of cyclic AMP which are known to occur in this tissue (see Chapter 8) would be expected to produce a substantial increase in phosphofructokinase activity. Other tissues where an effect of cyclic AMP on phosphofructokinase has been observed, but where the physiological significance is more doubtful, include cardiac muscle (Mansour, 1966), skeletal muscle (Lowry and Passonneau, 1966; Shen et al., 1968), and indeed every mammalian tissue studied (Lowry and Passonneau, 1964). On the other hand, cyclic AMP has been reported to have no effect on phosphofructokinase from higher plants (Dennis and Coultate, 1966) or unicellular organisms (Shen et al., 1968; Baumann and Wright, 1968).

E. Other Enzymes

Many other enzymes are probably affected by cyclic AMP. One of the most important examples may be triglyceride lipase, which is discussed again in Chapter 8. Little is known about the mechanism of lipase activation, except that in adipose tissue a requirement for ATP and Mg^{2+} has been established (Rizack, 1964). There is evidence that the stimulatory effect of cyclic AMP is not restricted to adipose tissue, but probably occurs in a variety of tissues, including liver (Bewsher and Ashmore, 1966) and muscle (Garland and Randle, 1963; Kruger et al., 1967). Whether the protein kinase studied by Corbin and Krebs (1969) participates in the activation of adipose tissue lipase remains to be seen.

The effect on fructose-1,6-diphosphatase has been studied by Mendicino and his colleagues (1966). They found that incubation of kidney extracts with ATP and Mg^{2+} led to a reversible inactivation of fructose diphosphatase, and that this process could be accelerated by the addition of cyclic AMP. Here again the mechanism could be basically similar to that involved in phosphorylase activation. Whether or not this effect of cyclic AMP is of any physiological significance is a separate question which remains to be answered, and whether it occurs in other tissues is likewise unknown. Reference to Fig. 5.7 shows that to whatever extent the effect occurs *in vivo*, it would complement the stimulatory effect on phosphofructokinase to favor glycolysis, just as the inhibitory effect on glycogen synthetase complements phosphorylase activation to favor the breakdown of glycogen to glucose-1-phosphate.

Sarkar (1969) observed that purified fructose-1,6-diphosphatase could be inhibited by 5'-AMP but not by equimolar concentrations of other nucleotides, thus raising the possibility that the effect observed by Mendicino et al. (1966) could have been caused by conversion of the ATP to 5'-AMP. Further work will be needed to clarify this point.

IV. EFFECTS ON PROTEIN SYNTHESIS

Within recent years it has become clear that cyclic AMP is capable of stimulating the synthesis or induction of a number of enzymes. Those known to be so affected at the time of this writing are listed in Table 5.I. Although the study of the mechanisms involved is only now beginning, a few general points can be made.

First, the effect of cyclic AMP is very selective. In the mammalian liver and in bacteria, where the effect has been studied in the most detail, the general tendency of cyclic AMP is to inhibit the overall rate of amino acid incorporation into protein (Pryor and Berthet, 1960; Perlman and Pastan, 1968). Of the many proteins which are produced by these cells, only a relatively few are affected by cyclic AMP.

Another important point is that the effect is highly specific for cyclic AMP. A variety of bases, nucleosides, and other nucleotides and phosphorylated derivatives have been tested in the liver (Yeung and Oliver, 1968b) and *E. coli* (Perlman and Pastan, 1968; Pastan and Perlman, 1970), and to date none has been found to share the activity of cyclic AMP.

A third point is that cyclic AMP is apparently capable of stimulating protein synthesis by several mechanisms. For example, cyclic AMP seems to stimulate the synthesis of β-galactosidase in *E. coli* by acting at the level of gene transcription, i.e., by stimulating the DNA-directed synthesis of *lac* messenger RNA (Perlman and Pastan, 1968; Ullmann and Monod, 1968). This effect can be regarded as a reversal of catabolite repression. Goldenbaum and Dobrogosz (1968) observed that cyclic AMP was much more effective when glucose was the source of the repressor than when glucose-6-phosphate was added, and suggested that cyclic AMP might act by simply reducing the rate at which glucose was converted to glucose-6-phosphate. However, such an effect has not been demonstrated. Pastan and Perlman (1968) found that in mutants not subject to transient repression by glucose, the cyclic AMP effect could not be demonstrated. In the parent organism and also in revertants of this strain, however, cyclic AMP overcame the transient but not the permanent repression of β-galactosidase synthesis. They suggested that the transient repression of the *lac* operon by glucose was mediated by the *lac* promotor, and that this might be the site of action of cyclic AMP. Chambers and Zubay (1969), working with a cell-free system, have obtained evidence to suggest that the effect of cyclic AMP is to promote the initiation rather than the elongation of RNA chain formation. Possibly cyclic AMP performs this function by virtue of its status as a high-energy phosphate compound (Greengard *et al.*, 1969b).

Langan's suggestion that histone phosphorylation might be important in higher organisms (Langan, 1968) is also attractive. As mentioned previously (Section III, B), a protein kinase capable of phosphorylating histones, and which is stimulated by cyclic AMP, has been identified in *E. coli* (Kuo and Greengard, 1969a). However, since histones are not thought to play an important role in regulating gene activity in prokaryotic cells (Bullough, 1967), the real significance of this enzyme in bacteria is uncertain.

The cyclic AMP-induced stimulation of tryptophanase synthesis in *E. coli* seems to occur by an entirely different mechanism. In this case, Pastan and Perlman (1969a) obtained evidence to suggest that the effect of cyclic AMP was exerted at the level of translation, i.e., that it stimulated polypeptide synthesis in response to preformed messenger RNA.

Cyclic AMP stimulated the synthesis of the enzyme even after mRNA synthesis had been arrested by actinomycin or by the removal of inducer, and it was also established that cyclic AMP was not acting by stimulating the conversion of an inactive form of tryptophanase to an active form.

Cyclic AMP may also act by more than one mechanism in mammalian cells. In the liver, in all cases studied, it has been possible to prevent the effect of cyclic AMP with actinomycin D as well as cycloheximide, suggesting that the effect occurs at the transcriptional level. However, as Wicks (1969) has pointed out, this evidence is not conclusive because actinomycin D is known to interfere with the basal synthesis of hepatic tyrosine transaminase. Thus the mechanism by which cyclic AMP stimulates the synthesis of this and several other hepatic enzymes is still far from clear.

As discussed in Chapter 9, protein synthesis has also been implicated in the steroidogenic effect of cyclic AMP in several tissues. In these cases, the effect can be blocked by cycloheximide and puromycin but not by actinomycin. If protein synthesis is indeed involved in this effect of cyclic AMP, the effect must therefore be exerted at the translational level.

Several possibly relevant effects have been observed in cell-free systems. Khairallah and Pitot (1967) found that cyclic AMP was capable of stimulating the release of nascent protein from liver polysomes. The significance of this effect is unclear, however, because in the partially purified system used in these experiments cyclic AMP was effective even in the absence of added ATP and GTP, which are normally required. On the other hand, these experiments were performed under conditions which were optimal for the complete system, and it is possible that the situation may be clarified by future studies employing more physiological conditions. That cyclic AMP itself was active, rather than some metabolite, was suggested by the potentiating effect of theophylline and also by the demonstration that 5'-AMP was inactive. Lissitzky *et al.* (1969) have more recently observed a stimulatory effect of cyclic AMP on the incorporation of amino acids into protein by thyroid polysomes. This effect was not produced by 5'-AMP or any of the other nucleotides tested.

One other effect which can be mentioned in this section, although it is clearly different from any of the others, is the stimulation of DNA synthesis which occurs in the parotid gland and possibly elsewhere in response to isoproterenol. Although an effect of exogenous cyclic AMP on this process has not so far been demonstrated, the available evidence suggests very strongly that the effect of isoproterenol is mediated by cyclic AMP (Malamud, 1969). This is a delayed response which seems ultimately to involve stimulation of DNA polymerase activity, although an earlier event may involve RNA and protein synthesis (Sasaki *et al.*, 1969). Net RNA synthesis is not increased, however, over and above that which can be accounted for by cell proliferation.

V. OTHER EFFECTS OF CYCLIC AMP

None of the other effects of cyclic AMP listed in Table 5.I is understood in any kind of detail, and our only purpose in this section is to call attention to a few features which may be common to some of them.

A feature common to the release of insulin from the pancreas (Lacy, 1967), of several hormones from the anterior pituitary gland (McCann and Porter, 1969), and of amylase from the parotid gland (Amsterdam *et al.*, 1969) is that they all seem to involve emiocytosis. This is the process by which granules containing the protein to be released fuse with the plasma membrane, following which the membrane at that point breaks, allowing the contents of the granules to be released into the extracellular space. The chemical events responsible for this phenomenon are unknown, although the process is known to require calcium (Douglas, 1968; Simpson, 1968). Perhaps the knowledge that the process is stimulated by cyclic AMP in at least some cells will be helpful in leading to a better understanding of it in all cells.

Another group of responses which require calcium are *opposed* by cyclic AMP. These include the contraction of smooth muscle, the aggregation of blood platelets, the conspersion of melanin granules in melanophores, and the rate of discharge of cerebellar Purkinje cells. It is appealing to think that some common denominator may underlie the action of cyclic AMP in these seemingly diverse systems. However, an additional note of complexity is introduced by some actions which do not seem to fit this pattern. For example, the aggregation of some cells (the cellular slime mold) is initiated by cyclic AMP instead of being prevented, as in blood platelets. Similarly, Purkinje cells in the heart seem to differ from those in the cerebellum in that their rate of firing is increased by cyclic AMP. It is evident that a good deal of research will have to be done before the mechanism of action of cyclic AMP in these and other systems is understood.

In summary, it is not possible at the present time to explain all of the known effects of cyclic AMP in terms of only one or a few basic actions. The effect which has been studied in the most detail, the stimulation of phosphorylase activation in skeletal muscle, is known to involve the stimulation of a protein kinase, which catalyzes the phosphorylation and hence activation of phosphorylase kinase. This latter enzyme then catalyzes the phosphorylation and hence activation of phosphorylase itself. A similar sequence of events seems to be involved in the effect of cyclic AMP on glycogen synthetase, except that in this case phosphorylation leads to a less active instead of a more active form of the enzyme, at least in some cells. The phosphorylation of a protein may be important in many of the other known effects of cyclic AMP, but in most cases too little is known about the basic processes involved to permit more than speculation on this point.

VI. ADDENDUM

An important development since this chapter was written has been the identification of the protein kinase that catalyzes the phosphorylation of phosphorylase b kinase (Section III, A, 1) and other proteins (Section III, B) with glycogen synthetase kinase (Section III, C). In other words, it has now been shown that the same enzyme catalyzes the phosphorylation of glycogen synthetase, on the one hand, and phosphorylase kinase, on the other. The evidence for this has been presented by Schlender, Wei, and Villar-Palasi (1969), Soderling and Hickenbottom (1970), and Reimann and Walsh (1970). Thus, starting from the site of action of cyclic AMP, there are three enzymes involved in the breakdown of glycogen whereas only two are involved in glycogen synthesis. The control of glycogen metabolism by cyclic AMP therefore seems to be "asymmetrical." The probability that this would prove to be the case was actually discussed several years ago by Appleman, Birnbaumer, and Torres (1966b).

Substantial progress has been made toward understanding the mechanism by which cyclic AMP increases protein kinase activity. It now seems clear that protein kinases which are sensitive to cyclic AMP are composed of regulatory and catalytic subunits, with the regulatory subunit tending to inhibit the activity of the catalytic subunit. Cyclic AMP binds only to the regulatory subunit, thereby effecting a functional dissociation between the subunits such that catalytic activity is no longer inhibited. Thus the apparent stimulation by cyclic AMP seems in fact to be a type of deinhibition. Earlier evidence for this was presented by Gill and Garren (1970) and Tao et al. (1970), and Reimann et al. (1971) using the muscle enzyme have now reported a complete separation of the two types of subunits. The ability of the regulatory subunit to selectively bind cyclic AMP has been utilized by Gilman (1970) and also by Walton and Garren (1970) as the basis of an assay for cyclic AMP.

The cyclic AMP-binding protein studied by Gill and Garren (1969) was later shown to have protein kinase inhibitory activity (Gill and Garren, 1970), and it seems likely that the binding proteins found in other tissues (e.g., Cheung, 1970b; Salomon and Schramm, 1970) may behave similarly. A cyclic AMP-binding protein required for the transcription of *lac* mRNA was isolated from *E. coli* (Emmer *et al.*, 1970; Zubay *et al.*, 1970). Its relation to the protein kinase previously obtained from *E. coli* (Kuo and Greengard, 1969a) is unknown, but the protein itself was found to lack kinase activity (Pastan and Perlman, 1970).

The possibility that all of the physiologically important effects of cyclic AMP in eukaryotic cells might result from increased protein kinase activity has still not been ruled out. Protein kinases from a variety of mammalian tissues were found to have generally similar properties (Kuo *et al.*, 1970), so that, on the basis of this hypothesis, specificity would have to depend upon different cells having different substrates for the enzyme. Potential substrates in addition to the proteins mentioned earlier have been found in brain membrane fractions (Weller and Rodnight, 1970), a neurotubular preparation (Goodman *et al.*, 1970), a fraction of bacterial RNA polymerase containing sigma factor activity (Martelo *et al.*, 1970), and rabbit ribosomes (Kabat, 1970). Further in line with this hypothesis, it was found that when broken cell preparations of rat adipose tissue were supplemented with a protein kinase from rabbit skeletal muscle, lipase activation in response to cyclic AMP could be easily demonstrated (Corbin *et al.*, 1970b; Huttunen *et al.*, 1970).

An interesting finding by Kuo *et al.* (1970) was that when Mg^{2+} in the reaction mixture was replaced by Ca^{2+}, the reduced protein kinase activity was not increased by cyclic AMP, but was instead further inhibited. Another interesting finding, discussed further in Chapter 11, was that lobster muscle contains a protein kinase preferentially sensitive to cyclic GMP (Kuo and Greengard, 1970). This enzyme was separated from the cyclic AMP-sensitive kinase by DEAE chromatography.

Rasmussen (1970) attempted to link cyclic AMP, protein kinases, calcium ions, ATP, microtubules, microfilaments, and secretory vesicles in a unifying hypothesis of intercellular communication. Such hypotheses are bound to be unsatisfactory at present because of our inadequate understanding of how these various known elements are related to each other in different types of cells. Future hypotheses should attempt to explain why cyclic AMP and calcium ions act together to affect some processes but oppositely to affect others, and why cyclic AMP affects superficially similar processes differently in different cells. These are among the questions we pondered during the course of writing this book, but for which satisfactory answers are not yet available.

Earlier in this chapter (Section II) we wrote that the effects of the acyl derivatives of cyclic AMP had generally been found to be qualitatively similar to those of cyclic AMP itself, when applied to intact cells and tissues. A number of exceptions to this generalization have now been reported, and most of these exceptions seem to fit the same pattern. The dibutyryl derivative, for example, usually produces the same effect as the hormone or hormones which increase the endogenous level of cyclic AMP in a given type of cell. In an increasing number of cases, however, the application of cyclic AMP itself has been found to produce the opposite effect. This has now been demonstrated with skeletal muscle (Chambaut *et al.,* 1969), isolated fat cells (Kitabchi *et al.,* 1970), HeLa cells (Hilz and Tarnowski, 1970), melanophores (Hadley and Goldman, 1969b), and intestinal smooth muscle (Bowman and Hall, 1970). One possible explanation of this apparent pattern, not previously considered, is that cyclic AMP might be converted in some cells to an antimetabolite of cyclic AMP, and that the presence of the acyl derivative tends to prevent this conversion. The inhibitor studied by Murad *et al.* (1969b) may be of interest in this regard. This material, which resembles cyclic AMP chemically and seems to be produced from cyclic AMP, antagonizes the phosphorylase-activating effect of cyclic AMP. Although its formation appears to parallel that of cyclic AMP under most conditions, the N^6-monobutyryl derivative was found to prevent its formation in a liver system. Although not established at the time of this writing, it seems possible that this factor may play an important physiological role in regulating the action of cyclic AMP.

CHAPTER 6

The Catecholamines

I. INTRODUCTION

The catecholamines are of special interest among hormones because they produce so many different effects in a variety of organs and species. They are of special interest from the point of view of this monograph because many of their effects are mediated by an increase in the level of cyclic AMP while others are mediated by a decrease.

The chemical structures of the three catecholamines which have been studied by pharmacologists in the greatest detail are shown in Fig. 6.1. It will be seen that all are derivatives of β-phenylethylamine in which phenolic hydroxyl groups are present in the 3- and 4-positions. Epinephrine is found principally in the secretory cells of the adrenal medulla, and functions primarily as a circulating hormone. Norepinephrine is also found in adrenal medullary cells, and may also be secreted to some extent as a circulating hormone. It plays a more important role, however, as the principle neurotransmitter of the sympathetic nervous system, and probably plays a similar role in the central nervous system. Isoproterenol is a synthetic catecholamine which is not known to occur naturally. As will be seen, however, this drug has been very useful as an experimental tool.

Another naturally occurring catecholamine is dopamine. This com-

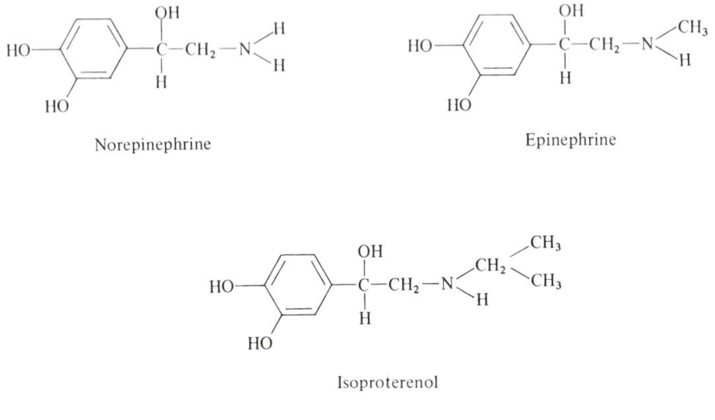

Norepinephrine   Epinephrine

Isoproterenol

Fig. 6.1. Chemical structures of three catecholamines.

pound has been known for many years as an intermediate in the biosynthesis of the catecholamines, but only recently has its possible importance as a neurohumoral substance been appreciated. Dopamine may play an especially important role as a transmitter within the CNS, and many of its actions, like those of epinephrine and norepinephrine, may be mediated by changes in the intracellular level of cyclic AMP. To date, however, this possibility has not been carefully studied.

The discussion in the remainder of this chapter will have to be restricted largely to those effects of the catecholamines in which the possible role of cyclic AMP has been experimentally investigated. For other aspects of the subject, readers are referred to the proceedings of the Second Symposium on Catecholamines (Acheson, 1966) and to a more recent review by Himms-Hagen (1967). Many references to the earlier literature can be obtained from a review by Ellis (1956). A useful summary of the various functional effects of the catecholamines has been provided by Innes and Nickerson (1965).

II. ADRENERGIC RECEPTORS

That the catecholamines could interact with at least two different types of receptors was known even before it was known that the adrenal medulla contained two catecholamines. Dale (1906) found that medullary extracts caused an increase in blood pressure when injected into anesthetized cats. However, when the cats were first treated with ergotoxine, a mixture of ergot alkaloids, then the injection of adrenalin caused a fall in blood pressure. Dale reasoned that one type of receptor normally predominated, and was responsible for the pressor effect. These receptors could be blocked by ergot alkaloids, thereby unmasking the effect of the other receptors, which were responsible for the fall in blood pressure. Adrenergic receptors were later identified in a variety of tissues and species in addition to mammalian vascular smooth muscle. In a classic monograph, Clark (1937) emphasized that the properties of these receptors might be different in different tissues.

Several attempts were made to explain the diverse actions of epinephrine in terms of a single type of adrenergic receptor, perhaps the best known of these being that of Cannon and Rosenblueth (1937). They suggested that in some tissues the hormone-receptor interaction led to the formation of a substance called "sympathin E," which was responsible for the "excitatory" effects of epinephrine (such as the pressor effect observed by Dale). The same type of interaction in certain other cells might lead to the formation of "sympathin I," which was postulated as the mediator of the "inhibitory" effects of epinephrine (such as the fall in blood pressure produced in the presence of ergotoxine). Cannon and Rosenblueth conceded that both types of effects could at times be produced in the same cells in certain lower species, but doubted that this ever occurred in mammals. Another troublesome observation was that while the ergot alkaloids seemed capable of blocking most of the "excitatory" effects of epinephrine, they did not prevent the positive inotropic and chronotropic effects in the heart. Subsequent arguments about the sympathin theory, which was featured in many textbooks of the period, were complicated by such side issues as whether or not there were two physiologically important catecholamines. Cannon and Rosenblueth argued in favor of one catecholamine (epinephrine) and the subsequent identification of norepinephrine as the sympathetic neurotransmitter (for a review of these developments, see von Euler, 1966) contributed to the downfall of the sympathin theory.

The dual receptor theory was later resurrected and placed on a firm basis by Ahlquist (1948). He studied a number of adrenergic responses in several species and showed that they could be classified into two relatively distinct groups, according to the order of potency of a series of catecholamines and related compounds in producing them. With reference to the structures shown in Fig. 6.1, epinephrine and norepinephrine were much more potent than isoproterenol in producing the responses of one group. These responses, which included most but not all of the responses which Cannon and Rosenblueth had referred to as "excitatory," could be prevented by some of the ergot alkaloids and several other drugs known at the time. Ahlquist suggested that the receptors mediating these responses might provisionally be referred to as α-receptors.

The other group of responses, including most of the responses which Cannon and Rosenblueth had referred to as "inhibitory," but including also the positive inotropic and chronotropic responses in the heart, differed from those mediated by α-receptors in that isoproterenol was a more potent agonist than either epinephrine or norepinephrine. Ahlquist suggested that the receptors mediating these responses, which could not be blocked by drugs known at the time, might be tentatively referred to as β-receptors.

Ahlquist's method of classifying these receptors did not receive widespread acceptance until Powell and Slater (1958) introduced dichloroisoproterenol (DCI), the first of a series of drugs capable of preventing the responses thought to be mediated by β-receptors. Moran and Perkins (1958) confirmed the adrenergic blocking activity of this compound, and suggested that it be referred to as a β-adrenergic blocking agent. This terminology is now very common. Subsequently many other β-adrenergic blocking agents, most of which can be regarded as structural analogs of isoproterenol, were synthesized and made available (for reviews, see Biel and Lum, 1966; Moran, 1967). It thus became possible to distinguish α- and β-receptors not only on the basis of the order of potency of agonists, but also on the basis of the type of blocking agent which could prevent the response in question.

Some confusion was engendered by the failure to realize that not all of the receptors of one group were identical to all other receptors of that group. It is now recognized that α- and β-receptors each constitute a class or family of receptors, rather than a set of identical entities. In this regard they resemble that class of flowering plants known as roses, which are all recognizable as roses but which are not all identical. We are tempted to apply the possibly more scientific analogy of isozymes (Davis et al., 1967), except that this might imply that we know more about these receptors than we do. What is known is that the ratio of the potencies of different agonists may vary considerably from one tissue or species to another (e.g., Ellis et al., 1967; Furchgott, 1967; Lands et al., 1967; Lum et al., 1966; van Rossum, 1965) and also that the various blocking agents are not equally potent in all tissues (e.g., Clark, 1937; Levy, 1966; Levy and Wilkenfeld, 1969; Moran, 1967; Robison et al., 1967a). The relative potencies of agonists do not necessarily correlate with the relative potencies of blocking agents. Some of these differences may of course be the result of such factors as differences in the rates of uptake and metabolism and the like (Furchgott, 1967), but in most cases probably reflect real, albeit subtle, differences in the patterns of forces which we define as the receptors for these compounds. Lands et al. (1967, 1969) have attempted to classify β-receptors into two major subgroups, but it is not clear that this is either valid or useful.

Another point which has to be kept in mind in attempting to classify these receptors is that both α- and β-adrenergic blocking agents may have many actions which are not related to their ability to block adrenergic receptors. The β-blocking agents are perhaps less troublesome in this regard, since they are closely related chemically to the catecholamines themselves, but some of them are known to possess local anesthetic and quinidinelike properties, among other possible actions (Lucchesi et al., 1967; Parmley and Braunwald, 1967). The α-adrenergic blocking agents constitute a diverse assortment of compounds, including some but not all of the ergot alkaloids, imidazole derivatives such as phentolamine, and the long-lasting β-haloalkylamine derivatives such as phenoxybenzamine (Innes and Nickerson, 1965; Furchgott, 1966; Ghouri and Haley, 1969), and they have a correspondingly wide array of actions. The chemical structures of the β-blocking agents are such that they would be expected to act as competitive antagonists of the catecholamines. The α-blocking agents may also compete with the catecholamines for a common site, and in the case of the β-haloalkylamine derivatives, which are alkylating agents, the evidence for this is very strong (Furchgott, 1966). Phentolamine and the ergot alkaloids may also interact directly with the α-receptor, but other primary sites of action have not been ruled out.

An embarrassing example to illustrate the pitfalls in this area is provided by an earlier discussion of the possible nature of adrenergic receptors in rat adipose tissue (Robison et al., 1967a).* In that case, it was known that the lipolytic response to the catecholamines could be prevented by both types of blocking agents, prompting us to suggest that the receptors involved might be some sort of cross between the α- and β-receptors which occurred in other tissues. It was later established that the α-blocking agents were acting nonspecifically to interfere with the response to all lipolytic agents, including the dibutyryl derivative of cyclic AMP (Aulich et al., 1967), and it is now recognized that the adrenergic receptors mediating lipolysis in this tissue are reasonably typical β-receptors (Finger et al., 1966).

With the foregoing notes of caution in mind, it has been possible to classify most of the known responses to the catecholamines according to whether they are mediated by α-receptors or β-receptors. As a general rule, responses mediated by α-receptors are such that the naturally occurring catecholamines are much more potent agonists than is isoproterenol, and their effects can be antagonized by concentrations of α-blocking agents which do not block β-receptors. Conversely, isoproterenol is generally a more potent β-receptor agonist than either of the naturally occurring catecholamines (although there are some exceptions to be discussed later), and these effects are competitively antagonized by one or more of the β-adrenergic blocking agents in doses which do not interfere with responses to other hormones or α-receptors.

Cyclic AMP entered this picture as soon as it was discovered (Chapter 1; see also Sutherland and Rall, 1960, for a more detailed review) as the mediator of the hepatic glycogenolytic effect of epinephrine. This coincided closely in time with the introduction of DCI as the first β-adrenergic blocking agent. After it was established that epinephrine increased the level of cyclic AMP in the liver by stimulating adenyl cyclase, Murad et al. (1962) showed that this effect in dog liver preparations could be competitively antagonized by DCI, and also that the order of potency of

---

*Several other errors of interpretation in that essay will not be specifically acknowledged, on the grounds that what was speculation was clearly marked as such.

agonists (Fig. 6.2) was the same as their order of potency as glycogeno-lytic agents, which was in turn the same as in most of the β-receptor responses studied by Ahlquist. The significance of this was partially obscured at the time by the finding that the stimulatory effect of the cate-cholamines on adenyl cyclase could also be inhibited to some extent by ergotamine, primarily an α-adrenergic blocking agent, and also by the fact that Ahlquist had not included the hepatic glycogenolytic response in his original study.

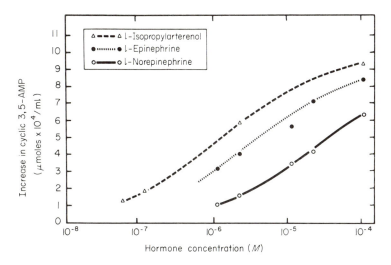

Fig. 6.2. The effect of increasing catecholamine concentration on the formation of cyclic AMP by dog liver adenyl cyclase. Washed particulate preparations were incubated with ATP and Mg^{2+} for 12 min at 30°C, in the presence of caffeine and a catecholamine. Forma-tion of cyclic AMP under these conditions in the absence of a catecholamine was 3.6×10^{-4} μmole/ml. From Murad *et al.* (1962).

Subsequently, however, adenyl cyclase activity in response to cate-cholamines and adrenergic blocking agents was studied in a great variety of tissues, as listed in Table 6.I. Although some of these tissues have been studied in much more detail than others, it gradually became clear that the stimulation of adenyl cyclase was a β-receptor effect rather than an α-receptor effect. The order of potency of agonists has generally been the same as the order of potency which Ahlquist used to initially define the β-receptor, and in all cases studied it has been possible to competi-tively antagonize the response with a β-adrenergic blocking agent. Since these properties were retained even in highly washed particulate prep-arations, the results have justified an earlier conclusion (Sutherland and

TABLE 6.I. Cells and Tissues in Which Adenyl Cyclase
is Stimulated by Catecholamines

Source of enzyme	Reference
Liver	Murad *et al.*, 1962
Spleen	Klainer *et al.*, 1962
Kidney	Melson *et al.*, 1969
Cardiac muscle	Murad *et al.*, 1962
Skeletal muscle	Posner *et al.*, 1965
Smooth muscle	Triner *et al.*, 1970
Brain	Klainer *et al.*, 1962
Pineal gland	Weiss and Costa, 1968a
Adipose tissue	Butcher *et al.*, 1968
Pancreatic islets	Turtle and Kipnis, 1967b
Avian erythrocytes	Davoren and Sutherland, 1963a
Amphibian erythrocytes	Rosen and Rosen, 1968
Human leukocytes	Scott, 1970
Parotid gland	Malamud, 1969
Lung	Unpublished observations

Rall, 1960; Sutherland, 1965) that at least some adrenergic receptors might be closely related to adenyl cyclase.

To state our present position in advance of the evidence, which will be presented in some detail in the pages to follow, we now believe that the adrenergic β-receptor is probably an integral component of the adenyl cyclase system in those tissues where β-receptors occur, and that β-adrenergic effects in general result from an increase in the intracellular level of cyclic AMP. This tentative conclusion will be reconsidered in Section VIII. An incidental corollary of this hypothesis is that in most tissues "sympathin I" can be identified with cyclic AMP. A difference between sympathin I as it was originally conceived and as we have identified it here is that it mediates not only the "inhibitory" effects of epinephrine but some of the "excitatory" effects as well, not to mention many of the effects of a number of other hormones.

Within recent years it has been found that the catecholamines also have the ability to reduce the level of cyclic AMP in some cells. The receptors responsible for this type of effect, where this has been studied, have been found to have the characteristics of adrenergic α-receptors. The possibility that all α-receptor effects might be mediated by a fall in the intracellular level of cyclic AMP was discussed previously (Sutherland, 1965; Robison *et al.*, 1967a) largely without benefit of definitive data. It will be reconsidered in Section VIII following. The data presently available suggest that just as sympathin I corresponds in most cases to cyclic AMP itself, sympathin E may correspond to the lack of it.

We should now like to review those effects of the catecholamines in which the participation of cyclic AMP has been studied.

III. EFFECTS ON CARBOHYDRATE METABOLISM

The hyperglycemia which follows the injection of epinephrine is the result of at least three separate effects: (1) the direct effect on the liver. This results not only in a net increase in the rate of conversion of glycogen to glucose, but also in a stimulation of the rate at which glucose is synthesized from noncarbohydrate precursors; (2) a direct effect on muscle. This results in an increased rate of conversion of muscle glycogen to lactic acid, which is released into the blood to produce hyperlacticacidemia. Part of this lactic acid is reconverted by the liver to glucose. Thus gluconeogenesis is stimulated by epinephrine not only by its direct effect on the liver but by the increased availability of substrate; and (3) a direct effect on the pancreas to suppress the release of insulin which would normally occur in response to the increase in blood glucose levels. This leads to an inhibition of the peripheral utilization of glucose, thus enhancing and prolonging the hyperglycemia which results from the direct effect of epinephrine on the liver. Glucose utilization by muscle may also be inhibited as a result of the increased rate of breakdown (and decreased rate of synthesis) of muscle glycogen.

Other effects, such as the central effect studied by Gangarosa and DiStefano (1966), may also contribute to the hyperglycemia produced by the catecholamines. Other factors, including the release of free fatty acids from adipose tissue (Chapter 8) and the presence or absence of steroids and other hormones, may also influence the response to the catecholamines. Clearly the hyperglycemic response in the whole animal may be the result of a complex series of interactions. We will now take a closer look at some of the individual components of this response.

A. The Response in the Liver

Epinephrine increases the rate of glucose production by the liver by stimulating both glycogenolysis and gluconeogenesis. The factors involved in the stimulation of hepatic glycogenolysis were, of course, discussed in Chapter 1, since it was the effort to understand these factors that led to the discovery of cyclic AMP. It is now known that the conversion of liver glycogen to glucose occurs as the result of phosphorylase activation (Sutherland and Rall, 1960) as well as the inactivation of glycogen synthetase (De Wulf and Hers, 1968b). These processes were discussed in Chapter 5. Hepatic gluconeogenesis will be discussed in Chapter 7.

That these hepatic effects of epinephrine are mediated by cyclic AMP seems well established. Epinephrine stimulates adenyl cyclase in broken cell preparations from liver (Makman and Sutherland, 1964; Bitensky et al., 1968) and also stimulates phosphorylase activation in these preparations so long as the particles containing adenyl cyclase are not removed by centrifugation (Rall et al., 1957). The ability of cyclic AMP to stimulate the activation of liver phosphorylase is still used in the assay of cyclic AMP (see Appendix). The effects in broken cell preparations of the other catecholamines and related derivatives, of glucagon, and of adrenergic blocking agents, are in agreement with their effects at higher levels of organization, to the extent that they have been studied (Murad et al., 1962; Makman and Sutherland, 1964). A problem to which we will return in the concluding section of this chapter is that the properties of hepatic adrenergic receptors differ considerably from one species to another. In some species, such as dogs and cats, these receptors resemble β-receptors found elsewhere (Murad et al., 1962). Although the response to catecholamines can be inhibited by ergotamine, the β-adrenergic blocking agents are more effective. This departure from the properties of "typical" β-receptors reaches something of an extreme in such species as the rat (Ellis et al., 1967; Arnold and McAuliff, 1968), where norepinephrine may be even more potent as an agonist than isoproterenol, at least in livers from adult animals.

The accumulation of cyclic AMP in intact tissue in response to epinephrine has been studied in rabbit liver slices (Sutherland et al., 1965) and also in the isolated perfused rat liver (Robison et al., 1967b). The effect of epinephrine on cyclic AMP and glucose production in rabbit liver slices is shown in Fig. 6.3. The responses in the perfused rat liver were more rapid but similar in the sense that the increase in cyclic AMP preceded the release of glucose into the perfusate. In collaboration with Exton and Park (unpublished observations) it was found that the response in the rat liver was biphasic. The addition of epinephrine at a concentration of $2 \times 10^{-7} M$ to a recirculating system, for example, caused a prompt threefold increase in the concentration of cyclic AMP, which after a few seconds fell to a level about twice that of the control livers. The steady-state level of cyclic AMP was dependent upon the dose of epinephrine, and correlated with the effect on glucose production. The reasons for the biphasic response to epinephrine, which has also been seen in other tissues, are poorly understood. In the liver this differs from the response to glucagon, where cyclic AMP increases monophasically to a maximum

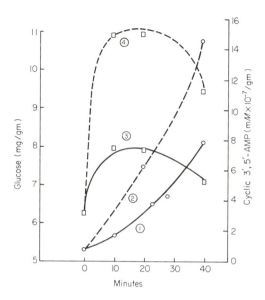

Fig. 6.3. Effect of epinephrine on cyclic AMP and glucose levels in rabbit liver slices. The slices were preincubated for 15 min at 37°C in a phosphate buffer, and then transferred to fresh media for the indicated times. Curve (1) represents glucose in the absence of epinephrine, while curve (2) shows the effect of 4×10^{-5} M epinephrine added at time zero. Curves (3) and (4) represent the corresponding values for the level of cyclic AMP. From Sutherland *et al.* (1965).

which is many times greater than that seen with maximally effective doses of epinephrine. An effect of epinephrine on phosphodiesterase has never been observed in a broken cell preparation, although it is conceivable that under the conditions of the intact cell such an effect may occur. Other possible explanations for the biphasic response suggest themselves, including the possibility that cyclic AMP levels in different cells respond to epinephrine in opposite directions. Whether the self-limiting nature of the hepatic response to epinephrine infusions in the dog (Altszuler *et al.,* 1967) can be related to a biphasic effect on cyclic AMP formation remains to be seen.

The hepatic effects of epinephrine can be mimicked by exogenous cyclic AMP. The activation of phosphorylase and the inactivation of glycogen synthetase have been demonstrated in homogenates (Sutherland and Rall, 1960; De Wulf and Hers, 1968a), tissue slices (Henion *et al.,* 1967; Reynolds and Haugaard, 1967), the isolated perfused rat liver (Northrop and Parks, 1964b; Levine, 1965), and in intact animals (De Wulf and Hers, 1968b). Hyperglycemia in response to exogenous cyclic AMP has been seen in all species studied, including rats (Northrop and

Parks, 1964a), cats (Ellis *et al.*, 1967), dogs (Henion *et al.*, 1967), and humans (Levine, 1968a), and at least part of this response can be attributed to the stimulation of hepatic glycogenolysis (Northrop and Parks, 1964a; Bergen *et al.*, 1966). Higher concentrations are required in intact tissues than are required in broken cell preparations, as discussed in Chapter 5 (Section II). The acyl derivatives of cyclic AMP are more potent glycogenolytic agents than cyclic AMP itself in intact tissues and whole animals (Henion *et al.*, 1967; Bieck *et al.*, 1969; Levine and Lewis, 1969). The effect of cyclic AMP is not shared by other adenine nucleotides (Bergen *et al.*, 1966).

The hyperglycemic response to catecholamines or exogenous cyclic AMP is enhanced in the presence of theophylline (Northrop and Parks, 1964a; Triner and Nahas, 1966). Potentiation by theophylline can also be demonstrated in the perfused liver by a suitable selection of dosages (J. H. Exton *et al.*, 1969, unpublished observations). This is easily demonstrated if the level of cyclic AMP is chosen as the end response; it is obscured at the level of glycogenolysis by the tendency of the methylxanthines to stimulate phosphatase activity (Sutherland, 1951a; De Wulf and Hers, 1968a), which opposes the kinase-activating effect of cyclic AMP (see Chapter 5, Sections III,A,4 and III,B).

Epinephrine also stimulates the rate of gluconeogenesis by the liver. *In vivo* this is partly a mass action effect due to the increased blood lactate levels which result from the stimulation of muscle glycogenolysis. However, even in the perfused rat liver (Exton and Park, 1968b) or in liver slices (Rinard *et al.*, 1969), where the substrate concentration can be maintained below saturating levels, epinephrine still stimulates the rate of conversion to glucose. As discussed again in the next chapter, potential gluconeogenic substrates other than lactate include glycerol, pyruvate, several amino acids, and the endogenous protein of the liver itself.

It was thought for a time that the stimulation of gluconeogenesis might require a higher level of cyclic AMP than was required for the stimulation of phosphorylase activation (Sutherland and Robison, 1966). Although phosphorylase activation in response to both glucagon and the catecholamines was well known, it had been reported that only glucagon could stimulate gluconeogenesis in the isolated rat liver (Miller, 1965). This was compatible with our later finding (Robison *et al.*, 1967b) that glucagon could raise the concentration of cyclic AMP in the liver to higher levels than could even supramaximal concentrations of epinephrine. However, it was later shown that the dose response curves for glycogenolysis and gluconeogenesis in response to exogenous cyclic AMP were similar, and that doses of glucagon and epinephrine which produced equal effects on glycogenolysis also stimulated gluconeogenesis equally, providing blood flow through the liver was held constant (Exton and Park, 1968b). The catecholamines have a powerful tendency to produce hepatic vasoconstriction, and the resulting ischemia may lead to glycogen breakdown by a mechanism not involving phosphorylase activation (Levine, 1965). In the type of system used in the experiments described by Miller (1965), vasoconstriction in response to catecholamines can be prevented by including an α-adrenergic blocking agent in the perfusion medium (Heimberg and Fizette, 1963).

It now seems likely, as Sokal and his colleagues have emphasized for many years (Sokal *et al.*, 1964; Sokal, 1966), that the concentrations of epinephrine which normally circulate (on the order of 10^{-9} to 10^{-8} M in most mammals) are too low to affect the liver directly. By contrast, sympathetic stimulation via the splanchnic nerve may at times be very important, since this leads to a rapid increase in phosphorylase activity and glycogen breakdown (Shimazu and Fukuda, 1965; Shimazu and Amakawa, 1968a). The suggestion that this effect of sympathetic stimulation might be mediated by the release of some factor other than norepinephrine (Shimazu and Amakawa, 1968b) was not warranted, in our opinion. It is not surprising that the liver responds to nerve stimulation more rapidly than to the intravenous injection of norepinephrine, and it is also not surprising that a dose of DCI which inhibited the smaller response to injected norepinephrine did not prevent the response to nerve stimulation. For one thing, the local concentration of norepinephrine following nerve stimulation may have been much greater than that obtained by intravenous administration.

Another hepatic response to epinephrine which has been studied to some extent is potassium release. This also occurs in response to glucagon (see Chapter 7) and exogenous cyclic AMP (Tsujimoto *et al.*, 1965; Exton and Park, 1968b; Northrop, 1968), suggesting that it is probably mediated, at least in part, by cyclic AMP. The response is transient, even when the hormones are perfused, and is usually over at a time when the output of glucose is still proceeding.

Careful measurements have indicated that the efflux of potassium is a very early event following the arrival of epinephrine or glucagon at the liver, either coincident with or perhaps slightly preceding the efflux of glucose (Craig and Honig, 1963; Elliott *et al.*, 1965; Exton and Park, 1968b). Craig and Honig (1963) suggested that the loss of potassium might result from an early step in the sequence of reactions leading to phosphorylase activation, perhaps from the interaction of the hormones with adenyl cyclase. This no longer seems probable, however, if only because the response is also produced by exogenous cyclic AMP.

Studies of the effects of adrenergic blocking agents on the hyperkalemic response in whole animals (Galansino *et al.*, 1960; Ellis and Beckett, 1963; Ellis *et al.*, 1967) have yielded conflicting results which are difficult to interpret. The ability of certain α-adrenergic blocking agents to inhibit the response to epinephrine could be understood partly on the basis of inhibition of adenyl cyclase stimulation (Murad *et al.*, 1962) and partly on the basis of inhibition of the action of cyclic AMP (Northrop, 1968). Glucagon would be less susceptible to this latter effect because it can increase the concentration of cyclic AMP in the liver to higher levels. Even assuming that the potassium measured in the experiments with whole animals came entirely from the liver, which may not have been the case, the observations with adrenergic blocking agents do not imply the participation of adrenergic α-receptors. Harvey *et al.* (1952) studied a series of fifteen α-adrenergic blocking agents in the cat and rabbit, and found that all fifteen had at least some ability to suppress epinephrine-induced hyperglycemia. There was no relationship, however, between this effect and the ability to suppress the epinephrine-induced pressor response. Even on the basis of experiments with the isolated perfused liver, the relation of the effect on potassium efflux to the other effects of cyclic AMP is poorly understood. Friedmann and Park (1968) found that epinephrine (and glucagon and cyclic AMP) also increased the rate of efflux of labeled calcium from the liver, but the significance of this effect is also poorly understood.

Other hepatic effects of the catecholamines which are probably mediated by cyclic AMP include stimulation of ketogenesis (Haugaard and Haugaard, 1954) and the induction of several hepatic enzymes (Wicks, 1968, 1969). These effects are discussed in more detail in the next chapter, in connection with the similar effects of glucagon.

One study has been reported on the effect of epinephrine on hepatic fine structure. Ashford and Porter (1961, 1962), using the isolated perfused rat liver, showed that the addition of either epinephrine or glucagon to the perfusate caused a dilation of the cisternae of the smooth endoplasmic reticulum (which also appeared to become more uniformly distributed throughout the cell after the addition of the hormones), and an increase in the number of lysosomes or autophagic vacuoles. Since these effects were seen with both glucagon and epinephrine, it seems likely that cyclic AMP was involved, although it is far from clear in what way. Possibly these structural changes are related to the depletion of liver glycogen, as suggested by Millonig and Porter (1960). Ashford and Burdette (1965) observed similar changes after prolonged hypoxia, which may lead to the depletion of glycogen by a mechanism not involving phosphorylase activation. The effect of glucagon treatment on the physical properties of lysosomes was recently studied by Deter and DeDuve (1967). The lysosomes were found after glucagon to be more sensitive to various types of stress, possibly because of an increase in size. Although glycogen levels were not measured in these experiments, the finding that the lysosomal changes did not begin until about 30 min after the intraperitoneal injection of glucagon would be consistent with effects secondary to glycogen depletion. However, a direct effect of cyclic AMP on lysosomes has not been ruled out.

B. The Response in Muscle

1. Skeletal Muscle

The catecholamines cause an increase in the level of cyclic AMP in skeletal muscle in all mammalian and amphibian species studied (Butcher *et al.*, 1965; Posner *et al.*, 1965; Lundholm *et al.*, 1967; Craig *et al.*, 1969). This is presumably the result of adenyl cyclase stimulation, as demonstrated in response to epinephrine in washed particulate preparations from dog skeletal muscle (Sutherland and Rall, 1960; Klainer *et al.*, 1962).

The distribution of adenyl cyclase in skeletal muscle seems to differ according to the functional properties of the muscle. Rabinowitz *et al.*, (1965) and Seraydarian and Mommaerts (1965) found that in rabbit skeletal muscle most of the adenyl cyclase activity was localized in the mitochondrial and microsomal fractions, apparently paralleling the distribution of relaxing factor activity. This activity is thought to derive largely from fragments of the sarcoplasmic reticulum or T system, which may be regarded as a continuation of the cell membrane (Franzini-Armstrong, 1964). Of course, the distribution of an enzyme may differ according to the homogenization procedure and other factors. However, when we compared different cat muscles under identical conditions, we found (unpublished observations) that the distribution of adenyl cyclase in the anterior tibialis muscle was similar to that reported by Rabinowitz and his colleagues for rabbit skeletal muscle. In the soleus, by contrast, and also in cardiac muscle, most of the adenyl cyclase activity sedimented in the low-speed fractions containing fragments of the cell membrane. The anterior tibialis muscle of the cat is a good example of a muscle composed almost entirely of "white" muscle fibers (also referred to as "fast" or "phasic" fibers), while the soleus consists predominantly of "red" muscle fibers ("slow" or "tonic" fibers). The different distribution of adenyl cyclase is only one of a large number of metabolic differences which have been noted between these two types of muscle, and it is not entirely clear how these differences may be related to the different functional properties of the muscles (Section VI,B,1).

The increased levels of cyclic AMP in muscle lead to an increase in the rate of glycogenolysis and a decrease in the rate of glycogen synthesis, as described in Chapter 5. The principal end-product of glycogenolysis in muscle is lactic acid, which may be oxidized by the tissue or released into the blood. Glucose is not released by muscle primarily because there is little or no glucose-6-phosphatase in this tissue (Cori, 1931; Cori and Welch, 1941). Epinephrine inhibits the uptake of glucose by muscle partly as a result of its glycogenolytic effect and partly, in whole animals, by suppressing the release of insulin from the pancreas.

The lactic acid released from skeletal muscle is probably the principal cause of the hyperlacticacidemia which occurs in response to epinephrine. Some of this lactic acid is returned via the bloodstream to the liver, where it is converted to glycogen and then to blood glucose, or to glucose directly. (As mentioned previously, the increased rate of gluconeogenesis seen after epinephrine is not entirely a mass action effect due to the increase in blood lactate levels, since epinephrine stimulates gluconeogenesis from lactate and other precursors even in the isolated perfused liver.) Thus much of the blood glucose in catecholamine-induced hyperglycemia is derived, however indirectly, from muscle glycogen. This conversion of muscle glycogen to blood lactate to liver glycogen to blood glucose (some of which may be used in the resynthesis of muscle glycogen) has been referred to as the Cori cycle, after its discoverer (Cori, 1931). It should be noted that not all of the lactate produced in this process is taken up by the liver, for a considerable amount of it may be directly utilized by other peripheral tissues, including the heart (Green and Goldberger, 1961).

The adrenergic receptors subserving the metabolic effects of the catecholamines in skeletal muscle are characteristically β-receptors, in that isoproterenol is much more potent than norepinephrine as an agonist, and also in that these effects can be blocked by β-adrenergic blocking agents but not by α-blocking agents (Hornbrook and Brody, 1963b; Ali et al., 1964; Dhalla, 1966; Estler and Ammon, 1966; Mayer et al., 1967). Adenyl cyclase from skeletal muscle has not been carefully studied from the point of view of its sensitivity to different catecholamines and adrenergic blocking agents. It seems reasonable to assume, however, based on studies of phosphorylase activation (Haugaard and Hess, 1965; Mayer et al., 1967), that the adenyl cyclase system in skeletal muscle, as in the other tissues where catecholamines stimulate the formation of cyclic AMP, has the characteristics of a β-receptor. Glucagon, which does not stimulate glycogenolysis in skeletal muscle, likewise has no effect on adenyl cyclase in this tissue (Sutherland and Rall, 1960). To this extent it can be said that our first criterion has been satisfied, although it would clearly be desirable in the future to include studies with other catecholamines and adrenergic blocking agents.

That cyclic AMP levels in skeletal muscle *in situ* increase in response to epinephrine was first demonstrated by Posner *et al.* (1965). In these experiments, in which the gastrocnemius muscles were removed from rats and frogs, 1 min and 12 min, respectively, after the intracardiac injection of epinephrine, the concentration of cyclic AMP was increased two- to threefold over the resting control levels, and phosphorylase *b* kinase activity and the amount of phosphorylase in the *a* form were also increased. These observations were in line with the view that phosphorylase may be an important rate-limiting enzyme in catecholamine-induced glycogenolysis in skeletal muscle. In similar experiments in cats (unpublished observations), we found that cyclic AMP levels were increased by epinephrine in both red (soleus) and white (tibialis anterior) muscles. Electrical stimulation of skeletal muscle leads to the activation of phosphorylase by a mechanism which involves neither an increase in the level of cyclic AMP nor the activation of phosphorylase *b* kinase (Drummond *et al.*, 1969). This effect may be secondary to the increase in free intracellular Ca^{2+} which probably occurs in response to this type of stimulation (Sandow, 1965).

A few studies have been reported on the effects of exogenous cyclic AMP in skeletal muscle. Edelman and his colleagues (Edelman *et al.*, 1966; Edelman and Schwartz, 1966) found that exposure of rat diaphragm muscle to high concentrations of cyclic AMP in a bicarbonate buffer caused a small but significant increase in the rate of glucose uptake by the tissue. This was similar to the effect of epinephrine in a phosphate buffer in the presence of 8-hydroxyquinoline (Sutherland, 1952). Epinephrine usually decreases glucose uptake when muscles are incubated in a phosphate buffer, but not when a bicarbonate buffer is used (Herman and Ramey, 1960). The dibutyryl derivative of cyclic AMP was studied in Clauser's laboratory (Chambaut *et al.*, 1969), and was found to inhibit both glucose uptake and glycogen synthesis. An interesting finding was that the derivative also inhibited the stimulatory effect of insulin on galactose uptake. The inhibition of glucose uptake was probably related at least in part to the stimulation of glycogenolysis and accumulation of hexose monophosphates. Ui (1965) found that omission of Ca^{2+} from the medium reduced the inhibitory effect of epinephrine, as would be expected from the requirement of Ca^{2+} in the activation of phosphorylase (Chapter 5).

The formation of lactic acid by muscle is influenced by a number of factors, and it is still not clear in all cases how epinephrine (or cyclic AMP) acts to increase it. Phosphorylase activation (Sutherland, 1951b;

Haugaard and Hess, 1965) is undoubtedly important in many cases. This effect, which involves the activation of phosphorylase *b* kinase secondary to the stimulation of a protein kinase by cyclic AMP (Krebs *et al.*, 1966; Walsh *et al.*, 1968; see also Chapter 5), is itself subject to a variety of environmental influences. For example, Lundholm *et al.* (1966) reported that in the isolated rat diaphragm phosphorylase activation in response to epinephrine was greatly reduced if the Na^+ in the medium was replaced by K^+. Further study disclosed that the level of cyclic AMP under these conditions was actually greater than normal, and that the effects of epinephrine and high K^+ were at least additive (Lundholm *et al.*, 1967). It was then found that replacing Na^+ with either K^+ or choline prevented phosphorylase activation in response to exogenous cyclic AMP as well as to catecholamines (Lundholm *et al.*, 1969). The basis for this effect of sodium deprivation is still not entirely clear. Possibly the activity of phosphorylase phosphatase is so high under these conditions that even maximal activation of phosphorylase kinase cannot shift the equilibrium toward phosphorylase *a*. There is some evidence from liver studies (Cahill *et al.*, 1957a) and studies with the isolated perfused rat heart (Ellis and Vincent, 1966) to suggest that this may be the case.

Muscle glycogen synthetase activity is decreased in response to epinephrine, as would be expected from the known effect of cyclic AMP on the activity of glycogen synthetase kinase (Huijing and Larner, 1966b; see also Chapter 5). Danforth (1965) showed that epinephrine decreased the percentage of glycogen synthetase in the I form at all glycogen levels in mouse skeletal muscle. Glycogen synthetase is inherently less stable at lower glycogen levels (Larner, 1966), which may account for the observation that epinephrine may at times decrease *total* glycogen synthetase activity as well as the percentage of the enzyme in the I form (Belocopitow, 1961; Craig and Larner, 1964).

In frog skeletal muscle, increased lactate production in response to epinephrine occurs only after a lag period, during which time there is a considerable accumulation of glucose-6-phosphate. This can be accounted for in part by phosphorylase activation (Helmreich and Cori, 1966) and probably also by reduced glycogen synthetase activity. However, the activity of phosphofructokinase seems to be unaffected (Ozand and Narahara, 1964). Mammalian skeletal muscle differs from frog muscle in that lactate production is increased more rapidly by epinephrine (Svedmyr, 1965), and in this case there is evidence that phosphofructokinase activation may be involved, in addition to the effects on phosphorylase and glycogen synthetase activity (Beviz *et al.*, 1967). There is at present no evidence that phosphofructokinase from mammalian skeletal muscle can exist in two forms, such as were demonstrated to occur in the liver fluke (Stone and Mansour, 1967a), but the possibility does not seem to have been carefully studied. In addition to a possible effect on the conversion of a less active to a more active form, a direct allosteric effect (Shen *et al.*, 1968) may at times be important.

The mechanism of what may be an important physiological effect of epinephrine (Cannon and Nice, 1913; Bowman and Nott, 1969) was suggested by experiments with fatigued frog muscle (Danforth and Helmreich, 1964). Electrical stimulation for 30 sec caused an increase in phosphorylase *a* from less than 5% of the total to about 70%. After a 5-min rest period, however, by which time phosphorylase *a* had returned to its basal level, a second stimulus was capable of increasing phosphorylase *a* to only about 15% of the total. This effect of fatigue was completely abolished by pretreatment with epinephrine. Exposure to low concentrations of epinephrine, in the 10^{-8} *M* range, had no effect on phosphorylase *per se*, but greatly increased the rate and degree of phosphorylase activation in response to electrical stimulation (Helmreich and Cori, 1966). These experiments suggested, what might also be inferred from experiments with purified enzymes (Chapter 5), that the effects of calcium and cyclic AMP on phosphorylase *b* kinase are synergistic. The defatiguing action of epinephrine *in vivo* may of course be partly related to facilitation of neuromuscular transmission (Section VI,B,2).

An interesting puzzle for future investigation is the mechanism by which epinephrine stimulates lactate production in mice which are genetically deficient in skeletal muscle phosphorylase *b* kinase (Lyon *et al.*, 1967). Although these mice may be at a disadvantage under certain conditions (Danforth and Lyon, 1964), they seem by and large to be reasonably healthy animals, and certainly are not comparable to patients with glycogen storage disease. They respond to epinephrine with an increase in blood lactate levels comparable to those seen in normal mice (Lyon and Porter, 1963), and recently it was shown (Lyon and Mayer, 1969) that skeletal muscle levels of cyclic AMP also rise in a relatively normal fashion. Possibly phosphorylase *b* in these animals is sufficiently sensitive to allosteric control (Morgan and Parmeggiani, 1964b) that the conversion to the active form is not necessary. Inspection of the data of Lyon and Porter (1963) suggests that the effect of epinephrine in these mice could be explained by an inhibition of glycogen synthetase activity coupled with an increase in the activity of phosphofructokinase.

Theophylline in a concentration of 10^{-4} *M* had no effect on phosphorylase activity in incubated rat diaphragm muscles, but significantly potentiated the phosphorylase activating effect of epinephrine (Hess *et al.*, 1963). The lack of effect of theophylline by itself suggests that its potentiating effect was due to phosphodiesterase inhibition. At higher concentrations theophylline actually reduced the percentage of phosphorylase in the *a* form, and its potentiating effect under these conditions was less. The effect of the higher concentration of theophylline on phosphorylase was probably the result of stimulation of phosphorylase phosphatase activity (Sutherland, 1951a; De Wulf and Hers, 1968a).

Danforth and Helmreich (1964) studied the effect of caffeine (10^{-2} M) on anaerobic frog skeletal muscle. Caffeine produced a contracture in these experiments which was associated with an increase in the percentage of phosphorylase in the *a* form. This effect could conceivably have resulted from phosphodiesterase inhibition, but the rapidity of the effect suggests that it could also have been caused by calcium. Contracture induced by caffeine is known to be associated with an increased influx of calcium (Sandow, 1965).

The effects of the methylxanthines on lactate production *in vivo* are of course complicated by actions other than those which occur in skeletal muscle, including a tendency to stimulate the release of catecholamines (Strubelt, 1969; Berkowitz *et al.*, 1970). The hyperlacticacidemic response to the combination of epinephrine and theophylline in dogs was nevertheless found to be greater than the sum of the responses to either agent given separately (Triner and Nahas, 1966).

Evidence suggesting an activation of skeletal muscle lipase has been presented by Garland and Randle (1963), who showed an increase in the intracellular level of free fatty acids in rat diaphragm in response to epinephrine. The possible effect of cyclic AMP on skeletal muscle lipase has not been investigated, but, in view of studies in other tissues (Rizack, 1964; Bewsher and Ashmore, 1966; Kruger *et al.*, 1967), a stimulatory effect would not be surprising. Such an effect might provide an explanation for the high citrate levels which occur in the muscles of mice with hereditary muscular dystrophy (Taussky *et al.*, 1962). Gordon and Dowben (1966) have reported that the levels of catecholamines and cyclic AMP in skeletal muscle of mice with hereditary muscular dystrophy are higher than in normal littermates, and both epinephrine and exogenous free fatty acids are known to cause increased citrate levels in the perfused rat heart (Williamson, 1965; Bowman, 1966). The significance of these findings for the pathogenesis of muscular dystrophy is presently unclear.

2. Cardiac Muscle

Carbohydrate metabolism in cardiac muscle seems to be basically similar to that in skeletal muscle, but some differences in the overall response to catecholamines have been noted and will be discussed in this section. The contribution of the heart to the hyperlacticacidemic response to epinephrine is quite small, and in fact, under normal conditions, the heart tends to use more lactate than it produces (Green and Goldberger, 1961). The principle function of the heart is to pump blood, and most of the metabolic reactions which occur in the heart seem to have evolved to serve this function, rather than to help maintain homeostasis in the organism as a whole.

Although there is some evidence that α-receptors exist in heart muscle (Govier, 1968; James *et al.*, 1968), most of the cardiac metabolic effects of the catecholamines which have been studied are thought to be mediated by β-receptors (Mayer *et al.*, 1967). These are distinguished from many other β-receptors by their relatively greater sensitivity to norepinephrine. They also seem to be less susceptible to blockade by the methoxamine derivatives than the receptors in many tissues (Burns *et al.*, 1964; Levy, 1966; Robison *et al.*, 1967a). These characteristics seem to prevail in the hearts of all species studied. They obviously do not apply, of course, to those species whose hearts respond to serotonin instead of catecholamines, as in certain molluscs (Rozsa, 1969).

Cardiac adenyl cyclase has been studied in washed particulate preparations from a variety of species, including dog (Murad *et al.*, 1962; Robison *et al.*, 1967a), cat (Sobel *et al.*, 1968; Levey and Epstein, 1969a), rat (Murad and Vaughan, 1969), guinea pig (Sobel *et al.*, 1969a), and human (Levey and Epstein, 1969a), and all of the characteristics which distinguish cardiac β-receptors seem to apply equally well to it. In preparations from dog heart, for example, isoproterenol was found to be approximately eight times more potent than either epinephrine or norepinephrine in stimulating adenyl cyclase (Murad *et al.*, 1962). These relative potencies are in good agreement with the results of experiments in which phosphorylase activation in intact tissue has been measured in response to catecholamines (Haugaard and Hess, 1965). Isoproterenol may give the impression *in vivo* of being an even more potent agonist than it really is, because it is taken up by nervous tissue to a lesser extent than the naturally occurring catecholamines (Hardman *et al.*, 1965). When reuptake is prevented, as by the administration of cocaine, then the relative potencies in organized tissue agree very closely with the values obtained in broken cell preparations. The effects of β-adrenergic blocking agents also correspond with their effects in intact cells. For example, isopropylmethoxamine (IMA) was found to be much less potent than pronethalol in preventing the stimulation of adenyl cyclase by isoproterenol *in vitro* (Fig. 6.4), in agreement with its lower potency in preventing phosphorylase activation in the intact heart (Burns *et al.*, 1964; Kukovetz and Pöch, 1967a). The data in Fig. 6.4 show that

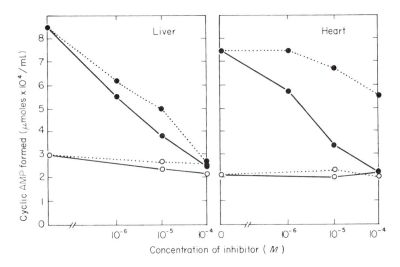

Fig. 6.4. Effect of pronethalol and isopropylmethoxamine (IMA) on adenyl cyclase activity of dog liver and heart. Washed particles were incubated with ATP and Mg^{2+} in the presence (solid circles) and absence (open circles) of 10^{-6} M isoproterenol. The amount of cyclic AMP formed in 15 min is plotted as a function of the concentration of pronethalol (solid lines) or IMA (dotted lines). From Robison *et al.* (1967a).

these two agents are roughly equipotent in dog liver preparations. These various correlations support the concepts that the β-receptor and adenyl cyclase are part and parcel of the same system, and that the cardiac metabolic effects of the catecholamines are mediated by cyclic AMP.

One discrepancy which has been noted between the *in vitro* and *in vivo* data is that high concentrations of ergotamine were found to be capable of inhibiting the effect of epinephrine on adenyl cyclase (Murad *et al.*, 1962). The ergot alkaloids are not noted for their cardiac effects, and indeed do not affect the heart in doses which block α-receptors completely (Innes and Nickerson, 1965). This discrepancy may not be too serious, however, for it seems entirely possible that ergotamine might inhibit the activation of cardiac adenyl cyclase *in vivo* as well as *in vitro* if present in sufficiently high concentrations (and in addition to whatever else it might inhibit at these concentrations).

Glucagon also stimulates cardiac adenyl cyclase, as shown in Fig. 6.5. This was demonstrated in rat tissue by Murad and Vaughan (1969) and in preparations from cat and human hearts by Levey and Epstein (1969a). Epinephrine and glucagon are seen from Fig. 6.5 to be roughly equipotent, in line with their effects in the intact heart, which are similar. Glucagon

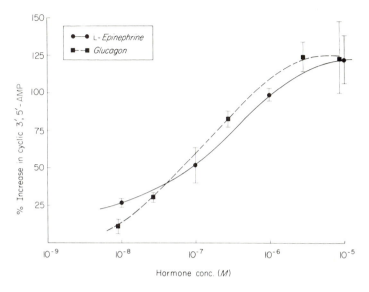

Fig. 6.5. Effects of epinephrine and glucagon on the activity of rat heart adenyl cyclase. Washed particulate preparations were incubated with ATP and Mg^{2+} for 10 min at 30°C in the presence of caffeine and the indicated concentration of hormone. Formation of cyclic AMP with no hormone added was 2.3×10^{-4} μmoles/ml. From Murad and Vaughan (1969).

is discussed in the next chapter, but its ability to stimulate cardiac adenyl cyclase is mentioned here because it adds to the evidence that the cardiac effects of the catecholamines are mediated by cyclic AMP. The stimulatory effect of glucagon was not affected by concentrations of propranolol which abolished the effect of epinephrine, and the effects of maximal concentrations of the two hormones were not additive (Murad and Vaughan, 1969). These results suggest that epinephrine and glucagon affect the same adenyl cyclase but by way of separate receptors.

Most of the adenyl cyclase activity in heart homogenates sediments out with the low-speed fraction, which is thought, by analogy with preparations from other tissues (e.g., Davoren and Sutherland, 1963b), to consist primarily of nuclear material and fragments of the plasma membrane. However, Entman *et al.* (1969a) found that cyclase activity in a rat heart homogenate was also associated with a microsomal fraction capable of accumulating calcium. This fraction is thought to consist largely of fragmented sarcoplasmic reticulum and perhaps also elements of the transverse tubular system, which can be thought of as extensions of the plasma membrane (Simpson and Oertelis, 1962). As in the preparations studied by others, the microsomal enzyme was found to be sensitive to stimulation by glucagon as well as by catecholamines (Entman *et al.,* 1969b).

The concentration of cyclic AMP in the beating heart rises very rapidly in response to epinephrine. This has been demonstrated in the isolated perfused working rat heart (Robison *et al.,* 1965),* in rat and rabbit hearts perfused by the Langendorff technique (Hammermeister *et al.,* 1965; Cheung and Williamson, 1965; Drummond *et al.,* 1966; Namm *et al.,* 1968), and in the rat heart *in situ* (Namm and Mayer, 1968). Descriptions of the time course of this response have varied somewhat from laboratory to laboratory, depending on how the experiments were done, but there is general agreement that the rise in cyclic AMP is the first response which can be detected. The response in the isolated perfused working rat heart is illustrated in Fig. 6.6. The activation of phosphorylase *b* kinase follows the rise in cyclic AMP with striking rapidity (Drummond *et al.,* 1966), and this is in turn followed by the activation of phosphorylase. As shown by Øye (1965), the increase in contractile force precedes the activation of phosphorylase by a measurable interval.

Glucagon also causes an increase in the cardiac level of cyclic AMP (R. H. Bowman *et al.,* 1967, unpublished observations; Mayer *et al.,* 1970), as expected from its effect on adenyl cyclase. The failure of La-Raia *et al.* (1968) to observe an effect of glucagon on the level of cyclic AMP in the rat heart may have been caused by the failure of these investigators to freeze the heart rapidly enough.

Among the results to be expected from the increased level of cyclic AMP in the heart, based on studies with broken cell preparations (see Chapter 5), would be an increase in the percentage of phosphorylase in the *a* form and a decrease in the percentage of glycogen synthetase in the I form, leading to a fall in the intracellular level of glycogen. Phosphorylase activation in response to catecholamines has been well documented (Haugaard and Hess, 1965; Mayer *et al.,* 1967), and will not be further discussed. Williamson (1966) observed a fall in glycogen synthetase activity and a fall in the level of glycogen in response to epinephrine in rat hearts perfused in the absence of glucose, as predicted from studies with cell-free systems (Huijing and Larner, 1966a,b). When glucose is present in the extracellular fluid, however, it would appear that other factors come into play to prevent this response. We were unable to de-

---

*Due to an oversight, the help of Dr. Lee C. Chumbley in carrying out this research was not acknowledged. Dr. Chumbley was at that time a medical student at Vanderbilt University. Despite this belated acknowledgment of it, we were and are grateful to him for his help.

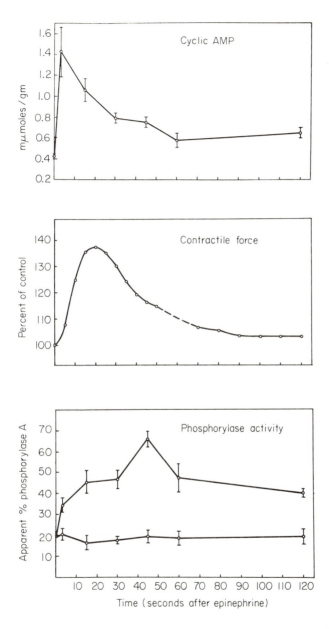

Fig. 6.6. Effect of a single dose of epinephrine in the isolated perfused working rat heart. Hearts were frozen at different times after injection of the hormone. Contractile force represents average response of the hearts frozen at 60 and 120 sec. Lower curve in phosphorylase panel shows lack of response to injected saline. From Robison *et al.* (1965).

tect a change in glycogen synthetase activity in response to epinephrine in the isolated perfused working rat heart (Robison *et al.,* 1965), but others have observed a transient *increase* in the activity of this enzyme both in the isolated heart (Williamson, 1966) and in the heart *in situ* (Williams and Mayer, 1966; Belford and Cunningham, 1968). Daw and Berne (1967) found that in cats the level of cardiac glycogen and the activity of glycogen synthetase were inversely correlated, and that this inverse relationship was unchanged by a variety of interventions. Surgical sympathectomy of the heart led to an increase in the level of cardiac glycogen and a fall in glycogen synthetase activity (Daw and Berne, 1967).

Studies of the effects of epinephrine in whole animals (e.g., Hornbrook and Brody, 1963a,b; Williams and Mayer, 1966) are, of course, complicated by effects occurring in organs other than the heart, and may be especially difficult to interpret when carried out in the presence of methylxanthines (e.g., McNeil *et al.,* 1969a). Even in the isolated perfused heart, however, there would appear to be a very effective mechanism or mechanisms for opposing the tendency of epinephrine or cyclic AMP to cause a fall in the level of cardiac glycogen (Williamson, 1966; Mayer *et al.,* 1967). The ability of epinephrine to stimulate glucose uptake by the heart (Smith *et al.,* 1967) may be important in this regard. Since this effect in the rat heart was not blocked by β-adrenergic blocking agents (Satchell *et al.,* 1968), there is no evidence that it is mediated by cyclic AMP.

Another metabolic effect of epinephrine is to cause an increase in the myocardial level of free fatty acids (Challoner and Steinberg, 1966). This is probably secondary to lipase activation by cyclic AMP, as demonstrated by Kruger *et al.* (1967) in broken cell preparations. Epinephrine also stimulates the myocardial uptake of exogenous free fatty acids (Masters and Glaviano, 1969). As suggested by Challoner and Steinberg (1966), fatty acid metabolism is probably the major factor responsible for the increased oxygen consumption (qO_2) seen in response to epinephrine in the arrested heart. Under normal conditions, the major part of the effect of epinephrine on qO_2 is probably secondary to increased work.

It is undoubtedly true that the overall metabolic response of the working heart to catecholamines is the result of at least two sets of effects, one set being the result of the direct effects of cyclic AMP on the various enzymes which are affected by this nucleotide, and the other set being secondary to the increased rate and force of contraction. These latter effects may also be caused by cyclic AMP (Section VI,B,3), but it is also true that increased work per se may lead to profound metabolic alterations which are independent of changes in the level of cyclic AMP. The metabolic effects to be expected as the result of increased work have been well studied by Morgan and Neely and their colleagues (Morgan *et al.,* 1965; Neely *et al.,* 1967a,b, 1969). These effects include an increase in the rate of glycogenolysis, an increase in the percentage of glycogen synthetase in the I form, an increase in the rate of transport of glucose and other substrates into the cells, an increase in the rate of glucose phosphorylation, marked increases in the levels of 5'-AMP and inorganic phosphate (probably sufficient to account for the apparent increases in the activities of phosphorylase *b,* phosphofructokinase, and hexokinase), an increase in the rate of production of lactate, and an increased rate of oxygen consumption. Cyclic AMP levels do not change at different work loads, as mentioned previously, and the percentage of phosphorylase in the *a* form also tends to remain constant. Evidence that phosphorylase activation in response to epinephrine is not secondary to increased contractile force has also been presented by others (Dhalla and McLain, 1967; Øye, 1967). The mechanism of the increased rate of glucose transport seen in response to increased work is not well understood, but the effect seems basically similar to the effects of insulin and anoxia (Morgan *et al.,* 1965).

As for the metabolic effects which occur in the heart in response to anoxia, these may be secondary to cyclic AMP to whatever extent the release of norepinephrine is involved. An increase in the level of cyclic AMP in the dog heart *in situ* in response to ischemia was demonstrated by Wollenberger *et al.* (1969), and it was found that this could be prevented by β-adrenergic blocking agents. Another component of the response to hypoxia cannot be blocked by β-adrenergic blocking agents (Mayer *et al.,* 1967; Wollenberger and Krause, 1968), and this involves a mechanism which is independent of changes in the level of cyclic AMP (Morgan *et al.,* 1965; Robison *et al.,* 1967a).

In view of the numerous metabolic adjustments which the heart is capable of making in response to the demand for increased contractile force per se, the question might legitimately be asked whether the additional metabolic effects produced by cyclic AMP in the heart confer any sort of survival value on the organism as a whole. This is an interesting but difficult question to try to answer. One intuitively feels that a hormone which orders an organ to do more work will be a better hormone if it simultaneously provides instructions for the necessary utilization of energy. In the case of the heart, however, these instructions may be superfluous. So long as a substrate and plenty of oxygen are available, the heart is capable of developing large amounts of tension for long periods of time, and it seems capable of doing this quite without the help of cyclic AMP.* Unlike skeletal muscle, where the utility of phosphorylase activation by cyclic AMP becomes apparent under various emergency conditions, there seems to be no theoretical need for phosphorylase activation in the heart. At the present time, one is pretty much reduced to the argument that phosphorylase activation does occur in the heart in response to sympathetic stimulation, and it seems reasonable to assume that this must be useful for something.

3. Smooth Muscle

The metabolic effects of epinephrine in smooth muscle seem basically similar to those which occur in other types of muscle. These include, characteristically, an increase in the level of cyclic AMP, an activation of phosphorylase, and an increase in the rate of production of lactic acid (Lundholm et al., 1966).

Bueding and Bulbring and their colleagues (Bueding et al., 1966; Bueding and Bulbring, 1967; Bueding et al. 1967) have studied the metabolic effects of small concentrations of epinephrine, in the 10^{-8} M range, on the smooth muscle of the guinea pig taenia coli. These concentrations of epinephrine, which are sufficient to cause the muscle to relax, also cause small but significant increases in the level of cyclic AMP, ATP, and creatine phosphate. These changes may occur without necessarily causing an activation of phosphorylase, and with no significant change in the intracellular levels of hexose monophosphates. It is possible that the relaxation is secondary to the increased levels of cyclic AMP and high energy phosphate compounds. The converse possibility, that the increased levels of these compounds are caused by the relaxation, has been ruled out.

The changes in cyclic AMP levels observed by Bueding et al. (1966) were very small relative to the large changes which have been seen in many other tissues. In the isolated rat uterus, under conditions where the presence of adrenergic α-receptors cannot be detected, the increase was larger (Butcher et al., 1965; Dobbs and Robison, 1968), as illustrated in Fig. 6.7. An interesting finding in the uterus was that the level of cyclic AMP rises and falls even in the continuous presence of a catecholamine. The order of potency of agonists in causing an increase in the uterine level of cyclic AMP is the same as that used by Ahlquist to initially define the

*See, however, Section VI,B,3 in this chapter and Section II,C,1 in Chapter 7.

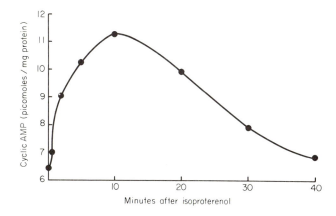

Fig. 6.7. Effect of isoproterenol on uterine level of cyclic AMP. Uteri from ovariecto-mized estrogen-primed rats were incubated in the presence of $5 \times 10^{-7} M$ isoproterenol, and frozen at the indicated times. From Dobbs and Robison (1968).

β-receptor, and the response can be blocked by β-adrenergic blocking agents.

Adenyl cyclase from smooth muscle has not been carefully studied, especially with respect to its subcellular localization, but it does exist (Klainer *et al.*, 1962; Dousa and Rychlik, 1968a). Adenyl cyclase activity in homogenates of rat uterus can be stimulated by catecholamines and sodium fluoride (Dobbs and Robison, 1968). These preparations also contain high phosphodiesterase activity which can be inhibited by methyl-xanthines.

Bueding *et al.* (1962) found that if samples of taenia coli which had been incubated were manipulated for a few seconds (as by drawing them over a hard surface) prior to homogenization, an apparent activation of phosphorylase occurred. This effect was significantly greater in samples which had been incubated in the presence of epinephrine (at concentra-tions which were later shown to increase the level of cyclic AMP) (Bued-ing *et al.*, 1966). However, if the samples were homogenized immediately after removal from the incubation medium, then no activation of phos-phorylase could be observed, even in tissues which had been exposed to epinephrine for up to 15 sec. This suggested the possibility that in smooth muscle cyclic AMP might be compartmentalized, and that manipulation of the tissue might tend to break down a barrier between the cyclic AMP and phosphorylase *b* kinase. Another possibility which could be con-sidered, especially in view of recent studies with the perfused rat heart

(Drummond *et al.*, 1966; Namm *et al.*, 1968), is that phosphorylase *b* kinase was indeed activated in the early experiments of Bueding and his colleagues, but that the concentration of free calcium (or possibly some other factor) in the vicinity of phosphorylase was too low to permit the phosphorylase *b* → *a* reaction to occur. Manipulation of the tissue may have altered the ionic environment sufficiently to allow the activated phosphorylase *b* kinase to exert its catalytic effect. The results obtained by Brody and Diamond (1967) were in agreement with those of Bueding *et al.* (1962) in that phosphorylase activation was not significant after 15 sec of exposure to epinephrine. However, after 30 sec there was a fivefold increase in the percentage of phosphorylase in the *a* form, and this was judged to be not artifactual. Both groups of investigators were in agreement that the activation of phosphorylase was not fast enough to account for the observed decrease in tension.

Several previously puzzling observations were clarified by the discovery by Diamond and Brody (1966b) that all interventions which increase the tone and motility of the isolated rat uterus lead at the same time to an activation of phosphorylase. That this type of phosphorylase activation is more likely to be a result than a cause of the enhanced contractility was suggested by experiments in which the time course of the change in phosphorylase activity was compared with the time course of a single contraction in spontaneously contracting uteri. The rise and fall in the level of phosphorylase *a* was found to parallel very closely the rise and fall of tension, with the change in phosphorylase following the change in tension by several seconds. Although the participation of cyclic AMP in these changes was not ruled out, this seems unlikely for several reasons. In no tissue is functional activity per se known to influence the level of cyclic AMP. Also, oxytocin and $BaCl_2$, which were among the uterine stimulants which Diamond and Brody found to also stimulate phosphorylase activation, were later found *not* to increase the uterine level of cyclic AMP (Dobbs and Robison, 1968). It would appear that in smooth muscle, as in skeletal muscle (Drummond *et al.*, 1969), phosphorylase activation can occur by a mechanism which is separate from the cyclic AMP-stimulated activation of phosphorylase *b* kinase. It should be noted that the type of phosphorylase activation noted by Diamond and Brody in the uterus may not occur in all types of smooth muscle. In vascular smooth muscle, for example, under conditions where epinephrine and other agents stimulated the muscle to contract, no change in phosphorylase activity could be observed (Lundholm *et al.*, 1966).

The possible effects of cyclic AMP on smooth muscle enzymes other than phosphorylase have not been well studied. Beviz and Mohme-Lundholm (1964, 1965) obtained indirect evidence suggesting an effect of epinephrine on phosphofructokinase in vascular smooth muscle. However, an effect of cyclic AMP on phosphofructokinase in broken cell preparations of taenia coli could not be demonstrated (Bueding *et al.*, 1966).

Studies of the relationship of cyclic AMP to the metabolic effects of the catecholamines in smooth muscle may be complicated by the presence of adrenergic α-receptors. Although this has not been demonstrated in smooth muscle, α-receptors in other tissues have been shown to mediate a fall in the intracellular level of cyclic AMP. In smooth muscle this would tend to cause contraction, which could in turn lead to various changes as summarized above for the uterus. More detailed studies in this area would be desirable.

C. The Response in the Pancreas

An important role for cyclic AMP in the regulation of insulin release has been suggested by a number of recent findings, as discussed in more detail in the next chapter. The most important physiological stimulus for insulin release is presumably an increase in the level of blood glucose, but this effect of glucose is shared by other substrates and is subject to modulation by a great variety of other factors (see Mayhew *et al.*, 1969, for a recent review). An inhibitory effect of epinephrine, via a direct effect on the pancreas, was long suspected as an important component of the hyperglycemic response to epinephrine (Ellis, 1956). With the development of newer and more sensitive methods for the measurement of insulin, this effect has now been established. This was first demonstrated by Coore and Randle (1964), using rabbit pancreas slices, and has since been demonstrated in a variety of species, *in vitro* and *in vivo*, in response to norepinephrine as well as epinephrine.

The finding by Coore and Randle that the inhibitory effect of epinephrine could be reversed by ergotamine suggested the participation of adrenergic α-receptors. This was later substantiated by Porte (1967a), who showed that the effect in humans could be blocked by phentolamine. During the course of these studies, Porte noted that in the presence of phentolamine the pancreas responded to epinephrine with an increase in insulin output, more than could be accounted for by the simultaneous increase in blood glucose levels. This, together with the observation that blood insulin levels were abnormally low following the administration of β-adrenergic blocking agents, suggested that β-receptors also might affect the release of insulin. Porte (1967b) then showed that isoproterenol was a potent stimulator of insulin release, and that this stimulatory effect could be blocked by the β-adrenergic blocking agents propranolol and butoxamine. Porte concluded that in the pancreas the stimulation of β-receptors led to the release of insulin, while α-receptor stimulation inhibited insulin release. This general conclusion is now well supported by experimental data (e.g., Malaisse *et al.*, 1967; Wong *et al.*, 1967; Senft *et al.*, 1968c; Gagliardino *et al.*, 1968).

Evidence that cyclic AMP was involved in these effects came in the meantime from other directions. Glucagon, which was of course known to be a powerful stimulator of adenyl cyclase in the liver, was found to also stimulate the release of insulin from the pancreas (Karam *et al.*, 1966; Samols *et al.*, 1966; Vecchio *et al.*, 1966). That this effect of glucagon was mediated by cyclic AMP was supported by the finding that the effect could be potentiated by methlxanthines (Lambert *et al.*, 1967a,b; Lacy *et al.*, 1968b; Allan and Tepperman, 1969) and mimicked by exogenous cyclic AMP (Sussman and Vaughan, 1967; Malaisse *et al.*, 1967). The ability of glucagon to cause an increase in the level of cyclic AMP in pancreatic islets was demonstrated by Turtle and Kipnis (1967b).

The possible relation of adrenergic receptors to adenyl cyclase in the pancreas was initially studied by Turtle, Littleton, and Kipnis (1967). These investigators showed that in rats the injection of theophylline by itself could cause an increase in the level of plasma insulin. When the theophylline was injected during the infusion of epinephrine, it still provoked a prompt eightfold increase in the level of plasma insulin. When α-adrenergic receptors were blocked by phentolamine, and the experiment repeated, then theophylline was found to provoke an extraordinary (30- to 40-fold) increase in the plasma insulin content. Conversely, when β-receptors were blocked by propranolol, then theophylline produced no effect whatever. Turtle and his colleagues interpreted these results as suggesting that α-receptors and β-receptors in the beta cells of the pan-

TABLE 6.II. Effect of Epinephrine and Adrenergic Blocking Agents on the Level of Cyclic AMP in Rat Pancreatic Islets[a]

Experimental conditions	Cyclic AMP (μmoles/gm)
Control	5570 ± 302
Epinephrine (5×10^{-5} M)	415 ± 12
Epinephrine + phentolamine	4450 ± 363
Epinephrine + propranolol	305 ± 20

[a]Islets were incubated for 60 min at 37°C in a bicarbonate buffer containing 5.6 mM glucose and 1 mM theophylline, in the presence or absence of the indicated agents. Cyclic AMP in the absence of theophylline was 355 ± 50 μmoles/gm. From Turtle and Kipnis, 1967b.

creas mediate divergent effects of epinephrine on the synthesis of cyclic AMP: α-receptors decreasing it and β-receptors increasing it. Turtle and Kipnis (1967b) later demonstrated the cyclic AMP-lowering effect of epinephrine in an isolated rat pancreatic islet preparation, and some of the data which they reported are reproduced in Table 6.II.

Thus the available evidence supports the hypothesis that the insulino-genic effects of glucagon and the catecholamines are mediated by cyclic AMP. The insulinogenic effect of the catecholamines, mediated by β-adrenergic receptors, is of course of doubtful physiological significance because under normal conditions the effect always seems to be obscured by the opposite effect mediated by α-receptors (Porte, 1969). It would appear that the normal physiological effect of the catecholamines is to inhibit the release of insulin, and the evidence suggests that this effect is secondary to a fall in the level of cyclic AMP in the pancreatic beta cells.

The presence of β-adrenergic receptors in the pancreas is nevertheless of some theoretical interest. As in other tissues where these receptors occur together with glucagon receptors, such as in the liver and heart of several species and also in rat adipocytes, they can be selectively blocked by β-adrenergic blocking agents (Porte, 1967b; Malaisse et al., 1967). Toman's Law, which states that enough of anything will inhibit anything (Chenoweth and Ellman, 1957), is of course always operative: higher concentrations of β-adrenergic blocking agents may inhibit insulin re-lease in response to all stimulatory agents, including glucagon and exo-genous cyclic AMP (Bressler et al., 1969).

Some interesting studies with diazoxide can be briefly mentioned here. Diazoxide is a hyperglycemic nondiuretic benzothiadiazine derivative closely related to the diuretic chlorothiazide. Following a careful investi-gation of the metabolic responses to this drug, Tabachnick et al. (1965) suggested, as one possible explanation of their results, that the hyper-glycemic effect of diazoxide might be related in part to a decreased rate of destruction of cyclic AMP in the liver. This suggestion was soon sub-stantiated by the work of Schultz et al. (1966), who demonstrated that in vitro diazoxide was capable of inhibiting phosphodiesterase activity in several tissues. This was later confirmed by Moore (1968), and may account for many of the observed effects of this agent, such as lipolysis (Nakano et al., 1968) and smooth muscle relaxation (Wilkenfeld and Levy, 1969). Its predicted effect in the pancreas, on this basis, would be to enhance the insulinogenic response to glucagon and β-adrenergic stim-ulation. In fact, however, the exact opposite occurs, for diazoxide is known to be a relatively potent inhibitor of insulin release (Wong et al., 1967).

Some light was shed on this problem by the finding that the inhibitory effect of diazoxide on insulin release could be reduced or prevented by the α-adrenergic blocking agent phentolamine (Blackard and Aprill, 1967; Senft *et al.*, 1968c; Porte, 1968). Malaisse and Malaisse-Lagae (1968) found that phenoxybenzamine did not prevent the effect of diazoxide *in vitro* in a dose which did prevent the similar effect of epinephrine, but Tabachnick and Gulbenkian (1968) observed a strong inhibitory effect of phenoxybenzamine after several hours pretreatment *in vivo*. These results suggest that adrenergic α-receptors are involved in the inhibitory effect of diazoxide. An indirect effect via catecholamine release has not been absolutely ruled out, but seems unlikely because the effect occurs *in vitro* and also *in vivo* in adrenalectomized and reserpinized animals (Blackard and Aprill, 1967; Tabachnick and Gulbenkian, 1968). It thus seems possible that diazoxide interacts directly with adrenergic α-receptors in pancreatic beta cells, and in this way causes a fall in the intracellular level of cyclic AMP. Direct measurement of cyclic AMP levels in pancreatic islets in response to diazoxide would be of interest.

Lambert *et al.* (1967a,b) found that the insulin-releasing effect of tolbutamide, like that of glucagon, was enhanced by caffeine in organ cultures of fetal rat pancreas, suggesting that tolbutamide might act by stimulating adenyl cyclase in pancreatic beta cells. Tolbutamide further resembled glucagon in that both agents acted synergistically with glucose to stimulate insulin release. Caffeine also enhanced the effect of glucose even in the absence of tolbutamide or glucagon, although caffeine itself was relatively ineffective. The relation of cyclic AMP to the insulinogenic action of glucose is poorly understood, but several observations suggest that it may be quite complex. In islets isolated from adult rat pancreas, exogenous glucagon stimulated insulin release slightly, but tolbutamide was completely ineffective (Lacy *et al.*, 1968b). On the other hand, tolbutamide was effective in the isolated perfused rat pancreas (Loubatieres *et al.*, 1967). The stimulatory effect of tolbutamide in this preparation could be antagonized by diazoxide or epinephrine.

The possible role of glucagon and cyclic AMP in the regulation of insulin release is discussed in more detail in the next chapter. The most important point to be made in this section is that the catecholamines interact with adrenergic α-receptors to cause a fall in the level of cyclic AMP in pancreatic islets, and inhibit the release of insulin. There is evidence that diazoxide may act by a similar mechanism.

D. Summary

The effects of the catecholamines on carbohydrate metabolism in vertebrates can be summarized briefly. These agents stimulate adenyl cyclase in the liver (presumably in hepatic parenchymal cells) and in all forms of muscle. The adrenergic receptors mediating this effect in muscle are characteristically β-receptors, in that isoproterenol is a more potent agonist than either of the naturally occurring catecholamines and their effects can be selectively blocked by β-adrenergic blocking agents. These characteristics are shared by the receptors in the liver, in many species, although they may deviate in others. In the rat liver, for example, isoproterenol may be a less effective agonist than either of the naturally occurring catecholamines, and the β-adrenergic blocking agents may be relatively ineffective as antagonists.

In the liver, the increased levels of cyclic AMP lead to an increased rate of glucose production by at least three mechanisms: the activation of phosphorylase, the inactivation of glycogen synthetase, and the stimulation of gluconeogenesis. The increase in phosphorylase activity and the decrease in glycogen synthetase activity combine to bring about a net conversion of glycogen to glucose. The conversion of other substrates to glucose is stimulated as a result of the increased rate of gluconeogenesis. Among these substrates is lactic acid produced as a result of muscle glycogenolysis.

The effects of cyclic AMP in muscle include phosphorylase activation and glycogen synthetase inactivation. The effect on phosphorylase in response to catecholamines is easily demonstrated in intact muscle, but the effect on glycogen synthetase may not always occur. Muscle seems to contain a highly effective mechanism or mechanisms for opposing the action of cyclic AMP on glycogen synthetase activity. The end result of glycogenolysis in muscle is the production of lactic acid instead of glucose, because muscle does not contain glucose-6-phosphatase.

In the pancreas, the characteristic effect of the catecholamines is to lower the level of cyclic AMP in islet tissue (presumably in the beta cells) and thereby inhibit the release of insulin which would otherwise occur in response to glucose. This effect is mediated by adrenergic α-receptors. Islets also contain β-receptors, which tend to mediate the opposite effect on cyclic AMP and insulin release, but under most conditions the α-receptors always seem to predominate.

An important result of this combination of actions is that the catecholamines tend to shunt glucose from the liver to the brain. The glucose which is released from the liver tends not to be utilized by muscle partly because of the stimulation of muscle glycogenolysis and partly because of the suppression of insulin release. The uptake of glucose by the central nervous system seems to be much less dependent on insulin than it is in muscle and other peripheral tissues.

The suppression of insulin release also facilitates the lipolytic response to the catecholamines, as discussed in more detail in Chapters 7 and 8. We can note here, however, that the increased blood levels of free fatty acids which result from this action, and which can be utilized as a source of energy by most peripheral tissues, tend to reduce further the peripheral utilization of glucose.

IV. THE CALORIGENIC EFFECT

The calorigenic effect of the catecholamines, the increased metabolic rate which characteristically occurs in response to sympathoadrenal discharge, may at times be one of the most important physiological effects of these hormones (Cannon, 1939). For example, normal rats adjust to an ambient temperature of 4°C by increasing their metabolic rate, and are able to exist under these conditions for long periods of time. Under the same conditions, rats in which adrenergic function has been completely blocked lose heat at the rate of about 1400 cal/hr, and die in 3 hr at a body temperature of about 15°C (Brodie *et al.*, 1966).

Usually measured as an increase in the rate of oxygen consumption, the calorigenic effect can be mimicked by appropriate doses of exogenous catecholamines. The mechanism of the effect is poorly understood, but probably represents a complex resultant of a variety of other effects, each of which may contribute to a greater or lesser extent at different ages and under different conditions. It is therefore not possible to define the role of cyclic AMP in this response with any degree of certainty, although a few general observations will be made. Readers interested in a more detailed analysis of the problem are referred to a review by Himms-Hagen (1967).

If each of the effects which contribute to the calorigenic effect is mediated by an increase in the level of cyclic AMP in the cells responsible for those effects, then it might be thought that the injection of exogenous cyclic AMP would stimulate calorigenesis. That this does not happen (Strubelt, 1968) is understandable because cyclic AMP also mimics a variety of other hormonal responses, some of which interfere with those characteristically produced by the catecholamines (Bieck *et al.*, 1969). Although exogenous cyclic AMP did not produce a calorigenic response, theophylline did (Strubelt, 1968).

The importance of adrenergic β-receptors in calorigenesis, as opposed to α-receptors, was illustrated by Estler and Ammon (1969). They found that pretreatment with any of several β-adrenergic blocking agents greatly reduced the survival rate of mice exposed to cold temperatures. Such heat-conserving mechanisms as vasoconstriction and piloerection, mediated by adrenergic α-receptors, are probably enhanced by β-blockade, but are inadequate at low temperatures in the absence of increased calorigenesis.

The response to theophylline depends upon the catecholamines. The hyperglycemic response in adult rats was completely abolished by the combination of adrenal medullectomy and reserpinization (Strubelt, 1969), and the calorigenic response was prevented by propranolol (Cockburn et al., 1967). These results can be understood partly in terms of catecholamine release and partly in terms of phosphodiesterase inhibition. Both actions are probably involved in the effects of theophylline in vivo.

Although the catecholamines seem to be of major importance as calorigenic hormones in most mammals, it should be noted that other hormones may at times participate in the response to cold, in addition to the various permissive hormones which may be required in all cases. In some species, such as birds, these hormones may be even more important than the catecholamines (Wekstein and Zolman, 1969).

One important component of the calorigenic response, especially in younger animals, is the mobilization of free fatty acids (FFA) in brown adipose tissue. This tissue differs from white adipose tissue in that it has a much greater capacity for oxidative metabolism. The brown color of the tissue is in fact mainly a reflection of its high content of cytochromes. Thus when lipolysis is stimulated in this tissue most of the FFA which are produced are not released into the bloodstream but are instead metabolized within the tissue itself. (The glycerol which is produced is released, however, and this provides a useful means for monitoring the rate of lipolysis.) Some of the acyl-CoA esters produced by FFA metabolism are oxidized via the Krebs cycle, but most are resynthesized to triglycerides, utilizing α-glycerophosphate synthesized from glucose. This accounts in large part for the increased utilization of glucose which occurs in this tissue in response to lipolytic hormones. This seems like an extraordinarily wasteful process from the point of view of energy metabolism, but from the point of view of heat production it seems ideally suited for the task. Thus Joel (1966) has shown that although the lipolytic response to norepinephrine is not strikingly different in the two types of adipose tissue, the increased oxygen consumption may be almost 40-fold greater in brown adipose tissue, in response to equivalent concentrations of norepinephrine. Hull and Segall (1965) calculated that in newborn rabbits over 90% of the heat produced in response to cold exposure or to the infusion of catecholamines could be accounted for by brown adipose tissue.

Heim and Hull (1966) found that larger doses of propranolol were required to block the calorigenic response to cold exposure than were required to block the response to infused catecholamines. To explain this result, several possibilities could be considered. It is possible that the most important physiological mechanism for stimulating adenyl cyclase in brown adipose tissue is the release of norepinephrine from sympathetic nerve endings (Correll, 1963). Possibly adenyl cyclase near nerve endings is in some manner more protected from exogenous influences (including both circulating hormones and adrenergic blocking agents) than adenyl cyclase in other parts of the cell. Perhaps a more likely explanation is that the local concentration of norepinephrine engendered by nerve stimulation is higher than can be achieved by infusing exogenous norepinephrine. Propranolol did not prevent the calorigenic response to infused glucagon or ACTH at all, in line with its inability to prevent the stimulation of adenyl cyclase by these hormones in brown adipose tissue (see Chapter 8).

Heim and Hull concluded from their experiments that the increased blood flow which occurs in brown adipose tissue in response to lipolytic hormones is primarily a result of increased tissue metabolism rather than a direct effect of the hormones on vascular smooth muscle. This may be related to the release of adenosine, as suggested by Berne (1964). The increased blood flow through brown adipose tissue may be of very great significance for the organism as a whole, since it facilitates the transfer of heat from the adipose tissue to the rest of the body.

Reed and Fain (1968a) showed that all of the effects of epinephrine in brown fat cells could be mimicked by the application of the dibutyryl derivative of cyclic AMP. The stimulation of respiration but not lipolysis was later found to require potassium (Reed and Fain, 1968b). On the basis of these and other studies, it was suggested that the calorigenic effect might be secondary to the uncoupling of oxidative phosphorylation by free fatty acids. Apparently the mitochondria of brown fat cells are more sensitive to this type of action than are mitochondria found elsewhere.

Another factor which might contribute to the calorigenic response is hyperlacticacidemia secondary to muscle glycogenolysis, especially since it is known that the injection of lactic acid does lead to an increase in oxygen consumption (Svedmyr, 1966). However, there are several observations which suggest that increased lactate metabolism may play only a minor role in the calorigenic response. Both glucagon and norepinephrine, for example, are as effective as epinephrine in stimulating oxygen consumption, at least in some species (Davidson *et al.*, 1960; Steinberg, 1966), and yet these agents have only negligible effects on blood lactate levels.

Other factors which might be important would include the increased blood levels of glucose and FFA which occur in response to sympathetic stimulation. Although neither parameter correlates well with calorigenesis under all circumstances (Steinberg, 1966; Kennedy and Ellis, 1969), they almost surely contribute to the overall effect.

Finally, in this section, it should be noted that the formation and metabolism of cyclic AMP is itself a calorigenic process, quite apart from the various processes which are stimulated by increased cellular levels of the nucleotide. The net result of the adenyl cyclase and phosphodiesterase reactions in the presence of adenylate kinase is that for every mole of ATP converted to cyclic AMP 2 moles of ADP are produced. Although normally only a small fraction of the available ATP is metabolized by this route, it may be increased to substantial proportions as a result of sympathoadrenal discharge. It is possible that this accounts for a larger portion of the calorigenic response than is presently realized.

V. SOME OTHER METABOLIC EFFECTS

A. Cholesterol and Fatty Acid Metabolism

Cyclic AMP inhibits the conversion of acetate to cholesterol in broken cell preparations from liver (Berthet, 1960), which may account for the tendency of epinephrine and glucagon to reduce blood cholesterol levels in acute experiments (Comsa, 1950; Davidson *et al.*, 1960).

Prolonged administration of epinephrine (e.g., subcutaneous administration in oil) tends to cause an *increase* in the plasma concentration of cholesterol (Shafrir *et al.*, 1959; Steinberg, 1966). The mechanism of this effect is poorly understood, and the role of cyclic AMP in it has not been studied.

Haugaard and Stadie (1953) showed that epinephrine and glucagon could depress the incorporation of acetate into fatty acids in liver slices, and this can be mimicked by cyclic AMP in broken cell preparations (Berthet, 1960). Some of the other hepatic effects of epinephrine, such as ketogenesis, are discussed in the next chapter in connection with the similar effects of glucagon.

The effect of epinephrine on hepatic bile secretion is biphasic, and Kasture *et al.* (1966) presented evidence to suggest that the initial increase is mediated by β-receptors while the more prolonged fall is an *alpha* effect. Homsher and Cotzias (1965) observed only a decrease in bile flow following the intravenous injection of cyclic AMP in rats, while Levine and Lewis (1967) could see no change in the isolated perfused liver. Papavasiliou *et al.* (1968) observed a rather striking fall in the rate of biliary excretion of manganese in response to epinephrine, glucagon, or exogenous cyclic AMP.

B. Inorganic Ions

A variety of changes in the blood and tissue levels of several inorganic ions have been noted in response to the catecholamines. In most cases the relation of cyclic AMP to these changes has not been investigated, while in some cases the change may be the result of some other effect of cyclic AMP. For these reasons the various effects of catecholamines on inorganic ions will be mentioned only briefly in this section.

The effects of epinephrine and cyclic AMP on potassium release from the liver were mentioned in Section III,A. Following the initial increase in the plasma potassium level, the catecholamines and glucagon also cause a relatively prolonged fall in potassium (and phosphate) levels. These changes may be secondary to hyperglycemia, since the same response is obtained by the infusion of glucose (Cahill *et al.*, 1957b). Most of the uptake of these ions is thought to occur in muscle, and may be secondary to the accumulation of hexose monophosphates (Cori and Welch, 1941).

The catecholamines cause a fall in the level of cyclic AMP in the toad bladder (Turtle and Kipnis, 1967b) and thus inhibit the increased permeability to water and sodium produced by vasopressin (Handler *et al.*, 1968b). This effect is mediated by adrenergic α-receptors. A similar effect may occur in the rabbit ileum (Field and McColl, 1968), except that in that case the principal effect of α-agonists may be to inhibit permeability to anions (Field *et al.*, 1968). Both α- and β-receptors may occur in the epithelial cells of the frog skin (Jard *et al.*, 1968; Watlington, 1968a,b), so that the effect of the catecholamines will depend upon which type of receptor predominates. This may depend on the species and also on the time of the year (see also Handler *et al.*, 1968b). The role of cyclic AMP in the regulation of membrane permeability is discussed in more detail in Chapter 10, in connection with the actions of vasopressin.

C. Phospholipid Turnover

A stimulatory effect of epinephrine on the incorporation of phosphate into phospholipids has been observed in liver (DeTorrontegui and Berthet, 1966), heart (Gaut and Huggins, 1966), salivary gland (Sandhu and Hokin, 1967), and brain tissue (Hokin, 1969). This effect is of special interest because there is now some evidence to suggest that it is not related to cyclic AMP. Although this point was not tested in the heart or brain studies, DeTorrontegui and Berthet could not mimic the effect with cyclic AMP in liver slices, and Sandhu and Hokin found the dibutyryl derivative to be ineffective in parotid gland slices.

DeTorrontegui and Berthet also studied the effect of glucagon in their experiments, and found that it did not share the ability of epinephrine to stimulate the incorporation of phosphate into phospholipids in rat liver slices. This was taken as further evidence that the effect was not mediated by cyclic AMP. It was also found that the effect of epinephrine could be blocked by ergotamine. Several additional observations, which are to some extent puzzling, were also made by De Torrontegui and Berthet. Although the addition of epinephrine clearly stimulated ^{32}P incorporation into all of the phospholipids which were separated, including phosphatidylcholine, the incorporation of labeled choline was not stimulated. In fact, both glucagon and epinephrine caused a small but significant inhibition of the incorporation of choline. Epinephrine did stimulate the rate of incorporation of ^{14}C-inositol into phospholipid, however, and this effect was not shared by either glucagon or high concentrations of cyclic AMP.

DeTorrontegui and Berthet suggested that the phospholipid effect might be an event preceding and possibly related to the stimulation of adenyl cyclase. This would be in line with the blocking effect of ergotamine, assuming rat liver adenyl cyclase is similar in this respect to the enzyme from dog liver (Murad *et al.*, 1962). This suggestion would also be compatible with the observations of Sandhu and Hokin (1967), and should probably be explored in more detail. It is also possible, of course, that the effects of catecholamines on phospholipid metabolism are mediated by α-receptors, in which event they would not be expected to be shared by either glucagon or exogenous cyclic AMP.

At the same time, the possibility that the phospholipid effect might have been caused by cyclic AMP should not be lost sight of. Several years ago a metabolic effect of epinephrine not mediated by cyclic AMP would have come as no surprise. In the light of present knowledge, however, such an effect does seem anomalous, and several other possibilities should be considered. One is that the phospholipid effect occurs in cells which are not permeable to cyclic AMP. In liver homogenates, for example, about 25% of the cells are derived from reticuloendothelial tissue. It has been recognized that hepatic parenchymal and enthothelial cells may be differentially sensitive to glucagon and the catecholamines and that they may likewise be differentially permeable to exogenous cyclic AMP (Sutherland and Robison, 1966; DeTorrontegui and Berthet, 1966), although neither possibility has been carefully investigated. The dibutyryl derivative of cyclic AMP *does* stimulate the incorporation of phosphate into phospholipid in thyroid gland slices, at least in some species (Pastan and Macchia, 1967), in which regard it mimics the effect of TSH.

A confusing observation in the thyroid has been that epinephrine *inhibits* phosphate incorporation into phospholipids in this tissue (Altman *et al.*, 1966). However, it is possible that this effect was due to oxidation products of epinephrine, especially since adrenochrome is known to be formed during incubation of epinephrine with thyroid tissue (Pastan *et al.*, 1962). This could also be considered as a possible explanation of the effects on phospholipid metabolism in the liver and other tissues. Adreno-chrome is known to have a number of metabolic effects, including un-coupling of oxidative phosphorylation (Park *et al.*, 1956), stimulation of glucose oxidation (Pastan *et al.*, 1962), and inhibition of actomyosin ATPase (Inchiosa, 1967), and in all of these instances it is much more potent than epinephrine. The conditions used in the experiments of DeTorrontegui and Berthet (incubation for 30 min in a bicarbonate buffer) would have been highly conducive to the formation of adrenochrome and other oxidation products.

In summary, neither the mechanism nor the significance of the effects of the catecholamines on phospholipid turnover has been established. The relation of cyclic AMP to these effects, if any, is obscure.

D. Glucose Oxidation

All possible effects of epinephrine on the rate of glucose oxidation have been reported, varying with the nature of the tissue, the species, and the experimental conditions. There are so many steps which might be affected between the initial uptake of glucose and its ultimate conversion to CO_2 and water that the mechanism of the effect is in many cases not clear.

The stimulatory effect of epinephrine in thyroid slices was mentioned in the preceding section. This effect was studied in some detail by Pastan *et al.* (1962), who concluded that the effect was in fact caused by adreno-chrome. Adrenochrome was found to be capable of causing a rapid in-crease in the intracellular level of NADP, which was judged to be rate-limiting in the conversion of glucose to CO_2 in this tissue. Epinephrine was active only after a lag period, during which time it was partially con-verted to adrenochrome. Adrenochrome acted immediately and was effective in much lower concentrations than epinephrine.

In contractile tissue the most commonly observed effect of epinephrine is an increase in the rate of glucose oxidation. This may primarily be a consequence of the increased contractile force brought about by epinephrine, but might in some cases be the result of increased glucose uptake secondary to some other action. In the perfused rat heart, for example, epinephrine stimulated glucose uptake but this effect was not blocked by doses of propranolol which did block the other effects of epinephrine (Satchell *et al.*, 1968). The inhibitory effect of epinephrine on glucose uptake was discussed in Section III,B,1, and in those cases it might be expected that the rate of glucose oxidation would be decreased.

In the liver, both epinephrine and glucagon inhibit the oxidation of glucose. This was demonstrated by Joseph R. Williamson and his colleagues (1966), among others, and could be secondary to any of several of the hepatic effects of cyclic AMP. However, the mechanism of the effect has not been carefully studied.

In adipose tissue, epinephrine and other lipolytic hormones tend to increase the rate of glucose oxidation (see Blecher *et al.*, 1969, and references cited therein). This may be the end result of a rather complicated mixture of effects, and may not occur at all in some species (Galton and Bray, 1967). One important factor is probably the increased concentrations of FFA which these hormones produce, since their effects on glucose metabolism can be duplicated by the addition of exogenous FFA to the medium (Himms-Hagen, 1967). When these hormones are added to adipose tissue in the absence of albumin, which functions as an external FFA acceptor, their effect may be to inhibit glucose oxidation. As discussed by Himms-Hagen, this is most likely the result of excessive accumulation of FFA within the tissue. The free fatty acids may themselves be toxic or may be converted to other more toxic compounds, such as lipid peroxides.

Another component of the effect of these hormones on glucose metabolism may be mediated by cyclic AMP but independently of its stimulatory effect on lipolysis. Blecher (1967) found that concentrations of the dibutyryl derivative of cyclic AMP too low to stimulate the release of FFA from isolated fat cells did stimulate glucose oxidation. This could have been the result of increased glucose uptake secondary to the stimulation of phosphofructokinase (Denton and Randle, 1966). Higher concentrations of the derivative (in the absence of albumin in the medium) inhibited glucose oxidation. This was probably secondary in large part to the accumulation of intracellular FFA, although other effects might have played a part. For example, Ho and Jeanrenaud (1967) obtained suggestive evidence for an inhibitory effect on the pentose shunt pathway, and Bray (1967) observed an inhibitory effect on 6-phosphogluconate dehydrogenase in a cell-free system.

Still another component of the effect of epinephrine on glucose oxidation may be unrelated to either lipolysis or cyclic AMP. Bray and Goodman (1968) found that a variety of procedures known to prevent the formation of cyclic AMP in adipocytes in response to epinephrine (see Chapter 8) did not prevent the stimulation of glucose utilization. Goodridge and Ball (1965) found that while both epinephrine and glucagon were effective lipolytic agents in pigeon adipose tissue, only epinephrine caused an increase in the rate of oxygen consumption. These observations suggest that a substantial part of the effect of epinephrine on glucose oxidation in fat cells, as in thyroid tissue (Pastan *et al.*, 1962), may be attributable to oxidation products of epinephrine. Ho *et al.* (1965) found that adrenochrome could antagonize the lipolytic effect of epinephrine to some extent, but did not measure its effect on glucose utilization.

The results of recent experiments by Blecher *et al.* (1969) are difficult to interpret, partly because lipolysis and glucose oxidation were measured under such different conditions and partly because of the unusually high concentrations of drugs and hormones used. On the basis of the observations which have been made so far, it seems unlikely that the catecholamines have any physiologically important action in adipose tissue other than to stimulate adenyl cyclase, although such an action or actions may eventually be established.

In summary, the catecholamines have various effects in various tissues on the uptake and utilization of glucose. The mechanism of these effects and their physiological significance are obscure in most cases.

E.   Effects on Salivary Glands

The catecholamines have several interesting effects on salivary gland function. Although sympathetic stimulation of the parotid gland does not markedly alter the volume of secretion, it greatly augments the level of amylase in the saliva. This effect appears to be mediated by adrenergic β-receptors because isoproterenol is a more potent agonist than either of the naturally occurring catecholamines and their effects can be blocked by β- but not by α-adrenergic blocking agents (Pohto, 1968; Katz and Mandel, 1968; Yamamoto *et al.*, 1968). The relatively high potency of norepinephrine suggests that these receptors are similar to the β-receptors in cardiac and adipose tissue.

In line with this, Schramm and his colleagues (Bdolah and Schramm, 1965; Babad *et al.*, 1967) have presented evidence that the effect on amylase secretion is mediated by cyclic AMP. The addition of cyclic AMP itself to parotid gland slices was not effective, but the addition of either the monobutyryl or dibutyryl derivative of cyclic AMP was found to produce a very striking increase in the rate of amylase secretion, mimicking the effects of the catecholamines. Theophylline and caffeine were also found to be effective in this system, and potentiation of the effect of epinephrine by theophylline was observed. More recently it has been shown that catecholamines stimulate adenyl cyclase activity in broken cell preparations from the parotid (Malamud, 1969; Schramm and Naim, 1970) and increase the level of cyclic AMP in these glands *in vitro*. The latter effect is illustrated in Fig. 6.8, which also shows the potentiating effect of theophylline.

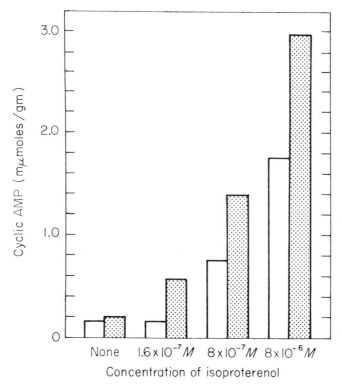

Fig. 6.8. Effect of increasing concentrations of isoproterenol on the level of cyclic AMP in quartered rat parotid glands incubated in the presence (shaded bars) or absence (open bars) of 1 mM theophylline.

Babad *et al.* (1967) found that approximately 50% of the amylase within the cells of control slices was contained within zymogen granules, and most of the enzyme which was released into the medium in response to epinephrine could be accounted for by loss from this fraction. This is now known to involve exocytosis (Amsterdam *et al.*, 1969), or what has also been termed "emiocytosis" or "reverse pinocytosis," but how this process is stimulated by cyclic AMP is completely unknown. Babad and his colleagues showed that it was dependent on oxidative metabolism, since the response to epinephrine was abolished either by replacing oxygen in the medium with nitrogen or by the addition of DNP. Of several exogenous substrates tested, β-hydroxybutyrate was found to be the most effective. The release of amylase in response to epinephrine was not dependent on this substrate, but was better maintained if it was present. The response probably does depend on Ca^{2+}, as is known to be the case in several other tissues (see Chapter 5), although this may be difficult to demonstrate in the parotid because of the unusually high levels of calcium which are normally present (Dreisbach, 1967).

Another effect which has been observed in salivary glands is increased cell proliferation, which is especially striking in response to isoproterenol. Selye *et al.* (1961) administered isoproterenol to rats twice daily for 17 days, and observed a six-fold increase in the weight of the salivary glands. It was pointed out that a similar selective parenchymatous growth of the salivary glands may occur as a result of malnutrition. Barka (1965) showed that the injection of a single large dose of isoproterenol in rats could produce, 24 hr later, a tenfold increase in DNA synthesis in the parotid and submaxillary glands. This increase in DNA synthesis was followed a few hours later by mitotic activity, suggesting that isoproterenol stimulates cell proliferation in salivary glands in a manner comparable to liver regeneration after partial hepatectomy. Baserga (1966) confirmed these findings in mice, and in addition showed that the effect of iso-proterenol could be prevented by pretreatment with a variety of agents, including actinomycin D and puromycin. These findings suggest that the effect on DNA synthesis is preceded by a complex sequence of events including the synthesis of new protein. Other effects which have been measured include a decrease followed by an increase in the level of glycogen (Malamud and Baserga, 1968a), a transient fall in uridylate kinase activity (Malamud and Baserga, 1968b), and a transient increase in peptide hydrolyase activity coinciding with the increased protein syn-thesis (Barka and Van der Noen, 1969; Sasaki *et al.*, 1969).

Evidence that these effects are mediated by cyclic AMP is now be-ginning to accumulate. The receptors involved appear to be β-receptors because isoproterenol is much more effective as an agonist than either of the naturally occurring catecholamines and its effect can be completely prevented by β- but not by α-adrenergic blocking agents (Fukuda, 1968; Schneyer, 1969). The administration of theophylline to rats produced a significant increase in salivary gland weight and potentiated such siala-denotrophic stimuli as tooth extraction and bulk feeding (Wells, 1967). Potentiation of isoproterenol-stimulated DNA synthesis in the parotid gland was reported by Malamud (1969). An effect of the dibutyryl derivative of cyclic AMP has not been reported at the time of this writing, but an apparently similar effect on DNA synthesis in thrombocytes has been reported by MacManus and Whitfield (1969).

The almost complete ineffectiveness of exogenous norepinephrine in stimulating DNA synthesis might argue that the receptors mediating this response are different from those involved in the cyclic AMP-mediated stimulation of amylase secretion, but this is not necessarily the case. Norepinephrine is particularly ill-suited for producing chronic effects because it is so rapidly taken up by neuronal tissue. Isoproterenol, on the other hand, might be capable not only of increasing the concentration of cyclic AMP in the salivary glands to higher levels, but of maintaining these levels for long enough periods of time to initiate whatever change is ultimately responsible for the enhanced rate of DNA synthesis. The physiological analog of this would of course be prolonged stimulation of the sympathetic nerves.

On the other hand, the possibility that different β-receptors are involved in these two responses should not be disregarded. Malamud and Baserga (1967) have made the interesting observation that after a single injection of labeled isoproterenol in mice, 26% of the radioactivity within salivary glands appeared to be localized within the nuclei (versus only 2 to 3% in other tissues). Whether this unusual localization is significant or simply fortuitous remains to be seen.

F. The Pineal Gland

Information about the pineal gland and its function was reviewed recently by Wurtman, Axelrod, and Kelly (1968). Sympathetic stimulation or the application of exogenous norepinephrine and other catecholamines leads to an increase in the pineal content of melatonin. This results, at least in part, from the stimulation of the rate of synthesis of melatonin from serotonin and of serotonin from tryptophan. Since adenyl cyclase in the pineal gland can be stimulated by catecholamines (Weiss and Costa, 1967, 1968a) and since the effects of the catecholamines on the synthesis of serotonin and melatonin can be mimicked by the dibutyryl derivative of cyclic AMP (Shein and Wurtman, 1969), it would appear that these effects are mediated by cyclic AMP.

Isoproterenol was not markedly more potent than either epinephrine or norepinephrine in stimulating pineal adenyl cyclase in broken cell preparations (Weiss and Costa, 1968a). However, since the stimulation could be prevented by β- but not by α-adrenergic blocking agents, the receptors involved appear to be reasonably typical β-receptors.

Weiss and Costa (1967) took advantage of the fact that the pineal gland is innervated exclusively by sympathetic fibers to study the phenomenon of denervation supersensitivity. The finding that the adenyl cyclase response to norepinephrine was greater in broken cell preparations from denervated glands than it was in control preparations was initially interpreted as evidence that adenyl cyclase in denervated cells was more sensitive to stimulation by norepinephrine. However, it was later found (Weiss, 1969b) that the response to fluoride was also greater in denervated tissue, suggesting that there was more adenyl cyclase present as a result of denervation. An implication of this finding was that norepinephrine (or perhaps some other factor) released from nerve endings ordinarily acts to inhibit the rate of synthesis of adenyl cyclase. This seems likely in view of the fact that several weeks were required before increased adenyl cyclase could be observed. It would appear that denervation supersensitivity, at least in this particular instance, involves an increase in the total number of receptors, rather than an increased sensitivity of each receptor already present.

Environmental lighting also affects the formation of melatonin within the pineal, and it is thought that the effects of light are mediated by the sympathetic nervous system (Wurtman et al., 1968). In the rat, and probably in all species that tend to be most active at night, the level of melatonin and the activity of hydroxyindole-O-methyltransferase (HIOMT), the enzyme which catalyzes the conversion of serotonin to melatonin, rise during periods of darkness and fall during periods of light. This diurnal rhythm can be abolished by bilateral superior cervical ganglionectomy. Apparently sympathetic stimulation in the rat is reduced by light, since exposure to constant light reduces HIOMT activity to very low levels. In line with this, Weiss (1969b) found that exposing rats to continuous light for several days resembled denervation in that it caused an apparent increase in the amount of pineal adenyl cyclase, or at least an increase in the magnitude of the response to either fluoride or norepinephrine.

An effect of cyclic AMP on HIOMT activity has not been demonstrated, although it seems reasonable to assume that an increase in the activity of this enzyme was responsible for the observed increase in the rate of conversion of serotonin to melatonin (Shein and Wurtman, 1969).* It is conceivable that HIOMT exists in two forms which are interconvertible by a mechanism similar to that which operates to control phosphorylase activity (see Chapter 5), although it is also possible that the synthesis of new enzyme protein is involved. In support of the latter possibility,

---

*See, however, Section IX.

Wurtman *et al.* (1968) have pointed out that inhibitors of protein synthesis do inhibit the increase in HIOMT activity which normally occurs in response to continuous darkness.

Shein and Wurtman (1969) reported that one effect of norepinephrine in the pineal which was not mimicked by exogenous cyclic AMP was an increase in the rate of uptake of tryptophan. A possibility which should probably be considered is that this apparent effect of norepinephrine was in fact produced by oxidation products, and therefore may not be physiologically important.

G. Renin Production

Cyclic AMP may play an important role in the regulation of renin release. Direct stimulation of the renal nerves and infusion of catecholamines are both effective stimuli to renin production *in vivo* (Gordon *et al.,* 1967), and washed particulate prepara'ions of rat kidney do contain a catecholamine-sensitive adenyl cyclase (E. J. Landon and G. A. Robison, *unpublished observations*), which could be derived from renin-producing cells. Evidence that the effects of the catecholamines on renin release were not entirely (if at all) the result of vascular changes was provided by Michelakis *et al.* (1969), who showed that the catecholamines also stimulated net renin production *in vitro,* and that this effect could be mimicked by exogenous cyclic AMP. Further studies to define the role of cyclic AMP in regulating the production and release of renin are in progress.

VI. SOME FUNCTIONAL EFFECTS OF ADRENERGIC STIMULATION

A. Introductory Remarks

The classification of the effects of the catecholamines into those which are metabolic and those which are something else (often referred to as "functional" or "mechanical" effects) is recognized as highly artificial. Metabolic effects, such as those discussed in the preceding sections of this chapter, can be measured chemically. This is most often done with the aid of an instrument designed for the purpose, so that the final observation is generally of the position of a pointer on a dial. The other effects differ in that the chemical events responsible for them are not well understood, and consequently are measured by changes in the physical properties of the tissues themselves.

In the following paragraphs we will briefly discuss those functional effects of the catecholamines in which the participation of cyclic AMP has been implicated. We can hope that the recognition that cyclic AMP is involved in some of them will lead eventually to an improved understanding of the chemical nature of the responses themselves.

B.   Effects in Muscle

1.   Skeletal Muscle

Some of the metabolic effects which occur in skeletal muscle in response to the catecholamines were discussed in Section III,B,1, where it was pointed out that the distribution of adenyl cyclase seems to differ in muscles with different functional properties. This is interesting because the catecholamines also produce different effects in these muscles, even though the overall increase in the intracellular level of cyclic AMP in response to a large dose of epinephrine is similar in both cases. These effects were reviewed recently by Bowman and Nott (1969), and will be summarized very briefly here. In the tibialis, a fast-contracting muscle, epinephrine causes an increase in twitch tension, associated with a prolongation of the active state, and a decrease in the rate of relaxation. Incomplete tetanic contractions are increased both in tension and degree of fusion. In the soleus and other slow-contracting muscles, by contrast, epinephrine causes a reduction in twitch tension and an increase in the rate of relaxation. Incomplete tetanic contractions are in this case reduced by epinephrine both in tension and in the degree of fusion. The soleus resembles cardiac muscle in its response to epinephrine in that the duration of the active state is decreased in both cases. A major difference is that twitch tension is increased by epinephrine in the heart, whereas this effect does not occur in slow skeletal muscle.

It should be emphasized that these different effects of the catecholamines on fast and slow muscles are direct effects on the muscles themselves, and are not secondary to changes in blood flow or neuromuscular transmission. In line with the supposition that these effects are mediated by cyclic AMP, Bowman and Raper and their colleagues (Bowman and Raper, 1967; Raper, 1967; Bowman and Nott, 1969) have established that they are initiated by an interaction with adrenergic β-receptors. The receptors in slow-contracting fibers are more sensitive to epinephrine than are those in fast-contracting muscles, such that the ability of epinephrine to affect the latter may not be physiologically important. Although the possible participation of cyclic AMP in these responses has yet to be investigated in any kind of detail, the observation that adenyl cyclase appears to be distributed differently in muscles which respond differently to catecholamines remains an interesting starting point for future studies.

The well-known defatiguing action of epinephrine may be partly related to phosphorylase activation, as discussed previously in Section III,B,1. This effect is almost certainly mediated by cyclic AMP. Another component of this action may be related to the facilitation of neuromuscular transmission. This may also be mediated by cyclic AMP, as discussed in the next section.

2. Neuromuscular Transmission

The catecholamines also affect the function of skeletal muscle by facilitating neuromuscular transmission, in addition to their direct effects on the muscles themselves. This effect of the catecholamines, which is evidenced by their anticurare effect and also by their ability to augment the repetitive firing and twitch potentiation produced by neostigmine and other drugs, is presumably due to their ability to increase the release of acetylcholine from nerve endings. Breckenridge et al. (1967) reported that this effect of epinephrine could be enhanced by theophylline and inhibited by propranolol, leading them to suggest that this might be a β-receptor effect mediated by cyclic AMP. This was supported by Goldberg and Singer (1969), who showed that the effect could be mimicked by the dibutyryl derivative of cyclic AMP. Goldberg and Singer also showed that the derivative (like epinephrine and theophylline) increased the frequency but not the magnitude of the miniature endplate potentials, supporting an earlier conclusion that the effect of epinephrine was exerted on nerve rather than muscle.

The nature of the adrenergic receptors involved in this response is controversial, because Bowman and Raper (1967) found that norepinephrine was a more potent agonist than isoproterenol, and that in vivo the response could be blocked by α-adrenergic blocking agents but not by β-blocking agents. The evidence that the response is mediated by cyclic AMP suggests that these receptors may resemble those in the livers of some species, where the receptors mediate the stimulation of adenyl cyclase but have many of the characteristics classically associated with α-receptors. Studies of the formation of cyclic AMP in nerve endings in response to catecholamines have not been reported.

3. Cardiac Muscle

Among the more striking effects of the catecholamines are the positive chronotropic and positive inotropic effects which occur in the hearts of all mammalian species studied. The chronotropic response has not been

widely studied from the biochemical point of view, at least in part because of the small number of cells involved in it relative to the total
mass of the heart, but a substantial amount of information has accumulated
regarding the inotropic effect. The evidence that this effect is mediated
by cyclic AMP will be summarized in this section.

A relationship between the inotropic effect and the activation of phosphorylase was noted at a fairly early stage (Kukovetz et al., 1959;
Haugaard and Hess, 1965). After comparing the relative potencies of a
series of catecholamines on both parameters, Kukovetz and his colleagues concluded that "the positive inotropic action of epinephrine and
related compounds is a process closely associated with activation of
phosphorylase." The possibility was even suggested that the relationship
was a direct one, the inotropic effect following as a result of the increased
activity of the enzyme. This was later shown to be improbable because
the two processes could be dissociated to at least some extent under
several sets of conditions. Mayer et al. (1963) and Drummond et al.
(1964c) found that with small doses of epinephrine substantial increases
in the force of contraction could be obtained without any measurable
change in the activity of phosphorylase. Other investigators (Øye, 1965;
Robison et al., 1965; Cheung and Williamson, 1965) found that when the
two responses were studied as functions of the time after epinephrine
administration, the inotropic response invariably preceded the phosphorylase response (see also Fig. 6.6). It thus seemed unlikely that
changes in phosphorylase activity were responsible for the observed
increases in the force of contraction, and the converse possibility, that
phosphorylase activation might be secondary to the increased contractile
activity, has also been ruled out (see Section III,B,2). However, in view
of the numerous correlations which have been noted between the two responses, and which led to the conclusion that both responses were mediated by adrenergic β-receptors in the first place, it seems reasonable to
suppose that a common intermediate might be involved. An obvious
candidate, because of its known role in the activation of phosphorylase,
would have to be cyclic AMP. As discussed in Section III,B,2, the activation of phosphorylase kinase seems to be closely related to the positive
inotropic response.

A new consideration was introduced into this area by Govier's finding
(Govier, 1968) that a small but measurable component of the inotropic
response to epinephrine and norepinephrine is due to α-receptor activation. This may explain at least part of the dissocation which has been observed between the inotropic response and the activation of phosphorylase. In the isolated perfused guinea pig heart, for example, Kukovetz and
Pöch (1967a) were unable to dissociate these two responses by the use of

the β-agonist isoproterenol, but were able to do so with low concentrations of norepinephrine, which is much more active than isoproterenol as an α-agonist. Another example of α- and β-receptors mediating superficially similar responses occurs in intestinal smooth muscle, where the response is relaxation (Section VI,B,4). Such observations pose an obvious challenge for the theory that β-receptor effects are mediated by an increase in the level of cyclic AMP while α-receptor effects are mediated by a decrease in the level of cyclic AMP. One possibility which could be considered is that in some tissues the cyclic AMP is located in two separate pools, or compartments, and that a decrease in the size of one pool may lead to an effect which superficially resembles the effect of increasing the size of the other pool. Evidence that cyclic AMP is compartmentalized in epithelial cells in the toad bladder has been provided by several groups of investigators (Petersen and Edelman, 1964; Argy et al., 1967).

The hypothesis that cyclic AMP is at least involved in the β-receptor component of the inotropic response to the catecholamines is supported by several lines of evidence, in addition to the studies dealing with phosphorylase activation. Thus, the order of potency of a series of catecholamines in stimulating dog heart adenyl cyclase was the same as their order of potency as inotropic agents in vivo (Murad et al., 1962). The stimulation of adenyl cyclase in vitro, like the inotropic response in the intact heart, can be competitively antagonized by β-adrenergic blocking agents (Murad et al., 1962; Robison et al., 1967a; Levey and Epstein, 1969a). The order of potency of the blocking agents, to the extent that they have been compared, seems to be the same both in preventing the formation of cyclic AMP and in opposing the inotropic response (Robison et al., 1965, 1967a; Drummond et al., 1966). Glucagon, which stimulates cardiac adenyl cyclase (Murad and Vaughan, 1969; Levey and Epstein, 1969a) and increases the level of cyclic AMP in the intact heart (Mayer et al., 1970), produces metabolic and functional changes which are the same or similar to those produced by the catecholamines (Farah and Tuttle, 1960; Cornblath et al., 1963; Kreisberg and Williamson, 1964; Williams and Mayer, 1966; Glick et al., 1968; Entman et al., 1969b), and these changes are unaffected by doses of β-adrenergic blocking agents which prevent the effects of catecholamines. Conversely, the choline esters, which inhibit the accumulation of cyclic AMP in vitro (Murad et al., 1962), also inhibit the positive inotropic response (Meester and Hardman, 1967; Levy and Zieske, 1969) and phosphorylase activation (Vincent and Ellis, 1963) in the intact heart. It is interesting to note that just as the catecholamines can increase the force of contraction without affecting phosphorylase, so also is it possible for acetylcholine to interfere with glycogenolysis without preventing the inotropic response. These various correlations between *in vitro* and *in vivo* effects suggest that cyclic AMP plays an important role in regulating cardiac function. In particular, they support the hypothesis that the positive inotropic response to the catecholamines is mediated by cyclic AMP.

The rapidity with which the level of cyclic AMP rises in response to epinephrine in the intact heart has already been mentioned (see Fig. 6.6). When epinephrine is recirculated through the isolated perfused working rat heart, the level of cyclic AMP rises and then falls to an intermediate level which is dose-dependent (Robison *et al.*, 1967a). This is illustrated in Fig. 6.9. Attempts to demonstrate similar changes in the dog heart *in situ*, using a biopsy sampling technique, were initially unsuccessful because of the difficulties involved in rapidly freezing the sample (Namm and Mayer, 1968), but a successful effort has since been reported (Wollenberger *et al.*, 1969). It was established that doses of β-adrenergic block-

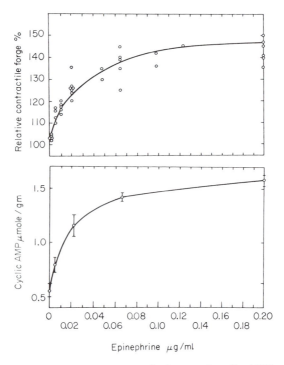

Fig. 6.9. Effect of epinephrine on contractile force and cyclic AMP concentration in isolated perfused working rat hearts. Cyclic AMP was measured in extracts of hearts which had been frozen 5 min after the addition of epinephrine. Points plotted in the upper panel represent peak systolic pressure developed in response to epinephrine (as percentage of control pressure before addition of epinephrine). From Robison *et al.* (1967a).

ing agents sufficient to prevent the inotropic response also prevented the rise in cyclic AMP (Robison *et al.*, 1965; Drummond *et al.*, 1966).

Some interesting results with the adrenergic blocking agent *N*-iso-propylmethoxamine (IMA) can be mentioned. This agent was found to be capable of preventing most of the metabolic responses to the catecholamines in the dog, in doses which had no effect on the inotropic response (Burns *et al.*, 1964). When IMA was recirculated through the isolated perfused working rat heart, at the highest concentration which could be used without producing severe cardiotoxicity, it was found to have no effect on the cyclic AMP response, the phosphorylase response, or the inotropic response to a relatively large dose of epinephrine (Robison *et al.*, 1965; Sutherland and Robison, 1966). This result was initially surprising (we had expected to see blockade of the cyclic AMP and phosphorylase responses but not the inotropic response) and led to a serious consideration of the idea that β-adrenergic responses in general might be mediated by cyclic AMP. Kukovetz and Pöch (1967a) obtained similar results in the isolated perfused guinea pig heart. However, by increasing the ratio of the concentration of IMA to that of the agonist, they were able to prevent both the inotropic response and the activation of phosphorylase. IMA thus appears to be classifiable as a β-adrenergic blocking agent, but as such to be less potent in some tissues (including the heart) than in others. We had arrived at a similar conclusion on the basis of our *in vitro* studies with adenyl cyclase (Robison *et al.*, 1967a; see also Fig. 6.4).

More recently, Shanfeld and Hess (1969) have reported that IMA is capable of producing a significant inhibition of the rise in cyclic AMP which occurs in the rat heart in response to a low dose of norepinephrine, without significantly reducing the inotropic response. Similarly, Mayer and Moran (Moran, 1966) noted a transient inhibition by IMA of the phosphorylase-activating effect of epinephrine in the dog heart *in situ,* with no apparent effect on the inotropic response. These observations can perhaps be understood as manifestations of the α-receptor component of the positive inotropic response (Govier, 1968), especially since the dissociation noted by Mayer and Moran could not be seen when isoproterenol was used as the agonist.

The methylxanthines, which are capable of inhibiting phosphodiesterase from heart muscle (Butcher and Sutherland, 1962), are also capable of enhancing the positive inotropic response to the catecholamines (Rall and West, 1963). Theophylline was found to be more potent than caffeine in both respects. Although these findings can be taken as support for the hypothesis that the β-receptor component of the inotropic response to the catecholamines is mediated by cyclic AMP, they are complicated by the fact that the methylxanthines by themselves can under some conditions produce an inotropic response. Although it is possible that this response is at least partly the result of phosphodiesterase inhibition, it now seems clear that it cannot be understood entirely on this basis. The positive inotropic response to the catecholamines, which resembles the positive inotropic response associated with an increase in heart rate (the positive inotropic effect of activation, or PIEA) (Koch-Weser and Blinks, 1963; Koch-Weser et al., 1964), is distinctive in that it is associated with an increase in the rate of rise of tension, a decrease in the time to peak tension, an apparent increase in the rate of relaxation, and a decrease in the overall duration of systole (Sonnenblick, 1962). In contrast to this, the response to the methylxanthines (at least under some conditions) is associated with an increase in the time to peak tension, a decrease in the rate of relaxation, and an overall prolongation of systole (Olson et al., 1967). Caffeine has been found to be capable of interfering with the uptake of calcium by elements of the sarcoplasmic reticulum (Weber, 1968; Fuchs, 1969), and this could perhaps account for the delayed onset and slower rate of relaxation. This action might also be the basis for the negative inotropic effect of the methylxanthines (an effect which is especially prominent when the level of extracellular calcium is high) (DeGubareff and Sleator, 1965) and also for the contracture which may occur at lower temperatures (Price and Nayler, 1967). The relative potencies of the methylxanthines in producing this effect on calcium metabolism are unknown, although such knowledge might be very helpful in future analyses of the inotropic response to these agents.

Although it seems clear that at least part of the inotropic response to the methylxanthines is not related to phosphodiesterase inhibition, it is possible that under certain conditions the inotropic response may be primarily the result of this action. This thought is based largely on the many similarities which have been noted between the responses to the catecholamines and methylxanthines, such as the similar ionic requirements (Reiter, 1964; Nayler, 1967) and the similar way in which the responses are affected by ouabain at low temperatures (Logan and Cotten, 1967). Kukovetz and Poch (1967c) noted that imidazole, which stimulates phosphodiesterase activity *in vitro,* inhibited the positive inotropic response to the methylxanthines, as well as to small doses of the catecholamines. Although the extent to which cyclic AMP participates in the cardiac effects of the methylxanthines and imidazole remains to be established, the observations which have been made to date are at least compatible with the view that cyclic AMP mediates the response to the catecholamines, however much they are complicated by actions not related to altered phosphodiesterase activity.

Positive inotropic and chronotropic responses were reported in unanesthetized dogs (Levine and Vogel, 1966) and human patients (Starcich *et al.,* 1967; Levine *et al.,* 1968) in response to the injection of cyclic AMP. Analysis of the human data led Levine and his colleagues to conclude that these were direct effects on the heart, rather than reflex effects, and this was substantiated in the isolated perfused guinea pig heart by Kukovetz (1968). In these latter experiments, high concentrations of the dibutyryl derivative were found to produce positive inotropic and chronotropic effects as well as phosphorylase activation. Our own attempts to mimic the cardiac effects of the catecholamines in the rat heart with cyclic AMP and several derivatives were unsuccessful (Robison *et al.,* 1965, 1967a), possibly because we used insufficiently high concentrations. Studies of the cardiovascular activity of these compounds will in any event be complicated by effects not related to the cyclophosphate moiety (see Chapter 5, Section II). Adenosine and its phosphorylated derivatives are potent vasodepressor substances and are generally toxic to the heart. James (1965, 1967) found that all of them, including cyclic AMP, produced a negative chronotropic effect when perfused through the sinus node in dogs, although cyclic AMP was less potent than the others. It seems possible, therefore, that the effect of exogenous cyclic AMP on the heart will always be the resultant of its effect as a second messenger for the catecholamines and of the opposite effect associated with the adenosine moiety. To the extent that this is the case, exogenous cyclic AMP might never duplicate the cardiac effects of the catecholamines exactly, although the effects of a mixture of isoproterenol *plus adenosine* might be mimicked quite closely. As for which component might predominate, this might depend very critically on the conditions of the experiment, and species differences may also be important.

The mechanism by which cyclic AMP might act to increase the force of myocardial contraction is unknown. An important feature of the positive inotropic response to the catecholamines is that it requires the participation of some labile factor or effect which is produced as the result of cardiac activity, and which is responsible for the well-known staircase phenomenon. Koch-Weser and Blinks (1963) have referred to this as the positive inotropic effect of activation (PIEA), and have suggested, on the basis of several lines of evidence, that the catecholamines may act by increasing the rate of formation of this factor (Koch-Weser *et al.,* 1964). The catecholamines do not increase the force of contraction if the interval between beats is sufficiently prolonged (i.e., if sufficient time is allowed for the accumulated PIEA to decay). In this regard the catecholamines differ from the cardiac glycosides, which stimulate the heart at all frequencies of contraction with no apparent effect on the PIEA. It can be noted that calcium also stimulates at all frequencies, and indeed, at any given frequency, the force of myocardial contraction is almost directly related to the concentration of extracellular calcium (Sonnenblick, 1962; Reiter, 1964). Contraction seems to represent another basic process in which calcium plays an essential role, with cyclic AMP being relegated to its accustomed role as a regulator of the process.

Some interesting results have recently been reported by Entman, Levey, and Epstein (1969b). They found that either epinephrine or glucagon increased the accumulation of Ca^{2+} by a microsomal fraction of dog myocardium in the presence of ATP and Mg^{2+}. This fraction had previously been shown (Entman et al., 1969a) to contain adenyl cyclase activity which could be stimulated by epinephrine or glucagon. The effects of epinephrine but not of glucagon could be blocked by propranolol, and the effect on Ca^{2+} accumulation could be mimicked by cyclic AMP. The response to cyclic AMP was measurably more rapid than to the hormones, as would be expected if the hormones were acting by way of cyclic AMP.

These in vitro results may be very important, and are of interest from several different points of view. The requirement for ATP and Mg^{2+} supports the view that many and perhaps all of the physiologically important actions of cyclic AMP are the result of the phosphorylation of a protein. Some such process as the phosphorylation and dephosphorylation of a protein, especially one important in regulating Ca^{2+} accumulation by the sarcoplasmic reticulum, could well account for the rise and fall of the PIEA. This process might depend on some critical concentration of cyclic AMP. We can recall here, as discussed in more detail in the previous chapter, that the protein kinase which Walsh et al. (1968) isolated from skeletal muscle was dependent on cyclic AMP, even though the reaction which this kinase catalyzed could occur to some extent in the absence of cyclic AMP. Failure of this hypothetical mechanism in the heart, either because of a fall in the level of cyclic AMP below its critical level or because of a lesion at some later step, could be an important factor in the etiology of some forms of heart failure.

Although it was reemphasized by Koch-Weser and Blinks (1963), the importance of the PIEA in cardiac function has not always been appreciated. Perhaps this is because changes in heart rate which are compatible with life do not have much of an effect on the force of each systolic contraction. However, the changes which can be produced experimentally in vitro are very dramatic indeed, and teach us (or should teach us) that the force of contraction as measured in situ is to a very large extent the result of the accumulated PIEA. The properties of mammalian cardiac muscle are such that if it were not for the PIEA, the force of contraction would be insufficient to allow the heart to function as a pump. Even slight failure of the mechanism for generating the PIEA might seriously compromise the heart's ability to carry out this function.

Evidence that the lesion responsible for cardiac failure may at times occur at the level of cyclic AMP formation was provided by Sobel and his colleagues (1969a). These investigators found that total adenyl cyclase activity (i.e., activity measured in the presence of fluoride) was less in homogenates of hearts from guinea pigs with congestive heart failure than it was in control homogenates. Heart failure was induced in these animals by chronic constriction of the ascending aorta, which procedure had no apparent effect on cardiac phosphodiesterase activity. It could be predicted from these experiments that the level of cyclic AMP in the failing hearts would be abnormally low, although this remains to be determined.

Just as abnormally low levels of cyclic AMP may be detrimental to cardiac function, abnormally high levels may also be damaging. Chronic administration of high doses of isoproterenol has long been known to lead to cardiomegaly (Stanton *et al.,* 1969), and cardiac necroses have been detected even in response to a single injection of isoproterenol (Dorigotti *et al.,* 1969). These effects can be prevented by β-adrenergic blocking agents. High levels of cyclic AMP in the heart may have effects on DNA synthesis similar to those which occur in salivary glands (Section V,E), and as demonstrated in response to cyclic AMP in thrombocytes (MacManus and Whitfield, 1969).

The cardiac glycosides may at times cause the release of stored catecholamines, which may contribute to their inotropic effect (Kukovetz and Poch, 1967b). However, these drugs also have a basic action of their own which is very different from that of the catecholamines (Koch-Weser and Blinks, 1963), and there is no evidence that cyclic AMP is involved in it in any way.

One other positive inotropic agent we can mention is the fluoride ion, which is of special interest because of its ability to stimulate adenyl cyclase in broken cell preparations. It is very unlikely, however, that the inotropic effect of this agent can be related to a change in the level of cyclic AMP. As mentioned previously, an effect of fluoride on the formation of cyclic AMP in an intact cell has never been demonstrated. Also, the response to fluoride is very different from that to the catecholamines: it can be seen only at slow frequencies of contractions (Berman, 1966) and is associated not with an increase in the rate of tension development but rather with a prolongation of the active state (Reiter, 1965).

In summary, a great deal of evidence has accumulated during the past few years to suggest that the positive inotropic response to β-adrenergic stimulation is mediated by an increase in the intracellular level of cyclic

AMP. The sequence of events by which such a change could lead to an increased force of contraction is unknown, but recent experiments suggest that an altered ability of the sarcoplasmic reticulum to accumulate calcium may be involved. A normal level of cyclic AMP may be required for normal cardiac function.

4. Smooth Muscle

Smooth muscle usually contracts as the result of α-receptor activity and relaxes in response to β-adrenergic stimulation. The only well-established exception to this generalization occurs in intestinal smooth muscle, where α-receptors as well as β-receptors mediate relaxation (Ahlquist and Levy, 1959). Conclusions that vascular smooth muscle may at times contract in response to both types of receptor interaction have been shown to be incorrect (Abboud *et al.*, 1965; Somlyo and Somlyo, 1966). Intestinal smooth muscle seems to resemble cardiac muscle (Govier, 1968) in that in both tissues α- and β-receptors mediate responses which are at least superficially similar. These exceptions are obviously relevant to the hypothesis that the effects associated with α- and β-receptors are in fact secondary to decreased and increased levels of cyclic AMP, respectively. At the present time it can only be said that the reasons for these exceptions are not known. They do not readily fit in with *any* theory of adrenergic receptors, and clearly a great deal of additional research will be required before they are understood.

The hypothesis that smooth muscle relaxation in response to β-adrenergic agonists is mediated by an increase in the intracellular level of cyclic AMP is supported by several lines of evidence. Perhaps the most clearcut evidence has been obtained from studies with the isolated rat uterus (Dobbs and Robison, 1968). In this preparation the dibutyryl derivative of cyclic AMP produces a striking reduction in tone and motility, whereas the other adenine derivatives which have been tested, including adenosine, 5′-AMP, ADP, and ATP, are incapable of producing this effect, even at concentrations many times greater. The effect of the dibutyryl derivative of cyclic AMP, in the presence and absence of theophylline, was shown in Fig. 5.1. Theophylline by itself has very little effect on the rat uterus, but greatly enhances the effect of the dibutyryl derivative. Concentrations which are ineffective in the absence of theophylline may produce a striking effect in the presence of theophylline, and the relaxing effect is prolonged by theophylline in all cases. These observations suggest that in order to produce an effect in smooth muscle, the derivative must first be metabolized to cyclic AMP, which then becomes subject to attack by the phosphodiesterase. Since exogenous cyclic AMP itself is ineffective, these observations also suggest that the greater potency of the derivative, in the rat uterus at least, may be due principally to its greater ability to penetrate the cell membrane.

Studies with other smooth muscle preparations have generally been complicated by the fact that adenosine and adenine nucleotides other than cyclic AMP may have pronounced effects. In all cases studied, however, cyclic AMP or the dibutyryl derivative or both have been shown to cause relaxation, and the opposite effect has never been observed.* By contrast, the effects of the other nucleotides are quite variable. While they all tend to cause vascular smooth muscle to relax, at least in most vascular beds, their effects in other types of smooth muscle run the gamut from relaxation to no effect to contraction, and the effect of a given nucleotide on a given type of smooth muscle may differ strikingly from one species to another (L. Hurwitz et al., 1967, unpublished observations). The mechanism of action is unknown in all cases.

In most smooth muscle studies the dibutyryl derivative has been found to be a more potent relaxant than cyclic AMP itself, but one exception has been reported. In the rat ileum, cyclic AMP and the other adenine nucleotides were found to be roughly equipotent in producing relaxation, whereas the dibutyryl derivative was without effect even in concentrations as high as 10^{-2} M (Kim et al., 1968). It will be of interest to see if the smooth muscle in this segment of the intestine lacks the enzymes which are necessary for metabolizing the derivative to cyclic AMP.

Studies of the effects of isoproterenol and β-adrenergic blocking agents in the isolated rat uterus (Dobbs and Robison, 1968) have provided further support for the hypothesis that the relaxing effect of the catecholamines is mediated by cyclic AMP. Isoproterenol causes an increase in the intracellular level of cyclic AMP (Fig. 6.7), and this occurs at least as rapidly as the relaxing effect. The relative potencies of the other catecholamines in producing this effect are the same as their relative potencies as relaxing agents. Both the effects on cyclic AMP and on tone can be blocked by propranolol and other β-adrenergic blocking agents, and doses which are too small to prevent the effect on cyclic AMP likewise do not prevent relaxation. By contrast, even large doses of propranolol do not prevent relaxation in response to the dibutyryl derivative of cyclic AMP, as was noted also in the case of the guinea pig trachea (Moore et al., 1968).

Synergism between the catecholamines and theophylline is striking both in terms of cyclic AMP (Fig. 6.10) and at the functional level. Synergism at the functional level has also been observed with aminophylline (Levy and Wilkenfeld, 1968). Unlike theophylline itself, however, amino-

*See, however, Section IX.

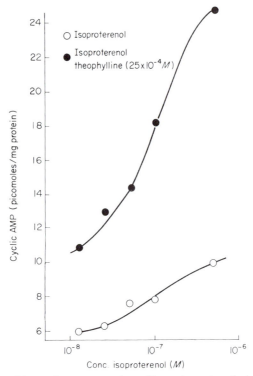

Fig. 6.10. Effect of increasing concentrations of isoproterenol on the level of cyclic AMP in isolated rat uteri in the presence and absence of theophylline ($2.5 \times 10^{-4}\,M$). Theophylline was added 5 min and isoproterenol 2 min before freezing. From Dobbs and Robison (1968).

phylline has considerable relaxing activity even in the absence of added catecholamines, and it is not entirely clear that this effect can be related to an effect on the level of cyclic AMP. The same may be true of nitroglycerine, which was also found to enhance the relaxing effect of norepinephrine.

Some of the interesting species differences which pertain to the uterus, especially as they relate to the effects of the ovarian hormones, were discussed previously (Robison *et al.,* 1967a). The biochemical bases for these differences are still unknown, but are currently being investigated.

Other smooth muscle preparations have been studied less extensively from the point of view of cyclic AMP formation, but most of the available information supports the view that relaxation in response to the catecholamines and also to theophylline is mediated by cyclic AMP. It was mentioned previously that theophylline by itself has very little effect on the tone and motility of the rat uterus, and it has a correspondingly slight effect on the intracellular level of cyclic AMP. By contrast, theophylline does cause the longitudinal smooth muscle from the guinea pig ileum to relax, and a dose of the drug which was judged to be just sufficient to produce a maximal effect was found to cause a doubling of the level of cyclic AMP in this preparation (L. Hurwitz *et al.*, 1967, unpublished observations). Wilkenfeld and Levy (1969) found that both theophylline and diazoxide enhanced the relaxing effect of isoproterenol in the rabbit ileum, while imidazole produced the opposite effect. These investigators also found that relaxation in response to phenylephrine, which acts primarily by way of α-receptors, was unaffected by these agents. This could perhaps be taken as evidence that this α-receptor response is not related to a fall in the level of cyclic AMP, although this possibility should probably be kept open pending further data. Bueding *et al.* (1966) showed that small doses of epinephrine, just sufficient to cause relaxation, produced a significant increase in the level of cyclic AMP in the guinea pig taenia coli.

It was suggested (Bartelstone and Nasmyth, 1965; Bartelstone *et al.*, 1967; Laborit and Weber, 1967) that vasoconstriction in response to α-receptor stimulation might be related to an increase rather than a decrease in the level of cyclic AMP in vascular smooth muscle. However, the indirect evidence upon which this suggestion was based was highly equivocal, and should probably be reinterpreted. For example, since it was known that both vasopressin and norepinephrine were capable of stimulating adenyl cyclase in some tissues, the finding that these hormones acted synergistically to produce vasoconstriction led Bartelstone and Nasmyth to suggest that this effect might be secondary to an increase in the level of cyclic AMP. This seems unlikely for several reasons. While it is understandable that the effects of two hormones which stimulate adenyl cyclase in the same cell could at times be additive, it is difficult to see how they could be more than additive. It has been found that oxytocin stimulates the rat uterus to contract without altering the level of cyclic AMP, and that this effect of oxytocin can be antagonized by cyclic AMP (Dobbs and Robison, 1968). It is at least conceivable that vasopressin causes vascular smooth muscle to contract by a similar mechanism, which, if norepinephrine acts by lowering the level of cyclic AMP, would explain the synergism between these agents.

Bartelstone *et al.* (1967) found that small concentrations of theophylline could under certain conditions increase the contraction of the isolated rat aortic strip in response to norepinephrine, a finding which again seemed to indicate that the response could be mediated by an increase in the level of cyclic AMP. Under most conditions, however, the methylxanthines antagonize the smooth muscle contracting effects of the catecholamines.

Sollman and Pilcher (1911) found that caffeine reduced or abolished the vasoconstrictor response to epinephrine in the isolated perfused dog kidney, while producing little or no vasodilatation by itself. They concluded that while many drugs may dilate the renal vessels directly, "caffeine acts only by preventing constriction." Kohn (1932) found that theophylline could actually convert the vasoconstrictor action of epinephrine in the dog hind limb to a vasodepressor effect. The antagonism by the methylxanthines of the vasopressor effects of epinephrine so impressed Fredericq and Melon (1922) that they considered them to be adrenergic blocking agents. These observations are not only compatible with the cyclic AMP hypothesis but can be explained by it.

The paradoxical effect noted by Bartelstone and his colleagues occurred with concentrations of theophylline below 10^{-6} M. These concentrations would be too low to inhibit phosphodiesterase activity in most tissues which have been studied, although the possibility cannot be dismissed that smooth muscle phosphodiesterase may be unusually sensitive to inhibition by the methylxanthines. For example, Ariens (1967) found that equally low concentrations of theophylline enhanced the relaxant effect of isoproterenol on the rabbit jejunum. It nevertheless seems likely that the effect noted by Bartelstone and his colleagues was the result of some action other than phosphodiesterase inhibition. Studies of the effects of vasoactive drugs and hormones on cyclic AMP levels in vascular smooth muscle are obviously needed, especially in view of their potential significance for studies of the etiology and treatment of hypertension.

Relaxation of tracheal smooth muscle in response to the dibutyryl derivative of cyclic AMP has been demonstrated (Moore *et al.,* 1968). In addition, some measurements of cyclic AMP in the lung have been made, and Fig. 6.11 illustrates the synergistic effect of epinephrine and caffeine. It was not established in these experiments that the increase in cyclic AMP occurred entirely or even primarily in smooth muscle, although it seems quite likely that this was the case. The use of the methylxanthines and β-adrenergic agonists to produce bronchodilatation in asthmatic patients is well known.

It seems appropriate at this point to mention the interesting theory proposed by Szentivanyi (1968), according to which bronchial asthma is regarded as the result of a defect at the level of the adrenergic β-receptor. This theory was based on a number of lines of evidence, including the similarity between the effects of β-adrenergic blockade and the injection of *Bordetella pertussis* vaccine. Both treatments sensitize certain strains of rodents to various stimuli, including anaphylactic shock and the exogenous administration of histamine and other autacoids. The

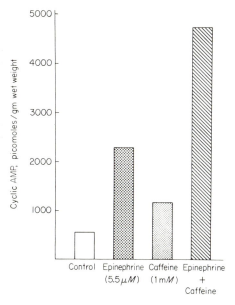

Fig. 6.11. Effect of epinephrine and caffeine on cyclic AMP levels in rat lung pieces. Lungs were removed from anesthetized rats, cut into sections, and incubated in Krebs-Ringer phosphate buffer at 30°C for 20 min. Tissue was transferred to fresh media containing the appropriate additions and incubated for an additional 20 min. Incubation was terminated by dropping the lung pieces into an operating Waring blendor containing 18 ml of 0.1 N HCl and a known quantity of tritiated cyclic AMP. Homogenates were purified and cyclic AMP assayed as described in the Appendix.

similarities between pertussis sensitization and β-adrenergic blockade do not extend to all tissues (Gulbenkian *et al.,* 1968), but it does seem possible that certain allergic phenomena might be associated with the production of a factor which could inhibit adenyl cyclase in certain cells. Since adenyl cyclase in so many cells is normally stimulated by β-adrenergic agonists, it is easy to see how the effects of this hypothetical factor could resemble the effects of β-adrenergic blockade. Reduced adenyl cyclase activity could account not only for supersensitivity to the effects of histamine but for a greater release of histamine as well. Lichtenstein and Margolis (1968) found that isoproterenol and theophylline acted synergistically to inhibit antigen-induced histamine release from human leukocytes, which contain adenyl cyclase activity which can be stimulated by catecholamines (see Section VI,D below). This inhibitory effect on histamine release could be mimicked by the dibutyryl derivative of cyclic AMP.

It is probable that many of the details of Szentivanyi's theory will have to be modified in the light of newer knowledge, but the hypothesis itself has already served a useful purpose as a stimulus for research. Experiments in which cyclic AMP was injected into mice (Cronholm and Fishel, 1968) were inadequate as a test of this hypothesis because of the rapidity with which injected cyclic AMP disappears from the blood. However, measurements of cyclic AMP formation in tissues from pertussis-sensitized animals may provide some useful insights into the nature of atopic allergy.

Returning to smooth muscle, the mechanism by which cyclic AMP causes relaxation (or prevents contraction) is completely unknown. As pointed out by Somlyo and Somlyo (1968), there are many possible ways by which the contraction and relaxation of smooth muscle could be affected. Bulbring and Bueding and their colleagues (Bulbring *et al.*, 1966; Bueding *et al.*, 1967) have emphasized the importance of changes in membrane excitability in responses to the catecholamines, although it should be noted that relaxation in response to these hormones may occur even in muscles which have been completely depolarized. Schild (1967) has postulated a role for calcium in both the contracting and relaxing effects of the catecholamines, but how the translocation or action of calcium might be affected by cyclic AMP in smooth muscle has yet to be investigated. Blumenthal and Brody (1965) found that a number of smooth muscle relaxants, including isoproterenol and theophylline, eventually became ineffective when bronchial smooth muscle was depleted of glycogen by repeated additions of epinephrine to the bath. Upon the readdition of glucose, however, relaxation could again be induced. Whether this is a clue to mechanism or not remains to be seen, since substrates other than glucose may be capable of supporting the same function. Bueding and Bulbring (1967) reported that β-hydroxybutyrate was an effective substrate in the glycogen-depleted taenia coli.

In summary, there are several lines of evidence which support the hypothesis that smooth muscle relaxation in response to β-adrenergic agonists occurs as the result of an increase in the intracellular level of cyclic AMP. The possibility that the opposite effect of α-adrenergic agonists occurs as the result of a fall in the level of cyclic AMP has not been investigated. The mechanism by which cyclic AMP acts to reduce the activity of smooth muscle is completely unknown.

C. Effects in Nervous Tissue

Although the role of cyclic AMP in the central nervous system has been the subject of a certain amount of speculation (e.g., Hechter and

Halkerston, 1964), there is to date very little in the way of experimental data to support this. The gray matter of the CNS has the highest activities of adenyl cyclase and phosphodiesterase of any mammalian tissue studied to date (Sutherland *et al.*, 1962), although this may to some extent be a function of the age of the animal (Schmidt *et al.*, 1970; see also Fig. 4.4).

Insofar as adrenergic stimulation is concerned, studies with slices (Kakiuchi and Rall, 1968a) and broken cell preparations from the cerebellum of several species (Klainer *et al.*, 1962; P. Greengard *et al.*, 1968, unpublished observations) have shown that adenyl cyclase from these sources possesses the characteristics associated with adrenergic β-receptors. In the rat brain, these receptors are not detectable for at least the first 3 days postpartum (Schmidt *et al.*, 1970), as illustrated in Fig. 6.12. These experiments were carried out with chopped brain preparations, which are thought to consist primarily of intact cells. Broken cell preparations from most brain areas other than the cerebellum have been unsatisfactory in that they have not yielded consistent and repro-

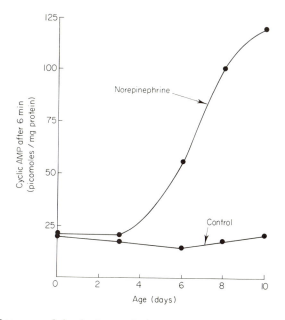

Fig. 6.12. Response of developing rat brain to norepinephrine. Brains were removed from rats of different ages and divided along the longitudinal fissure, with one-half serving as the control for the other. Chopped tissue slices were prepared and incubated essentially as described by Kakiuchi and Rall (1968a). Data represent cyclic AMP levels after incubating for 6 min in the presence or absence of 10^{-5} M norepinephrine. From Schmidt *et al.* (1970).

ducible effects in response to the catecholamines or any other hormone (see, for example, Weiss and Costa, 1968b). It seems likely that the hormonal sensitivity of adenyl cyclase may be different in different parts of the brain, or even in different parts of a single neuron. Adrenergic α-receptors may exist in the brain, but this point has not been carefully studied.

Some interesting results have been reported by Rall and his co-workers. Kakiuchi and Rall (1968a) found that both norepinephrine and histamine were effective in enhancing the accumulation of cyclic AMP in slices of rabbit cerebellum. Exposure to 5×10^{-5} M norepinephrine led to a rapid increase of more than tenfold in the level of cyclic AMP. This began to decline after a few minutes to an apparent equilibrium level of about twice that of control slices. The addition of theophylline potentiated the actions of norepinephrine and histamine, but did not prevent the decline in the level of cyclic AMP following the initial peak. That independent receptors were involved in the responses to norepinephrine and histamine was suggested by several observations. First, the effects of the two agents were additive when maximal concentrations were added together. Second, it was found that exposure of the slices to one hormone for 35 min would effectively prevent a response to the readdition of that hormone but not to the other. Finally, it was found that the response to histamine could be almost completely prevented by concentrations of phenoxybenzamine or diphenhydramine which did not alter the response to norepinephrine, while DCI could prevent the response to norepinephrine in concentrations which did not inhibit the response to histamine. Chlorpromazine inhibited the effects of both agents.

Kakiuchi and Rall (1968b) also studied the effects of norepinephrine and histamine in slices of rabbit cerebral cortex. In these slices histamine was found to be more effective than norepinephrine in increasing cyclic AMP levels, and the decline following the initial increase was less rapid.

Sattin and Rall (1967) found that exposure of slices of guinea pig cerebral cortex to 40 mM KCl for 4 min led to a tenfold increase in the intracellular level of cyclic AMP. As mentioned previously, a similar effect of high levels of extracellular K^+ was noted in the isolated rat diaphragm (Section III,B,1). Sattin and Rall (1970) also found that several nucleotides were capable of stimulating the production of cyclic AMP in guinea pig cortical slices. Preliminary experiments have indicated that the nucleotides must be metabolized to adenosine in order to be effective. Adenosine acts synergistically with norepinephrine and histamine, and its action can be antagonized by theophylline. The mechanisms involved in these interactions are currently being investigated by Rall and his associates.

As mentioned previously, De Robertis *et al.* (1967) found adenyl cyclase from rat brain cortex to be localized primarily in the mitochondrial and microsomal fractions, with the highest specific activity in those sub-fractions containing nerve endings. It seems possible, therefore, that adenyl cyclase in the brain may be located primarily in synaptic regions, and extrapolation from studies with pineal gland homogenates (see Section V,F) suggests the likelihood of a postsynaptic site. However, the studies on neuromuscular transmission (Section VI,B,2) suggest that adenyl cyclase may also occur presynaptically.

The injection of the dibutyryl derivative of cyclic AMP into the lateral ventricle of several species, including rats, cats, and rabbits, leads to central stimulation which is manifested by an increased respiratory rate and other signs (S. J. Strada *et al.,* 1968, unpublished observations; Krishna *et al.,* 1968c). These include convulsions which may recur inter-mittently for periods in excess of 24 hr. It should be noted that these effects are not the same as those of exogenous catecholamines admin-istered by the same route, for the catecholamines are unusually inert under these conditions. The role of cyclic AMP in the central actions of the catecholamines, or of agents such as amphetamine which are thought to act by way of the catecholamines, is therefore unclear. The central excitatory effects of the methylxanthines are suggestive, but here again there is no clear evidence that these effects are mediated by cyclic AMP.

Siggins, Hoffer, and Bloom (1969) found that the microelectrophoretic application of norepinephrine reduced the discharge frequency of Perkinje cells in the rat cerebellum. This effect could be enhanced by theophylline and mimicked by exogenous cyclic AMP but not by other nucleotides. The effects of norepinephrine and theophylline but not of cyclic AMP could be prevented by several prostaglandins, which are known to be capable of preventing the accumulation of cyclic AMP in several other tissues (see Chapters 8 and 10). These results are consistent with the view that this inhibitory effect of norepinephrine is mediated by cyclic AMP. They suggest that an important role of cyclic AMP within the central nervous system may be to modulate the rate of neuronal firing.

D. Hematological Effects

The shortening of the bleeding time which is produced by epinephrine is probably the result of a combination of factors, including vasocon-striction, platelet aggregation, and an increase in the rate of production of factor VIII. Of these, only platelet aggregation has been studied from the standpoint of the participation of cyclic AMP.

An important role for cyclic AMP in the regulation of platelet function has been established on the basis of several lines of evidence. It has been known for some time that the catecholamines were capable of potentiating the platelet-clumping activity of ADP and certain related nucleotides (Marcus and Zucker, 1965; Ardlie *et al.*, 1966; Born, 1967). This action of the catecholamines may be of considerable pathological as well as physiological significance (Mustard and Packham, 1970; Thomas *et al.*, 1968). ADP is by far the most effective of the naturally occurring nucleotides which have been tested, while adenosine and nucleotides which can be metabolized to adenosine tend to antagonize this effect of ADP (Marcus and Zucker, 1965; Rozenberg and Holmsen, 1968). Exogenous cyclic AMP also opposes platelet aggregation (Marcus and Zucker, 1965) and is less effective in this regard than the dibutyryl derivative (Marquis *et al.*, 1969).

The prostaglandins also inhibit platelet aggregation (Weeks *et al.*, 1969) and appear to do so by causing an increase in the intracellular level of cyclic AMP. These agents stimulate adenyl cyclase activity in homogenates prepared from human platelets (Wolfe and Shulman, 1969) and cause a dose-dependent increase in the level of cyclic AMP in intact platelets (Robison *et al.*, 1969; see also Fig. 10.6). The order of potency of the prostaglandins in stimulating the formation of cyclic AMP is the same as their order of potency as inhibitors of platelet aggregation. Theophylline acts synergistically with prostaglandin E_1 to increase the level of cyclic AMP and to inhibit aggregation (Cole *et al.*, 1970). The stimulatory effects of the prostaglandins on platelet adenyl cyclase have recently been confirmed by Marquis *et al.* (1969), who also demonstrated the inhibitory effect of imidazole on the accumulation of cyclic AMP in broken cell preparations. Imidazole has previously been shown to enhance platelet aggregation (Rossi, 1968).

The catecholamines may induce platelet aggregation in the presence of concentrations of ADP which are too low to produce an effect by themselves, and tend to abolish the inhibitory effect of adenosine (Ardlie *et al.*, 1966). Epinephrine is more potent than norepinephrine, while isoproterenol is ineffective. The effects of the catecholamines can be selectively blocked by α-adrenergic blocking agents (Born, 1967), and can also be inhibited by theophylline (Ardlie *et al.*, 1967) and the prostaglandins (Cole *et al.*, 1970). The ability of epinephrine to lower the level of cyclic AMP in human platelets in the presence of high concentrations of prostaglandin E_1 and theophylline, together with the reversal of this effect by phentolamine, is illustrated in Fig. 6.13. The reduced levels of cyclic AMP in these experiments were still too high to permit aggregation, but the results of preliminary studies by Cole *et al.* (1970) have indicated that in the presence of lower concentrations of PGE_1 the cyclic AMP-

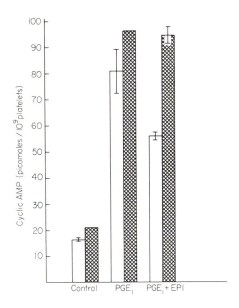

Fig. 6.13. Effects of prostaglandin E₁(PGE₁) and epinephrine (EPI) on the level of cyclic AMP in human blood platelets in presence (shaded bars) and absence (open bars) of phentolamine. Platelets were incubated for 10 min at 37°C in platelet-rich plasma containing 2 mM theophylline. Phentolamine (5×10^{-4} M), epinephrine (5×10^{-5} M), and PGE₁ (100 μg/ml) were added in that order at intervals of 1 min, and incubation continued for an additional 10 min. The mixtures were then centrifuged for 5 min at 4°C, and the pellets rapidly frozen. From Robison *et al.* (1969).

lowering effect of the catecholamines and their effects on platelet aggregation are well correlated. These observations suggest that platelet aggregation in response to the catecholamines is the result of a decrease in the intracellular level of cyclic AMP secondary to the stimulation of adrenergic α-receptors.

Another effect which the catecholamines have in blood is to cause an increase in the level of factor VIII, the antihemophilic factor. The demonstration by Ingram and Vaughan Jones (1966) that this effect is mediated by β-receptors suggests that it may be mediated by cyclic AMP, but this has not been investigated directly.

The catecholamines also inhibit the antigen-induced release of histamine from human leukocytes. This effect can be enhanced by theophylline and mimicked by the dibutyryl derivative of cyclic AMP (Lichtenstein and Margolis, 1968), suggesting that it is mediated by an increase in the intracellular level of cyclic AMP. It was shown some years ago by R. E. Scott, then a medical student at Vanderbilt University, that homogenates prepared from human leukocytes contain adenyl cyclase activity which can be stimulated by epinephrine (Table 6.III). It was later suggested by Wolfe and Shulman (1969) that the adenyl cyclase apparently derived from leukocytes might be accounted for by contamination from platelets, but Scott's data show that this cannot be the case. Epinephrine did not stimulate platelet adenyl cyclase but had a substantial effect in the leukocyte preparation.

TABLE 6.III. Effects of Epinephrine, Two Prostaglandins (E₁ and F$_{1\beta}$), and NaF on Adenyl Cyclase Activity in Homogenates of Washed Leukocytes and Platelets from Normal Human Subjects[a]

	Cyclic AMP formed in 15 min	
Treatment	Leukocytes (nmoles/10^6 cells)	Platelets (nmoles/10^9 cells)
Control	0	0
Epinephrine ($2 \times 10^{-4}\ M$)	7.0 ± 1.5	0
PGE₁ (200 μg/ml)	15.0 ± 3.0	6.5 ± 2.5
PGF$_{1\beta}$ (200 μg/ml)	3.5 ± 1.0	1.0 ± 0.5
NaF (10 mM)	8.5 ± 1.5	3.5 ± 1.0

[a]Incubations were carried out at 37° for 15 min in 40 mM tris buffer (pH 7.3) containing 2 mM ATP, 3 mM Mg^{2+}, and 6.7 mM caffeine. Results are mean values of triplicate determinations on 20 individual specimens. From unpublished experiments by R. E. Scott, Vanderbilt University, 1967. See also Scott, 1970.

It can be inferred from the data of Lichtenstein and Margolis (1968) that the adrenergic receptors in leukocytes are β-receptors, although this point has not been carefully studied. Another question which remains to be answered is whether or not epinephrine might be capable of inhibiting the effect of PGE₁ on platelet adenyl cyclase. The mechanism by which α-receptors mediate a fall in the level of cyclic AMP in intact platelets is presently unclear.

An interesting feature of the response of leukocytes to cyclic AMP is that it represents an inhibition of the release of material thought to be contained within intracellular granules. In other cells in which this type of process has been studied (e.g., amylase release from salivary glands, insulin release from pancreatic beta cells, TSH release from the anterior pituitary, and apparently also the release of acetylcholine at the neuromuscular junction), the evidence suggests that cyclic AMP promotes or accelerates the process. Why the release of histamine from leukocytes should be so different in this regard is presently unclear.

Adenyl cyclase could not be detected in red blood cells from dogs (Sutherland *et al.,* 1962) or humans (Wolfe and Shulman, 1969) but may exist in the erythrocytes of some mammalian species (Sheppard and Burghardt, 1969). It definitely exists in the cell membranes of the nucleated erythrocytes of birds (Davoren and Sutherland, 1963b; Øye and Sutherland, 1966) and frogs (Rosen and Rosen, 1969). The formation of cyclic AMP in avian erythrocytes required the presence of a catecholamine, and the receptors involved were found to have the characteristics classically associated with adrenergic β-receptors (Davoren and Sutherland, 1963a). An interesting finding was that these cells pumped cyclic AMP into the intracellular medium against a concentration gradient, when stimulated by catecholamines. This process could be inhibited by probenecid and also by caffeine; this represents another of the many effects of caffeine which cannot be understood in terms of phosphodiesterase inhibition. Fluoride did not stimulate the formation of cyclic AMP in intact erythrocytes but did stimulate adenyl cyclase activity in hemolyzed ghosts or broken cell preparations (Øye and Sutherland, 1966).

An interesting finding in frog cells was that adenyl cyclase in preparations from both tadpole and adult erythrocytes could be stimulated by fluoride, whereas the catecholamines were effective only in adult preparations (Rosen and Rosen, 1968). As in avian erythrocytes, the receptors mediating this effect had the characteristics of β-receptors. Perhaps the development of these receptors during metamorphosis is similar to their development in the rat brain beginning a few days after birth (Schmidt *et al.,* 1970). Further studies along these lines should lead to a better understanding of the nature of hormone receptors and their relation to adenyl cyclase.

The role of cyclic AMP in nucleated erythrocytes, if any, is presently unknown. Perhaps these cells have some as yet undiscovered function which is regulated by cyclic AMP. Another possibility, based on their ability to pump cyclic AMP into the extracellular fluid, is that they represent an intermediate target cell for circulating catecholamines. The cyclic AMP which they presumably release *in vivo* might then go on to affect certain other cells, although which ones is quite unclear.

E. Effects on Melanophores

The control of physiological color changes in amphibians has recently been reviewed by Novales and Davis (1969), and readers are referred to their review for a comprehensive discussion of the subject. The effects of melanocyte-stimulating hormones (MSH) and melatonin are discussed in Chapter 10 (Section III). The effect of MSH in the dorsal skin of *Rana pipiens* is to cause an increase in the level of cyclic AMP, which in turn leads to the dispersion of melanin granules within melanophores and hence to skin darkening. The principal point to be made in this section is that this effect of MSH is antagonized by the catecholamines *via* adrenergic α-receptors, as illustrated in Figs. 6.14 and 6.15. Figure 6.14 illustrates the antagonism between MSH and norepinephrine at the level of cyclic AMP, while the data of Fig. 6.15 show that the cyclic AMP-lowering effect of norepinephrine can be prevented by either phentolamine or dihydroergotamine. This effect of the α-adrenergic blocking agents is specific in that the similar effect of melatonin is not blocked by them (Abe *et al.*, 1969b).

Although the melanophores of *Rana pipiens* do not contain adrenergic β-receptors, other amphibian species may contain both types of receptors (Novales and Davis, 1969), and the same appears to be true of fish (Fujii and Novales, 1969) and reptiles (Hadley and Goldman, 1969a). Some reptiles appear to contain only β-receptors. To whatever extent β-receptors predominate over α-receptors, then to that extent the catechola-

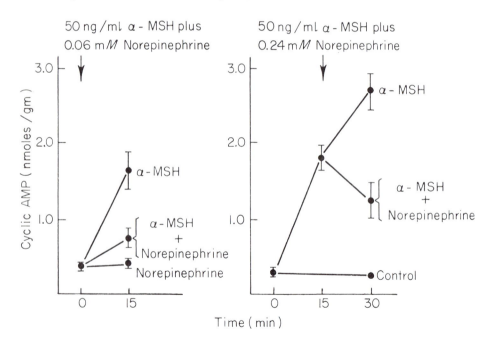

Fig. 6.14. Effects of MSH and norepinephrine on cyclic AMP levels in dorsal frog skin *in vitro*. In one experiment (left hand panel) norepinephrine and MSH were added simultaneously, while in the other experiment norepinephrine was added 15 min after MSH. From Abe *et al.* (1969a).

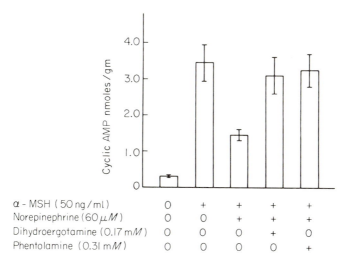

α - MSH (50 ng / ml)	O	+	+	+	+
Norepinephrine (60 μM)	O	O	+	+	+
Dihydroergotamine (0.17 mM)	O	O	O	+	O
Phentolamine (0.31 mM)	O	O	O	O	+

Fig. 6.15. Effects of MSH, norepinephrine, and α-adrenergic blocking agents on the level of cyclic AMP in dorsal frog *in vitro*. Sectioned frog skins were frozen 30 min after addition of MSH or MSH plus norepinephrine. Phentolamine or dihydroergotamine added 15 min before norepinephrine. From Abe *et al.* (1969b).

mines will tend to mimic rather than oppose the action of MSH. The physiological or evolutionary significance of these species differences has not been established.

Although there may be some points of similarity between exocytosis (Lacy *et al.*, 1968a) and the dispersion of melanin granules (Fujii and Novales, 1969), over and above the fact that both processes are stimulated by cyclic AMP, there are also some differences. For example, calcium is required for exocytosis whereas it opposes melanin dispersion (or facilitates consperion, depending on how one chooses to view the problem). The mechanism of action of cyclic AMP is poorly understood in either case.

VII. OTHER FACTORS INFLUENCING ADRENERGIC RESPONSES IN-VOLVING CYCLIC AMP

A. Permissive Hormones

Several hormones are permissive with respect to the catecholamines, which is to say that in their absence certain of the effects of the catecholamines are either not produced or are diminished. The adrenocortical and thyroid hormones may be especially important in this regard (Ramey and Goldstein, 1957; Waldstein, 1966). It would appear that many of the effects of these hormones can be related to their ability to selectively stimulate the synthesis of new enzyme protein (Hechter and Halkerston, 1965; Tata, 1967). Either as a result of this or in addition to it, these hormones may at times influence the formation or action of cyclic AMP. The overall effect may be very complex, and not necessarily the same in all cells.

To illustrate the complexity that may be encountered in this area, we can briefly mention some experiments on the effects of the glucocorticoids in the liver. It had been shown several years ago (Friedmann *et al.*, 1967) that glucagon did not stimulate gluconeogenesis in perfused livers from adrenalectomized rats. Although the ability of glucagon to stimulate the formation of cyclic AMP in these livers was not markedly different from controls, the sensitivity of the gluconeogenic process to stimulation by cyclic AMP was greatly reduced. This sensitivity could be restored almost to normal by a single injection of dexamethasone 30 min prior to sacrifice.

In collaboration with Exton and Park, we later studied the effects of adrenalectomy and dexamethasone on the hepatic response to epinephrine. To our surprise, we found that livers from adrenalectomized rats responded to epinephrine with a greater than normal increase in the level of cyclic AMP. Conversely, administration of the glucocorticoid suppressed the cyclic AMP response to epinephrine. Thus we have an example of a hormone reducing the ability of epinephrine to stimulate the formation of cyclic AMP and at the same time increasing the sensitivity of a system which responds to cyclic AMP. The physiological or ontological significance of this is presently quite obscure, although it may explain some of the differences which have been noted in rats of different ages. For example, adult rat livers seem to be much less sensitive to the catecholamines than are livers from younger rats (Wicks, 1969; Holt and Oliver, 1969a), and this could be the result of a reduced ability to form cyclic AMP in response to the catecholamines, secondary to an effect of the glucocorticoids. Presumably this effect represents a loss in the number or sensitivity of adrenergic receptors, although this remains to be established.

An increase in the sensitivity of various responding systems to the action of cyclic AMP could account for some of the other effects of the glucocorticoids, such as on cardiac metabolism and function (Hess *et al.*, 1969; T. B. Miller *et al.*, 1969, unpublished observations). Some of the earlier observations suggesting that one of the effects of the corticosteroids was to increase the sensitivity of adrenergic receptors to stimulation by catecholamines (Ramey and Goldstein, 1957) could perhaps also be understood on this basis. Whether or not the glucocorticoids ever increase the ability of catecholamines to stimulate the formation of cyclic AMP remains to be determined.

Just as some of the effects of the catecholamines may depend upon steroid hormones, it is possible that some of the effects of steroid hormones may depend upon the catecholamines. Szego and Davis (1967, 1969a) found that the injection of estradiol in rats led to an increase in the level of cyclic AMP in the uterus, and that this could be prevented by pretreatment with the β-adrenergic blocking agent propranolol. Whether or not the catecholamines or cyclic AMP can be implicated in the physiologically important actions of estrogen in the uterus remains to be seen. Some results reported by Barnea (1969) were suggestive in this regard. It is known that some of the uterine effects commonly associated with estrogenic activity can be mimicked *in vitro* by the application of exogenous cyclic AMP (Hechter *et al.*, 1967). Although these effects were shared by other nucleotides, the dibutyryl derivative of cyclic AMP was found to be more effective in stimulating amino acid uptake (Griffin and Szego, 1968). Possibly estradiol renders adenyl cyclase more sensitive to stimulation by catecholamines or other agents, and may permit cyclic AMP to be formed at sites within the uterus at which it would not otherwise be formed. It is also possible, as suggested by Hechter and his associates, that cyclic nucleotides other than cyclic AMP may participate in the action of estrogen.

The influence of the thyroid on the lipolytic response of adipose tissue to the catecholamines and other hormones has been studied, and it is generally agreed that this is one response (as opposed to many other adrenergic responses) which does depend on thyroid status. Adipose tissue from thyroidectomized rats is relatively insensitive to lipolytic hormones while that from hyperthyroid rats is hypersensitive (Fisher and Ball, 1967; Vaughan, 1967; Krishna *et al.*, 1968b). Current evidence suggests that the thyroid effect may involve an enhanced accumulation of cyclic AMP in response to the hormones, possibly because of a greater amount of adenyl cyclase (Krishna *et al.*, 1968b). Growth hormone may also enhance the accumulation of cyclic AMP in adipose tissue in response to lipolytic hormones (Moskowitz and Fain, 1970; Goodman, 1968a; Eisen and Goodman, 1969), although here again many details remain to be understood (Goodman, 1968b).

The relation between the thyroid gland and the sympathetic nervous system vis-à-vis cardiovascular function seems almost as controversial today as it was at the beginning of the century, possibly reflecting the complex nature of this relationship. Levey and Epstein (1969b) have reported that thyroxine and triiodothyronine are capable of stimulating adenyl cyclase in washed particulate preparations of cat heart. Since the effects of maximal concentrations of thyroxine and norepinephrine were additive, it would appear that there are at least two adenyl cyclase systems present in myocardial tissue, one sensitive to stimulation by the catecholamines and glucagon, and the other sensitive to stimulation by thyroxine. This may explain a good deal of the similarity between the cardiac manifestations of hyperthyroidism and sympathetic stimulation, although the overall relationship between these systems may be much more complex.

Our principal purpose in this section has been to indicate some of the ways in which cyclic AMP may participate in the actions of several permissive hormones. Serious research in this area is in a comparatively early stage, and our present understanding is meagre. However, what seems to be a very complicated picture now will probably seem much less so as a result of future research.

B. The Ionic Environment

A number of ions may influence the formation or action of cyclic AMP in responses to the catecholamines. Although no general principles have emerged from studies in this area, we can summarize some of the observations which have been made.

The hydrogen ion may be especially important from the clinical standpoint, since many of the effects of the catecholamines are depressed by acidosis. Several studies suggest that the effect of pH may be exerted primarily on the formation rather than the action of cyclic AMP. Poyart and Nahas (1968) showed that while lipolysis in response to lipolytic hormones was reduced by lowering the pH, the response to the dibutyryl derivative was unaffected. Similarly, Reynolds and Haugaard (1967) studied phosphorylase activation in rat liver slices, and found that a decrease in pH inhibited the response to epinephrine but not to exogenous cyclic AMP. These observations suggest that hydrogen ions inhibit adenyl cyclase activity in intact cells, as had been demonstrated in broken cell preparations (Sutherland *et al.*, 1962; Vaughan and Murad, 1969).

Potassium ions also interfere with some of the effects of the catecholamines, although the nature of the effect may differ from one type of cell to another. In the rat heart, for example, potassium interferes with the ability of epinephrine to stimulate the formation of cyclic AMP (Namm *et al.*, 1968), but in rat skeletal muscle (Lundholm *et al.*, 1967) and guinea pig brain cortex (Sattin and Rall, 1967) excess potassium increases the level of cyclic AMP and does not prevent the effect of epinephrine. These effects of K^+ could be direct effects on the cells involved or they might be secondary to some other action. In addition to these effects on the formation of cyclic AMP, potassium may also promote or interfere with many of the processes which are affected by cyclic AMP. The antilipolytic effect of ouabain is probably mediated in large part by an effect on potassium transport (Ho *et al.*, 1967; Kypson *et al.*, 1968; Fain, 1968c; Vulliemoz *et al.*, 1968).

Many cyclic AMP-mediated responses do not occur in the absence of calcium ions, and in all cases studied this seems to be because the process being stimulated requires Ca^{2+}, as in the case of phosphorylase activation. Calcium ions tend to inhibit adenyl cyclase in broken cell preparations, and there is no known example of Ca^{2+} being required in order for a catecholamine to stimulate adenyl cyclase. The catecholamines thus differ from ACTH, which does seem to require Ca^{2+} in order to stimulate adenyl cyclase (Birnbaumer and Rodbell, 1969; Bär and Hechter, 1969d).

It is probable that many factors other than ions or permissive hormones influence the formation or action of cyclic AMP in one or more of the effects of the catecholamines. One possibility which comes to mind is the inhibitory factor studied by Murad (1965).* However, the importance of this and other factors in cellular function has not been investigated.

VIII. GENERAL CONSIDERATIONS AND SUMMARY

We can begin to summarize the information presented in this chapter by saying that a large number of the effects of the catecholamines are mediated by an increase in the level of cyclic AMP within the cells which are responsible for these effects. This increase in the level of cyclic AMP seems to be secondary, in all cases studied, to stimulation of adenyl cyclase. A separate group of effects are mediated by a fall in the intracellular level of cyclic AMP. The mechanism of this opposite effect on the level of cyclic AMP has not been established, but may involve an inhibitory effect on adenyl cyclase. The catecholamines share with the prostaglandins (see Chapter 10, Section IX) the distinction of being the only hormones known which produce some of their effects by way of an increase in the level of cyclic AMP and others by way of a decrease.

The effects of the catecholamines which are mediated by an increase in the level of cyclic AMP include several effects on carbohydrate metabolism (summarized previously in Section III,D), some additional metabolic effects described in Sections IV and V, and also the lipolytic effect

*See also Chapter 5, Section VI.

discussed in Chapter 8. In addition, a group of functional effects which cannot be measured chemically also seem to be mediated by cyclic AMP. These effects include, in all probability, the positive inotropic response in the heart and the relaxation of smooth muscle.

The four effects which seem to be mediated by a fall in the level of cyclic AMP are the inhibition of insulin release by pancreatic beta cells, reduced permeability to water and electrolytes in several epithelial tissues, stimulation of platelet aggregation, and inhibition of melanophore dispersion in frog skin and probably other species. The latter effect could be expressed more positively as stimulation of melanophore conspersion, leading to skin lightening.

As we look at these two groups of effects in more detail, an interesting pattern begins to emerge. Most of the effects which are mediated by an increase in the level of cyclic AMP are mediated by adrenergic β-receptors, while those which involve the opposite effect on the level of cyclic AMP are mediated by α-receptors. The stimulation of adenyl cyclase by catecholamines, even when studied in highly washed particulate preparations, and even when the *in vitro* effect has not been related to any known change in cell function, has always been found to be mediated by β-receptors. In operational terms, this means that isoproterenol has generally been found to be more potent as an agonist than either of the naturally occurring catecholamines, and that the effect can be prevented by β-adrenergic blocking agents. One hypothesis which these findings have led to is that adrenergic β-receptors and adenyl cyclase are very closely related, at least in those cells in which β-receptors occur. We might go further and suggest that in these cells the β-receptor and adenyl cyclase are integral components of the same system, which boils down to a special case of the more general hypothesis discussed in Chapter 2, according to which the receptors for *all* hormones which stimulate adenyl cyclase were regarded as parts of an adenyl cyclase system. In the present case, this amounts to redefining adrenergic β-receptors as those receptors which mediate the stimulation of adenyl cyclase by catecholamines.

An obvious corollary of this hypothesis is that all β-adrenergic effects without exception are mediated by increased levels of cyclic AMP in the cells responsible for these effects. To say the same thing in negative terms, the implication is that no β-adrenergic effects ever occur which are not mediated by cyclic AMP. Since not all β-adrenergic effects have been studied from this point of view, the hypothesis cannot be said to have been established. However, since a large variety of such effects have been studied, and since no exceptions have emerged, the hypothesis is well supported by experimental data. If and when this hypothesis is established beyond question it will probably lose most of its value, but at present it has a certain amount of predictive value. Thus, whenever we see an effect of the catecholamines which can be produced by smaller or equimolar concentrations of isoproterenol, and which can be prevented by doses of a β-adrenergic blocking agent which do not prevent cell function per se, then we can predict with a reasonable degree of certainty that that is an effect which is mediated by an increase in the intracellular level of cyclic AMP. Should this prove to be the case, then it diminishes our ignorance somewhat.

Turning now to effects mediated by adrenergic α-receptors, the evidence that four of these effects are associated with a fall in the intracellular level of cyclic AMP provides the experimental foundation for the hypothesis that *all* such effects are mediated by a fall in the level of cyclic AMP. The evidence supporting this hypothesis is not as strong as that supporting the corresponding hypothesis that all β-adrenergic effects are mediated by an increase in the level of cyclic AMP. This is partly because there are so many α-adrenergic effects which have not been studied from this point of view — the contraction of smooth muscle is perhaps the most important of these — and partly because of uncertainty about the mechanisms by which α-receptors mediate a fall in the level of cyclic AMP. What is perhaps the simplest possibility in this regard is illustrated in Fig. 6.16. In this hypothetical model α-receptors and β-receptors are envisioned as specific patterns of forces located on separate regulatory subunits, such that interaction with α-receptors leads to a decrease in

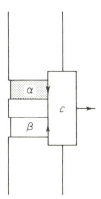

Fig. 6.16. Possible model of the membrane adenyl cyclase system as related to adrenergic receptors. The hypothetical regulatory subunit of which the α-receptor is a part is bonded to the catalytic subunit such that an effective interaction with a catecholamine leads to a decrease in enzyme activity. The subunit containing the β-receptor is connected in such a way that interaction with a catecholamine leads to an increase in enzyme activity. Receptors for other hormones could be located on the same or different subunits.

adenyl cyclase activity while β-receptors mediate an increase in the activity of the same catalytic subunit. Variations of this type of model would include one in which both types of receptors were located on the same regulatory subunit, and another in which the two types of receptors were parts of entirely separate systems.

The type of scheme illustrated by Fig. 6.16 may well symbolize an aspect of reality, and indeed this even seems likely in the case of the adrenergic β-receptor. As for α-receptors, however, it has to be emphasized that an inhibitory effect on adenyl cyclase activity by a catecholamine in a broken cell preparation has never been observed. Although the possibility that this may occur has not been carefully investigated, the lack of evidence that it does occur suggests that other possibilities should be considered. For example, Bloom and Goldman (1966) and Belleau (1967) have speculated that the adrenergic α-receptor may be a part of a membrane ATPase system, such that interaction with a catecholamine leads to a stimulation of the enzyme. If such a system were to compete for the same substrate pool utilized by adenyl cyclase, then it is easy to imagine how the stimulation of such a system could reduce the level of cyclic AMP without affecting the catalytic potential of adenyl cyclase in any way. Consideration of this hypothetical example illustrates the point that if the α-receptor is part of a system which indirectly influences the formation or accumulation of cyclic AMP, then it could also mediate effects which are unrelated to changes in cyclic AMP. Since research in this area seems to be entering an active stage, further speculation seems inappropriate.

Some puzzling observations were mentioned previously, and can be briefly discussed in this section. The observations that both types of adrenergic receptors mediate superficially similar effects in the heart and in certain segments of intestinal smooth muscle are not compatible with the model shown in Fig. 6.16. It is therefore ruled out as a model of universal validity, even though it does seem adequate for most cells in which both types of receptors occur. Perhaps in most such cells both types of receptors do influence the same pool of cyclic AMP, whereas in certain cardiac and smooth muscle cells they are separated and thus influence different pools. Is it possible that a decrease in the level of cyclic AMP in one compartment could lead to an effect which superficially resembles that produced by an increase in the level of cyclic AMP in the other? The effect would be an increase in contractile force in the case of the heart, and relaxation in the case of intestinal smooth muscle. Rather than speculate about this further, we will leave it as a research problem which will undoubtedly be investigated within the near future.

We mentioned previously that there were no known exceptions to the generalization that all β-adrenergic effects were mediated by an increase in the intracellular level of cyclic AMP. We can now consider the converse, and ask whether all of the effects of the catecholamines which are mediated by an increase in the level of cyclic AMP are also mediated by β-receptors. By and large, to the extent that it has been studied, this seems to be true, although there may be some apparent exceptions. Perhaps the most striking of these occurs in the liver. In some species, such as dogs and cats, the receptors mediating the hepatic metabolic effects of the catecholamines seem to be quite similar to β-receptors in other tissues. Although it is true that ergotamine is capable of inhibiting the stimulation of adenyl cyclase by catecholamines, in doses which do not interfere with the effect of glucagon, it is also true that the β-adrenergic blocking agents are more effective. The fact that isoproterenol is a more potent agonist than either of the naturally occurring catecholamines suggests further that the hepatic adrenergic receptors in these species should be classified as reasonably typical β-receptors. By contrast, the properties of hepatic adrenergic receptors in certain other species, such as the rat and apparently also in the human, may diverge considerably from those classically associated with β-receptors. Thus, the β-adrenergic blocking agents may be much less effective as antagonists than they are in other tissues, and isoproterenol may at times be even less effective as an agonist than norepinephrine. It is understandable that the hepatic adrenergic receptors in these species have often been classified as α-receptors. We feel inhibited in discussing this problem in more detail because of the recent evidence that in immature rats these receptors do have the characteristics classically associated with β-receptors. Therefore, the question we should probably be asking is not why the hepatic adrenergic receptors in the rat differ from those in other species, but what happens in the course of development and ageing to cause them to *become* different. In other words, although we have been thinking about the possible relation between adrenergic receptors and adenyl cyclase for several years, we now feel that a detailed discussion of this question may be premature.

It is possible that the deliberate effort which we have made to fit the cyclic AMP data summarized in this chapter into the context of some earlier concepts, such as the concept of α- and β-receptors, may in the final analysis prove to be ill-founded. On the other hand, the relation between changes in the level of cyclic AMP and the receptors mediating these effects does in general seem to be very striking, and we feel it has not been entirely a waste of time to call attention to it. The hypothesis that all α-adrenergic effects are mediated by a fall in the level of cyclic AMP, just as all β-adrenergic effects seem to be mediated by an increase in the level of cyclic AMP, has already been useful as a working hypothesis, and should continue to be useful as additional effects are studied. It is possible that in the future, however, adrenergic receptors will have to be classified in some other manner.

In final summary of this chapter, we should say that a variety of adrenergic responses are mediated by an increase in the intracellular level of cyclic AMP, while certain others are mediated by a fall in the level of cyclic AMP. Increased levels in response to catecholamines occur as the result of stimulation of adenyl cyclase, and in most cases the receptors mediating this effect have properties which in the past have been associated with adrenergic β-receptors. The mechanism by which the catecholamines act to produce a fall in the level of cyclic AMP is unknown, but the receptors mediating this effect, to the extent that they have been studied, have characteristics which in the past have been associated with adrenergic α-receptors. Whether all of the physiologically important effects of the catecholamines are mediated by one or the other of these divergent effects on the level of cyclic AMP remains to be seen.

XI. ADDENDUM

A number of papers published since this chapter was written support the view that β-adrenergic effects in general are mediated by cyclic AMP. These include studies of skeletal muscle (Walaas and Walaas, 1970), cardiac muscle (Skelton et al., 1970; Kukovetz and Pöch, 1970a; Shinebourne and White, 1970; Krause et al., 1970), smooth muscle (Andersson and Mohme-Lundholm, 1969; Kawasaki et al., 1969; Bowman and Hall, 1970; Kukovetz and Pöch, 1970b; Lamble, 1970; Mitznegg et al., 1970; Somlyo et al., 1970; Triner et al., 1970), melanophores (Hadley and Goldman, 1969b; Goldman and Hadley, 1969a,b, 1970a,b; Van de Veerdonk and Konijn, 1970),* and nucleated erythrocytes (Riddick et al., 1969; Rosen and Erlichman, 1969; Rosen et al., 1970). The formation of cyclic AMP in vascular smooth muscle has still not been carefully studied, despite the potential clinical importance of such studies.

The statement in Section VI,B,4 that smooth muscle contraction had never been seen in response to exogenous cyclic AMP was incorrect. This had been observed by Bartelstone et al. (1967) and subsequently by Bowman and Hall (1970). These exceptions fit the pattern discussed previously in the addendum to Chapter 5.

Some interesting studies have been carried out with human fat cells. Burns and Langley (1970) found that these cells, in contrast to adipocytes

---

*See also Chapter 10, Section III.

from rats and certain other species, contained α- as well as β-adrenergic receptors. The α-receptors could be unmasked by the use of β-adrenergic blocking agents, and under these conditions the catecholamines were found to inhibit lipolysis. It was later shown (Burns *et al.,* 1970) that the divergent effects on lipolysis were paralleled by divergent effects on the level of cyclic AMP. Inhibition of lipolysis in human fat cells thus seems to be another α-adrenergic effect mediated by a fall in the intracellular level of cyclic AMP.

Whether α-adrenergic effects in general result from a fall in the level of cyclic AMP per se still seems doubtful. Salzman and Neri (1969) observed a fall in the level of cyclic AMP in platelets in response to epinephrine, and Marquis *et al.* (1970) reported a reduced rate of conversion of labeled ATP to cyclic AMP under similar conditions, in line with observations mentioned in Section VI,D. A fall in the level of cyclic AMP in pancreatic islets was mentioned in Section III,C as the most likely cause of the α-receptor-mediated inhibition of insulin release. More recently, however, Feldman and Lebovitz (1970) reported that epinephrine could also inhibit insulin release in response to exogenous cyclic AMP (added as the dibutyryl derivative), and that this could be prevented by α-adrenergic blocking agents. Similarly, Novales and Fujii (1970) observed aggregation in fish melanophores in response to epinephrine in the presence of exogenous cyclic AMP. It would thus appear that epinephrine may interfere with the action as well as the formation of cyclic AMP in some cells. A possibility which should probably be considered is that adrenergic α-receptors may be part of a system which catalyzes the conversion of cyclic AMP to an antimetabolite of cyclic AMP. The resulting lack of cyclic AMP might be the most important factor under some conditions, whereas the presence of the antimetabolite might be more important under others. Although highly conjectural at present, this possibility seems worth investigating.

The relation of receptors to adenyl cyclase in rat liver preparations was studied by Marinetti and his colleagues (Ray *et al.,* 1970; Tomasi *et al.,* 1970). The results differed from those reported earlier in that basal activity in the absence of epinephrine or glucagon was relatively high and was inhibited instead of stimulated by fluoride. The addition of Ca^{2+} was found to inhibit stimulation by glucagon but to enhance the effect of epinephrine. The binding of Ca^{2+} to the membrane fraction was increased by epinephrine but decreased by glucagon (Ray *et al.,* 1970). Fractionation on Sephadex G-200 disclosed that the hormones were preferentially bound to a fraction different from that containing most of the adenyl cyclase activity (Tomasi *et al.,* 1970), but whether the sites to which the hormones were bound can be related to the stimulation of adenyl cyclase remains to be seen. Some interesting recent studies by Bitensky *et al.* (1970) on the effects of steroid hormones on hepatic adenyl cyclase are summarized in Chapter 10 (Section XII). The recent studies by Rodbell and his colleagues, suggesting an important role for guanine nucleotides in the activation of adenyl cyclase, were mentioned briefly in the addendum to Chapter 4. The rat liver preparations used in these latter experiments were sensitive to glucagon but not to catecholamines.

Some additional observations on the role of cyclic AMP in the parotid gland can be mentioned. Grand and Gross (1969) showed that epinephrine stimulated protein synthesis in this gland independently of its effect on amylase secretion, and that both effects could be mimicked by the dibutyryl derivative of cyclic AMP. Subsequent experiments (Grand and Gross, 1970) pointed to an effect at the translational level as the cause of the increased protein synthesis. It is not clear how or if these observations can be related to the pronounced stimulatory effect of isoproterenol on DNA synthesis in salivary glands. Salomon and Schramm (1970) observed that cyclic AMP was bound specifically and with high affinity to a microsomal fraction obtained from parotid gland homogenates, and further experiments suggested that the binding site was derived primarily from the smooth endoplasmic reticulum. Whether this binding site represents part of a protein kinase system (see Chapter 5, Section VI) remains to be seen.

Some recent experiments with cultured pineal glands suggest a different interpretation from that offered in Section V,F. Serotonin is converted to melatonin in the pineal gland by a two-step process: *N*-acetyltransferase catalyzes the conversion of serotonin to *N*-acetylserotonin, and this is then converted to melatonin under the catalytic influence of HIOMT. Based on their analysis of labeled intermediates, Klein *et al.* (1970a) suggested that the earlier step might be a more likely site of action for cyclic AMP than HIOMT. This was substantiated by the demonstration (Klein *et al.*, 1970b) of a six- to tenfold stimulation of *N*-acetyltransferase in response to norepinephrine or the dibutyryl derivative of cyclic AMP, with little or no change in HIOMT activity. It would appear, therefore, that cyclic AMP stimulates melatonin synthesis not by changing the properties of HIOMT but rather by making more substrate available to it. As expected from previous observations, Ebadi *et al.* (1970) were able to show that light does indeed alter the level of cyclic AMP in the rat pineal gland: the level was six times higher after 10 hours of light than after 10 hours of darkness.

A variety of observations on the formation of cyclic AMP in brain tissue have been reported. Contributions not mentioned in Section VI,C include those by Kakiuchi *et al.* (1969), Shimizu *et al.* (1969, 1970a,b,c), Palmer *et al.* (1969, 1971), Ditzion *et al.* (1970), and Paul *et al.* (1970a). Kakiuchi and his colleagues showed that electrical stimulation of slices of guinea pig cerebral cortex led to an elevenfold increase in the level of cyclic AMP within 10 minutes, and this was confirmed by Shimizu *et al.* (1970c). The effects of electrical stimulation and adenosine were similar in that both enhanced the effects of either norepinephrine or histamine and both were inhibited by theophylline. Shimizu *et al.* (1970b) had previously shown that several depolarizing agents, of which batrachotoxin was the most potent, also increased cyclic AMP formation and acted synergistically with the biogenic amines. An interesting correlation was observed by Palmer *et al.* (1971): drugs which resemble chlorpromazine pharmacologically were found to share chlorpromazine's ability to inhibit the rise in cyclic AMP that occurs in brain slices in response to norepinephrine; chemically related compounds which lack tranquilizing activity had no effect on cyclic AMP under these conditions. Ditzion *et al.* (1970) confirmed the earlier finding by Breckenridge (1964) that decapitation leads to a large increase in the brain level of cyclic AMP; the brains of rats killed by an overdose of ether or by exsanguination under pentobarbital anesthesia also contained cyclic AMP levels higher than those seen after rapid freezing. Schmidt *et al.* (1970) showed that the postdecapitation rise in cyclic AMP occurred in rats of all ages, thus supporting the conclusion by Kakiuchi and Rall (1968b) that brain catecholamines are probably not involved in this effect. Paul *et al.* (1970a) found that convulsant doses of theophylline caused an increase in brain cyclic AMP levels if the rats were killed by rapid freezing, but a decrease (compared to controls) if the rats were killed by decapitation. Other stimulants, including amphetamine, had no effect on whole brain cyclic AMP levels under either set of conditions. Many of the foregoing observations were discussed at a symposium entitled "Role of Cyclic AMP in Neuronal Function," the proceedings of which were published (Greengard and Costa, 1970). Most of them are difficult to interpret because of the extreme heterogeneity of brain tissue, and the role of cyclic AMP in neuronal function unfortunately remains obscure.

Conceivably related to the subject of this chapter were the reports that urinary excretion of cyclic AMP was below normal in depressed patients and above normal in manic patients (Paul *et al.,* 1970b,c; Abdulla and Hamadah, 1970). It was later found, however, that pooled samples of cerebrospinal fluid from patients with affective disorders contained cyclic AMP levels which were not different from those seen in normal patients (Robison *et al.,* 1970). Factors affecting the urinary excretion of cyclic nucleotides are discussed in Chapter 11.

CHAPTER 7

Glucagon and Insulin

I. INTRODUCTION

The two principal hormones produced by the endocrine pancreas are *glucagon* and *insulin*. Both of these hormones are polypeptides, with molecular weights of approximately 3450 and 6000, respectively. Glucagon seems to be produced primarily in the *alpha* cells of the islets of Langerhans (Foa and Galansino, 1962; Orci *et al.*, 1969), although it may also be produced in gastrointestinal tissue to some extent (Sutherland and DeDuve, 1948; Makman and Sutherland, 1964; Orci *et al.*,

1968; Valverde *et al.*, 1968). Insulin seems normally to be produced only in the pancreatic *beta* cells (Lacy, 1967; Steiner *et al.*, 1969). The two hormones are conveniently discussed together not only because of the close proximity of their sites of origin but also because they are frequently physiological antagonists of each other. They are further related in that glucagon is capable of stimulating the release of insulin, and indeed glucagon may even be required in order for insulin release to occur.

It is now known that some of the effects of glucagon are mediated by an increase in the intracellular level of cyclic AMP, while some of the effects of insulin are the result of a fall in the level of cyclic AMP. These divergent effects on the level of cyclic AMP may account for much of the antagonism between these hormones, especially in hepatic and adipose tissue. It seems possible that most and perhaps all of the known effects of glucagon will eventually be understood in terms of its ability to alter the intracellular level of cyclic AMP. By contrast, only some of the effects of insulin can be accounted for on this basis.

II. GLUCAGON

Glucagon was at one time referred to as the hyperglycemic factor of the pancreas, or HGF, a name based on its most prominent pharmacological effect when injected into animals.* Although there is evidence to suggest that glucagon may play an important role in the maintenance of blood sugar levels, at least in some species, the exact nature of this role is still controversial (e.g., Sokal, 1966; Foa, 1968; Ohneda *et al.*, 1969; Buchanan *et al.*, 1969a). There are several reasons for this. First, it has been difficult to detect changes in the blood level of glucagon under conditions where such changes might reasonably be expected to occur. Second, despite the powerful tendency of exogenous glucagon to stimulate hepatic glucose production, it has been found that the hormone is also capable of acting directly on the pancreatic *beta* cells to stimulate the release of insulin. Both effects seem to be mediated by cyclic AMP, and the question arises as to which of them is physiologically most significant. This will be discussed in more detail in Section II,A,4.

As mentioned previously, many of the effects of glucagon are similar to the effects of β-adrenergic stimulation. This is, of course, understandable in a general sense because these effects are mediated by cyclic AMP,

---

*The earlier literature on glucagon, as well as several aspects of the subject not included in this monograph, have been reviewed by Foa and Galansino (1962) and Berthet (1963). These earlier reviews were recently updated by Foa (1968).

but this does not explain why the effects of isoproterenol and glucagon seem to overlap more often than the effects of other hormones which stimulate adenyl cyclase. Perhaps the unknown patterns of forces which we identify with the β-receptor and those which constitute the glucagon receptor overlap to some extent, or else exist in very close proximity to each other. It is also possible that the similarities between the effects of glucagon and isoproterenol are more apparent than real, and that when all of the hormones that stimulate adenyl cyclase are studied in greater detail, especially at concentrations far removed from their normal physiological levels, they will be found to overlap just as much or more. Certainly it is true that the relative potencies of isoproterenol and glucagon differ drastically from one tissue to another. In the liver, for example, in all species studied, glucagon is more potent than any of the catecholamines. In striated muscle, by contrast, glucagon is so much less potent than the catecholamines that from a physiological standpoint it can be considered inert. In certain other tissues, however, as in the liver, it is possible that both glucagon and the catecholamines play important physiological roles.

A. The Hyperglycemic Response

The hyperglycemic response to glucagon differs in several respects from the response to the catecholamines (Chapter 6, Section III). Although the hepatic effects are similar, the effect of glucagon on muscle glycogenolysis is less pronounced, and at physiological levels does not occur at all. A second difference is that whereas the normal effect of the catecholamines is to inhibit the release of insulin, glucagon acts to stimulate this process. The hepatic and pancreatic components of the response to glucagon are discussed separately in the following sections.

1. The Response in the Liver

The overall hepatic response to glucagon resembles that of epinephrine, which is understandable if cyclic AMP functions as a second messenger for both hormones. The nature of this response was discussed in some detail in the last chapter (Section III,A), and the evidence that cyclic AMP mediates the response to glucagon specifically can be summarized briefly.

First, glucagon stimulates adenyl cyclase activity in washed particulate preparations from the livers of all species studied (Makman and Sutherland, 1964; Bitensky *et al.*, 1967, 1968; Pohl *et al.*, 1969; Marinetti *et al.*, 1969). In preparations from cats and dogs, glucagon was found to be much more potent than the catecholamines (Makman and Sutherland, 1964), in line with data obtained at higher levels of organization (Robison *et al.,* 1967b; Exton and Park, 1968b). The stimulatory effect of glucagon on hepatic adenyl cyclase activity is illustrated in Fig. 7.1. This effect has been reported to occur at all levels of ATP and Mg^{2+}, and to persist even through dialysis. According to Bitensky and his colleagues (1967), there is a factor or factors in liver extracts capable of reversing the effect of glucagon on adenyl cyclase. Inhibition of this effect by adrenergic blocking agents has not been observed, and it is possible that the slight inhibitory effect of DCI, which was seen at the level of glucose production (Ashmore *et al.,* 1962), was exerted at a step beyond the formation of cyclic AMP.

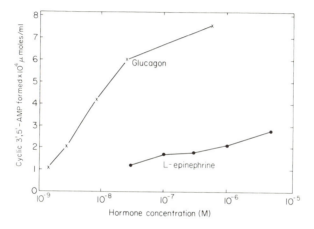

Fig. 7.1 Effects of glucagon and epinephrine on the activity of cat liver adenyl cyclase. Washed particulate preparations were incubated with ATP and Mg^{2+} for 15 min at 30°C. in the presence of caffeine and the indicated concentration of hormone. Cyclic AMP formation was not detectable in the absence of hormone. From Makman and Sutherland (1964).

Glucagon causes a prompt increase in the intracellular level of cyclic AMP when added to the isolated perfused rat liver (Robison *et al.,* 1967b; Exton and Park, 1968b; Exton *et al.,* 1969b, unpublished observations), and the dose-dependent nature of this response is illustrated in Fig. 7.2. By measuring the amount of cyclic AMP released into the perfusate, Lewis *et al.* (1970) were able to see effects of glucagon at concentrations even lower than those shown in this figure. Stimulation of cyclic AMP accumulation is the earliest hepatic effect of glucagon that has been measured, and precedes glycogenolysis, gluconeogenesis, potassium efflux, amino acid uptake, urea production, and the increase in tyrosine aminotransferase activity, all of which effects can be mimicked by exogenous cyclic AMP (Bergen *et al.,* 1966; Exton and Park, 1968a,b; Friedmann and

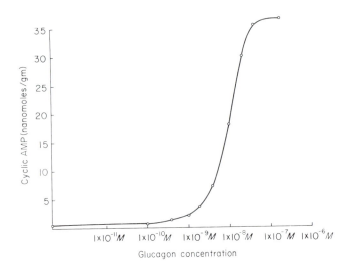

Fig. 7.2 Effect of increasing concentrations of glucagon on the level of cyclic AMP in the isolated perfused rat liver. Livers were freeze-clamped after 4 min exposure to the indicated concentration of glucagon. From Exton *et al.* (1969b, unpublished observations).

Park, 1968; Mallette *et al.,* 1969a; Wicks *et al.,* 1969; Chambers and Bass, 1970). It is interesting to compare the dose-response curve illustrated in Fig. 7.2 with that shown in Fig. 7.3, the dose-response curve for glucose release. It can be seen that the concentration of glucagon which produces a maximal stimulation of glucose production is quite small relative to the concentration needed to produce a maximal effect on the level of cyclic AMP. The concentration of glucagon which produces a maximal effect on glucose production is close to the highest blood glucagon level that has been reported to occur under physiological conditions (Ohneda *et al.,* 1969). This means that under normal conditions *in vivo* the hepatic receptors for glucagon are probably never saturated, and that the majority of them (those responsible for the increment in the level of cyclic AMP which occurs when the concentration of glucagon is increased above the normal physiological maximum) constitute something of a reserve. According to classical receptor theory (Ariens, 1966) these would be referred to as spare receptors.

For any given effective concentration of glucagon, the level of cyclic AMP in the liver remains high so long as the glucagon is present in the perfusing medium, and begins to fall toward the control level as soon as the hormone is removed. This is illustrated in Fig. 7.4.

The hepatic effects of glucagon in combination with a methylxanthine or other phosphodiesterase inhibitor have not been carefully studied, but

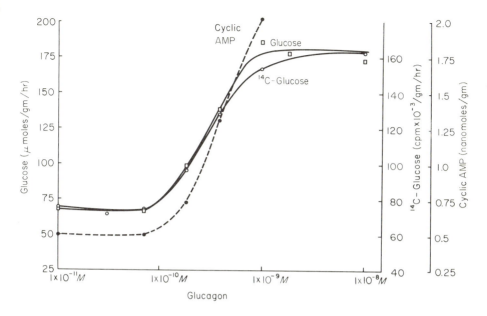

Fig. 7.3. Effect of increasing concentrations of glucagon on glucose production and ^{14}C-glucose synthesis from ^{14}C-lactate by the isolated perfused rat liver. Livers from fed rats were perfused for 20 min in the presence of the indicated concentration of glucagon. Also plotted in this figure is the foot of the dose-response curve shown in Fig. 7.2. From unpublished experiments by Exton and Park.

synergism has been observed, both in terms of the level of cyclic AMP and with respect to the stimulation of glucose output (J. H. Exton and G. A. Robison, 1969, unpublished observations). The latter effect may have been related primarily to stimulation of gluconeogenesis, since there is a tendency for the glycogenolytic effect of cyclic AMP to be antagonized by the phosphatase-stimulating action of the methylxanthines. This has been demonstrated for liver phosphorylase phosphatase (Sutherland, 1951a) and also for the glycogen synthetase system (DeWulf and Hers, 1968a). Whether the same phosphatase serves both systems is unknown.

All of the known hepatic effects of glucagon, as mentioned previously, have been mimicked by exogenous cyclic AMP or one of its derivatives. Other adenine nucleotides are largely without effect when added to the perfused liver (Bergen *et al.*, 1966), but recently cyclic GMP has been found to be capable of mimicking at least some of the hepatic effects of glucagon without altering the level of cyclic AMP (Glinsmann *et al.*, 1969). The effects of cyclic GMP are discussed in more detail in Chapter 11.

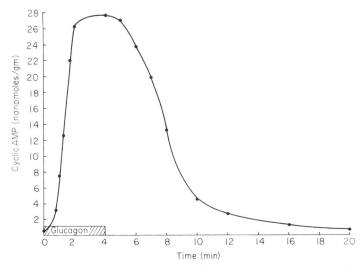

Fig. 7.4. Effect of glucagon on the level of cyclic AMP in the isolated perfused rat liver. Livers from fed rats were perfused for 20 min with a bicarbonate buffer containing albumin and red blood cells. Starting at time zero, glucagon (5 μg/ml) was infused at a rate of 0.1 ml/min for 4 min. Perfusion with glucagon-free medium was continued for an additional 16 min. Livers were frozen at various intervals and cyclic AMP determined in purified extracts. From Exton *et al.* (1969b, unpublished observations).

Some of the differences between the hepatic effects of glucagon and those of the catecholamines were mentioned in the last chapter, and can be reiterated here. The greater potency of glucagon has been seen in all species studied and at all levels of organization, from broken cells *in vitro* (Makman and Sutherland, 1964) to the intact animal (Sanbar, 1968). Bitensky and his associates (1968) have reported that it is possible to separate two adenyl cyclase systems from rat liver homogenates, one sensitive to glucagon and the other to catecholamines. Since both systems seemed to be sensitive to similar concentrations of their respective hormones, Bitensky *et al.* suggested that the previously observed differences in potency might have been the result of a preparative artifact. This does not seem likely, however, because the greater potency of glucagon has been observed in the perfused liver as well as in broken cell preparations. Assuming that the findings of Bitensky *et al.* can be confirmed, a reconciliation of these findings with earlier data could be brought about if it were found that the liver contained more glucagon-sensitive cyclase systems than catecholamine-sensitive ones. It is also possible that there is a factor or factors in liver which tend to enhance the effect of glucagon or inhibit the effect of the catecholamines or both. Other possibilities, including the possibility that different cell types may be involved (see below), will undoubtedly have to be considered in the course of future research in this area.

With reference to their effects on the level of cyclic AMP in the perfused rat liver, another difference between these agents is that the biphasic response seen with the catecholamines is not seen in response to glucagon. It would appear that either the receptors for these hormones are related to adenyl cyclase in different ways, or else that the catecholamines exert an additional effect which tends to suppress the enhanced formation of cyclic AMP. Glucagon, it may be noted, has in no tissue been found to be capable of lowering the level of cyclic AMP, whereas this has been seen in response to the catecholamines in a number of tissues.

The possibility that glucagon and the catecholamines might affect different types of hepatic cells was mentioned previously (Sutherland and Robison, 1966), and recent histochemical evidence suggests that this may in fact be the case (Reik *et al.,* 1970). Surprisingly, this evidence suggests that glucagon-sensitive adenyl cyclase may be located primarily in the reticuloendothelial cells. There has been a tendency to assume that most of the hepatic effects of glucagon occur in parenchymal cells, but this had never been established. The histochemical method suggesting the opposite was based on the precipitation of pyrophosphate by lead, and its validity is open to question on the grounds that the lead nitrate used would be expected to inhibit adenyl cyclase. The results are nevertheless of great interest, and emphasize the need for further research in this area. The effect of isoproterenol in these experiments was restricted almost exclusively to parenchymal cells.

Some early observations by Cornblath (1955) on phosphorylase activation in liver slices, showing additivity of the effects of glucagon and ephedrine but not of glucagon and epinephrine, have not been satisfactorily explained. The ability of certain anabolic steroids to prevent the hyperglycemic response to glucagon without affecting the response to epinephrine (Landon *et al.,* 1962) is also poorly understood. Perhaps the suggestion that glucagon and the catecholamines affect different hepatic cells to different degrees (Reik *et al.,* 1970) will be helpful in future experiments designed to study these and other phenomena.

It should be noted that these hormones may be capable of exerting one or more of their hepatic effects independently of an effect on the level of cyclic AMP. This has never been observed in the case of glucagon, but in the case of epinephrine there is evidence that the effect on phospholipid metabolism is not mediated by an increase in cyclic AMP (since it could be reproduced neither by glucagon nor exogenous cyclic AMP). This was discussed in Section V,C in the preceding chapter.

Three important components involved in the stimulation of hepatic glucose release in response to glucagon are phosphorylase activation, inhibition of glycogen synthetase activity, and stimulation of gluconeogenesis. Phosphorylase activation in response to glucagon was noted at an early date, and indeed it was this observation that led eventually to the discovery of cyclic AMP (see Chapter 1). Phosphorylase activity and glucose production are closely correlated in glycogen-containing livers (Weintraub *et al.*, 1969). Inactivation of glycogen synthetase in response to glucagon in several species has also been noted (Bishop and Larner, 1967; De Wulf and Hers, 1968b); indeed it now seems likely that glycogen synthetase kinase and the protein kinase that catalyzes the activation of phosphorylase kinase are one and the same enzyme (Schlender *et al.*, 1969; Soderling and Hickenbottom, 1970; Reimann and Walsh, 1970). Since these responses were discussed in some detail in Chapters 5 and 6, they will not be discussed further here. Our discussion of gluconeogenesis was deferred until the present chapter, however, and we can proceed with it now.

2. Gluconeogenesis

Gluconeogenesis is the process by which glucose is formed from a variety of small molecules. The hormonal control of gluconeogenesis has been studied in the most detail in liver, although the process also occurs to a lesser extent in the kidney. Known precursors for hepatic gluconeogenesis include blood lactate, pyruvate, and several amino acids (of which alanine appears to be the most important), as well as endogenous liver protein. Gluconeogenesis from all of these precursors can be stimulated by glucagon, the catecholamines, and exogenous cyclic AMP (Exton *et al.*, 1966; Menahan and Wieland, 1967; Exton and Park, 1968a,b; Mallette *et al.*, 1969a), and it therefore seems likely that the hormonal effects are mediated by cyclic AMP. The mechanisms by which cyclic AMP might act, however, are poorly understood.

Considering first gluconeogenesis from lactate and pyruvate, it is now well recognized that the thermodynamics of the situation make a simple reversal of glycolysis extremely unlikely. The chief energy barriers tending to make glycolysis a one-way process occur between pyruvate and phosphoenolpyruvate, between fructose-1, 6-diphosphate and fructose-6-phosphate, and between glucose-6-phosphate and glucose.

The conversion of pyruvate to phosphoenolpyruvate in gluconeogenesis involves at least two enzyme systems. Pyruvate carboxylase catalyzes the formation of oxaloacetate from pyruvate and CO_2. This reaction occurs in mitochondria and requires the presence of acetyl-CoA as well as ATP and Mg^{2+}. Oxaloacetate is then released into the cytosol and is converted to phosphoenolpyruvate under the catalytic influence of phosphoenolpyruvate carboxykinase. This reaction utilizes GTP and leads to the regeneration of CO_2.

Fructose-1,6-diphosphate is converted to fructose-6-phosphate by the action of fructose-1,6-diphosphatase, while the breakdown of glucose-6-phosphate to glucose is catalyzed by glucose-6-phosphatase.

Exton and Park and their colleagues have studied the control of gluconeogenesis in the isolated perfused rat liver. They found that saturating concentrations of fructose or dihydroxyacetone led to a rate of gluconeogenesis about twice that observed with saturating concentrations of lactate or pyruvate. Glucagon and the catecholamines (and exogenous cyclic AMP) stimulated gluconeogenesis from lactate or pyruvate but not from fructose or dihydroxyacetate (Exton and Park, 1968a). These observations led to the conclusion that the step or steps limiting the rate of gluconeogenesis from lactate or pyruvate most likely occurred between the conversion of pyruvate to phosphoenolpyruvate, and that the stimulatory effect of cyclic AMP was probably exerted, at least in part, at this level. This conclusion was supported by studies involving the measurement of steady state levels of cytoplasmic intermediates, showing a crossover point between pyruvate and phosphoenolpyruvate (Exton and Park, 1969). This is illustrated in Fig. 7.5.

Whether cyclic AMP affects pyruvate carboxylase or phosphoenolpyruvate carboxykinase or neither or both of these enzyme systems is still unclear. Experiments by Adam and Haynes (1969) support the possibility that cyclic AMP might act by increasing the rate of transfer of pyruvate into the mitochondria. It is conceivable that this process, rather than the activity of either converting enzyme, could at times be rate-limiting in the conversion of pyruvate to phosphoenolpyruvate.

It should be noted that glucagon does increase the *amount* of phosphoenolpyruvate carboxykinase in the liver (Shrago *et al.,* 1963). This can be mimicked by cyclic AMP (Yeung and Oliver, 1968b; Wicks, 1969), but it is a relatively slow and long-lasting effect which seems clearly to involve the synthesis of new protein. Yeung and Oliver showed that this effect was not shared by 5'-AMP or cyclic GMP or any of the other nucleotides which they examined, and suggested that cyclic AMP might act in this instance to derepress the synthesis of the carboxykinase. As discussed in Section II,B, cyclic AMP also stimulates the synthesis of several other hepatic enzymes, but the effect is not general. For example, Yeung and Oliver showed that the amounts of pyruvate kinase and fructose diphosphatase were unchanged by cyclic AMP.

Although the ability of cyclic AMP to stimulate enzyme synthesis is

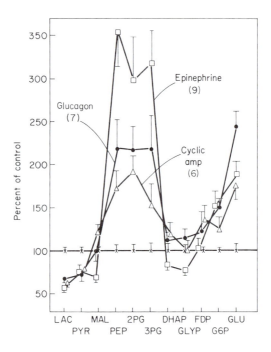

Fig. 7.5. Effects of glucagon, epinephrine, and exogenous cyclic AMP on levels of intermediates in the isolated perfused rat liver. Livers from fasted rats were perfused for 1 hr without substrate and then for another hour with 20 mM lactate plus glucagon ($10^{-8}M$), epinephrine ($10^{-6}M$), or cyclic AMP ($10^{-3}M$). Number of livers shown in parentheses; a control was run with each experimental liver. Reading from left to right, intermediates measured were lactate, pyruvate, malate, phosphoenolpyruvate, 2-phosphoglycerate, 3-phosphoglycerate, dihydroxyacetone phosphate, glycerol-1-phosphate, fructose-1, 6-diphosphate, glucose-6-phosphate, and glucose. Note that in all cases pyruvate is below the control level whereas phosphoenolpyruvate is above it. From Exton and Park (1969).

probably important for the development and maintenance of the gluconeogenic capacity of the liver, it seems unlikely that it can account for the rapid stimulation of gluconeogenesis which occurs in response to glucagon or cyclic AMP in the isolated perfused rat liver. John R. Williamson *et al.* (1966) suggested that this rapid effect might be secondary to the stimulation of hepatic lipolysis (Bewsher and Ashmore, 1966; Claycomb and Kilsheimer, 1969). The idea here was that metabolism of the resulting free fatty acids would lead to an increased intracellular level of acetyl-CoA, which might in turn increase the rate of pyruvate carboxylase activity. Fritz (1967) has postulated a role for carnitine in this process. It has now been shown, however, that gluconeogenesis can be stimulated markedly by glucagon with no detectable change in the level of acetyl-

CoA (Menahan *et al.,* 1968). It has also been shown that perfusion with fatty acids may lead to large increases in ketone body production with no change in the rate of gluconeogenesis, under conditions where glucagon and cyclic AMP stimulate gluconeogenesis maximally (Exton *et al.,* 1969a; Menahan and Wieland, 1969b). In addition, it has been demonstrated that under conditions where exogenous fatty acids do stimulate gluconeogenesis, the effects of maximally effective concentrations of fatty acids and of glucagon are at least additive (Ross *et al.,* 1967; Exton *et al.,* 1969a). These and other observations (e.g., Adam and Haynes, 1969) make it unlikely that the stimulation of hepatic lipolysis is an obligatory step in the gluconeogenic effect of glucagon.

The stimulation of hepatic gluconeogenesis seems to be a longer lasting effect than the stimulation of glycogenolysis (Sokal, 1966). This makes good sense from the teleonomic point of view, and might suggest that the basic process could be similar to phosphorylase activation, which is to say it might involve some process such as the phosphorylation of a protein, but that the recovery process has a longer time constant than that associated with the phosphorylase phosphatase reaction.

Lewis *et al.* (1970) have shown that in perfused livers from normal rats, gluconeogenesis may be slightly more sensitive than glycogenolysis to the stimulating effect of glucagon. This may depend critically on the hormonal state of the animal. Exton *et al.* (1966) had found that perfused livers from adrenalectomized animals were relatively insensitive to the gluconeogenic action of glucagon. It was later shown that the formation of cyclic AMP in response to glucagon in these livers was normal, and that the reduced sensitivity extended to the action of exogenous cyclic AMP (Friedmann *et al.,* 1967). Sensitivity to glucagon or cyclic AMP could be restored by the administration of dexamethasone *in vivo* or even by adding the steroid to the perfusate *in vitro*. It was also established that dexamethasone did not stimulate gluconeogenesis from lactate in the perfused liver when glucagon was absent from the perfusate. These observations suggested that the glucocorticoids stimulate hepatic gluconeogenesis, at least in part, by enhancing the sensitivity of whatever system is ultimately affected by cyclic AMP. This direct effect of the steroids would complement the tendency of these hormones to increase the rate of supply of gluconeogenic substrates to the liver, and together these effects could account for a substantial part of their total gluconeogenic effect. Whether any of the other permissive effects of steroid hormones can be related to sensitization of a system or systems responsive to cyclic AMP remains to be seen. These hormones could also act by stimulating a reaction that becomes rate-limiting only after the cyclic AMP-sensitive step has been accelerated.

A requirement for Ca^{2+} in renal gluconeogenesis has been deduced from studies with isolated segments of renal tubules (Nagata and Rasmussen, 1970). In this preparation, parathyroid hormone rather than glucagon stimulates cyclic AMP formation and gluconeogenesis (see also Chapter 10). Parathyroid hormone stimulated the formation of cyclic AMP whether calcium was added to the external medium or not, but neither parathyroid hormone nor exogenous cyclic AMP stimulated gluconeogenesis from lactate in the absence of external Ca^{2+}. This requirement for Ca^{2+} probably applies to hepatic gluconeogenesis as well, and it is of interest that an effect of glucagon or exogenous cyclic AMP on hepatic Ca^{2+} flux has been demonstrated (Friedmann and Park, 1968). However, whether cyclic AMP stimulates gluconeogenesis by altering calcium permeability, or whether calcium is simply required in order for another action of cyclic AMP to be expressed, is unknown at present.

Considering next gluconeogenesis from amino acids, it is clear that in this case there might be other steps in addition to those between pyruvate and glucose which could at times be rate-limiting. Mallette *et al.* (1969a,b) found that when isolated livers were perfused with a mixture of amino acids at various multiples of their normal plasma concentrations, glucose production approached saturation at three times the normal levels and was about half-maximal at the normal concentrations. As in the case of gluconeogenesis from other substrates, therefore, changes in the blood level of amino acids, and especially of alanine (Felig *et al.*, 1970), would be expected to have an important influence on the overall rate of gluconeogenesis *in vivo*. In addition, Mallette *et al.* (1969a) found that gluconeogenesis and urea production could be stimulated by glucagon, as well as by epinephrine and exogenous cyclic AMP, at all amino acid concentrations studied.

There is now evidence that part of this stimulatory effect may result from the stimulation of amino acid uptake by the liver. Mallette *et al.* (1969b) showed that when the perfusate concentration of alanine was high, under which conditions transamination to pyruvate seemed to be rate-limiting, glucagon or cyclic AMP caused an increase in the intracellular concentration as well as a decrease in the extracellular concentration of this amino acid. Also, Chambers *et al.* (1970) have observed that glucagon is capable of stimulating the rate of uptake of α-aminoisobutyric acid, an amino acid which is not utilized by the liver either for protein synthesis or for gluconeogenesis. This effect, like all of the other hepatic effects of glucagon which have been studied, could be mimicked by the dibutyryl derivative of cyclic AMP.

The glucocorticoids also stimulate gluconeogenesis from amino acids (Eisenstein, 1967; Rinard *et al.*, 1969), although in this case a permissive effect for the actions of glucagon and the catecholamines has not been demonstrated. Presumably such an effect will be demonstrated when physiological concentrations of hormones are used. An even more important effect of glucocorticoids *in vivo* may be the stimulation of protein catabolism in peripheral tissues, leading to an increase in the rate at which alanine and other precursors are made available for hepatic conversion to glucose (Dunn *et at.*, 1969).

Another potential bottleneck in the conversion of amino acids to glucose, in addition to amino acid transport and the conversion of pyruvate to phosphoenolpyruvate, may occur at the level of transamination. Mallette *et al.* (1969a) showed that with alanine as the substrate, transamination could indeed be the rate-limiting step under certain conditions. They were unable to demonstrate a clear-cut stimulation of this step in response to glucagon or cyclic AMP, but the possibility of such an effect was not ruled out. The stimulatory effect of cyclic AMP on the induction of hepatic tyrosine transaminase was mentioned in Chapter 5, and it is at least conceivable that alanine transaminase may be affected similarly. The effect of glucagon on tyrosine transaminase is a rapid effect whose time course coincides closely with that of the hyperglycemic effect (Civen *et al.*, 1967). Its rapidity notwithstanding, this effect seems clearly to involve the *de novo* synthesis of enzyme protein (Wicks *et al.*, 1969). The induction of hepatic enzyme synthesis is discussed again in Section II,B following.

Still other enzymes could be rate-limiting when gluconeogenesis from endogenous liver protein is considered. Stimulation of this process, which in the adult rat could account for as much as 700 mg of glucose daily (Sokal, 1966), may at times be an important physiological effect of glucagon. This effect is perhaps best demonstrated by the use of perfused livers from 24-hr fasted rats with no exogenous carbon source added to the perfusion fluid. Under these conditions glucose and urea are produced in an almost one-to-one relationship, and the production of both metabolites is stimulated to an equivalent degree by physiological levels of glucagon (Miller, 1965; Sokal, 1966; Menahan and Wieland, 1967). This effect can also be produced by catecholamines, providing care is taken to maintain blood flow constant (Heimberg *et al.*, 1964; Mallette *et al.*, 1969a), and can be mimicked by exogenous cyclic AMP (Menahan and Wieland, 1967; Mallette *et al.*, 1969a; Chambers and Bass, 1970). The site of action of cyclic AMP is not entirely clear. In addition to the mass action effect that might be expected to follow stimulation of gluconeogenesis from endogenous amino acids, there is evidence that cyclic AMP also accelerates the conversion of protein to amino acids by a more direct mechanism. Inhibition of incorporation of amino acids into protein when liver slices were exposed to cyclic AMP was noted many years ago (Pryor and Berthet, 1960), but whether this was the result of a direct inhibitory effect on the protein synthesizing machinery or a stimulatory effect on the rate of proteolysis or both or neither of these effects was an unanswered question. The bulk of current evidence is on the side of enhanced proteolysis. For example, Mallette *et al.* (1969b) found that while glucagon increased the uptake by the perfused liver of certain gluconeogenic amino acids, it actually increased the output of some other amino acids, including valine. Since valine is neither synthesized nor catabolized by the liver to any significant extent, this result suggested that proteolysis had been stimulated. This interpretation is in line with the effect of insulin, which lowers the level of cyclic AMP in the liver (Jefferson *et al.,* 1968) and seems to inhibit proteolysis (Glinsmann and Mortimore, 1968). It is at least conceivable that this action of cyclic AMP involves the synthesis or release of proteolytic enzymes from lysosomes (Deter and De Duve, 1967; Vaes, 1968b).

In summary, the liver is capable of utilizing a variety of carbohydrate and noncarbohydrate precursors for the production of glucose. These precursors include lactate, pyruvate, and alanine and other amino acids present in blood, as well as endogenous liver protein. Gluconeogenesis from all of these precursors can be stimulated by physiological concentrations of glucagon, and in all cases the effect can be mimicked by exogenous cyclic AMP. In the case of gluconeogenesis from lactate, the rate-limiting step or steps are between pyruvate and phosphoenolpyruvate, and cyclic AMP accelerates this conversion. In the cases of gluconeogenesis from amino acids and protein, still other reactions may be rate-limiting. Cyclic AMP increases the rate of uptake of exogenous amino acids and seems also to increase the rate of breakdown of endogenous protein. In addition, the rate of transamination may be increased. The molecular mechanisms by which cyclic AMP produces these effects are poorly understood.

3. The Response in the Pancreas

The stimulatory effect of glucagon on the release of insulin was mentioned briefly in the last chapter, in connection with the pancreatic effects of the catecholamines. This effect of glucagon has now been studied by a number of investigators, both *in vivo* and *in vitro*, and there is no longer any doubt that it represents a direct effect of glucagon on the pancreas. The available evidence suggests strongly that this effect is mediated by cyclic AMP. Although it has not been possible to make direct measurements of the level of cyclic AMP in pancreatic beta cells, a stimulatory effect of glucagon on the formation of cyclic AMP by pancreatic islets has been demonstrated (Turtle and Kipnis, 1967b). The stimulatory effect of the hormone on insulin release can be enhanced by caffeine or theophylline (Lambert *et·al.*, 1967a,b, 1969a; Lacy *et al.*, 1968b; Allan and Tepperman, 1969) and mimicked by exogenous cyclic AMP (Sussman and Vaughan, 1967; Malaisse *et al.*, 1967; Feldman and Lebovitz, 1970). Although 5'-AMP and other adenine nucleotides may stimulate insulin release when injected *in vivo* (Levine *et al.*, 1970), 5'-AMP was not effective when applied to the isolated perfused rat pancreas (Sussman and Vaughan, 1967).

Glucose itself has long been recognized as an important stimulus for insulin release (Mayhew *et al.*, 1969). The greater effectiveness of oral over intravenously administered glucose can be understood in terms of the ability of glucose to stimulate the release of several gastrointestinal hormones (Unger *et al.*, 1967, 1968), which in turn stimulate insulin release by an unknown mechanism (Buchanan *et al.*, 1969b). Still it is clear from *in vitro* studies (Coore and Randle, 1964) that glucose also stimulates the pancreas directly. Little is known about the mechanism of this action of glucose, although metabolism to at least the stage of glucose-6-phosphate seems to be required (Samols *et al.*, 1966; Kilo *et al.*, 1967; Montague *et al.*, 1967; Kuzuya and Kanazawa, 1969; Lambert *et al.*, 1969a).

The general observation with adult pancreatic tissue has been that cyclic AMP and agents that stimulate its endogenous accumulation do not stimulate insulin release unless a certain minimum concentration of glucose (on the order of 100 mg%) is also present (Malaisse *et al.*, 1967; Lacy *et al.*, 1968b). Glucose, on the other hand, does not under these conditions require the addition of a hormone or cyclic AMP in order to stimulate insulin release (Coore and Randle, 1964; Buchanan *et al.*, 1969b). On the basis of these findings, it seemed reasonable to suppose that cyclic AMP might act either by increasing the rate at which glucose was converted to an active metabolite (or transported into a special compartment) or else by promoting the action of the active metabolite, without actually being required in order for the process to occur.

Recently, however, a series of experiments with fetal pancreatic tissue have cast some doubt on this proposition. Using organ cultures of fetal rat pancreas, which contain fewer alpha cells and therefore less glucagon than corresponding preparations of adult tissue (Foa, 1968; Orci *et al.*, 1969), Vecchio *et al.* (1966) found that glucagon could stimulate the release of insulin even in the absence of added glucose. The addition of glucose by itself was relatively ineffective, and this was also found by Milner (1969) using isolated segments of fetal rabbit pancreas. However, when glucagon and glucose were added together (Lambert *et al.*, 1967a, 1969b), the amount of insulin released was much greater than the sum of the responses to either agent added separately. The response to either glucagon or glucose or both could be further enhanced by caffeine (Lambert *et al.*, 1969a). Pyruvate could replace glucose in all of these situations, although on a molar basis it was less potent than glucose (Lambert *et al.*, 1969a).

The striking synergism between glucose and glucagon which is apparent in these preparations from fetal tissue, but which is less apparent in the adult pancreas, suggests that cyclic AMP generated in beta cells in response to glucagon released from neighboring alpha cells may play an essential role in the insulin-releasing mechanism (Lambert *et al.*, 1967b). The finding that glucose inhibits the release of glucagon from adult pancreatic islets (Vance *et al.*, 1968; Chesney and Schofield, 1969) does not argue strongly against this interpretation, since it seems possible that enough glucagon is always present under the conditions of these experiments to permit insulin release in response to glucose. The possibility that a defect in the cyclic AMP-generating mechanism may be involved in the pathogenesis of human diabetes mellitus is discussed in Section III,C.

A group of amino acids, of which lysine and arginine seem to be most effective (Lambert *et al.*, 1969b), are also capable of stimulating the release of insulin. Since these amino acids also stimulate the release of glucagon (Chesney and Schofield, 1969), it is conceivable that their effect on insulin release is mediated by glucagon, at least in part. The finding that theophylline also stimulates glucagon release (Chesney and Schofield, 1969) raises the interesting possibility that cyclic AMP may regulate the release of glucagon as well as of insulin. Experiments designed to test this hypothesis have not been reported.

The mechanism by which cyclic AMP stimulates the release of insulin is at least as poorly understood as the mechanism by which glucose acts. Like many other cyclic AMP-regulated processes (see Chapter 5), the release of insulin is an exocytotic process (Lacy, 1967) that requires calcium (Hales and Milner, 1968). It is at least conceivable that the phosphorylation of one or more membranal proteins in response to cyclic AMP leads to fusion of the insulin-containing granules with the interior of the plasma membrane of the beta cell, following which rupture of the membrane at that site occurs. Another idea (Lacy *et al.*, 1968a) is that microtubular function within the beta cell may be affected by cyclic AMP and calcium. Roth *et al.* (1970) have suggested an interesting possible mechanism by which conformational changes in the microtubular elements could lead to the movement of granules and other intracellular structures.

Grodsky and Bennett (1966) found that increasing the concentration of potassium from 4 to 8 mM stimulated insulin release even in the absence of glucose. This is of interest in view of the ability of K^+ to increase cyclic AMP levels in other tissues (Lundholm *et al.*, 1967; Sattin and Rall, 1970; Shimizu *et al.*, 1970b), but whether a similar effect occurs in pancreatic beta cells is unknown.

Despite intensive study over the years (Butterfield and Van Westering, 1967), the mechanism by which tolbutamide stimulates the release of insulin is still obscure. The finding that tolbutamide and caffeine acted synergistically in fetal rat tissue (Lambert *et al.*, 1967a) raised the possibility that tolbutamide and glucagon might act in a similar fashion, i.e., by stimulating adenyl cyclase. It was later established, however, that whereas glucose was considerably more effective than caffeine in enhancing the glucagon effect, caffeine was more effective than glucose in enhancing the tolbutamide effect (Lambert *et al.*, 1969b). The possibility that only part of the effect of caffeine in these cells is related to phosphodiesterase inhibition has to be kept in mind, especially in view of the calcium requirement. Caffeine affects the uptake of calcium by muscle, for example, by a mechanism which seems not to involve a change in the level of cyclic AMP (Sandow, 1965; Weber, 1968). Recognition of this possibility does not explain the action of tolbutamide, but may be helpful in the interpretation of future experiments with these drugs.

We might also mention in this section that several other hormones, including ACTH, gastrin, secretin, and pancreozymin, are capable of stimulating the release of insulin in several species (Lebovitz and Pooler, 1967; Unger *et al.*, 1967). The physiological significance of the insulinogenic action of these hormones is uncertain; future studies with gastrin, which is thought to be produced in the delta cells in pancreatic islets, may be of special interest in this regard. However, whether any or all of these hormones affect the beta cells directly, and, if so, whether they act by stimulating adenyl cyclase, are questions that remain for future study. Potentiation by phosphodiesterase inhibitors (Lebovitz and Pooler, 1967; Allan and Tepperman, 1969) is suggestive but not conclusive.

In addition to these hormones, Unger *et al.* (1968) have studied a glucagonlike material obtained from extracts of canine jejunum which also appears to have insulin-releasing properties. This material resembles glucagon in the sense that it cross-reacts with glucagon antibodies, but differs from pancreatic glucagon, and also from the gastrointestinal material studied by Makman and Sutherland (1964; see also Valverde *et al.*, 1968), in that it does not stimulate hepatic adenyl cyclase, and consequently does not produce hyperglycemia. Based on chromatographic studies, this material appears to have a molecular weight at least twice that of glucagon, and Unger and his colleagues have suggested that it may be an aggregate of glucagon, or possibly glucagon complexed to some other molecule. Although the biological significance of this material is unknown, its tendency to react in the glucagon immunoassay appears to account for several conflicting reports in the earlier literature relating to blood glucagon levels.

The insulinogenic activity of growth hormone seems to involve the stimulation of protein synthesis, including the synthesis of insulin itself (Martin and Gagliardino, 1967). It has been suggested that the protein component of adenyl cyclase may be among the proteins whose rate of synthesis is stimulated by growth hormone in adipose tissue (Fain, 1968a; Moskowitz and Fain, 1970), but whether such an effect occurs in pancreatic beta cells is unknown.

The tendency of epinephrine and norepinephrine to inhibit insulin release was discussed in Chapter 6 (Section III,C). Another biogenic amine capable of inhibiting the release of insulin, at least in some species, is serotonin. Using pieces of hamster pancreas *in vitro*, Feldman and Lebovitz (1970) found that serotonin inhibited insulin release in response to all agents tested, including the dibutyryl derivative of cyclic AMP. Interestingly, cytochemical evidence suggests that serotonin is localized in beta cells within the same granules in which insulin is stored (Jaim-Etcheverry and Zieher, 1968).

4. Possible Significance of the Insulinogenic Response to Glucagon

Since physiological concentrations of glucagon were known to be capable of stimulating the production and release of glucose by the liver, it seemed reasonable for many years to assume that glucagon might function primarily as a hormone of glucose need (Foa and Galansino, 1962; Sokal, 1966; Unger, 1966). Although questioned for a time, this earlier interpretation now seems to be reasonably well supported by experimental data. Thus, increasing the concentration of glucose in the incubation medium inhibits the release of glucagon from isolated pancreatic islets (Vance *et al.*, 1968; Chesney and Schofield, 1969), and infusing glucose intravenously leads to a fall in the level of glucagon in the pancreaticoduodenal vein of intact animals (Ohneda *et al.*, 1969). Conversely, insulin-induced hypoglycemia is associated with an increase in the level of pancreatic vein glucagon (Ohneda *et al.*, 1969; Buchanan *et al.*, 1969a). Decreased levels of circulating immunoreactive glucagon in response to intravenous glucose (Shima and Foa, 1968) and increased levels in response to starvation (Aguilar-Parada *et al.*, 1969) have also been detected in human volunteers. These and other observations suggest that glucagon could function as a hormone of glucose need, possibly as part of a relatively simple feedback system. A fall in the level of blood glucose reaching the pancreas would stimulate the release of glucagon, which could then travel to the liver to stimulate hepatic glucose production. The resulting increase in blood glucose would then return to the pancreas to inhibit the further release of glucagon.

The seemingly contradictory finding that glucagon could also stimulate the release of insulin was dismissed by some investigators on the grounds that most of the early experiments demonstrating this effect had made use of pharmacological rather than physiological doses of glucagon. It was later shown, however, that the intraportal injection of very small doses of glucagon (too small to produce a detectable increase in circulating glucagon, by methods which could detect an increase in response to hypoglycemia) did cause an increase in insulin levels (Ketterer *et al.,* 1967; Buchanan *et al.,* 1969a). It therefore seems possible that even circulating levels of glucagon, not to mention the higher levels likely to be in contact with the pancreatic beta cells, could at times have an effect on the rate of release of insulin. The experiments with fetal pancreatic tissue, summarized in the preceding section, suggest that the presence of glucagon released from alpha cells may play an important role in regulating the release of insulin from neighboring beta cells.

The ability of glucagon to stimulate the release of insulin from the pancreas as well as glucose from the liver seemed all the more puzzling in view of the evidence (summarized in Section III,A of this chapter and Section III,A of the next chapter) that insulin produced some of its effects in hepatic and adipose tissue by suppressing the accumulation of cyclic AMP. The picture developed of one hormone, glucagon, stimulating adenyl cyclase in at least two types of cells. This action in the pancreas led to the release of a hormone which tended to oppose the action in the liver.

Upon reflection, it is possible to see that this unusual arrangement could have a certain amount of survival value for the animal in which it occurs. In the first place, it should be noted that the low blood glucose levels that presumably constitute the normal stimulus for pancreatic glucagon release in adult animals might be insufficient to permit the concomitant release of insulin. Because of the important requirement of a minimum glucose level for insulin release, glucagon and glucose could participate in the feedback system discussed previously, despite the tendency of glucagon to stimulate insulin release at higher levels of blood glucose. Higher levels of glucose returning to the pancreas would of course tend to stimulate the release of insulin at the same time that glucagon release was being suppressed, and this might have the beneficial effect of dampening the oscillations in blood glucose levels that might otherwise occur. It is probably that under most conditions the level of cyclic AMP in the liver is a function of the ratio of glucagon to insulin; subtle shifts in the balance between these two hormones are probably more important than large changes in the absolute level of either one of them.

It is also possible that enough insulin could at times be released with glucagon to affect muscle and adipose tissue without preventing the hepatic effects of glucagon, because muscle and adipose tissue (at least in rats) are more sensitive to insulin than is the liver. The possibility that this might occur is not altered by the fact that the liver is also exposed to higher concentrations of insulin than are the other tissues. To the extent that this does occur, then glucagon would function to shunt glucose from the liver to other peripheral tissues, thus differing from the catecholamines which tend to shunt glucose from the liver to the brain (Chapter 6, Section III,D).

The extent to which gastrointestinal glucagon is released in response to oral glucose is presently unknown. Observations that circulating levels of immunoreactive glucagon are increased in response to glucose feeding (e.g., Shima and Foa, 1968) are complicated by the putative presence of the cross-reacting material studied by Unger *et al.* (1968). The release of this material may account, at least in part, for the long-standing observation that orally administered glucose is a more effective stimulus for insulin release than comparable or even greater hyperglycemia produced by parenteral glucose. Other gastrointestinal hormones that are released in response to feeding, and which stimulate the pancreas without affecting the liver, include gastrin, secretin, and pancreozymin. It is even possible that if glucagon itself is released from the gastric mucosa in response to feeding, it stimulates the release of enough insulin to inhibit its own action in the liver. Although the capacity of glucagon to increase the hepatic level of cyclic AMP exceeds the capacity of insulin to reduce it, this may not be very meaningful when only physiological concentrations of the hormones are considered.

The release of any or all of these insulinogenic hormones may have played an important role in assuring our own survival. The conditions facing early man and his predecessors, like those facing most wild animals today, would surely have included prolonged periods of fasting interspersed with relatively infrequent (and brief) periods of plenty. Those animals which had developed a mechanism for releasing insulin during these periods (thereby facilitating the conversion of glucose to glycogen, free fatty acids to triglycerides, and amino acids to protein) would have had a considerable advantage over those animals which lacked such a mechanism. They would have been able to travel longer distances for longer periods without food, and, when they finally did come upon a potential meal, would have had a reserve source of energy (mobilized in all probability by cyclic AMP generated in response to catecholamine release) to provide the strength to make the kill. Those animals which lacked this mechanism would more often have *been* the meal, and would have had correspondingly fewer opportunities to reproduce themselves.

The exact nature of the hormonal response to feeding may of course depend very critically on the nature of the food products ingested. Certain amino acids, for example, are capable of stimulating the release of glucagon from pancreatic islets *in vitro* (Chesney and Schofield, 1969), but whether this is accompanied by the release of insulin may depend upon the prevailing glucose level. Aguilar-Parada *et al.* (1969) found that in human subjects the infusion of arginine caused a greater increase in plasma glucagon levels after 3 days of starvation than it did in fed controls. Conversely, the rise in insulin was less in the starved subjects than it was in the controls. A high ratio of glucagon release to insulin release would be an appropriate initial response of a starving animal to a meal rich in protein but low in carbohydrate. Although it might deplete the liver of its remaining glycogen, this would eventually be restored because of the increased rate of gluconeogenesis. The glucose produced from the ingested amino acids would be converted to glycogen when blood glucose reached a certain level partly because of a direct effect of glucose on the liver (Buschiazzo *et al.*, 1970) and partly because of the glucose-induced release of insulin. Presumably the initial ratio of glucagon to insulin release will be more balanced in response to a more balanced meal. The importance of subtle shifts in the ratio of glucagon to insulin in regulating the level of cyclic AMP in the liver was mentioned previously, and it seems likely that this will be further emphasized by future studies of the hormonal response to feeding.

The foregoing remarks have been based entirely on the results of metabolic studies of mammalian tissues, and it should be noted that the relative importance of changes in the rate of release of glucagon and insulin may be quite different in other species. For example, it is well known that in most mammalian species pancreatectomy leads to hyperglycemia, with the administration of insulin being required in order to prevent death. In birds, pancreatectomy has just the opposite effect; in these species death is associated with hypoglycemia, and life can be maintained by the administration of glucagon (Sokal, 1966). It would appear that insulin may be relatively less important in birds than it is in mammals, and also that glucagon serves some of the same functions in birds that are served in mammals by the catecholamines. This latter interpretation is strengthened by the finding (Grande and Prigge, 1970) that in birds glucagon is a more potent lipolytic agent than are the catecholamines, whereas just the opposite is true in most mammalian species studied. The impression created by the results of these and other studies is that the role of cyclic AMP in a given system may be relatively constant from one species to another, but that the hormonal mechanisms regulating cyclic AMP may differ considerably. Obviously many details of the relation between hormones and cyclic AMP remain to be discovered, even in mammals, but the situation seems to offer an especially interesting challenge for the comparative endocrinologist.

Although we have made a retrospective attempt in this chapter to show that glucagon's ability to stimulate adenyl cyclase in the pancreas may be important, or at least not as anomalous as it was once thought to be, our success is open to question. The evolutionary pattern would seem simpler if glucagon did not have this ability or if another pancreatic hormone were found to share this ability. Gastrin would seem to be a prime candidate for this role, but, in the one study of which we are aware in which gastrin was added to isolated pancreatic islets (Buchanan *et al.,* 1969b), it did not stimulate the release of insulin. However, since glucagon was also relatively ineffective in these experiments, the possibility remains that the islets were so heavily contaminated with endogenous gastrin that an effect of the added hormone could not have been seen. Further studies seem warranted, and studies with fetal pancreatic tissue may be especially revealing.

In summary, glucagon stimulates the formation of cyclic AMP in liver cells and also in pancreatic islets. In the liver, this leads to an increased rate of glucose production. In the pancreas, it causes an increase in the rate of release of insulin. Recent demonstrations that glucose inhibits the release of glucagon from islets support the earlier concept that glucagon may function primarily as a hormone of glucose need. Its action in the liver may therefore be more important than its action in the pancreas. Other observations suggest that glucagon may also play an important role in the regulation of insulin release, even though it would seem more appropriate, teleonomically, if this role could be played by a different hormone. A better understanding of the normal physiological role of glucagon will probably emerge from the results of future research.

B. Other Hepatic Effects

Glucagon produces several other effects in the liver in addition to those discussed in Section II,A. There is evidence that all of these effects are mediated by cyclic AMP, as described briefly in the following paragraphs.

1. Ion flux

Glucagon stimulates the efflux of potassium ions from the liver (Galansino *et al.,* 1960; see also Foa, 1968). Since this effect is shared by epinephrine and can be mimicked by exogenous cyclic AMP (Northrop, 1968; Exton and Park, 1968b), it is presumably mediated by cyclic AMP. The transient increase of plasma K^+ that occurs in dogs in response to injected glucagon can be blocked by doses of dihydroergotamine (DHE) which do not prevent the hyperglycemic response (Galansino *et al.,* 1960). Higher concentrations of DHE may interfere with other effects of both glucagon and exogenous cyclic AMP (e.g., Northrop, 1968; Chambers and Bass, 1970), but little is known about the mechanism of action. Some other puzzling aspects of research in this area were mentioned in Chapter 6 (Sections III,A and V,B).

Glucagon, epinephrine, and exogenous cyclic AMP are also capable of increasing the rate of efflux of labeled calcium from the isolated perfused rat liver (Friedmann and Park, 1968). The physiological significance of these effects on hepatic ion flux is unknown, and their relation to the other hepatic effects of cyclic AMP is poorly understood.

2. Ketogenesis

Although the rate of ketone body formation is determined largely by the concentration of free fatty acids in the blood, and hence regulated indirectly by changes in the level of cyclic AMP in adipose tissue (Chapter 8), cyclic AMP also stimulates ketogenesis more directly by acting within the liver. The ability of exogenous cyclic AMP to mimic the ketogenic effect of glucagon in liver slices (Haugaard and Haugaard, 1954) was demonstrated many years ago by Berthet (1959), but the mechanism is still obscure.

It has been suggested that the ketogenic effect may be secondary to the stimulation of hepatic lipolysis (Bewsher and Ashmore, 1966; Claycomb and Kilsheimer, 1969), but the greater effect seen in the presence of increasing concentrations of exogenous free fatty acids (Heimberg *et al.,* 1969) would argue against this interpretation. Heimberg and his colleagues suggested that the stimulation of fatty acid oxidation might be a more important factor. In line with this, Regen and Terrell (1968) studied the metabolism of labeled acetate by the isolated perfused rat liver, and found that while glucagon increased ketogenesis about threefold, it increased the incorporation of label into ketone bodies only 1.7-fold. This fall in the specific activity of ketone bodies produced in the presence of glucagon was interpreted as evidence for an increased rate of turnover of acetyl-CoA. From a consideration of label incorporation into Krebs cycle intermediates in relation to the specific activity of acetyl-CoA, Regen and Terrell concluded that the initial steps of the Krebs cycle must have been stimulated about twofold under these conditions.

Berthet (1960) had previously reported a reduction in the incorporation of label into fatty acids in response to glucagon and cyclic AMP in liver slices, and Regen and Terrell showed that this corresponded to the reduction in specific activity of acetyl-CoA.

Berthet (1959) also found that glucagon and cyclic AMP reduced the incorporation of label into cholesterol in liver slices, but this effect could not be demonstrated in the isolated perfused rat liver (Regen and Terrell, 1968). There is agreement that glucagon does markedly inhibit the release of cholesterol and cholesterol esters into the perfusate (Penhos et al., 1967; Gorman et al., 1967; Regen and Terrell, 1968), and it is conceivable that under some conditions this could contribute to an inhibition of cholesterol synthesis. Regen and Terrell showed that in the perfused liver glucagon also inhibited the release of newly synthesized fatty acids in the form of lipoprotein-bound triglycerides. This effect of glucagon was presumably mediated by cyclic AMP, since Heimberg et al. (1964) had previously shown the same effect in response to norepinephrine and have more recently demonstrated the effect of exogenous cyclic AMP (Heimberg et al., 1969).

That ketogenesis and gluconeogenesis may each be stimulated independently of the other (Menahan and Wieland, 1969b; Exton et al., 1969a), was cited previously as evidence that an effect on fatty acid metabolism was probably not important in the gluconeogenic response to glucagon. The ability of cyclic AMP to stimulate ketogenesis by the liver probably contributes to the production of diabetic ketoacidosis, even though its action in adipose tissue to stimulate the release of free fatty acids may be a more important factor.

3. Histone Phosphorylation

The ability of glucagon to stimulate histone phosphorylation in the liver (Langan, 1969b) is of interest for several reasons, and represents one of the few potentially important effects of this hormone to have been discovered *after* its mechanism of action had been studied in a cell-free system.

Following the discovery by Walsh, Perkins, and Krebs (1968) of the cyclic AMP-sensitive protein kinase that catalyzes the phosphorylation of muscle phosphorylase kinase, Langan (1968) searched for and found a similar enzyme in rat liver that catalyzed the phosphorylation of several nuclear histones. There is evidence that histones may function as repressors of DNA template activity in eukaryotic cells (Bullough, 1967) and the possibility had long been considered that phosphorylation of these proteins might render them less effective in this capacity, thereby leading to derepression of gene activity. Langan's discovery that histone phosphorylation could be stimulated by cyclic AMP thus suggested a possible mechanism by which glucagon might act to stimulate the synthesis of several hepatic enzymes (see Section 4 following).

The existence of a relatively specific phosphatase for these histones was later established (Meisler and Langan, 1969). The degree of phosphorylation of at least some histones therefore seems to represent a balance between the activities of a kinase and a phosphatase, analogous to the mechanism by which the activities of glycogen phosphorylase and glycogen synthetase are regulated (see Chapter 5). It is even possible that the kinase studied by Langan is the same enzyme that functions in these other systems, but this had not been established at the time of this writing.

Langan later showed that the kinase-catalyzed phosphorylation of lysine-rich histone (the fl fraction) occurred primarily on a single specific serine residue. A tryptic phosphopeptide containing this residue could be isolated from tryptic digests of rat liver, and it was found that the degree of phosphorylation at this site was increased dramatically following the intraperitoneal injection of cyclic AMP or its dibutyryl derivative. Equivalent doses of 5'-AMP were essentially inactive (Langan, 1969a).

As mentioned, the effects of hormones were not studied until later (Langan, 1969b). The degree of phosphorylation was increased some 25-fold following the injection of glucagon, and this was not prevented by pretreating the rats with either cycloheximide or actinomycin D. By contrast, two hormones which do not alter the level of cyclic AMP in the liver, hydrocortisone and ACTH, had no effect on histone phosphorylation. The results of dose-response and time course studies indicated that even at a maximum glucagon caused the phosphorylation of only about 1% of the fl histones present in the liver.

An apparently discrepant finding was that insulin, which lowers the level of cyclic AMP in the liver, also stimulated histone phosphorylation (Langan, 1969b). Although it was not established in these experiments that this was a direct effect of insulin on the liver, it is of interest that insulin does share with glucagon the ability to stimulate the synthesis of tyrosine transaminase in the liver *in vitro* (Wicks, 1969), even though it opposes most of the other hepatic effects of glucagon. Should it be shown that insulin does stimulate histone phosphorylation at the same time that it lowers the level of cyclic AMP in the liver, this may provide an important clue to the mechanism of action of insulin. It may be especially interesting to learn whether insulin and glucagon stimulate the phosphorylation of the same or different histones, and whether insulin may oppose the effect of glucagon on some.

4. Hepatic Enzyme Induction

The net effect of glucagon or cyclic AMP on hepatic protein metabolism is to inhibit protein synthesis or stimulate proteolysis or both. These effects were discussed previously in Section II,A,2 in connection with the stimulation of gluconeogenesis. Cyclic AMP is also capable of selectively stimulating the synthesis of certain liver enzymes, as discussed in the following paragraphs.

a. Tyrosine Transaminase. This enzyme is extraordinary in several respects. It seems to have an unusually rapid rate of turnover and seems to be sensitive to an unusual variety of environmental influences. Despite intensive biochemical study, the physiological significance of changes in the amount or activity of this enzyme is still unclear (Kim and Miller, 1969).

The level of tyrosine transaminase activity in the livers of fetal rats *in utero* is normally very low, but begins to increase dramatically immediately following birth (Greengard, 1969a; Holt and Oliver, 1969a). The factors responsible for this postnatal rise in activity are incompletely understood, but the *de novo* synthesis of enzyme protein seems clearly to be involved, and cyclic AMP seems to play an important role.

It is generally agreed that the normal postnatal rise in rat liver tyrosine transaminase activity can be increased by the injection of either glucagon, epinephrine, or exogenous cyclic AMP (Greengard, 1969b; Holt and Oliver, 1969a). In the fetus 3 to 4 days before term, Greengard (1969b) found that only cyclic AMP was capable of inducing activity. Glucagon and epinephrine were not effective until later (1 to 2 days before term), suggesting that either adenyl cyclase or the receptors for glucagon and epinephrine do not develop until this time. Hydrocortisone had no effect when injected into the fetus (Greengard, 1969b), but adrenalectomy prevented the normal postnatal rise in activity (Holt and Oliver, 1969a). In contrast to the impressive results obtained by Greengard (1969b), Holt and Oliver (1969a) saw no effect of either cyclic AMP or epinephrine when injected into fetal rats *in utero,* although the results with glucagon seem similar. The higher and more variable baseline values in the experiments of Holt and Oliver may account for this discrepancy. Insulin inhibits the normal postnatal rise in activity when injected at birth, but increases activity when injected into older animals (Holt and Oliver, 1969a).

All of these agents stimulate the synthesis of tyrosine transaminase in older animals (Greengard, 1969b; Wicks *et al.,* 1969) or in fetal rat liver obtained at term and maintained in organ culture for 42 hr (Wicks, 1968, 1969). Greengard's erroneous conclusion that glucagon and cyclic AMP become ineffective in older rats derives from the fact that her measurements were made 5 hr after injection. Cyclic AMP and hormones that stimulate its formation in the liver act rapidly but transiently, so that after 5 hr the level of tyrosine transaminase has returned to normal (Fig. 7.6). The action of hydrocortisone is slower in onset but more prolonged.

There seem to be at least three separate mechanisms by which hormones stimulate the induction of this enzyme. Evidence that glucagon and the catecholamines act by way of cyclic AMP includes the similar time course of their effects, lack of additivity when maximally effective concentrations are added together, and potentiation by theophylline (Wicks, 1968, 1969). In addition, it was established by Wicks (1969) that

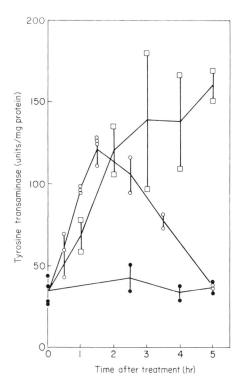

Fig. 7.6. Induction of tyrosine transaminase in rat liver by single or repeated injections of dibutyryl cyclic AMP. Adrenalectomized adult rats were fasted overnight before the experiments were done. Each point represents a single determination. Rats were given 5 mg of dibutyryl cyclic AMP in a single dose at zero time (open circles) or else 2 mg at zero time and again at each hour for the remaining time (open squares). Closed circles refer to untreated controls. From Wicks *et al.* (1969).

the hormones did stimulate adenyl cyclase in homogenates prepared from fetal rat liver at term. The mechanism by which cyclic AMP acts is unknown. Langan's suggestion that histone phosphorylation may be involved (Section B,3 above) is an appealing hypothesis, and seems to be supported by the inhibitory effect of actinomycin D. However, Wicks and his colleagues feel that a posttranscriptional site of action has not been ruled out, and further studies are indicated. Cyclic AMP stimulates protein synthesis in bacteria (Chapter 12) at both the transcriptional and translational levels.

Hydrocortisone acts by a different mechanism, as indicated by the entirely different time course of its inductive effect (Wicks, 1968) and by synergism when added with either glucagon or cyclic AMP (Wicks, 1969; Wicks et al., 1969). In addition, Granner et al. (1968) found that in hepatoma cells which did not respond to either glucagon or exogenous cyclic AMP, hydrocortisone was as effective as it was in normal liver cells. Holt and Oliver (1969b) found that several forms of tyrosine trans-aminase exist in liver, and presented evidence that hydrocortisone and cyclic AMP affected different forms. Hydrocortisone may be permissive for the effect of cyclic AMP, but may in addition stimulate the synthesis of a form not affected by cyclic AMP. As mentioned previously, hydro-cortisone does not stimulate histone phosphorylation in the liver (Langan, 1969b).

Still a third mechanism for the induction of tyrosine transaminase is indicated by the stimulatory effect of insulin (Wicks, 1969). Although the time course of this response resembles that seen with glucagon, insulin cannot act by way of cyclic AMP because it lowers the level of cyclic AMP in the liver (Jefferson et al., 1968). The response to insulin in com-bination with either glucagon or cyclic AMP was greater than that with each inducer alone, but was less than additive (Wicks, 1969). These results are compatible with the hypothesis that insulin stimulates the formation of another second messenger, possibly a derivative of cyclic AMP, which, in the case of tyrosine transaminase at least, may act simi-larly to cyclic AMP. It seems possible that this second messenger could stimulate the phosphorylation of one or more histones which are different from the ones affected by cyclic AMP (Langan, 1969b), and it is also possible that insulin ultimately affects a different form of tyrosine trans-aminase (Holt and Oliver, 1969b). It is obvious that much additional research remains to be done in this area.

Pyridoxine seems to affect tyrosine transaminase partly by an effect on enzyme synthesis (Holt and Oliver, 1969a) and partly by forming pyridoxal 5'-phosphate, which acts as a cofactor (Black and Axelrod, 1969). Black and Axelrod found that norepinephrine and other catecholamines could compete with pyridoxal 5'-phosphate to inhibit tyrosine transaminase *in vitro,* but whether any physiological significance can be attached to this effect remains to be seen. Inhibition by norepinephrine *in vivo* (Black and Axelrod, 1969) could have been related to vasoconstriction (Heimberg and Fizette, 1963).

In addition to the changes already mentioned, rat liver tyrosine transaminase activity undergoes a daily rhythm which depends on the feeding schedule and other factors (Fuller and Snoddy, 1968; Fuller *et al.,* 1969). The possible role of cyclic AMP in regulating this rhythm is unknown at present.

b. Phosphopyruvate Carboxylase. The effect of glucagon on this enzyme was mentioned previously in the section on gluconeogenesis, and its regulation seems simpler than the complex pattern of mechanisms seen to govern tyrosine transaminase activity. Activity is normally absent in fetal rat livers, but appears rapidly after birth (Yeung and Oliver, 1968a). Activity can be induced *in utero* by glucagon or catecholamines or by the injection of exogenous cyclic AMP or its dibutyryl derivative (Yeung and Oliver, 1968b). These agents are also effective in fetal rat liver maintained in organ culture (Wicks, 1969) and also in adult rats *in vivo* (Wicks *et al.,* 1969). Nucleotides other than cyclic AMP are ineffective (Yeung and Oliver, 1968b). Although it seems likely that the *de novo* synthesis of enzyme protein is involved in these effects (Yeung and Oliver, 1968; Wicks, 1969), this has not been unequivocally established. It is conceivable that the phosphorylation of one or more nuclear histones (Langan, 1969a) may be involved.

In contrast to its pronounced effect on tyrosine transaminase, hydrocortisone seems not to exert an important influence on the level of phosphopyruvate carboxylase (Wicks, 1969). In line with its ability to lower the level of cyclic AMP in the liver, insulin opposed the effect of glucagon. That this effect of insulin may not have been due entirely to its effect on cyclic AMP levels per se was indicated by its ability to also antagonize the effect of dibutyryl cyclic AMP (Wicks, 1969). Conceivably insulin stimulates the conversion of cyclic AMP to another derivative, which may in some systems compete with cyclic AMP.

c. Serine Dehydratase. Regulation of this enzyme in rat liver has been especially well studied by Jost and his colleagues (Jost *et al.,* 1968, 1969, 1970). Increased levels have been demonstrated in response to glucagon, epinephrine, or cyclic AMP injected into rats *in vivo* (Jost *et al.,* 1969; Wicks *et al.,* 1969) and *de novo* synthesis of the enzyme has been unequivocally established (Jost *et al.,* 1970). Inhibition of the effect of cyclic AMP by actinomycin D raises the possibility of an effect at the transcriptional level, and is compatible with Langan's suggestion of an involvement of histone phosphorylation. As in the case of phosphopyruvate carboxylase, hydrocortisone has little or no effect on the level of serine dehydratase (Wicks *et al.*, 1969; Jost *et al.*, 1970).

The mechanism by which glucose suppresses the synthesis of this enzyme (Jost *et al.*, 1968) has not been completely defined. Several observations, including reversal of the glucose effect by cyclic AMP (Jost *et al.*, 1969), suggested that the mechanism might be analogous to that of catabolite repression in bacteria (Chapter 12). However, unlike the situation in *E. coli* (Makman and Sutherland, 1965), glucose has not been shown to have an effect on the level of cyclic AMP in the liver (Buschiazzo *et al.*, 1970).

The rise in hepatic cyclic AMP levels that followed the injection of a mixture of amino acids (Jost *et al.*, 1970) may have been secondary to the stimulation of glucagon release (Chesney and Schofield, 1969; Aguilar-Parada *et al.*, 1969). In support of this suggestion, the results of preliminary experiments with D. R. Regen indicate that tryptophan, at least, does not affect the level of cyclic AMP when added to the isolated perfused rat liver. The ability of tryptophan and other amino acids to stimulate the synthesis of serine dehydratase cannot be understood entirely in terms of cyclic AMP in any event, since their effect on cyclic AMP was less than that of glucagon whereas their effect on serine dehydratase was greater (Jost *et al.*, 1970).

d. Glucose-6-Phosphatase. The level or activity of this enzyme can be increased by glucagon, epinephrine, or cyclic AMP when injected into fetal rats *in utero* (Greengard, 1969b), but not when injected after birth or when applied to fetal rat liver in organ culture (Wicks, 1969). Other aspects of the regulation of this enzyme have been discussed by Greengard (1969a).

e. Other Enzymes. The rates of synthesis of many other enzymes may eventually be found to be regulated by cyclic AMP. For example, an effect of glucagon on the induction of aminolevulinic acid synthetase in embryonic chick liver has been reported (Simons, 1970). The effect of glucagon on the apparent activity of this enzyme (15 ± 10 units vs. 8 ± 3 in control embryos) was judged significant, whereas the effect of dibutyryl cyclic AMP (10 ± 3) was not. Further work seems indicated. Among the enzymes that seem clearly *not* to be affected by glucagon or cyclic AMP are tryptophan pyrrolase (Wicks *et al.*, 1969) and reduced NADP dehydrogenase (Greengard, 1969a,b).

5. Hepatic Fine Structure

The effects of glucagon on structural changes within the liver were mentioned in the last chapter, in connection with the hepatic effects of the catecholamines. Since the effects of epinephrine and glucagon are similar, they are probably mediated by cyclic AMP, although an effect of exogenous cyclic AMP had not been reported at the time of this writing. Deter and De Duve (1967) found that, starting fairly suddenly about 30 min after the intraperitoneal injection of a large dose of glucagon, rat liver lysosomes showed an increase in mechanical fragility, an increase in osmotic sensitivity, and changes in sedimentation pattern indicative of an increase in size. These changes may in turn be responsible for the autophagic vacuole formation noted previously by others (Millonig and Porter, 1960; Ashford and Porter, 1962; Ashford and Burdette, 1965). The significance of the enhanced cellular autophagy in response to glucagon and epinephrine is unknown. It could be secondary to some effect such as glycogen depletion, but a more direct effect has not been ruled out. A possible relation between the lysosomal effect and the protein catabolic effect has been suggested (Mortimore and Mondon, 1970).

6. Hepatic Effects Summarized

Glucagon stimulates hepatic adenyl cyclase and thereby increases the level of cyclic AMP in the liver. This is in turn responsible for a variety of effects. Activation of phosphorylase and inhibition of glycogen synthetase activity combine to cause a net breakdown of liver glycogen, and this, together with stimulation of gluconeogenesis, leads to an increased production of glucose. Gluconeogenesis from a variety of substrates (e.g., lactate, pyruvate, amino acids, and endogenous liver protein) may be stimulated by cyclic AMP, partly because of an increased rate of conversion of pyruvate to phosphoenolpyruvate. Amino acid uptake, net proteolysis of endogenous protein, and urea production may all be increased. Associated with these effects are a rapid but transient increase in the efflux of potassium and calcium, an increase in the rate of production of ketone bodies, and a decreased rate of secretion of triglycerides and cholesterol. Cyclic AMP also stimulates histone phosphorylation and increases the rate of synthesis of several hepatic enzymes.

C. Effects in Other Tissues

1. The Positive Inotropic Effect

The positive inotropic effect of glucagon was first observed and studied by Farah and Tuttle (1960). Since this effect occurs only at concentrations of glucagon that exceed normal plasma levels by several orders of magnitude (Aguilar-Parada *et al.*, 1969), it is not likely to be of any physiological significance. It has been of interest from the therapeutic standpoint, however, because of the possibility that glucagon administration may result in fewer unwanted side-effects than are seen with catecholamine therapy (Parmley *et at.,* 1968; Brogan *et al.,* 1969). The evidence that this effect of glucagon is mediated by cyclic AMP will be summarized in this section.

Glucagon stimulates adenyl cyclase in broken cell preparations of cardiac tissue from all species studied, including rats (Murad and Vaughan, 1969), cats (Levey and Epstein, 1969a), dogs (Entman *et al.,* 1969a), and humans (Levey and Epstein, 1969a). Figure 6.5 shows that glucagon is slightly more potent than epinephrine on a molar basis, and this has been seen with all species tested. This correlates well with the relative potencies of these agents in the isolated perfused rat heart (Mayer *et al.,* 1970) and with their cardiac effects in several species *in vivo* (Glick *et al.,* 1968; Lucchesi, 1968).

Murad and Vaughan found that concentrations of propranolol and DCI which completely prevented the effect of epinephrine on adenyl cyclase had no effect on the response to glucagon. This also correlates with observations made at higher levels of organization (Glick *et al.,* 1968; Lucchesi, 1968; Parmley *et al.,* 1968; Mayer *et al.,* 1970). Farah and Tuttle (1960) had reported that the inotropic response to glucagon in the dog heart-lung preparation could be blocked by a large dose of DCI. The possibility that this effect of DCI was related to its not inconsiderable agonist activity was supported by the finding (Lucchesi, 1968) that DCI was no longer capable of blocking the glucagon response after pretreatment with propranolol. This and other observations (e.g., Tuttle, 1970) suggest that cardiac adrenergic receptors may be related to the receptors for glucagon in a rather complex way.

Murad and Vaughan (1969) and Levey and Epstein (1969a) found that maximally effective concentrations of glucagon and epinephrine were not additive in stimulating cardiac adenyl cyclase *in vitro,* suggesting that the receptors for these hormones were related to (or were part of) the same adenyl cyclase system. Since the effect of glucagon was not reduced in the presence of epinephrine, these results do not provide any evidence for the relationship implied by the above mentioned observation by Lucchesi. It seems possible, however, that in the intact heart the interaction of an adrenergic agonist with its receptors might have an effect on the interaction of glucagon with its receptors, and vice versa, in a way not easily seen in a broken cell preparation. Entman *et al.* (1969a) found that glucagon and catecholamines could stimulate cardiac adenyl cyclase not only in low-speed fractions but also in fractions thought to represent fragments of sarcoplasmic reticulum. A separate adenyl cyclase that could be stimulated by thyroid hormones but not by catecholamines has also been reported to exist in cat heart homogenates (Levey and Epstein, 1969b).

Other evidence supporting the concept that the cardiac effects of glucagon are mediated by cyclic AMP has come from studies on the isolated perfused rat heart. It has long been known that the biochemical effects of glucagon in this preparation are essentially the same as those of the catecholamines (Cornblath *et al.*, 1963; Kreisberg and Williamson, 1964; Williams and Mayer, 1966), and recently Mayer *et al.* (1970) showed that these effects of glucagon are associated with an increase in the intracellular level of cyclic AMP. There was some indication that the effect of glucagon on cyclic AMP was less rapid than the effect of epinephrine, as had previously been observed in the liver (Exton and Park, 1968b), although in the heart this was not demonstrated in the same experimental series. The ability of glucagon to increase the level of cyclic AMP had also been observed at Vanderbilt (unpublished observations), although others were unable to see an effect of glucagon either in beating hearts (LaRaia *et al.*, 1968) or in slices (LaRaia and Reddy, 1969). The reasons for this failure are not obvious, especially since epinephrine did produce an effect. However, the effect of epinephrine in these experiments was smaller than that seen by other investigators (e.g., Robison *et al.*, 1965, 1967a; Drummond *et al.*, 1966), and it seems possible that the method of fixation or the relatively insensitive assay procedure or both were inadequate for the purpose. Still another possible explanation is brought to mind by the recent report by Gold *et al.* (1970) that glucagon did not stimulate adenyl cyclase from failing hearts even though epinephrine was active.

If the positive inotropic response to glucagon is mediated by cyclic AMP, as postulated in the last chapter for the catecholamines, then the response to glucagon should be very similar to that seen with the catecholamines. To date this has not been well studied, insofar as kinetic parameters are concerned, and the analysis of existing data is complicated by observations suggesting that adrenergic α-receptors may mediate a positive inotropic response (Govier, 1968) and a negative chronotropic response (James *et al.*, 1968) under certain conditions. However, even if the α-adrenergic component is important in the overall response to the naturally occurring catecholamines, the response to isoproterenol should resemble very closely the response to glucagon. Unfortunately, most of the detailed information we have about the inotropic response to the catecholamines has been obtained using one or another of the naturally occurring hormones. In any event, we can briefly summarize the information presently available.

Mayer *et al.* (1970) studied contractility and cyclic AMP levels in response to glucagon and epinephrine in the isolated perfused rat heart. Both hormones were approximately equipotent in altering both parameters, and additivity was demonstrated over a limited range of doses.

Rigorous correlation between the level of cyclic AMP and the degree of contractility was rendered difficult for several reasons, one being the relative insensitivity of the cyclic AMP assay, such that measurable changes in contractility and phosphorylase activity were produced by doses of the hormones that did not produce a detectable change in the level of cyclic AMP. However, even if the cyclic AMP assay could be made as sensitive and precise as the other methods, and even if it is assumed that the other effects are mediated by cyclic AMP, it is not certain, in a study of this type, that the correlation between cyclic AMP and the other effects would be perfect. Although it has often been supposed, by ourselves as well as others, that the contribution of nonmyocardial cells to the overall level of cyclic AMP in a sample of cardiac tissue might be negligible, this has not been proved and may not be valid. As one of many possible examples suggesting caution in this regard, we might point to experiments with MSH in the frog skin (Abe *et al.*, 1969a,b). This hormone produces a striking effect in skin melanophores but is not known to affect other epithelial cells in this tissue. Since melanophores are absent from ventral skin, and must constitute only a small fraction of the total mass of dorsal skin, it might have been predicted that MSH would produce no effect on the level of cyclic AMP in ventral skin and a relatively modest rise in dorsal skin. The predicted result with ventral skin was in fact obtained, but, far from a relatively modest rise in dorsal skin, supramaximal doses of MSH were found to increase the level of cyclic AMP by more than tenfold. This unexpectedly large effect suggests that if some cardiac cells contain adenyl cyclase that is differentially sensitive to glucagon and the catecholamines, even if most cardiac cells are the same, then it may be unreasonable to expect identical changes in contractility to be associated with identical changes in the overall level of cyclic AMP. This is assuming an ideal assay for cyclic AMP and also that glucagon and the catecholamines owe their inotropic activity exclusively to their ability to stimulate the formation of cyclic AMP. When the known α-receptor component of the catecholamines is taken into account, not to mention other possible actions of the hormones, which could modify their effect *on* cyclic AMP as well as the effects *of* cyclic AMP, then the outlines of a very serious problem emerge. Quantitative analysis of the effects of different hormones can be greatly facilitated if the affected cells can be isolated in a functional state (as has been done, for example, with adipocytes and certain blood cells), but this may be difficult in the case of contractile tissue.*

---

*See, however, Krause *et al.* (1970).

Acetylcholine inhibits the response to glucagon (Lucchesi, 1968) just as it inhibits the response to catecholamines (Meester and Hardman, 1967; Levy and Zieske, 1969). This supports the concept that these hormones act by a common mechanism, and can be understood in terms of the demonstrated ability of acetylcholine to inhibit cardiac adenyl cyclase (Murad *et al.*, 1962). A fall in the level of cyclic AMP in response to acetylcholine has now been demonstrated in rat heart slices (LaRaia and Reddy, 1969) and also in the perfused rat heart (George *et al.*, 1970).

Potentiation of glucagon by theophylline has not been demonstrated, and in fact, in the experiments of Lucchesi (1968), theophylline was found to inhibit the response to glucagon. However, the cardiac effects of the methylxanthines seem very complex (DeGubareff and Sleator, 1965; Olsen *et al.*, 1967), and are probably the result of several actions in addition to phosphodiesterase inhibition. Perhaps the most that can be expected at the present time is that theophylline will affect the inotropic responses to glucagon and the catecholamines similarly, which is to suggest that under conditions where theophylline inhibits the effect of glucagon, it will probably also inhibit the effect of isoproterenol. On the other hand, the previously mentioned tendency of DCI to prevent the response to glucagon, and the ability of propranolol to block the effect of DCI, suggests a complex relationship between adrenergic β-receptors and the receptors for glucagon. Tyramine (Lucchesi, 1968) and norepinephrine (Tuttle, 1970) have also been reported to interfere with the positive inotropic response to glucagon.

Glucagon did increase the rate of tension development in the isolated cat papillary muscle, in which respect it resembles the catecholamines, but did not consistently alter the time to peak tension in these experiments (Glick *et al.*, 1968). A reduction in the time to peak tension seems to be a characteristic feature of the inotropic response to the catecholamines, so that this observation does not support the view that glucagon acts by the same mechanism. However, this parameter is undoubtedly influenced by a variety of factors, such as the temperature and ionic environment, and before any firm conclusions are drawn on the basis of these data, isoproterenol and glucagon should probably be compared in the same series of experiments. Perhaps it will be found that in papillary muscles in which the active state is prolonged by glucagon, then so also will it be prolonged by isoproterenol. This will be an important and interesting series of experiments to perform.

Regarding the possible effect of glucagon on the PIEA, this has not to our knowledge been studied. We might predict, however, on the basis of the studies of Koch-Weser and Blinks (1963), and on the assumption that a major part of the positive inotropic response to the catecholamines is mediated by cyclic AMP, that glucagon would enhance the production of the PIEA and would have little or no effect on the force of the rested state contraction.

The ability of glucagon to stimulate Ca^{2+} uptake by cardiac sarcoplasmic reticulum (Entman et al., 1969b) was mentioned in the previous chapter in connection with the similar effect of the catecholamines and cyclic AMP. Whether this effect can account for the positive inotropic response, as suggested by Shinebourne et al. (1969), remains to be seen. It could certainly account for the greater rate of relaxation (Shinebourne and White, 1970).

The reduced adenyl cyclase activity seen in homogenates prepared from failing hearts (Sobel et al., 1969a) was also discussed in the preceding chapter. The recent report by Gold et al. (1970) that such preparations can be stimulated by catecholamines but not by glucagon is of great interest in this regard. We suggested previously that the ability of glucagon to stimulate cardiac adenyl cyclase was not likely to be of physiological significance, but this opinion may have to be modified.

Glucagon also produces a positive chronotropic effect (Ueda et al., 1965; Whitehouse and James, 1966) and increases the rate of A-V nodal discharge (C. Steiner et al., 1969; Lucchessi et al., 1969), but these effects have not been studied with a view to determining their biochemical basis.

2. Skeletal Muscle

Glucagon has no effect on skeletal muscle metabolism at physiological concentrations, and does not stimulate adenyl cyclase from skeletal muscle at concentrations which stimulate hepatic adenyl cyclase maximally (Sutherland and Rall, 1960). Reported effects of higher concentrations on muscle metabolism (e.g., Beatty et al., 1963; Bowman and Raper, 1964) may have been caused by contaminating insulin or other factors. Hyperlacticacidemia has occasionally been reported in response to injected glucagon, but the source of the lactate has not been studied.

3. Smooth Muscle

The effects of glucagon in smooth muscle have not been carefully studied. The hormone is known to be capable of reducing the tone and motility of intestinal smooth muscle (Farah and Tuttle, 1960; Dotevall and Kock, 1963), and may inhibit the tone of vascular smooth muscle as well. Glick et al. (1968) reported a small but significant decrease in peripheral vascular resistance in dogs in response to glucagon, and Kuschke *et al.* (1966) found that the injection of 1 mg i.v. in humans decreased sensitivity to the pressor activity of norepinephrine. The increased blood flow that occurs in several organs in response to glucagon (e.g., Shoemaker *et al.*, 1959; Goldschlager *et al.*, 1969) could be the result of a direct effect on smooth muscle or may reflect metabolic changes occurring in the surrounding tissues. Possible effects of glucagon on the level of cyclic AMP in smooth muscle have not been measured.

4. Adipose Tissue

Lipolysis is discussed in detail in the following chapter. We will only mention here that in avian adipose tissue glucagon is a more effective lipolytic agent than any of the catecholamines (Carlson *et al.*, 1964; Goodridge, 1964; Goodridge and Ball, 1965; Grande and Prigge, 1970). Glucagon stimulates adenyl cyclase and is therefore lipolytic in rat adipose tissue (Butcher *et al.*, 1968), but is probably not an important lipolytic hormone in most mammalian species (De Plaen and Galansino, 1966; Sokal *et al.*, 1966; Burns and Langley, 1968).

5. Calorigenesis

Glucagon exerts a calorigenic effect in several species, including dogs, rats, rabbits, and guinea pigs (Davidson *et al.*, 1960; Miller and Krake, 1963; Heim and Hull, 1966; see also Foa, 1968). As with the catecholamines, this response is probably the result of several effects, including stimulation of adenyl cyclase in brown adipose tissue (Joel, 1966; Cockburn *et al.*, 1967; Reed and Fain, 1968a,b). However, the finding that supramaximal doses of glucagon and epinephrine may at times be almost additive (Davidson *et al.*, 1960) suggests that at least some of the systems affected by these hormones to stimulate calorigenesis are different. Corticosteroids and the thyroid were shown to be permissive for both hormones.

6. Hypocalcemia

The injection of glucagon produces a relatively prolonged hypocalcemic effect (Paloyan *et al.*, 1967). This can be understood, at least in part, in terms of the ability of glucagon to stimulate the release of thyrocalcitonin (Care *et al.*, 1969; Avioli *et al.*, 1969). This effect can be mimicked by exogenous cyclic AMP (Care and Gitelman, 1968), suggesting that glucagon may be capable of stimulating adenyl cyclase in the parafollicular cells (Copp, 1969). The possibility that glucagon may increase the level of cyclic AMP in the thyroid gland does not seem to have been tested.

Another factor which may contribute to the hypocalcemic effect is decreased tubular reabsorption of Ca^{2+} (Pullman *et al.*, 1967). The ability of glucagon to inhibit the reabsorption of several other ions had been discovered earlier (Elrick *et al.*, 1959), and its ability to stimulate renal medullary adenyl cyclase was demonstrated by Marcus and Aurbach (1969).

Still a third factor may be involved. Williams *et al.* (1969) found that the injection of very large doses of glucagon produced a transient hypocalcemic effect in rats that had been either thyroidectomized or nephrectomized, and suggested that the effect was secondary to increased calcium uptake by bone. Glucagon did not stimulate adenyl cyclase in particulate preparations of fetal rat calvaria (Chase *et al.*, 1969a), but supraphysiological concentrations may stimulate skeletal adenyl cyclase elsewhere.

In view of the large doses that have been used to produce these various effects, it seems unlikely that glucagon plays an important role in calcium homeostasis.

7. Gastric Secretion

The inhibition of gastrointestinal motility by glucagon was mentioned previously. Glucagon also reduces the volume (although not necessarily the acidity) of gastric secretion, and, since these effects are produced by glucagon in doses similar to those that raise blood sugar, they may be physiologically important (Clarke *et al.*, 1960; Von Heimburg and Hallenbeck, 1964). On the other hand, Charbon *et al.* (1963) have reported that glucagon may at times enhance the stimulatory effect of histamine on gastric secretion. Zollinger and Ellison (1955) had earlier suggested a possible ulcerogenic role for glucagon, on the basis of their finding pancreatic islet cell tumors in several patients with peptic ulcers. These tumors are now known to contain gastrin (Zollinger *et al.*, 1962), oversecretion of which might well be expected to lead to ulcer formation. Gastrin, histamine, and the parasympathetic nervous system seem to interact in a highly complex manner in the regulation of gastric secretion, and it now seems possible that the prostaglandins (Way and Durbin, 1969) as well as glucagon may have to be added to this list of regulatory influences. The role of cyclic AMP in the control of gastric secretion is presently far from clear, and what evidence there is on this point is considered further in Chapter 10.

III. INSULIN

Insulin, together with the parasympathetic nervous system and several gastrointestinal hormones, constitutes an important part of what Hess (1948) referred to as the *trophotropic* system. The components of this system are called into play during rest or after eating to facilitate anabolic or assimilatory processes. At least part of the function of this system depends on the ability of insulin to inhibit the formation or action of cyclic AMP.

A. Hepatic Effects of Insulin

The ability of insulin to lower the intracellular level of cyclic AMP in the liver can be demonstrated under appropriate conditions. The normal level of cyclic AMP in the unstimulated liver is on the order of 400 to 700 picomoles/gm of tissue (wet weight). This amount of cyclic AMP, if evenly distributed throughout the intracellular water, would, on the basis of *in vitro* studies with cell extracts, lead to a concentration of the nucleotide sufficient to stimulate glycogenolysis and gluconeogenesis maximally and to virtually eliminate glycogen synthesis. That these conditions do not prevail suggests that in the resting liver cyclic AMP is present largely in a bound or sequestered form, or else that intracellular cyclic AMP is effectively antagonized by other metabolites. Perhaps both factors are involved. In any event, insulin does not seem to be capable of reducing hepatic cyclic AMP below this baseline level.* However, when the control level of cyclic AMP is increased, as in livers from alloxan diabetic rats, or in normal livers in the presence of glucagon or a catecholamine, then the cyclic AMP-lowering effect of insulin can be readily demonstrated (J. H. Exton *et al.,* 1969, unpublished observations; Lewis *et al.,* 1970).

Apparently under normal physiological conditions insulin exerts a continuous damping effect on the hepatic level of cyclic AMP. Exton and Park and their associates (Exton *et al.,* 1966; Exton and Park, 1968b) have suggested that glucose output by the liver may depend on a balance between agents such as glucagon and the catecholamines, which increase the level of cyclic AMP, and insulin which lowers it. This interpretation is now well supported by experimental evidence. Figure 7.7, from the

---

*The imprecision and variability of most assays for cyclic AMP make it difficult to detect slight changes above or below the normal baseline level, even though such small changes may be physiologically very important.

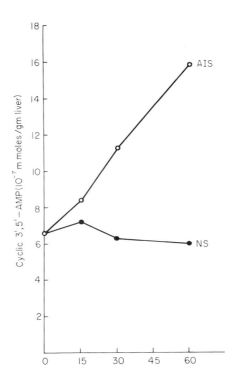

Fig. 7.7. Effect of anti-insulin serum (AIS) and normal serum (NS) on cyclic AMP levels in rat liver *in vivo*. Serum was injected intravenously into anesthetized rats and the livers rapidly frozen at the indicated times. From Jefferson *et al:* (1968).

work of Jefferson *et al.* (1968), shows the rise in hepatic cyclic AMP which occurs in normal rats in response to the injection of insulin antibodies. Cyclic AMP levels are also increased in livers from fasted rats and from rats made diabetic by the injection of alloxan. The data in Table 7.I shows that the administration of insulin to these latter animals leads to a prompt restoration of the normal level. As mentioned previously, insulin is also capable of suppressing the rise in cyclic AMP which normally occurs in response to glucagon or the catecholamines (Exton *et al.,* 1969, unpublished observations). A point which should perhaps be re-emphasized is that glucagon, especially, is capable of increasing the concentrations of cyclic AMP in the liver to levels much higher than are necessary to produce a maximal stimulation of glucose output. It is thus possible to choose experimental conditions under which the antagonism between glucagon and insulin can be easily demonstrated at the level of cyclic AMP, but such that the level of cyclic AMP is still so high that an

TABLE 7.I Effects of Alloxan and Insulin on Cyclic AMP Levels in Rat Liver *in Vivo*

Treatment[a]	Cyclic AMP (mμmoles/gm)
None	0.49 ± 0.06
Alloxan (48 hr)	0.96 ± 0.07
Alloxan (48 hr) + insulin (30 min)	0.49 ± 0.07
Alloxan (48 hr) + insulin (24 hr)	0.39 ± 0.05

[a]Rats were made diabetic by a single intravenous injection of alloxan monohydrate (60 mg/kg) and killed 48 hr later. Insulin was given 30 min before death (0.1 unit crystalline insulin i.v.) or beginning 24 hr after alloxan (2 units of protamine zinc insulin s.c. twice daily). From Jefferson *et al.* (1968).

effect on carbohydrate metabolism cannot be seen. Many of the conflicting reports in the earlier literature, which led some investigators to conclude that insulin did not have a direct effect on the liver, can probably be understood on this basis. Another reason for the earlier confusion would be the inability of insulin to reduce the level of cyclic AMP below a certain baseline level. If experimental conditions are such that the level of cyclic AMP is already at a minimum, then insulin may well appear to have no effect on the liver.

Although this point has not been carefully studied under all conditions, it would appear that the liver, at least in the rat, may be less sensitive to the effects of insulin than either adipose tissue or muscle. This difference in sensitivity might be physiologically important. For example, as suggested in Section II,A,4, the ability of glucagon to stimulate the release of insulin would be useful only to the extent that the released insulin did not interfere with the direct hepatic effects of glucagon. It seems possible that a quantity of insulin could at times pass through the liver without producing a local effect, and yet still lead to levels in peripheral blood sufficient to affect muscle and adipose tissue. At other times, such as during periods of glucose plethora, the amount of insulin released would be expected to be enough to produce a substantial effect on the liver. It should be noted that since the blood draining the islets of Langerhans empties directly into the portal circulation, the liver will always be exposed to higher concentrations of insulin than will the other tissues of the body.

Although it now seems likely that insulin has other effects, most of its known effects in the liver can be accounted for in terms of its ability to lower the intracellular level of cyclic AMP. Thus, on this basis, insulin should antagonize phosphorylase activation in response to glucagon and should increase hepatic glycogen synthetase activity, and that it does so was demonstrated by Bishop and Larner (1967), who studied both effects in the dog liver *in situ*. Insulin should certainly inhibit gluconeogenesis, and that it does so in the isolated perfused rat liver was shown by Exton and Park and their colleagues (Jefferson *et al.*, 1968; Exton and Park, 1968b). Inhibition by insulin of hepatic glucose release and urea production has also been observed by others (e.g., Miller, 1965; Steele, 1966; Mondon and Mortimore, 1967; Mortimore *et al.*, 1967; Glinsmann and Mortimore, 1968; Menahan and Wieland, 1969a; Mackrell and Sokal, 1969). Since cyclic AMP and cyclic AMP-generating hormones cause a net efflux of potassium from the liver, insulin should produce the opposite effect, and it does (Mortimore, 1961; Burton *et al.*, 1967; Exton and Park, 1968b; Glinsmann and Mortimore, 1968). Cyclic AMP inhibits the incorporation of labeled acetate into fatty acids in liver slices (Berthet, 1960), and the opposite effect of insulin was demonstrated many years ago (Brady and Gurin, 1950). The antiketogenic effect of insulin is of course well known, and is discussed further in Section III,C. That insulin increases the rate of incorporation of leucine into liver protein, both *in vitro* and *in vivo,* was shown by Penhos and Krahl (1963). This is the opposite of the effect of cyclic AMP (Pryor and Berthet, 1960), and probably reflects, at least in part, an inhibition of proteolysis (Mortimore and Mondon, 1970).

It is thus clear that many of the hepatic effects of insulin could result from its ability to lower the intracellular level of cyclic AMP in the liver. The mechanism by which insulin produces this effect is unknown, however. It could in theory involve an inhibition of adenyl cyclase or a stimulation of phosphodiesterase or both, but neither effect has been demonstrated in a broken cell preparation. Increased hepatic phosphodiesterase activity in response to insulin treatment *in vivo* was reported (Senft *et al.*, 1968d) but could not be confirmed (Müller-Oerlinghausen *et al.*, 1968; Menahan *et al.*, 1969). Similarly, Blecher *et al.* (1968) and Hepp *et al.* (1969) were unable to see an effect of insulin on phosphodiesterase activity in adipose tissue. It seems likely that in liver, as in adipose tissue, inhibition of adenyl cyclase or stimulation of phosphodiesterase may be a consequence of some more primary action of insulin. Whichever enzyme is ultimately shown to be affected, the effect may be difficult or even impossible to see in a cell-free system. A possibility which should definitely be considered is that insulin reduces the level of cyclic AMP in both of these tissues by stimulating the conversion of cyclic AMP to a metabolite other than 5'-AMP.

The nature of the primary action of insulin has been studied in most detail in tissues other than liver. It was known that in adipocytes (Kuo *et al.,* 1966) and muscle (Rieser, 1966; Weis and Narahara, 1969) the actions of insulin could be mimicked by trypsin. Kono (1969a) found that by incubating adipocytes with slightly higher concentrations of trypsin, the response of these cells to insulin (as measured both by the fall in cyclic AMP and by stimulation of glucose uptake) could be completely abolished. This treatment also led to a loss of the insulin-binding reaction that Crofford (1968) had postulated as the first step in the hormone-cell interaction. Although it was later found (Kono, 1969b; Rodbell *et al.,* 1970) that sensitivity to glucagon was also impaired by this treatment, the cells in other respects seemed to be normal. The ability of trypsin to selectivity interfere with the action of insulin in fat cells, observed also by Fain and Loken (1969), is reminiscent of the effect of trypsin treatment on the receptors for acetylcholine in the rabbit intestine (Lu, 1952). It was later found (Kono, 1969b) that continued incubation of the treated cells after the trypsin had been inactivated led to a restoration of the response to insulin. This restoration could be prevented by either puromycin or cycloheximide. These results suggest that an essential part of the insulin effector system is a rapidly renewable peptide element located on the surface of the cell, and that an initial action of trypsin on the system mimics the effect of insulin while subsequent proteolytic modification renders it unresponsive to the hormone. Additional strong evidence that insulin interacts with a receptor on the external side of the cell membrane was provided by Cuatrecasas (1969). How this interaction leads to a fall in the intracellular level of cyclic AMP is of course still obscure.

It was mentioned previously that insulin has other effects in the liver, in addition to lowering the level of cyclic AMP, and it is possible that some of the effects which appear to be the result of a lowered level of cyclic AMP may in fact involve an additional mechanism. For example, insulin increases glycogen synthetase activity in muscle in concentrations which do not affect phosphorylase activity (Craig and Larner, 1964) or the level of cyclic AMP (Goldberg *et al.,* 1967; Craig *et al.,* 1969). Longer incubation with higher concentrations of insulin may interfere with the rise in cyclic AMP and the activation of phosphorylase normally seen in response to epinephrine (Torres *et al.,* 1968; Craig *et al.,* 1969), but such conditions are apparently not necessary in order for insulin to activate glycogen synthetase. It has been suggested that one of the effects of insulin in muscle may be to reduce the sensitivity of glycogen synthetase kinase to the action of cyclic AMP (Villar-Palasi and Wenger, 1967; Larner *et al.,* 1968), although the recent evidence that this kinase is the same one that activates phosphorylase *b* kinase (Schlender *et al.,* 1969; Soderling and Hickenbottom, 1970; Reimann and Walsh, 1970) raises the question of how such a mechanism could operate without simultaneously interfering with phosphorylase activation.

Regardless of the mechanism, it is clear that if insulin does reduce the sensivitity of the glycogen synthetase system in muscle to the action of cyclic AMP, then the possibility of a similar effect in liver would have to be considered. That such an effect may occur was supported by the finding that insulin is capable of antagonizing the glucose-releasing effect of small concentrations of exogenous cyclic AMP in the isolated perfused rat liver (Glinsmann and Mortimore, 1968); in the case of potassium release, insulin was effective against even higher concentrations of cyclic AMP. Similarly, Wicks (1969) found that insulin could interfere with the ability of exogenous cyclic AMP to stimulate the synthesis of phosphopyruvate carboxykinase. All of these effects could in principle be understood in terms of the stimulation of phosphodiesterase activity. Another possibility, mentioned previously, is that insulin might stimulate the conversion of cyclic AMP to another metabolite, which could at times interfere with one or more of the actions of cyclic AMP.

As discussed previously in Section B,4, one hepatic effect of insulin that is especially difficult to explain in terms of a fall in the level of cyclic AMP is the stimulation of tyrosine transaminase synthesis (Wicks, 1969). This has been demonstrated in response to glucagon-free insulin *in vitro,* and seems to be essentially similar to the effect produced by exogenous cyclic AMP or by hormones that stimulate the endogenous formation of cyclic AMP. When insulin is combined with one of these other agents, the response is greater than to either of them added separately, but less than additive. Although the mechanism of this action of insulin is poorly understood, it is at least conceivable that the previously mentioned possibility, that insulin stimulates the conversion of cyclic AMP to another metabolite, could also be applied here. Although this hypothetical substance might function in most systems as an antimetabolite of cyclic AMP, it might in other systems have an effect similar to that of cyclic AMP. Various other possibilities could be considered, but all in the absence of hard data.

The claim that insulin increases the total amount of glycogen synthetase in the liver by a mechanism involving the synthesis of new protein (Steiner, 1966) has been criticized on the grounds that the assay conditions were inadequate (Mersmann and Segal, 1967). Therefore, the conclusions drawn from these experiments may or may not be valid. The possible role of cyclic AMP in certain other hepatic effects of insulin, such as inhibition of the release of inorganic phosphate, has yet to be studied.

B. Other Effects of Insulin

Evidence that insulin inhibits lipolysis in adipose tissue by reducing the level of cyclic AMP in adipocytes is summarized in the following chapter (Section III,A). As mentioned there and in the preceding section on the liver, the mechanism by which insulin acts to lower the level of cyclic AMP is unknown. The antilipolytic effect of insulin has been seen in adipose tissue of all mammalian species studied, but apparently does not occur in some avian species. This interesting difference between birds and mammals has not been studied from the standpoint of cyclic AMP.

Evidence that the stimulatory effect of insulin on glucose transport (Morgan *et al.*, 1965; Park *et al.*, 1968) is *not* related to a change in the intracellular level of cyclic AMP can be summarized here. Insulin had no effect on the level of cyclic AMP in skeletal muscle (Goldberg *et al.*, 1967; Craig *et al.*, 1969) or cardiac muscle (R. W. Butcher and H. E. Morgan, unpublished observations), in which tissues the effect on glucose transport is very pronounced. As mentioned above, insulin is not an anti-lipolytic hormone in birds, presumably because it does not lower the level of cyclic AMP in avian adipose tissue; it nevertheless produces a fall in the level of blood sugar in birds similar to that seen in mammals (Good-ridge, 1964; Grande, 1969). In mammalian adipocytes, the effect of insulin to lower the level of cyclic AMP correlates well with inhibition of lipolysis but not with stimulation of glucose uptake. For example, prostaglandin E_1, which is more potent than insulin in lowering the level of cyclic AMP in rat adipocytes (Butcher and Baird, 1968), is less potent than insulin in stimulating glucose utilization (Fain, 1968b; Blecher *et al.*, 1969). The observation that hormones which increase the level of cyclic AMP in adipose tissue do not inhibit the effect of insulin on glucose uptake (Rodbell *et al.*, 1968) would also argue against a role for cyclic AMP in the regulation of glucose transport. It can be added that the glucose transport system in liver seems to be completely unresponsive to insulin (Williams *et al.*, 1968).

The observations that exogenous cyclic AMP (but not 5'-AMP) caused a significant increase in the rate of glucose uptake by the rat diaphragm *in vitro* (Edelman *et al.*, 1966) whereas the dibutyryl derivative produced an inhibition that could be overcome by insulin (Chambaut *et al.*, 1969) are difficult to interpret at the present time. We suggested previously that insulin might lower the level of cyclic AMP in hepatic and adipose tissue by stimulating the conversion of cyclic AMP to a derivative of cyclic AMP. It should be noted that this hypothetical reaction would not necessarily lead to a fall in the level of cyclic AMP, since that could be prevented by an increase in the activity of adenyl cyclase relative to that of phosphodiesterase. It can be imagined that exogenous cyclic AMP might be converted to this metabolite in muscle relatively easily, even in the absence of insulin, whereas the dibutyryl derivative, perhaps by virtue of its resemblance to the natural product, might tend to inhibit this conversion. Chambaut and her colleagues had suggested that their data was compatible with the idea that insulin might stimulate the formation of a second messenger different from cyclic AMP "which could competitively antagonize the action of the latter compound on all the enzymatic systems involved in insulin action." The only addition we are making to this speculation is that the new second messenger might actually be derived *from* cyclic AMP. It can be noted that if the dibutyryl derivative inhibits this conversion in adipose tissue, as suggested here for skeletal muscle, it might explain the ability of insulin to overcome the lipolytic action of exogenous cyclic AMP (Goodman, 1969b) but not that of the dibutyryl derivative (Peterson *et al.*, 1968).

It is entirely possible, of course, that all of the various observations that seem so puzzling to us now will eventually be understood in relatively prosaic terms, without reference to reactions and other factors of which we presently know nothing. For example, much of our present reasoning is based on the apparent fact that insulin does not reduce the level of cyclic AMP in striated muscle. However, the possibility that insulin does alter the level of cyclic AMP in a small compartment in muscle cells, separate from the bulk of the cytoplasm, has not been ruled out.

It can be added that much of our present reasoning is also based on the supposition that all of the diverse effects that insulin is known to be capable of producing will eventually be understood in terms of a single primary action. The conviction that this is probably so is more akin to religious faith than to anything that could be called scientific. It will not be discussed further, therefore, except to say that our experience with cyclic AMP has probably strengthened rather than weakened it.

The stimulatory effect of insulin on muscle glucogen synthetase (Larner *et al.*, 1968) was mentioned in the preceding section on the liver. It was also pointed out there that insulin has a tendency to oppose the phosphorylase activating effect of epinephrine (Torres *et al.*, 1968). In addition to these effects and the effect on glucose transport, the interaction of insulin with muscle also leads to an increased rate of amino acid uptake (Kipnis and Noall, 1958), a net decrease in the rate of potassium efflux (Zierler, 1966), and stimulation of protein synthesis by muscle ribosomes (Wool *et al.*, 1968). Increased phosphoglucomutase activity has also been reported (Hashimoto *et al.*, 1967). No one of these effects seems explicable in terms of any of the others, and their possible relation to cyclic AMP has not been carefully studied.

C. Possible Role of Cyclic AMP in Diabetes

In view of the evidence that cyclic AMP plays a role in the release as well as some of the actions of insulin, a brief discussion of its possible role in diabetes seems warranted. Although the etiology and pathogenesis of its various forms are obscure, human diabetes mellitus has been commonly defined as a chronic hereditary disease characterized by a relative or absolute deficiency of insulin and complicated by various microangiopathic disorders.

At least two defects have been described that seem to be characteristic of the prediabetic state (Camerini-Davalos and Cole, 1970). One, described by Cerasi and Luft (1967), is a reduced rate of insulin release in response to the infusion of high concentrations of glucose. Cerasi and Luft (1969) have more recently found that this deficiency can be ameliorated by the injection of theophylline, leading them to propose, as the inherited factor in diabetes, or at least as a marker of this factor, a reduced ability of pancreatic beta cells to generate cyclic AMP. This suggestion will be considered further below.

Although this defect in the insulin-releasing mechanism seems to be a very early event in the pathogenesis of diabetes, detectable in early childhood (Cerasi and Luft, 1970), another lesion that shows up before changes in carbohydrate or lipid metabolism can be detected is a thickening of the basement membranes associated with a variety of epithelial cells. This has been especially well described by Siperstein and his colleagues for the basement membranes associated with muscle capillaries (Siperstein et al., 1970). Whether or not this can be related to the reduced insulin response in the glucose infusion test is unknown. It is of interest, however, that Lacy (1967) has reported that hyalinization or amyloidosis of pancreatic islets is commonly seen in diabetic patients. Although the amyloid is deposited between the two basement membranes that separate the islet cells from the capillary, rather than forming part of one or the other of the basement membranes proper, it nevertheless seems possible that a fundamentally similar process might be involved in the pathogenesis of these two lesions. It seems possible, in other words, that the amyloidosis seen in islets and the thickening of basement membranes seen elsewhere could be manifestations of the same genetic defect. Unfortunately, little is known about the site or mechanism of formation of the basement membrane (Misra and Berman, 1969), and much additional research in this area is needed.

From the standpoint of the possible role of cyclic AMP, several possibilities can be considered. One is that even if the reduced insulin secretion in diabetes results from a relative deficiency in the level of cyclic AMP in pancreatic beta cells, as suggested by Cerasi and Luft, adenyl cyclase and phosphodiesterase in these cells might be perfectly normal. As suggested by Lacy, the amyloid deposits within the islets might act as a barrier delaying the transfer of insulin from the beta cells to the bloodstream. Another possibility is that this barrier might impede glucagon or gastrin produced in adjoining cells from gaining access to the adenyl cyclase of the beta cells. It seems possible that either of these postulated defects could be overcome if the normal rate of formation of cyclic AMP could be increased, as by the addition of theophylline.

Beyond these considerations, and not necessarily exclusive of them, there is the distinct possibility that the cyclic AMP-forming mechanism is itself defective in diabetes, as suggested by Cerasi and Luft. The genetic defect might be manifested in the form of a defective adenyl cyclase system or by excessive phosphodiesterase activity. This defect might not be confined to the pancreatic beta cells, but might occur elsewhere and might even be related to the observed thickening of basement membranes. The expression of many genetic abnormalities seems to be restricted to a single tissue, as in the example of mice deficient in skeletal muscle phosphorylase b kinase (Lyon et al., 1967), but may in other cases be more widespread. Further speculation regarding the basement membrane would not be warranted because of our ignorance of the mechanisms regulating the formation and breakdown of this structure. It is interesting to note, however, that if this hypothetical defect also occurred in fat cells, it might explain the paradoxical obesity so often seen in association with diabetes mellitus (Bierman et al., 1968). The possibility that adipocytes from older diabetic patients contain an adenyl cyclase relatively insensitive to factors tending to stimulate it is a testable proposition, since human fat cells can be isolated and studied (Galton and Bray, 1967; Burns and Langley, 1968, 1970). Excess lipolysis leading to ketoacidosis would then occur primarily in the presence of severe insulin deficiency.

In view of the implication that the ratio of phosphodiesterase to adenyl cyclase in pancreatic beta cells may be too high in diabetes, it is of interest that Kupiecki (1969) found phosphodiesterase activity in the pancreas and adipose tissue of spontaneously diabetic mice to be below that of normal controls. This finding was consistent with the high levels of plasma insulin found in the diabetic animals. It is not clear, of course, that these mice provide a good model for most or even any forms of human diabetes mellitus, although this possibility has not been discounted. It is conceivable that in some forms of human diabetes, reduced phosphodiesterase activity is associated in some cells with an even more marked impairment of adenyl cyclase activity.

Whether all of the observations made by Cerasi and Luft are compatible with their hypothesis of a defective adenyl cyclase is uncertain. There are two peaks in the insulin response to glucose infusion, and it has been suggested (Cerasi and Luft, 1969; Grodsky *et al.*, 1970) that this may reflect two compartments of insulin, one for quick release and the other, a storage compartment, feeding into the first at a reduced rate. Only the first peak is missing in most diabetics. Theophylline restored this response in prediabetics but not in overt diabetics, and had no effect in normal subjects. These observations could perhaps be understood on the assumption that cyclic AMP is not rate-limiting in the glucose infusion test in normal subjects, whereas in advanced diabetes the cyclase might be so defective that inhibiting phosphodiesterase would be of little use. As emphasized previously, the effects of phosphodiesterase inhibition at the functional level will always depend upon the ongoing rate of adenyl cyclase activity.

Another finding by Cerasi and Luft (1970) was that chronic treatment with growth hormone greatly enhanced the insulin response to glucose in normal as well as prediabetic patients, with normals being more sensitive to growth hormone than prediabetics. These findings are of special interest in view of the demonstration by Moskowitz and Fain (1970) that growth hormone increases the accumulation of cyclic AMP in rat adipocytes *in vitro* by a mechanism that seems to involve protein synthesis. It seems possible, therefore, that one of the effects of growth hormone in these cells is to stimulate the rate of synthesis of one or more components of the adenyl cyclase system. If this is the basis of the insulinogenic activity of growth hormone, then the lower sensitivity to growth hormone in prediabetics might be taken as evidence for a genetically defective mechanism for synthesizing adenyl cyclase. On the other hand, since the experiments with theophylline suggested that cyclic AMP was not rate-limiting in the insulin response to glucose infusion, the results with growth hormone might indicate that the effect was to increase the amount or sensitivity of the system directly affected by cyclic AMP. That a higher dose of growth hormone was required in the prediabetics would still be compatible with the hypothesis of a defective adenyl cyclase system. It should be noted that even in adipose tissue, where an effect of growth hormone on cyclic AMP accumulation has been observed, the evidence suggests that growth hormone may also act to increase the sensitivity of lipolysis to cyclic AMP (Goodman, 1968b).

Regardless of whatever role cyclic AMP may play in the etiology and pathogenesis of human diabetes mellitus, it now seems clear that many of the metabolic derangements associated with this disease are the result of the excessive production of cyclic AMP in adipose and hepatic tissue, secondary to insulin deficiency. This presumption is, of course, based entirely on studies of experimental diabetes in laboratory animals (Exton and Park, 1968b; Sneyd *et al.*, 1968), but the applicability of these studies to the clinical situation seems obvious. It should be noted that a relative insulin deficiency could result not only from reduced tissue sensitivity to insulin, but also from increased sensitivity to (or excess production of) catecholamines or glucagon.

In adipose tissue, as discussed in more detail in the following chapter, insulin deficiency leaves unopposed the stimulatory action of the catecholamines on adenyl cyclase. The resulting increase in the level of cyclic AMP leads to excessive mobilization of fatty acids, and this is probably the principal factor causing diabetic ketosis. The ketogenic process is of course further enhanced by the higher levels of hepatic cyclic AMP that result from the unopposed action of glucagon and the catecholamines. By stimulating glycogenolysis and gluconeogenesis and suppressing glycogen synthesis, these higher levels of cyclic AMP also lead to excessive glucose production by the liver. This contributes to the characteristic hyperglycemia of diabetes, which would tend to occur in any event because of the reduced peripheral utilization of glucose. The continued high levels of cyclic AMP that probably occur in the face of chronic insulin deficiency would be expected to increase the amount as well as the activity of a number of hepatic enzymes, as discussed previously in Section B,4. This would certainly increase the gluconeogenic and perhaps also the ketogenic *capacity* of the liver.

We might briefly consider the question of why these metabolic abnormalities are not apparent in the prediabetic, even in the face of the demonstrated abnormality in the insulin releasing mechanism. At least two possibilities come to mind, on the basis of studies involving cyclic AMP. One, mentioned previously, is that adenyl cyclase may be relatively unresponsive in adipose and hepatic tissue as well as in pancreatic beta cells. To the extent that this is the case, the reduced rate of insulin release would be counteracted by the reduced sensitivity of adenyl cyclase in these peripheral tissues, and it is conceivable that such a balance could be maintained for a considerable period of time. Alternatively, it is conceivable that the reduced rate of insulin release is counterbalanced by a similar reduction in the rate of release of glucagon and/or reduced sympathetic tone. The point to be made is that many of the metabolic consequences of relative or absolute insulin deficiency will occur only to the extent that this deficiency leads to excessive production of cyclic AMP. This can be prevented, within limits, by a reduction in the level of stimulation of adenyl cyclase. Other factors which might play a role here, in addition to glucagon and the catecholamines, might include the prostaglandins (discussed in Section III,B in the following chapter and again in Chapter 10) and any of several permissive hormones, including growth hormone, the corticosteroids, and thyroxine.

Since human diabetes mellitus seems to be a hereditary disease (Camerini-Davalos and Cole, 1970), a real cure may have to involve some form of treatment such as the injection of a virus to correct the basic flaw. The prospect of using this type of therapy to induce auxiliary sites of insulin synthesis has been discussed in an interesting article by Sinsheimer (1969). Before this can be seriously considered, however, more will have to be learned about the basic defect involved. As we have tried to show in this section, for example, it is not entirely clear that the defective gene is directly related to the production or release of insulin. Perhaps continued study of the role of cyclic AMP in diabetes will suggest other avenues by which therapeutic progress can be made.

IV. ADDENDUM

We were able to incorporate a number of recent advances into the text of this chapter, before the galley proof stage, and some other pertinent findings have been mentioned in the addenda to preceding chapters. The interesting recent studies by Bitensky *et al.* (1970), on the differential effects of age and steroid hormones on the hormonal sensitivity of rat liver adenyl cyclase, are summarized in Chapter 10 (Section XII). A few additional recent findings can be mentioned here.

Concerning the mechanism of the cyclic AMP-lowering action of insulin, Ray *et al.* (1970) reported that insulin could inhibit adenyl cyclase activity when added to broken cell preparations of rat liver. However, the specificity of this effect was not established, and it seems doubtful that it could account for the effect of insulin in the intact liver. In partial confirmation of the work of Senft and Schultz and their colleagues (Section III,A), Loten and Sneyd (1970) observed a small but reproducible increase in phosphodiesterase activity (a higher maximum velocity of the low-K_m enzyme) in homogenates of insulin-treated fat cells. They concluded that this effect was probably sufficient to account for the effect of insulin in intact cells.

In line with observations of fetal pancreatic tissue obtained from experimental animals (Section II,A,3), glucagon was found to stimulate insulin release from isolated human fetal pancreatic islets in the absence of glucose, and at a time when glucose itself was ineffective (Espinosa *et al.,* 1970). An interesting lecture by Renold (1970), emphasizing the complexity of the mechanisms regulating insulin biosynthesis and release, has been published.

Finally, of interest from the standpoint of the possible role of cyclic AMP in diabetes, Unger and his colleagues (1970) were able to show that diabetic patients do contain abnormally high levels of plasma glucagon. Fasting levels were similar to those seen in healthy subjects, but the increase in response to arginine infusion was much greater while the fall in response to glucose infusion was less. Highest levels were seen in patients with severe ketoacidosis.

CHAPTER 8

Lipolysis in Adipose Tissue

I. INTRODUCTION

A glance at Table 8.I, listing some of the many agents which alter the rate of a single process, adipose tissue lipolysis, will point out why we have chosen to treat fat in a separate chapter. At present, the endocrine interrelationships in mammalian adipose tissue are perhaps the most complex in nature, and most of the hormones affecting fat have been shown to involve cyclic AMP, either directly or indirectly.

TABLE 8.I. Some Agents Affecting Lipolysis in Adipose Tissue

Effects on lipolysis	Agents	References
Fast increase	Catecholamines	Gordon and Cherkes (1958)
	ACTH	White and Engel (1958)
	TSH	Freinkel (1961)
	Glucagon	Hagen (1961)
	LH	Butcher *et al.* (1968)
	Secretin	Carlson (1968)
	Vasopressin	Vaughan (1964)
	Serotonin	Bieck *et al.* (1966)
	Methylxanthines	Vaughan and Steinberg (1963)
Slow increase	Growth hormone	Raben and Hollenberg (1959)
	Glucocorticoids	Jeanrenaud and Renold (1960)
	GH + glucocorticoids	Fain *et al.* (1966)
	Thyroid hormones	Debons and Schwartz (1961)
Decrease	Insulin	Jungas and Ball (1963)
	Prostaglandins	Steinberg *et al.* (1964)
	Nicotinic acid	Carlson (1963)
	5-Methylpyrazole-3-carboxylic acid	Gerritson and Dulin (1965)
	α-Adrenergic blockers	Schotz and Page (1960)
	β-Adrenergic blockers	Mayer *et al.* (1961)
	Nucleotides	Vaughan (1960); Dole (1961)

For the purposes of this chapter, lipolysis can be defined as breakdown of triglycerides to free fatty acids and glycerol (Fig. 8.1). The lipolysis of 1 mole of triglycerides yields 3 moles of free fatty acids (FFA) and 1 of glycerol. However, FFA are readily reesterified (as fatty acyl-CoA intermediates), so that even at very high rates of lipolysis, the net increase in FFA may be very small if the reesterification rate is high. Glycerol, however, is not extensively reutilized since only moderate amounts of glycerol kinase activity are found in fat. The rate-limiting step in lipolysis is at the level of the triglyceride lipase (TG lipase), and it is on this system that lipolytic hormones as well as cyclic AMP act. Lipolysis is also a process of considerable physiological significance. As Carlson (1968)

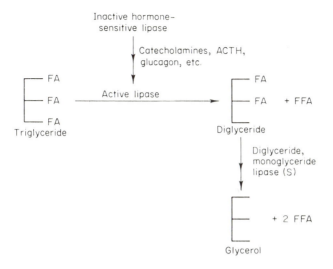

Fig. 8.1. Lipolysis of adipose tissue triglycerides.

recently stated, although the concentration of FFA in plasma is low (usually less than 5% of the total plasma lipids), the turnover of these compounds is so high that about 200 gm of FFA are released from fat per 24 hr by the lipolytic process, and the oxidation of these FFA by peripheral tissues could account for about 80% of the basal O_2 consumption in man. In point of fact, lipolysis would appear to be the lipid counterpart of glycogenolysis in carbohydrate metabolism, just as the whole FFA cycle is a mechanism for the synthesis, storage, rapid mobilization, transport, and utilization of FFA (Steinberg, 1966).

We cannot deal further with the endocrine physiology of adipose tissue because of space limitations, and must now concentrate on the role of cyclic AMP in fat. However, a number of excellent reviews on fat have been published, and the reader is referred to a partial listing of these at the end of this chapter.

II. THE ROLE OF CYCLIC AMP IN THE ACTIONS OF LIPOLYTIC HORMONES

A. The Catecholamines

1. Data Suggesting a Role of Cyclic AMP in Lipolysis

Considerable evidence had been obtained by others to suggest that cyclic AMP might mediate the lipolytic action of the catecholamines. The presence of an epinephrine-sensitive adenyl cyclase system in washed cell-free preparations of adipose tissue was reported by Sutherland and Rall (1960) and Klainer *et al.* (1962). Also, Vaughan (1960) had shown that phosphorylase activation (which might be taken as an indirect measurement of tissue levels of cyclic AMP) and lipolysis were both increased in fat pads incubated with catecholamines, ACTH, TSH, or glucagon. At that time serotonin appeared to present a dissociation in that it caused phosphorylase activation without increased lipolysis, but Vaughan and Steinberg (1965) subsequently noted that serotonin at higher concentrations also had lipolytic activity. Later, Bieck *et al.* (1966) found that serotonin was without lipolytic activity unless theophylline and/or an inhibitor of monoamine oxidase were included. Vaughan and Steinberg (1963) provided additional and very suggestive evidence when they reported that caffeine, an inhibitor of the phosphodiesterase, acted synergistically with epinephrine on lipolysis. This was the first report of the use of a methylxanthine on adipose tissue, and deserves mention because these compounds have been used extensively by many groups in studying many sorts of hormone actions both directly and indirectly. Finally, Rizack (1964) reported the activation of an epinephrine-sensitive lipase activity by cyclic AMP in cell-free preparations of fat pads incubated with ATP, Mg^{2+}, and caffeine.

2. Correlations between Cyclic AMP Levels and Lipolysis

Our first series of experiments was designed to examine the relationship between intracellular cyclic AMP levels and lipolysis (Butcher *et al.,* 1965). To do this, we had to use intact fat pads, since cell-free preparations which responded to hormones with increased lipolysis were not available. The results obtained may be summarized briefly: (1) The incubation of fat pads with increasing concentrations of epinephrine caused increased cyclic AMP levels and free fatty acid release (Fig. 8.2). The effects of epinephrine on cyclic AMP and lipolysis were linear at lower concentrations of the catecholamine, but at higher concentrations FFA release slowed somewhat, as if some step in the lipolytic mechanism other than cyclic AMP had become rate limiting (see Section II, D). (2) β-Adrenergic blocking agents (e.g., DCI) were weak agonists of FFA release and cyclic AMP levels, and antagonized the effects of epinephrine on both. (3) The effects of epinephrine on cyclic AMP levels were detectable earlier than the effects on lipolysis in whole fat pads in incubation or in perfusion. For example, cyclic AMP levels were maximal within 5 min after the addition of epinephrine to fat pads, while FFA release

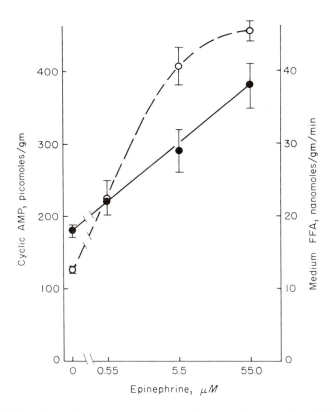

Fig. 8.2. The effects of varying concentrations of epinephrine on cyclic AMP levels and FFA release in epididymal fat pads *in vitro*. (From Butcher *et al.,* 1965.) Key: ●——●, cyclic AMP; 0---0, FFA.

was not significantly increased until 13 min. In addition, highly significant changes in cyclic AMP levels were detected 30 sec after epinephrine was introduced to perfused fat pads, or within 15 sec after addition to isolated fat cells (Butcher *et al.,* 1968). (4) The synergistic effects of methylxanthines and catecholamines on lipolysis, first reported by Vaughan and Steinberg (1963), also occurred when cyclic AMP levels were measured (Fig. 8.3). While neither 0.55 μM epinephrine nor 1 mM caffeine strongly stimulated cyclic AMP levels or lipolysis, the combination of the two produced striking responses.

Weiss *et al.* (1966) and Humes *et al.* (1968) have confirmed the effects of catecholamines on adipose tissue cyclic AMP levels and lipolysis, and also the synergistic action of catecholamines and methylxanthines on cyclic AMP levels.

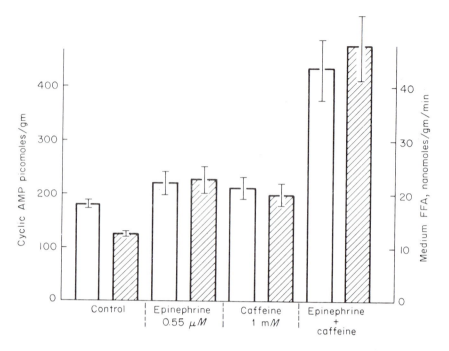

Fig. 8.3. Effects of epinephrine and caffeine on cyclic AMP levels and FFA release in epididymal fat pads *in vitro*. (From Butcher *et al.*, 1965.)

3. Effects of Exogenous Cyclic Nucleotides on Lipolysis

Like Vaughan (1960) and Dole (1961, 1962) we found that exogenous cyclic AMP inhibited rather than stimulated lipolysis in intact fat pads. Fortunately, the derivatives described in Chapter 3 by Professor Poster-nak were available and N^6-2'-O-dibutyryl-cyclic AMP, at high concentrations (between 3×10^{-4}–3×10^{-3} M), was a very effective lipolytic agent on intact epididymal fat pads either in incubation or perfusion, or on isolated fat cells (Butcher *et al.*, 1965). Cyclic AMP also had lipolytic activity when incubated with isolated fat cells, but was at least an order of magnitude less effective than the dibutyryl compound. The *in vitro* lipolytic activity of dibutyryl cyclic AMP has been repeatedly confirmed (Aulich *et al.*, 1967; Blecher *et al.*, 1968; Goodman, 1969b; Peterson *et al.*, 1968; Stock and Westermann, 1966; Weiss *et al.*, 1966).

Bieck *et al.* (1968) have studied the actions of dibutyryl cyclic AMP on lipolysis in the rat *in vivo*. They found that the intraperitoneal injection of the derivative (10 μmoles/kg) caused a prompt and sustained fall in plasma FFA and glycerol. However, this was apparently not a direct effect of dibutyryl cyclic AMP, for when fat pads were removed from rats injected with large amounts of the derivative and incubated *in vitro,* FFA release was significantly stimulated. The reasons for the antilipolytic effect of dibutyryl cyclic AMP *in vivo* are not completely clear. One possibility which Bieck *et al.* considered, and which seems likely, was that insulin (known to have antilipolytic activity) was being released directly in response to dibutyryl cyclic AMP or indirectly because of the hyperglycemia caused by the derivative. In alloxan diabetic animals, dibutyryl cyclic AMP did not lower plasma glycerol levels as it did in normals, and the inhibitory effect on plasma FFA was considerably diminished. Bieck *et al.* were quick to point out that at least in the context of the second messenger system, cyclic AMP is an intracellular compound, and that it was somewhat surprising that exogenous cyclic AMP will mimic the action of any hormone (cf. Chapters 2 and 5).

The sorts of problems which may be encountered in experiments with exogenous nucleotides were well illustrated by Mosinger and Vaughan (1967) who found that the lipolytic hormones and cyclic AMP optimally stimulated lipolysis in media that were very different (i.e., epinephrine was maximally effective in complete Krebs-bicarbonate buffer while cyclic AMP worked best in a drastically simplified system). In other words, the continued stimulation of adenyl cyclase by a hormone with the concomitant increase in intracellular cyclic AMP levels is a very different situation than the penetration of this very polar compound into the cell.

B.   Other Fast-Acting Lipolytic Hormones

ACTH and glucagon stimulated cyclic AMP levels in fat pads *in vitro* (Butcher *et al.,* 1966) and acted synergistically with methylxanthines on lipolysis (Hynie *et al.,* 1966) and cyclic AMP levels (Butcher and Sutherland, 1967).

The intact epididymal fat pad is less than ideal for studies of cyclic AMP levels because of the heterogeneity of the cell population found therein (Butcher and Baird, 1968). As a result, we shifted to the isolated fat cell (I.F.C.) preparation developed by Rodbell (1964). In this technique, fat pads are incubated with crude bacterial collagenase for short periods (20–40 min), resulting in the dissolution of some of the connective tissue of the fat pad. The I.F.C. are separated from the other cells of the fat pad by taking advantage of their low specific gravity. When centrifuged at low speed, the fat cells rise to the top of the centrifuge tube, and repeated washing yields an essentially homogenous population of I.F.C.

In addition to the homogeneity obtained with I.F.C., other advantages of the preparation include better sampling and also unimpaired access of hormones and other agents to the cells. There are disadvantages as well — one is a quantitative (but never qualitative) variability in the response of well-prepared I.F.C., and the other is the notorious difficulty in obtaining preparations of collagenase which will dissolve connective tissue without causing serious damage to the fat cell.

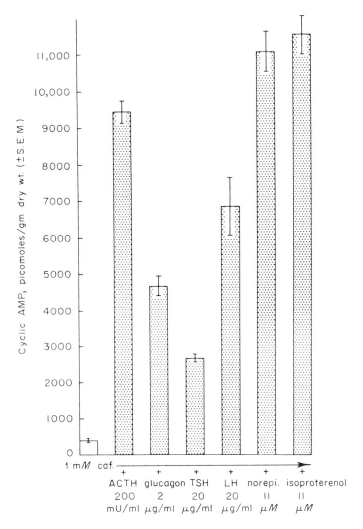

Fig. 8.4. Effects of 1 mM caffeine and a variety of hormones on cyclic AMP levels in isolated fat cells. (From Butcher *et al.*, 1968.)

As shown in Fig. 8.4, ACTH, glucagon, TSH, LH, and, as would be expected, norepinephrine and isoproterenol increased cyclic AMP levels in I.F.C. (Butcher *et al.*, 1968). Preparations which were without significant effect on cyclic AMP levels under the conditions used in these experiments included prolactin (5 μg/ml), pitressin (1 U/ml), and pitocin (1 U/ml). The adenyl cyclase of I.F.C. was more sensitive to ACTH than to epinephrine (Butcher *et al.*, 1968). Rodbell and his associates (1968) found essentially the same orders of potency with the adenyl cyclase activity of fat cell ghosts, and they found that glucagon was active at even lower concentrations than ACTH. However, in their experiments as in ours, the effectiveness of these hormones was not obviously related to sensitivity. As shown in Fig. 8.4, glucagon at supramaximal concentrations caused cyclic AMP levels which were only one-half those obtained with supramaximal ACTH or catecholamines.

C.   Adenyl Cyclase and the Fast-Acting Lipolytic Hormones

The lipolytic hormones appear to increase intracellular cyclic AMP levels by activating the adenyl cyclase system rather than by inhibiting the phosphodiesterase. Although final proof of this contention must await the preparation of a pure, hormone-sensitive adenyl cyclase from fat, two kinds of evidence support it. First, the hormones increase formation of cyclic AMP from exogenous ATP by cell-free preparations (Klainer *et al.*, 1962; Butcher and Sutherland, 1967; Rodbell *et al.*, 1968; Butcher and Baird, 1969). Although these preparations are still contaminated with phosphodiesterase, its influence is minimized. Second, the striking synergistic effects of the hormones and the methylxanthines on I.F.C. are strong evidence that the lipolytic hormones are acting on adenyl cyclase. In addition, these hormones have been without acute effects on the phosphodiesterase preparations available. Thus, all the available evidence supports the idea that the lipolytic hormones act on the adenyl cyclase system in fat, as they appear to do in tissues with which they are more commonly associated.

However, it was hard to visualize how a single adenyl cyclase system could be stimulated by molecules of such different structures as epinephrine, ACTH, and LH. Since the specificities of the adenyl cyclase systems of other tissues have been very exacting, it seemed possible that adipose tissue might contain two or more adenyl cyclases with different specificities. However, supramaximal concentrations of several hormones were as effective alone as in combination (Table 8.II). Rodbell *et al.* (1968) have obtained results qualitatively identical to ours using the adenyl cyclase activity of isolated fat cell ghosts. Thus these experiments

TABLE 8.II. Effect of Supramaximal Concentrations of Epinephrine, ACTH, and the Combination of the Two on Cyclic AMP Levels in I.F.C.[a]

Additions	Cyclic AMP picomoles/gm (dry weight)
1 mM Caffeine	300 ± 20
1 mM Caffeine + 110 μM epinephrine	10,510 ± 1,100
1 mM Caffeine + 400 mU/ml ACTH	9,500 ± 500
1 mM Caffeine + 110 μM epinephrine + 400 mU/ml ACTH	8,500 ± 350

[a]From Butcher, Baird and Sutherland, 1968.

did not support the multiple adenyl cyclase system theory. However, they are subject to many qualifications. For example, it seems possible that the maximal accumulation of cyclic AMP in fat cells may be regulated by factors other than adenyl cyclase activity (e.g., substrate availability or phosphodiesterase activity). In addition, since adenyl cyclase does vary from tissue to tissue, it may be expected that some variation in a given tissue may be encountered in the future, just as several enzymes are known to differ within a single tissue.

D. Cyclic AMP and Lipolysis

The relationship between intracellular cyclic AMP levels and the rate of lipolysis requires some comment. In adipose tissue, as in several other hormone actions where the physiological response and intracellular cyclic AMP levels have been measured simultaneously, the excursion of cyclic AMP required for maximal stimulation of the physiological response is smaller than the total possible excursion of cyclic AMP (cf. Chapters 5, 6, and 7). When glycerol release and intracellular cyclic AMP were plotted against one another, cyclic AMP was rate-limiting only between 180–300 picomoles/gm (Fig. 8.5). At cyclic AMP levels even as high as 1000 picomoles/gm the rate of lipolysis was not further increased. Although this curve was displaced when different experimental conditions were employed, the relationship was not qualitatively altered and was essentially independent of the agent used to increase cyclic AMP levels (Corbin et al., 1970a). On the other hand, Weiss et al. (1966) reported that I.F.C. responded to theophylline with higher lipolytic rates even though already maximally stimulated with norepinephrine or ACTH. Since these results are at variance not only with our earlier data but also with those of several others, one can only surmise that for some reason Weiss and

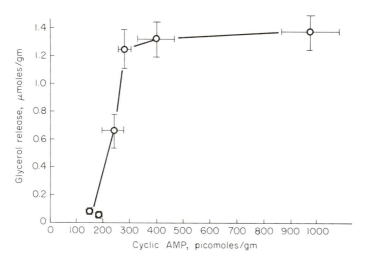

Fig. 8.5. Cyclic AMP levels and lipolytic rates.

his co-workers had not reached cyclic AMP levels capable of sustaining maximal lipolytic rates with the hormones alone.

The cyclic AMP present in I.F.C. incubated without hormones or phosphodiesterase inhibitors appears to be highly stable. As shown in Table 8.III, caffeine alone had no significant effect on cyclic AMP levels during 10-min incubations, suggesting that adenyl cyclase activity was low in the absence of added hormones.

TABLE 8.III. Synergism between Epinephrine and Caffeine on Cyclic AMP Levels in I.F.C.

Addition	Cyclic AMP picomoles/gm (dry weight)
NaCl (control)	410 ± 20
5.5 μM Epinephrine	850 ± 30
1.0 mM Caffeine	520 ± 20
5.5 μM Epinephrine + 1.0 mM caffeine	9800 ± 300

The addition of 5.5 μM epinephrine raised cyclic AMP levels significantly, but the elevation was small. However, the two agents in combination acted synergistically. Thus, there was a considerable phosphodiesterase activity in the fat cells, since even partial inhibition of the enzyme so greatly magnified the effect of epinephrine. Despite this, the basal levels of cyclic AMP in I.F.C. persist even after 4 hr of incubation without hormones or phosphodiesterase inhibitors. By contrast to the basal situation, when cyclic AMP levels are elevated in I.F.C., turnover appears to be rapid (cf. Fig. 8.8). In addition, the levels of cyclic AMP found in I.F.C. under basal conditions, although high enough to stimulate lipolysis, phosphorylase activation, etc., do not appear to do so to any great degree perhaps because of binding or sequestration (Butcher *et al.*, 1966; Sneyd *et al.*, 1968). On the other hand, the increased cyclic AMP found in fat cells exposed to lipolytic hormones must be in a form or location that is physiologically active, so that only a small rise relative to the basal content of the tissue stimulates lipolysis maximally.

The measurement of cyclic AMP at the low intracellular concentrations which are physiologically relevant is at best difficult. Methylxanthines, by providing amplification of the cyclic AMP response to hormones which stimulate adenyl cyclase, have facilitated studies not only of the hormones but especially of agents which act to decrease cyclic AMP levels. Although measurements of lipolysis in these experiments were of little value, since even in the presence of inhibitors cyclic AMP levels were far in excess of those required for maximal lipolytic rates, they were of great value for studies of the control of cyclic AMP levels. The extrapolation of these data at such high intracellular levels of cyclic AMP, to the lower levels, where cyclic AMP is rate-limiting on lipolysis, appear to be justified. For example, Corbin *et al.* (1970a) have found that the effects of insulin on cyclic AMP levels are reflected by changes in the rate of lipolysis when they are kept within the rate-limiting range. Also, they found that the effects of insulin on cyclic AMP were quantitatively similar at high and low levels of the cyclic nucleotide.

E. Lipase Activation

Studies on the mechanism by which cyclic AMP activates TG lipase have been surprisingly limited.* Rizack (1964, 1965) reported that cyclic AMP would, in the presence of ATP and Mg^{2+}, largely restore TG lipase activity which had been previously inactivated by incubation of fat pads for 3 hr prior to homogenization. The concentration of cyclic AMP required for a maximal effect on lipase activation in the homogenate system (2×10^{-5} M) was comparable to the intracellular concentration of cyclic AMP, although the range of concentrations over which cyclic AMP was effective was extremely narrow. ATP was required for the effect of cyclic AMP but CTP could, at twice the concentration, be substituted for ATP. Both ATP and CTP were inhibitory at slightly higher concentrations.

*See, however, Section VII.

Caffeine (6.7×10^{-4} M) stimulated the lipolytic system in the presence or absence of the nucleotides and Mg^{2+}, and at higher concentrations was inhibitory. At concentrations between 0.1 and 0.25 M Ca^{2+} could replace cyclic AMP.

Another cell-free lipolytic system was reported by Hales *et al.* (1968). Reasoning that since the cytoplasmic volume of fat cells is small in comparison to their total volume, Wade and his co-workers used very concentrated homogenates to avoid the possibly excessive dilution of cellular components which would occur with more conventional tissue/medium ratios.

The production of FFA from endogenous triglycerides in these concentrated homogenates was inhibited by the addition of Mg and ATP. Cyclic AMP (1 mM) had no effect on FFA production in the absence of Mg^{2+} and ATP, but in the presence of 3 mM Mg and ATP, cyclic AMP in concentrations ranging from 1–100 μM caused a restoration of the activity to that seen in the absence of Mg^{2+} and ATP. These effects of cyclic AMP were small (40–60% stimulation) but significant. Exposure of fat pads to epinephrine before homogenization produced threefold increases in lipase activity, and cyclic AMP did not stimulate FFA production in homogenates of stimulated pads although the ATP and Mg^{2+} inhibition remained. These authors also stated that they were unable to obtain effects of cyclic AMP in homogenates using Rizack's system.

To date, these reports are the only ones dealing with the mechanism of cyclic AMP on lipase activation. It seems reasonable to expect that the activation involves phosphorylation or dephosphorylation of the lipase and that this is probably enzymatically catalyzed. However, it is manifest that the system must be simplified and the lipase and the still hypothetical kinases, etc., studied in detail.

F. Species Variations

One of the more interesting aspects of the endocrinology of adipose tissue is that striking variations are encountered between species (see Rudman *et al.*, 1965). For example, while both epinephrine and ACTH stimulate lipolysis in rat epididymal fat pads or perirenal fat, perirenal fat from mature rabbits is stimulated by ACTH but not by epinephrine. Recently, Carlson *et al.* (1970 and unpublished observations) have found the effects of hormones on cyclic AMP levels in adipose tissues from man, dog, and rabbit paralleled their effects on lipolysis, and Bhakthan and Gilbert (1968) reported that exogenous cyclic AMP mimicked the action of epinephrine on cockroach fat bodies.

The rate of lipolysis of rat interscapular brown fat is stimulated by catecholamines, glucagon, ACTH, and TSH (Joel, 1966). In addition, cyclic AMP levels in brown fat are increased by incubation with epinephrine and caffeine and lowered by prostaglandin E_1 and insulin (Butcher and Baird, 1968, 1969; Beviz et al., 1968). Beviz et al. also reported that exogenous cyclic AMP stimulated lipolysis in brown fat.

III. AGENTS WHICH DECREASE CYCLIC AMP LEVELS AND LIPOLYSIS IN ADIPOSE TISSUE

A. Insulin

Gordon and Cherkes (1958), who first demonstrated the *in vitro* lipolytic effect of epinephrine, also reported that insulin antagonized the increased release of FFA engendered by epinephrine. However, the medium used in these experiments contained glucose, and the effect of insulin was due at least in part to its stimulatory effect on glucose transport: the increased availability of glucose leads to increased levels of α-glycerol phosphate, and hence to increased reesterification of the FFA produced by lipolysis. However, Jungas and Ball (1963) reported that insulin was capable of antagonizing the effects of epinephrine on lipolysis in fat pads even in the absence of glucose, proving that insulin could also inhibit lipolysis independently of its effects on glucose transport. Insulin was effective against low or moderate but not high concentrations of lipolytic agents (Jungas and Ball, 1963; Carlson and Bally, 1965), and Jungas and Ball suggested that insulin might in some way affect lipolysis through decreased cyclic AMP levels. Later, in collaboration with Drs. J. G. T. Sneyd and C. R. Park (Butcher et al., 1966), we found that insulin lowered cyclic AMP levels significantly and rapidly in fat pads (Fig. 8.6). The effect of insulin in suppressing cyclic AMP accumulation in fat appeared to be specific for native insulin. Neither bovine serum albumin or insulin denatured by incubation with dithiothreitol lowered cyclic AMP in the presence of caffeine and epinephrine, nor was any effect obtained with zinc ions in concentration of 0.2 or 1.0 μM (Butcher et al., 1966). Insulin lowered cyclic AMP levels in isolated fat cells at low concentrations, as little as 0.01 mU/ml causing significant decreases in cyclic AMP levels (Butcher et al., 1968). In addition, insulin lowered the increased levels of cyclic AMP found in fat pads incubated with glucagon or ACTH as well as epinephrine in intact fat pads or isolated fat cells. In these experiments, there was a poor correlation between the effects of insulin on cyclic AMP levels and lipolysis. This was because of the high background of inactive cyclic AMP (alluded to in Section II, D), the imprecision of the assay,

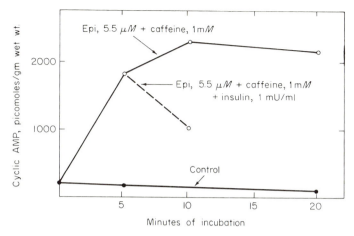

Fig. 8.6. Effect of insulin on cyclic AMP levels in epididymal fat pads. (From Butcher *et al.*, 1966.)

and the variations encountered due to the problems of sampling fat pads. Corbin *et al.* (1970a) later studied the effects of insulin on glycerol release and cyclic AMP levels in isolated fat cells. This preparation was preferable to fat pads because it is very sensitive to hormones and, as a homogeneous population of cells, is better from the standpoint of sampling. In the presence of mild lipolytic stimuli (e.g., 5.5 μM epinephrine), insulin caused significant decreases in both cyclic AMP and lipolysis, supporting the idea of a causal relationship between these two effects of insulin.

Unfortunately, the mechanism by which insulin lowers cyclic AMP levels remains unknown. This is in part because of the high turnover of cyclic AMP in fat cells exposed to hormones, and also because the effect of insulin on cyclic AMP cannot be obtained in the absence of cell structure, i.e., in cell-free preparations. Jungas (1966) reported that preliminary incubation of fat pads with insulin decreased the accumulation of ^{14}C-cyclic AMP from ^{14}C-ATP by homogenates. He also showed that phosphorylase activity was decreased and glycogen synthetase increased, both of which would be expected if intracellular cyclic AMP levels were lowered. The data of Jungas suggested that insulin might act by inhibiting adenyl cyclase, but did not eliminate stimulation of the phosphodiesterase. Senft and his co-workers (1968d) reported that alloxan treatment *in vivo* lowered phosphodiesterase activity in homogenates of rat liver and adipose tissue, and that the injection of insulin resulted in rapid increases in phosphodiesterase activity in both tissues. However, this has not as yet been confirmed by other investigators. We have not observed effects of insulin on the phosphodiesterase *in vitro,* and recently Blecher *et al.*

(1968) reported that insulin was without effect on adipose tissue phosphodiesterase activity under a variety of experimental situations, including administration of insulin *in vivo*, to intact tissues *in vitro*, or with broken cell preparations. However, it should be emphasized that these negative results cannot totally eliminate the phosphodiesterase. Blecher and his colleagues measured phosphodiesterase activity at concentrations of cyclic AMP far in excess of those found in cells, and if insulin were acting to alter the K_m rather than to change V_{max}, this would have been missed.* Also, it must be recognized that the intracellular distribution of the phosphodiesterase is complex, at least in some tissues (Butcher and Sutherland, 1962), and it does not seem inconceivable that insulin might in some way alter the distribution of the phosphodiesterase (or for that matter of cyclic AMP). Thus, while it seems possible that insulin lowers cyclic AMP via an effect on adenyl cyclase, this cannot as yet be stated with any certainty. An alternate possibility is that insulin might lower cyclic AMP levels indirectly. For example, insulin might promote the formation of another second messenger which in some way affects cyclic AMP levels, and which might even be a derivative of cyclic AMP. Still another possibility is that insulin might act primarily to alter the configuration of a constituent of the plasma membrane, e.g., lipoproteins, as suggested by Rodbell and Jones (1966). This change in the membrane could then alter the activities of either adenyl cyclase or of the phosphodiesterase.

Although it seems very likely that decreased cyclic AMP levels are the basis of the antilipolytic action of insulin, it is less likely that another important effect of insulin, that of enhanced carbohydrate transport into fat cells (Renold *et al.*, 1965), involves cyclic AMP. We have been unable to demonstrate effects of insulin on lowering cyclic AMP levels in the absence of stimulation by a lipolytic hormone (e.g., epinephrine), and Sneyd *et al.* (1968) found a number of circumstances which made it seem unlikely that decreased cyclic AMP levels were involved with the stimulation of transport. Similar conclusions have been drawn by Rodbell (1967b). Furthermore, although insulin does lower cyclic AMP levels in liver, to date we have been unable to detect any effect of insulin on striated muscle (perfused heart or diaphragm), where the effect of insulin on carbohydrate transport is very striking. Finally, the stimulation of glucose transport seems to be a feature of insulin action in adipose tissue of all species studied, including birds (Goodridge and Ball, 1965), in which species insulin is not antilipolytic (Grande, 1969). The latter observation might suggest that the interaction of insulin with its receptors (Kono,

*See also Loten and Sneyd (1970).

1969a,b) does not lead to a fall in the level of cyclic AMP in avian adipose tissue, although this has yet to be established.

B. The Prostaglandins

Our interest in the prostaglandins was aroused by two reports dealing with the "antihormonal" actions of these compounds. First, Steinberg and Vaughan working in collaboration with Professor Sune Bergström (Steinberg et al., 1964) reported that PGE_1 at very low concentrations antagonized the effects of the lipolytic hormones on both lipolysis and phosphorylase activation in the rat epididymal fat pad. Next, Orloff, Handler, and Bergström (1965) demonstrated that PGE_1 antagonized the actions of vasopressin and theophylline on water and ion movement in the toad bladder but did not antagonize the ability of cyclic AMP to mimic vasopressin. Therefore, it seemed reasonable to expect that PGE_1 would depress lipolysis via decreased levels of cyclic AMP.

Our first experiments were quite surprising. PGE_1 did in fact antagonize the effect of epinephrine on cyclic AMP levels in fat pads, but curiously, also increased cyclic AMP levels by itself. The increase in cyclic AMP, although small, was sufficient to have maximally activated lipolysis, but no activation of lipolysis occurred (Butcher et al., 1966). Therefore, it appeared that we had succeeded in dissociating cyclic AMP from lipolysis. However, a possible explanation for this was that PGE_1 was increasing cyclic AMP levels in cells other than fat cells. If this were the case, one would not expect stimulation of lipolysis, since the increased cyclic AMP would be irrelevant to lipolysis. Fortunately, this could be tested using I.F.C.

As shown in Fig. 8.7, PGE_1 did not cause increased cyclic AMP levels in isolated fat cells although the effect of epinephrine was antagonized (Butcher et al., 1967; Butcher and Baird, 1968). The loss of the ability of the tissue to respond to PGE_1 with increased cyclic AMP levels was not due to the collagenase treatment but rather to the separation of the fat cells from the rest of the fat pad. In addition, stimulatory effects of PGE_1 on cyclic AMP levels in the "stromovascular" component of the fat pad, i.e., that part of the fat pad which is found in the bottom of a tube after centrifugation at 100 g for 1 min, have been observed (Butcher and Baird, 1969).

PGE_1 antagonized the actions of not only the catecholamines but also of polypeptide lipolytic hormones including ACTH, glucagon, TSH, and LH on cyclic AMP accumulation in isolated fat cells (Butcher and Baird, 1968). This was in accord with the earlier observations by Steinberg et al.

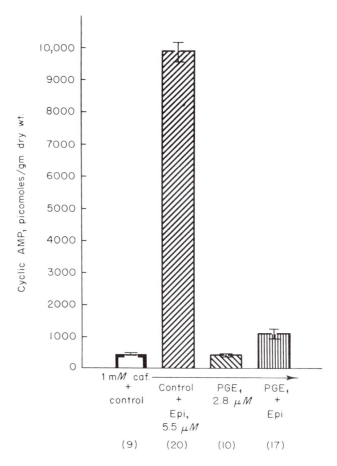

Fig. 8.7. Effects of PGE₁ on cyclic AMP levels in isolated fat cells. (From Butcher *et al.,* 1968.)

(1964). In addition, the cyclic AMP mechanism in isolated fat cells is very sensitive to PGE₁, for 50% inhibition of the effect of 5.5 μM epinephrine occurred with 0.004 μM PGE₁. Finally, as shown in Fig. 8.8, PGE₁ affected cyclic AMP levels very rapidly. Within 2 min after the addition of PGE₁ to fat cells incubated with 5.5 μM epinephrine and 1 mM caffeine, cyclic AMP levels were reduced to about one-half those incubated in the absence of the prostaglandin. Humes *et al.* (1968) obtained essentially identical results for the effects of PGE₁ on cyclic AMP levels in fat pads and also in isolated fat cells.

Unfortunately, the mechanism by which PGE₁ lowers cyclic AMP levels is unclear. Just as with insulin, we have been unable to elicit effects

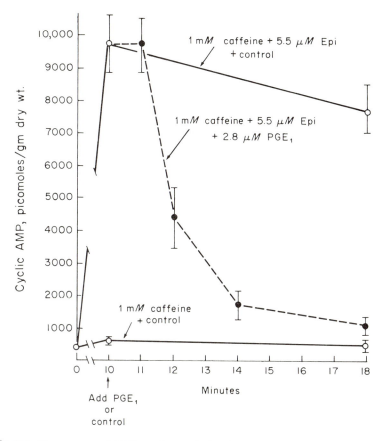

Fig. 8.8. Time course of PGE$_1$ action on isolated fat cells. (From Butcher and Baird, 1968.)

of PGE$_1$ on either adenyl cyclase or the phosphodiesterase in cell-free preparations of fat. Steinberg and Vaughan (1967) and Humes *et al.* (1968) reported that PGE$_1$ did not block the lipolytic action of exogenous dibutyryl cyclic AMP. This was somewhat equivocal, however, since in very limited studies dibutyryl cyclic AMP was not hydrolyzed by purified phosphodiesterase (Posternak *et al.*, 1962). However, Blecher *et al.* (1968) reported that adipose tissue homogenates degraded dibutyryl cyclic AMP at nearly the rate of cyclic AMP, although it was not clear if the phosphodiesterase was hydrolyzing the dibutyryl compound directly or if the derivative was first converted to free cyclic AMP which was then degraded. Thus, it is tempting to postulate that PGE$_1$ is probably not acting at the level of the phosphodiesterase and that it does not antagonize the effects of cyclic AMP on lipolysis. In addition, Steinberg and Vaughan

(1967) showed that PGE_1 was capable of antagonizing the lipolytic effect of theophylline except at the very highest concentrations of the methylxanthine. Again, this is compatible with an action of PGE_1 on the formation of cyclic AMP.

Stock *et al.* (1968) examined the nature of the antilipolytic action of PGE_1 and found that it competitively antagonized the actions of norepinephrine, ACTH, and theophylline. By contrast, the β-adrenergic antagonist Kö-592 acted competitively only against norepinephrine, and noncompetitively blocked ACTH, theophylline, and dibutyryl cyclic AMP. The inhibition by phentolamine was noncompetitive against all four stimulatory agents. In a well-reasoned kinetic analysis, Stock and his colleagues proposed that PGE_1 interferes with the binding of ATP to adenyl cyclase without reducing the rate of breakdown of the ATP-adenyl cyclase complex.

C. Other Antilipolytic Agents

Ouabain has been shown by Ho *et al.* (1967) to have insulin- and prostaglandinlike activity on lipid and carbohydrate metabolism in adipose tissue. The addition of $3 \times 10^{-4} M$ ouabain to isolated fat cells incubated in the absence of glucose caused a marked inhibition of the lipolytic effects of ACTH, epinephrine, or glucagon. Conversely, ouabain did not significantly antagonize the lipolytic effects of caffeine or dibutyryl cyclic AMP, suggesting that ouabain was acting at the level of the adenyl cyclase system. Direct evidence of this was also presented by Ho and his co-workers, for ouabain was found to cause about 50% inhibition of adenyl cyclase activity while the phosphodiesterase was unaffected.

The well-recognized actions of ouabain on the transport of Na^+ and K^+ suggested that ions might be involved, and Ho *et al.* showed that the omission of K^+ from the incubation medium produced an effect like that of ouabain not only on fat and carbohydrate metabolism but also on adenyl cyclase. The effects of ouabain on carbohydrate metabolism and of ions on the control of cyclic AMP and subsequent events are discussed in other sections of this chapter.

The antilipolytic actions of nicotinic acid are well recognized and have been thoroughly reviewed (Carlson and Bally, 1965). Krishna *et al.* (1966) reported in an abstract that nicotinic acid stimulated the activity of adipose tissue phosphodiesterase but this has not been reproduced in our laboratory or in others (Kupiecki and Marshall, 1968). Nonetheless, concentrations of nicotinic acid sufficient to slow lipolysis also lowered cyclic AMP levels in isolated fat cells incubated with catecholamines, ACTH, or glucagon (Butcher *et al.*, 1968; Butcher and Baird, 1969). Kupiecki and

Marshall have reported that 5-methylpyrazole-3-carboxylic acid (U-19425) inhibited lipolysis in a fashion identical to nicotinic acid, but that neither compound inhibited lipolysis stimulated by dibutyryl cyclic AMP, or affected the activity of the cyclic nucleotide phosphodiesterase, although imidazole (which has been recognized as a phosphodiesterase stimulator for several years) did activate the enzyme. 5-Methylpyrazole-3-carboxylic acid also lowered cyclic AMP levels (Butcher and Baird, 1969).

The actions of adrenergic blocking agents on lipolysis in adipose tissue were quite confusing for several years but have been simplified recently. The first demonstration of blockade of the actions of lipolytic hormones by adrenergic blockers was reported by Schotz and Page (1960). They showed that the α-adrenergic blockers phenoxybenzamine and phentolamine antagonized the lipolytic actions of epinephrine and ACTH. Later, Mayer et al. (1961) showed that the β-adrenergic blocker DCI was both a weak agonist and strong antagonist of epinephrine-stimulated lipolysis in the dog. The situation then became very complicated when Love et al. (1963) reported that β-adrenergic blockers antagonized not only the catecholamines but also at high concentrations the polypeptide lipolytic hormones, including ACTH, glucagon, and TSH.

There was no precedent for β-adrenergic blockers to antagonize the stimulatory effects of polypeptide hormones on adenyl cyclases, and therefore it sounded as though adipose tissue adenyl cyclase was indeed peculiar. However, when we tested the effects of β-adrenergic blockers on cyclic AMP levels in fat cells we found no antagonism of ACTH or glucagon although the action of epinephrine was abolished (Butcher et al., 1968). Earlier, Stock and Westermann (1966) had shown that the β-adrenergic blocker Kö-592 blocked catecholamines competitively and confirmed Love et al. in that much higher concentrations of the blocker were required to antagonize the polypeptide lipolytic hormones. In addition, Stock and Westermann found that this inhibition was noncompetitive. They also demonstrated that the blockade of the lipolytic actions of all of these hormones (including the catecholamines) by α-adrenergic blockers was non-competitive and that much higher levels were required. The whole quandry was satisfactorily resolved when Aulich et al. (1967) and Peterson et al. (1968) showed that at the very high concentrations of α- and β-adrenergic blockers required to antagonize the polypeptide hormones, these agents also antagonized the lipolytic action of exogenous cyclic AMP or of dibutyryl cyclic AMP. Thus, these apparently were not specific blocking effects on adenyl cyclase (as with β-blockade of catecholamines) but rather nonspecific effects which occurred after the formation of cyclic AMP.

Fassina (1967) reported that inhibitors of oxidative phosphorylation, including dinitrophenol and oligomycin antagonized the effects of catecholamines on lipolysis. She suggested that these inhibitors might be acting at the level of adenyl cyclase (either directly or by decreasing the availability of ATP for cyclization) or at the level of lipase activation, which is apparently ATP dependent.

Kuo and Dill (1968) have reported that the antibiotics valinomycin and nonactin inhibited the adenyl cyclase activity in isolated fat cells in the presence of theophylline and norepinephrine. These are "ionophorous" antibiotics, which have been reported to affect membranes and alter ion transport. Kuo and Dill reasoned that, since adenyl cyclase appears to be associated with the cell membrane, that these antibiotics might affect adenyl cyclase and hence secondarily inhibit lipolysis. Their method of assay of adenyl cyclase also warrants comment, in that they incubated fat cells with adenine-8-^{14}C and measured the accumulation of cyclic AMP-8-^{14}C in response to various lipolytic and antilipolytic agents (Kuo and DeRenzo, 1969).

D. Comments on Antilipolytic Substances

The diversity and complexity surrounding the antilipolytic agents have been only slightly relieved by the knowledge that most of them are acting through cyclic AMP. One of the most distressing aspects has been our inability to prove the site of action of insulin or the prostaglandins. However, certain generalizations, which are of no great help but still are of interest, can be offered. First, at least in terms of their antagonism of the effects of lipolytic hormones on cyclic AMP levels, insulin and the prostaglandins are by far the most potent. Half-maximal inhibition (of 5.5×10^{-6} M epinephrine) occurred at about 2×10^{-10} M for insulin and at about 2×10^{-9} M for PGE$_1$ and PGE$_2$. Nicotinic acid and 5-methylpyrazole-3-carboxylic acid were almost two orders of magnitude less effective, and propranolol produced 50% inhibition at about 4×10^{-6} M.

In addition, of all these antilipolytic agents, only the β-adrenergic blocking agents were specific (i.e., they would antagonize only catecholamines while the other inhibitors were effective against the polypeptide lipolytic hormones as well).

The β-adrenergic blocking agents are known to act, in part at least, by inhibiting adenyl cyclase. Therefore, if PGE$_1$, insulin, or nicotinic acid were acting to activate the phosphodiesterase, one might expect an effect of the two which would be greater than additive (potentiation or synergism). However, such was not the case. Combinations of propranolol with the other antagonists resulted in effects which were slightly less than additive and which caused displacements of the propranolol dose response curve which were parallel (Butcher and Baird, 1969).

Finally, of these agents which lower cyclic AMP levels in fat cells, the only type which worked in the absence of cellular structure are the β-adrenergic blocking agents (Butcher and Baird, 1969).

IV. PHOSPHODIESTERASE INHIBITORS

In most tissues, the activity of phosphodiesterase is far greater than that of adenyl cyclase (at least as measured in cell free systems). This difference may be in part artifactual, since the phosphodiesterase is considerably more stable than adenyl cyclase and because it is also easier to assay properly. In addition, measurements of phosphodiesterase activity are usually carried out at concentrations of cyclic AMP higher than those found in cells (the K_m is between 1×10^{-5} and 1×10^{-4} M). The reader is referred to Chapter 4 for a more detailed discussion of phosphodiesterase.

However, changes in the activity of the phosphodiesterase can result in very rapid and large changes in cyclic AMP levels in I.F.C. as shown in Fig. 8.9 (Butcher and Baird, 1969). Thus, it is obvious that the diesterase can play a key role in the control of cyclic AMP levels in I.F.C. The best known inhibitors of the phosphodiesterase are the methylxanthines, and unfortunately these are also very potent compounds in a variety of systems. For example, in the phosphorylase system alone, in addition to inhibiting the phosphodiesterase, methylxanthines activate phosphorylase phosphatase, and at higher concentrations inhibit phosphorylase itself and also inhibit adenyl cyclase. Thus, it seems likely that systems involving cyclic AMP may also be affected by methylxanthines at points other than the phosphodiesterase.

It is difficult to say exactly what degree of inhibition of the phosphodiesterase in I.F.C. is produced by a given concentration of the methylxanthines, for example, in experiments like those shown in Figs. 8.3 and 8.9. Davies (1968) reported that 2 mM caffeine inhibited fat phosphodiesterase 30%. The concentration of cyclic AMP he used was considerably in excess of the levels found in I.F.C. under basal conditions, and the degree of inhibition by caffeine would be higher if lower concentrations of cyclic AMP were used. In addition, very little is known about the relationship of phosphodiesterase activity as measured in homogenates and that which is actually effective in the cell. For example, Cheung

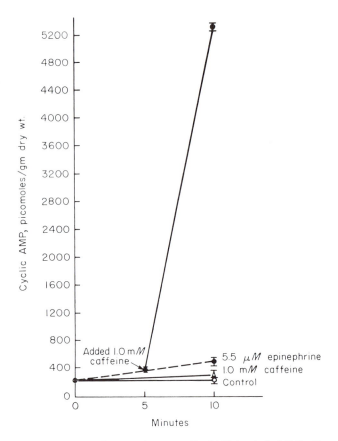

Fig. 8.9. Time course of caffeine action on cyclic AMP levels in I.F.C. (From Butcher and Baird. 1969.)

(1967) has shown that several cellular constituents including ATP and PP_i are phosphodiesterase inhibitors at concentrations which are physiologically meaningful.

Other inhibitors of phosphodiesterase have also been discovered. Appleman and Kemp have demonstrated that puromycin and puromycin aminonucleoside inhibit fat phosphodiesterase, and Davies (1968) has shown a variety of purine and pyrimidine derivatives to be inhibitors. Dalton et al. (1970) found that an imidazolidine derivative was more potent than theophylline in inhibiting phosphodiesterase in adipose tissue homogenates, but less potent in stimulating lipolysis. This might indicate different mechanisms of action, but could also reflect differential penetration into intact cells. When a series of xanthine derivatives were compared, Beavo *et al.* (1970b) found that their ability to inhibit adipose tissue phosphodiesterase correlated well with their relative potencies as lipolytic agents.

V. OTHER FACTORS CONTROLLING LIPOLYSIS WHICH MAY INVOLVE CYCLIC AMP DIRECTLY OR INDIRECTLY

A. Pituitary Function

Goodman and Knobil (1959) showed that hypophysectomy decreased the effects of epinephrine on FFA release. More recently, Goodman (1966) reported that the effects of ACTH, TSH, and fasting on lipolysis were reduced in adipose tissue from hypophysectomized rats.

An involvement of growth hormone (GH) in the regulation of lipolysis has been suspected for many years. Raben and Hollenberg (1959) showed that GH injection caused increased plasma FFA in rats, and Kovacev and Scow (1966) reported that GH increased FFA release by fat pads perfused *in vivo.* Also, under conditions where plasma growth hormone levels were high (for example, exercise and fasting), plasma FFA concentrations were elevated.

Fain and his co-workers (1966, 1967) have studied the lipolytic action of GH *in vitro.* Incubation of adipose tissue with GH in the presence of a glucocorticoid caused increased lipolysis but the effects of GH differed from the effects of the usual lipolytic hormones in several ways. First, GH was ineffective unless a glucocorticoid was also added. Second, the GH acted slowly as compared to the other lipolytic hormones. Effects of GH + dexamethasone were not significant until after 2 hr of incubation. Further, actinomycin D and puromycin blocked the lipolytic effect of GH + dexamethasone but not of ACTH or other "fast-acting" hormones. These results suggested that the mechanism of action of growth hormone and the glucocorticoid involved DNA-RNA-dependent protein synthesis.

However, Fain later found some similarities between the actions of GH + dexamethasone and the better established lipolytic hormones. Theophylline potentiated and insulin and nicotinic acid antagonized the effect of GH + dexamethasone suggesting that cyclic AMP might in some way be involved in the action of GH. Quite recently, Fain (1968a) reported that incubation of isolated fat cells with GH + dexamethasone for 4 hr resulted in five- to tenfold increases in the sensitivity of the lipolytic mechanism to norepinephrine or theophylline without affecting the maximal response and the lipolytic action of dibutyryl cyclic AMP was mildly depressed. The ability of GH + dexamethasone to increase the sensitivity of the fat cells to theophylline was prevented by the inclusion of cycloheximide in the incubation, and Fain suggested that GH might somehow be involved in the synthesis of a protein factor which contributed to the regulation of cyclic AMP levels in the isolated fat cell.

Goodman (1968a) reported that fat pads from hypophysectomized rats failed to respond to GH + dexamethasone *in vitro,* but that the addition of theophylline to the incubation medium restored the response to a level which was quantitatively and qualitatively similar to normal tissues incubated in the absence of theophylline. GH administrations to hypophysectomized rats for at least 2 days prior to sacrifice restored the response to normal. Goodman suggested that GH might somehow decrease the level of phosphodiesterase activity in the fat pad.

B. Glucocorticoids

Adipose tissue from adrenalectomized animals is much less sensitive to lipolytic hormones than that from normals (Jeanrenaud and Renold, 1960), and the lipolytic response appears to be another dependent upon a "permissive" effect of glucocorticoids. Corbin and Park (1969) recently found that although lipolysis (on the basis of DNA content) in fat pads from adrenalectomized rats was much less sensitive to epinephrine than those from controls, the cyclic AMP mechanism responded normally. This suggested that either the amount of lipase was reduced by adrenalectomy or, as appeared more likely, the activation mechanism was defective.

Jeanrenaud (1968) has shown that the glucocorticoids affect FFA mobilization. Prolonged incubation of fat pads with dexamethasone resulted in increased FFA release which was due mainly to decreased FFA reesterification and only at high concentrations of the steroid to increased lipolysis. He suggested that physiologically, the glucocorticoids may act by decreasing the availability of carbohydrates for reesterification rather than by the activation of lipolysis.

C. Thyroid Hormones

Fat pads from hyperthyroid rats are considerably more sensitive to lipolytic hormones than those from eu- or hypothyroid animals. This was first described by Debons and Schwartz (1961) who thought it possible that thyroid hormone somehow controlled the amount of lipase activity in adipose tissue. More recently, Brodie *et al.* (1966) concluded that this was not the case, for although fat pads from hyperthyroid animals were considerably more active lipolytically than from normals in the presence of norepinephrine or ACTH alone, the combination of norepinephrine and theophylline caused equal lipolytic maxima regardless of thyroid function. They also reported that homogenates of fat pads from hyperthyroid animals produced considerably more cyclic AMP from radioactive ATP in the presence of norepinephrine. They recently published another report in which they stated that adenyl cyclase activity (as measured in the presence of NaF) in fat pads from hyperthyroid rats was twice that in fat pads from euthyroid rats and 4 times that in hypothyroid rats, and that phosphodiesterase activity was unchanged (Krishna *et al.*, 1968b).

Fisher and Ball (1967) published a thorough study on the effect of thyroid status upon oxygen consumption and lipolysis in the fat pad. They, like Brodie *et al.*, concluded that the effect of thyroid status upon the lipolytic mechanism does not involve a change in the tissue content of lipase but rather an alteration in the mechanism of activation of the lipase, presumably involving cyclic AMP. Fisher and Ball also cited preliminary studies in which the incubation of fat pads from hypothyroid rats with triiodothyronine for 14 hr in the presence of a mixture of amino acids resulted in some enhancement of lipolytic rates. They felt that the action of the thyroid hormones on the lipolytic mechanism might be through a primary action on the synthesis of adenyl cyclase.

Mendel and Kuehl (1967) have reported that very high concentrations of triiodothyronine caused a significant inhibition of phosphodiesterase activity in cell-free preparations of fat pads. However, the concentrations of triiodothyronine used in these experiments were so high that it is difficult to assess the physiological significance of the observation.

D. Nutritional State

Changes in the nutritional state of the animal may profoundly affect the lipolytic mechanism. Fat pads removed from animals fasted for 72 hr and refed for 48 hr prior to sacrifice have an extremely high basal rate of lipolysis which is inhibited by insulin (Jungas and Ball, 1963). Recently, Corbin *et al.* (1970a) found that the cyclic AMP mechanism in fat pads from fasted and refed animals was likewise hypersensitive to lipolytic stimuli, and it appeared as though the balance between the adenyl cyclase and the phosphodiesterase may have been in some way altered so that the response of the tissue was greatly enhanced. Since the measurements of cyclic AMP were expressed on the basis of the DNA content rather than the wet weight of the tissue, it seems unlikely that this effect could be explained solely by a change of the triglyceride content of the tissue. Brodie *et al.* (1969) also observed increased adenyl cyclase activity in adipose tissue from fasted rats. Cyclase activity also appeared to be greater after cold exposure (Therriault *et al.*, 1969). The mechanisms responsible for these changes have not been established.

VI. SUMMARY

The very complicated control system involving cyclic AMP and adipose tissue is shown in Fig. 8.10. The catecholamines or polypeptide lipolytic hormones stimulate the adenyl cyclase system. The lipolytic

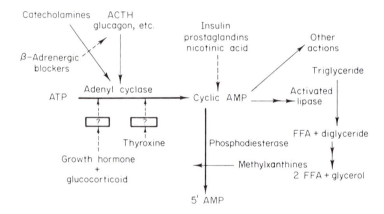

Fig. 8.10. The role of cyclic AMP in lipolysis.

hormones appear to be acting upon the same catalytic function (i.e., that part of the system which catalyzes the conversion ATP to cyclic AMP), in that the effects of supramaximal concentrations of the hormones on cyclic AMP levels were not additive. However, it seems very unlikely that the hormones are acting on the same regulatory function, not only because of their very different structure but also because only the catecholamines are specifically and competitively antagonized by β-adrenergic blocking agents. Insulin, the prostaglandins, and nicotinic acid lower cyclic AMP levels stimulated by lipolytic hormones, but the mechanisms by which they do so remain unclear. The methylxanthines cause increased intracellular levels of cyclic AMP by virtue of their ability to inhibit the cyclic nucleotide phosphodiesterase, and thus act synergistically with the lipolytic hormones. Small increases in intracellular cyclic AMP levels cause activation of a hormone-sensitive triglyceride lipase, which is the rate-limiting enzyme along the pathway of lipolysis. Finally, other factors may also affect lipolysis directly or indirectly through cyclic AMP; these include thyroid hormones and the combination of growth hormone and a glucocorticoid, which appear to modify either the concentration of adenyl cyclase or the hormone sensitivity of the system to the lipolytic hormones, ions, and the nutritional state of the animal.

VII. ADDENDUM

Adenyl cyclase activities in broken rat epididymal fat cell preparations have been further studied by several groups (Birnbaumer *et al.,* 1969; Birnbaumer and Rodbell, 1969; Vaughan and Murad, 1969; Bär and Hechter, 1969a,b,c,d). The results were generally as expected from data obtained by measuring lipolysis and cyclic AMP levels in intact fat cells. For example, β-adrenergic blocking agents antagonized the stimulation by catecholamines but not by polypeptide hormones. In addition, Birnbaumer and Rodbell (1969) found that an analog of ACTH, containing D-isomers of five amino acids, could antagonize the effect of ACTH on adenyl cyclase without affecting basal activity or the response to other hormones. Calcium ions appear to be required for ACTH but not for other hormones to stimulate adenyl cyclase (Bär and Hechter, 1969d). The ability of secretin to increase cyclic AMP levels in fat cells and fat cell homogenates (Butcher and Carlson, 1970) was in line with its high potency as a lipolytic agent (Carlson, 1968). Rodbell *et al.* (1970) reported that secretin was effective at lower concentrations than glucagon in stimulating fat cell adenyl cyclase.

R. H. Williams *et al.* (1968a) described a method for estimating the effects of hormones on cyclic AMP levels in I.F.C. by incubating them with radioactive ATP and measuring the cyclic AMP formed. In general, their results were similar to those obtained with cell-free preparations of fat cells, i.e., while the lipolytic hormones stimulated cyclic AMP accumulation and propranolol antagonized the catecholamines, insulin and the prostaglandins were without effect and sodium fluoride was effective. In view of the relative impermeability of intact cells to ATP and previous difficulties in demonstrating cyclization of exogenous ATP by avian erythrocytes (Øye and Sutherland, 1966) and I.F.C., (Table 2.III), it seems possible that what was in fact being measured was the adenyl cyclase activity of cell fragments.

Kuo and DeRenzo (1969) and Humes *et al.* (1969) described a new assay for intracellular cyclic AMP based on incubation of fat cells with adenine 8-C^{14} and isolating the cyclic AMP-8-C^{14} ultimately formed. Using this method they confirmed many of the earlier studies dealing with the effects of lipolytic hormones, phosphodiesterase inhibitors, and anti-lipolytic agents on cyclic AMP levels. In addition, Kuo and DeRenzo reported that certain proteases which mimicked insulin on lipolysis did not lower cyclic AMP levels and that filipin and pimaricin (polyene antibiotics which are anti-lipolytic) lowered intracellular cyclic AMP levels and markedly increased cyclic AMP in the media.

Kono's studies with trypsin-treated fat cells (see Chapter 7, Section III,A) were confirmed and extended by others. Fain and Loken (1969) found that the antilipolytic action of PGE_1 was unaffected by treatment which abolished the effect of insulin. Rodbell *et al.* (1970) found that glucagon did not stimulate adenyl cyclase in ghosts prepared from trypsin-treated fat cells. Trypsin treatment reduced the effects of secretin and ACTH by only 60% and 40%, respectively, and had no effect on the activation by epinephrine or fluoride.

Forn *et al.* (1970) reported that the adenyl cyclase activity in fat cell homogenates (measured in the presence of either norepinephrine or NaF) was much reduced in old rats as compared to younger animals, as was cyclic AMP accumulation in intact fat cells in response to norepinephrine. Conversely, phosphodiesterase activity was higher in the fat cells of the older rats. These data suggest that the diminished lipolytic sensitivity of older animals may be due to modifications of the cyclic AMP responsiveness in fat cells.

Krishna *et al.* (1970) confirmed an earlier observation that norepinephrine, acting by way of adrenergic β-receptors, could depolarize brown fat cell membranes. This is the opposite of the effect on membrane potential seen in vascular smooth muscle (Somlyo *et al.*, 1970). The effect in brown fat cells could not be reproduced by the dibutyryl derivative of cyclic AMP in concentrations up to 3×10^{-4} M, leading Krishna and his colleagues to suggest that depolarization might be involved in the stimulation of adenyl cyclase.

Some additional species differences have been reported. The reduction of cyclic AMP levels and inhibition of lipolysis mediated by adrenergic α-receptors in human fat cells (Burns *et al.*, 1970) was mentioned in the addendum to Chapter 6. Evidence that even the receptors for the same hormone may differ considerably from one species to another was presented by Ramachandran and Lee (1970). They found that the o-nitro-phenyl sulfenyl derivative of ACTH did not stimulate rat fat cell adenyl cyclase, but instead inhibited the effect of ACTH. In rabbit fat cells, by contrast, the derivative was as potent as ACTH in stimulating adenyl cyclase.

The lipolytic defect in several strains of genetically obese mice has been studied by Yen and his colleagues (Steinmetz et al., 1969; Yen et al., 1970) and by Enser (1970). Adipose tissue preparations from these mice are much less sensitive to lipolytic hormones and somewhat less sensitive to methylxanthines than are control preparations from non-obese litter-mates, but they respond normally to the dibutyryl derivative of cyclic AMP. Apparent adenyl cyclase and phosphodiesterase activities were both lower in homogenates prepared from the fat pads of obese mice (Enser, 1970). These results point to a defect in the adipocytes of these mice at the level of cyclic AMP formation.

Wing and Robinson (1968) reported that dibutyryl cyclic AMP could inhibit the activity of clearing factor lipase. This enzyme, which is thought to participate in the regulation of triglyceride uptake by fat cells, is also decreased by lipolytic hormones and methylxanthines, and increased in response to insulin or nicotinic acid. These observations suggest that the two enzyme systems, one involved with the mobilization of fatty acids and the other with storage, may be reciprocally controlled by cyclic AMP levels.

Goodman (1969b) reported that insulin antagonized the lipolytic effects of theophylline or exogenous cyclic AMP, but, if anything, enhanced the lipolytic action of dibutyryl cyclic AMP. Imidazole also antagonized the effects of exogenous cyclic AMP (Goodman, 1969a) but not of dibutyryl cyclic AMP (Nakano et al., 1970).

Moskowitz and Fain (1970) extended the studies on the effects of growth hormone and dexamethasone on isolated fat cells by reporting that the combination of hormones caused an increased accumulation of labeled cyclic AMP from radioactive adenine. This effect was apparent after a lag period of one hour and was blocked by inhibitors of protein synthesis.

Goodman (1970) reported that glucocorticoids alone or in combination with growth hormone enhanced the lipolytic response of adipose tissue to catecholamines. This effect appeared to require RNA and protein syn-thesis, and Goodman suggested that it might reflect induced modifications of cyclic AMP metabolism. Triiodothyronine had similar effects which, in the early hours after administration, were not additive with those of glucocorticoids and growth hormone. Another effect of glucocorticoids was described by Braun and Hechter (1970). Adenyl cyclase in fat cell ghosts obtained from adrenalectomized or hypophysectomized rats was found to respond poorly to ACTH, whereas the stimulatory effects of other hormones or fluoride were unchanged. Pretreatment with glucocorticoids selectively increased the response to ACTH, and this effect could be blocked by actinomycin D or cycloheximide.

Some progress has been made towards understanding the mechanism by which cyclic AMP activates lipase. Tsai *et al.* (1970) confirmed Rizack's earlier report that cyclic AMP could increase lipase activity in a cell-free system providing ATP and Mg^{++} were also present. The effect of cyclic AMP was larger and more reproducible when the incubation mixture was supplemented with skeletal muscle protein kinase (Corbin *et al.*, 1970b; Huttunen *et al.*, 1970). Since a similar protein kinase is known to exist in adipose tissue (Corbin and Krebs, 1969; see also Chapter 5), these observations support the view that lipase activation and phosphorylase activation may involve similar mechanisms. Whether the lipase is phosphorylated directly, or whether there is an intervening "lipase kinase," had not been established at the time of this writing.

In place of the list of references promised in the introduction to this chapter, readers interested in a more comprehensive account of the regulation of adipose tissue metabolism are referred to the monograph edited by Jeanrenaud and Hepp (1970).

CHAPTER 9

Cyclic AMP and Steroidogenesis

I. INTRODUCTION

Four endocrine glands, the adrenal cortex, the testicular interstitial cells of Leydig, the ovarian interstitial cells, and corpus luteal cells elaborate increased amounts of steroid hormones when stimulated by specific hypophyseal hormones (Fig. 9.1). The steroid hormones are synthesized from cholesterol by similar mechanisms. The rate-limiting step between cholesterol and the active hormones is the conversion of cholesterol to pregnenolone. In all these tissues, pregnenolone is converted to proges-

317

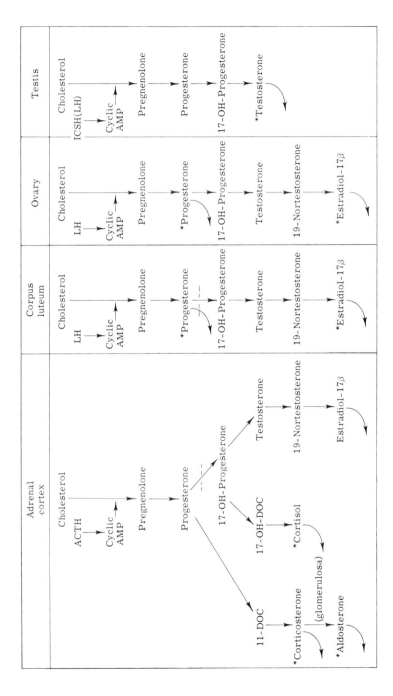

Fig. 9.1. Major steroidogenic pathways in the adrenal cortex and in the gonads. (*Indicates steroids synthesized and secreted in quantities sufficient to be considered as major steroid hormones.)

terone and in most cases, but not all, to 17-OH-progesterone. In the adrenal, 17-OH-progesterone is hydroxylated twice (at C-21 and C-11) by NADPH and O_2-dependent hydroxylases. Alternately, 17-OH-progesterone is (in lesser amounts) converted to testosterone and then to estradiol 17β and other estrogens. In the testis, ovary, and corpus luteum, however, the 21- and 11-hydroxylases are missing or vestigial. Thus, in the interstitial cells of the testis, testosterone is the major product of steroid hormone synthesis and is secreted as such. The ovary is somewhat more versatile, however, and can convert testosterone to estradiol-17β and other estrogens.

It seems clear that cyclic AMP is involved in the steroidogenic actions of the trophic hormones, and also that cyclic AMP is acting at least in part to accelerate the conversion of cholesterol to pregnenolone. Unfortunately, how cyclic AMP does this is not known because of the complexity of the reactions involved and the difficulty in establishing responsive cell-free systems in which steroidogenesis is responsive to ACTH or cyclic AMP.

II. THE ADRENAL CORTEX

A. The Haynes-Berthet Hypothesis

Cyclic AMP was first implicated in steroidogenesis by Haynes and Berthet (1957), who accumulated considerable data indicating that the activation of adrenal phosphorylase might be involved in the action of ACTH. Homogenates of beef adrenal cortex produced essentially no corticosteroid formation from endogenous precursors. However, the addition of fumarate markedly stimulated corticosteroid production, and the addition of NADP caused a further increase. Others had shown that NADP and fumarate were necessary for the 11β-hydroxylation of steroids, and that NADPH was the true stimulatory agent. Haynes' findings with the homogenates suggested that NADPH might drive the entire chain of reactions involved in steroidogenesis. Since the adrenal cortex was known to be well endowed with the enzymes of the hexosemonophosphate shunt relative to the enzymes of glycolysis Haynes reasoned that glucose-6-phosphate in the adrenal cortex would be metabolized primarily by the oxidative pathway, resulting in the generation of NADPH. This prediction was supported by experiments showing that the addition of glucose-6-P or glucose-1-P to homogenates stimulated steroidogenesis to the same extent as fumarate. However, when glycogen was added to the homogenates, steroidogenesis was not stimulated, suggesting that phosphorylase was the rate-limiting step between glycogen and glucose-6-P in the adrenal cortex. The addition of glycogen to homogenates fortified with purified liver phosphorylase resulted in a clear stimulation of steroidogenesis. They next found that slices of beef adrenal cortex incubated with ACTH contained considerably higher levels of active phosphorylase than did controls, and that the qualitative specificity of the phosphorylase response to ACTH exactly paralleled the specificity of the steroidogenic response of the slices. Conversely, ACTH did not stimulate phosphorylase activation in liver homogenates which were known to be responsive to glucagon and epinephrine. The striking similarities between the actions of ACTH on adrenal phosphorylase and epinephrine and glucagon on liver phosphorylase suggested that cyclic AMP might be involved in the adrenal response, as it was by now known to be in the liver.

Haynes (1958) found that particulate fractions of beef adrenal cortical homogenates were well endowed with adenyl cyclase, and in addition, that cyclic AMP levels in adrenal slices incubated with ACTH were higher than those in control slices. The specificity of the cyclic AMP response of the adrenal (like the phosphorylase response) paralleled the steroidogenic response in that hormones like epinephrine and glucagon were without effect on adrenal cyclic AMP concentrations. Finally, Haynes, in collaboration with Koritz and Péron (1959), showed that exogenous cyclic AMP stimulated steroidogenesis in rat adrenal fragments. Haynes assembled these data into a hypothesis for the action of ACTH which is illustrated in Fig. 9.2. ACTH activated adenyl cyclase, and the resulting rise in cyclic AMP caused phosphorylase activation. As phosphorylase was the rate-limiting enzyme in the pathway to G-6-P, the amount of G-6-P available for oxidation by G-6-P dehydrogenase was increased. This resulted in NADPH generation and the increased NADPH then drove steroidogenesis by increasing the rates of the hydroxylation reactions.

During the past decade, the relevance of phosphorylase activation to the triggering of increased steroidogenesis has been questioned, and at this time it seems unlikely that it is involved in any direct, causal fashion. However, the most important part of Haynes' hypothesis, that cyclic AMP mediates the steroidogenic effect, has received much supporting evidence.

B. Cyclic AMP Levels and Glucocorticoid Synthesis

1. Adenyl Cyclase

Studies with cell-free preparations of adrenal adenyl cyclase have been limited. Stimulatory effects of ACTH on cyclic AMP accumulation

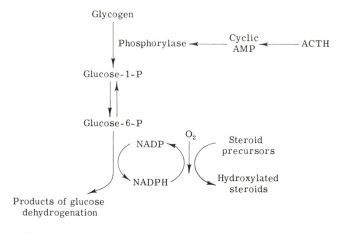

Fig. 9.2. The Haynes hypothesis. (From Haynes *et al.*, 1960.)

in homogenates of rat adrenals have been elicited (Grahame-Smith *et al.*, 1967). Similar data were reported by Taunton *et al.* (1969), who found that ACTH increased the conversion of ATP^{32} to cyclic AMP^{32} in particulate fractions of a functional mouse adrenal tumor. Other peptide hormones, including insulin, glucagon, and TSH, as well as epinephrine and PGE_2 were tested and found to be without effect. They did not detect effects of ACTH on cyclic AMP^{32} degradation and concluded that ACTH acted to stimulate the adenyl cyclase system rather than by inhibiting the phosphodiesterase.

2. Intact Adrenal Preparations

Quartered rat adrenals were incubated with graded concentrations of ACTH and the tissue cyclic AMP levels and medium corticosterone levels measured (Grahame-Smith *et al.*, 1967). Cyclic AMP levels in the rat adrenal were subject to rather dramatic excursions, and as the concentration of ACTH in the medium was increased, so the adrenal cyclic AMP increased and was accompanied by progressive stimulation of steroidogenesis until saturating levels of cyclic AMP were reached. The same sort of relationships were seen following the injection of ACTH into hypophysectomized rats *in vivo* (Fig. 9.3). It should be mentioned that the adrenal *in situ* is much more sensitive to ACTH (both in terms of cyclic AMP response and of the steroidogenic response) than *in vitro*.

The qualitative specificity of the cyclic AMP and steroidogenic responses were studied using certain structural analogs of ACTH which have differing steroidogenic potencies. As shown in Table 9.I, the potency of the ACTH analogs in producing stimulation of adrenal steroidogenesis

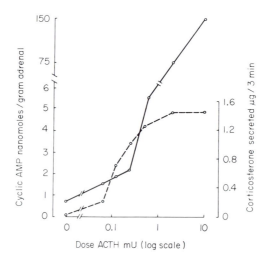

Fig. 9.3. The effects of various dosages of ACTH on adrenal cyclic AMP levels and corticosterone secretion in hypophysectomized rats *in vivo*. (From Grahame-Smith *et al.*, 1967.)

was reflected in their potencies in producing increases in adrenal cyclic AMP concentrations.

Finally, the temporal correlation between the ACTH and adrenal cyclic AMP in steroidogenesis was studied. The cyclic AMP response of rat adrenal quarters to ACTH preceded any detectable increase in medium corticosterone. In addition, when corticosterone secretion was maintained by continuous incubation with ACTH, adrenal cyclic AMP levels were also maintained. In systems in which the physiological pro-

TABLE 9.I. Effect of Analogs of ACTH on Adrenal Cyclic AMP and Corticosterone Secretion in Rat *in vivo*[a]

Analog	Intravenous dose (μg)	Adrenal cyclic AMP (nanomoles/gm tissue)[b]	Corticosterone secretion (μg/3 min)[b]
$\alpha^{25\text{-}39}$-ACTH	2	0	0
α-MSH[c]	2	2.63	0.34
N-α-acetyl-$\alpha^{1\text{-}24}$ACTH	0.06	4.17	0.78
$\alpha^{1\text{-}24}$-ACTH	0.004	2.12	0.39

[a] From Grahame-Smith *et al.* (1967)
[b] Minus 0.9% sodium chloride control.
[c] α-MSH, α-melanocyte-stimulating hormone.

cess stimulated by cyclic AMP is slowly inactivated, such a relationship may not be applicable or demonstrable. However, Grahame-Smith and his colleagues (1967) found that while quartered adrenals stimulated by incubation with ACTH maintained elevated steroidogenic rates for hours after rinsing and incubation in fresh media, the rate of steroidogenesis in adrenals exposed to cyclic AMP and then rinsed, returned to control levels almost immediately.

C. Methylxanthines

The steroidogenic response of the adrenal cortex has provided an example of how studies with methylxanthines may be misleading (see Chapter 2). Halkerston *et al.* (1966) reported that at almost all concentrations theophylline failed to act synergistically with ACTH on corticosterone production by rat adrenal quarters *in vitro* and that at most concentrations it inhibited the action of ACTH. However, they also found that theophylline was inhibitory to adrenal protein synthesis and reasoned that because of this it could not enhance steroidogenesis. More recently, Bieck and co-workers (1968) confirmed the finding of Halkerston *et al.*, and went on to show that theophylline antagonized the steroidogenic effects of cyclic AMP or dibutyryl cyclic AMP, which also suggested that the methylxanthines could act at a site other than the phosphodiesterase.

D. Exogenous Cyclic AMP

The ability of exogenous cyclic AMP (or its derivatives) to stimulate steroidogenesis in adrenal tissue from a variety of species has been confirmed many times (Birmingham *et al.*, 1960; Cushman *et al.*, 1966; Ferguson, 1963; Imura *et al.*, 1965; Karaboyas and Koritz, 1965; Studzinski and Grant, 1962). Two sets of experiments were of particular interest, however. In 1963, Ferguson reported that the steroidogenic action of cyclic AMP, like that of ACTH, was antagonized by puromycin, and he suggested that protein synthesis might in some way be involved in the actions of both.

Another report identified the site of action of cyclic AMP. Stone and Hechter (1954) had demonstrated that ACTH acted to stimulate steroidogenesis between cholesterol and pregnenolone. More recently, Karaboyas and Koritz (1965) made use of the steroidogenic activity of exogenous cyclic AMP in a series of experiments which defined the site of action of the compound in the steroidogenic mechanism. As shown in

TABLE 9.II. The Effect of ACTH and 3',5'-AMP on the Transformation of Corticoid Precursors to Rat Adrenal and Beef Adrenal Cortex Slices[a]

Tissue	Precursor	Product Isolated	Corrected cpm[b] in product			Ratios	
			Control	+ ACTH	+ 3',5'-AMP	ACTH/control	3',5'-AMP/control
Rat adrenal	[1-^{14}C]Acetate	Corticosterone	1,000	2,400	2,300	2.4	2.3
	[7α-^{3}H]Cholesterol	Corticosterone	1,330		3,580		2.7
	[7α-^{3}H]Cholesterol	Corticosterone	840	1,880		2.3	
	[7α-^{3}H]Δ^5-Pregnenolone	Corticosterone	2,420,000	2,650,000	3,380,000	1.1	1.4
	[7-^{3}H]Progesterone	Corticosterone	3,900,000	3,900,000	4,380,000	1.0	1.1
Beef adrenal	[1-^{14}C]Acetate	Cortisol	9,600	62,000	83,000	6.5	8.7
	[7α-^{3}H]Cholesterol	Cortisol	3,500	29,200	31,700	8.4	9.0
	[7α-^{3}H]Δ^5-Pregnenolone	Cortisol	770,000	760,000	850,000	1.0	1.1
	[7-^{3}H]Progesterone	Cortisol	890,000	870,000	810,000	1.0	0.9

[a] From Karaboyas and Koritz, 1965.

[b] Calculated as follows: Initial CPM X (constant specific activity/initial specific activity). This gives a measure of the radioactivity in the original sample due to corticosterone or cortisol.

Table 9.II, they studied the effect of ACTH or cyclic AMP on the incorporation of radioactive precursors into steroid hormones in rat adrenal quarters (where they measured corticosterone) and beef adrenocortical slices (where cortisol was the product measured). The specific activity of corticosterone synthesized in the rat adrenals incubated with 7-^{3}H-cholesterol when stimulated by either ACTH or cyclic AMP was increased 2.4-fold over the controls. In beef adrenal slices, ACTH and cyclic AMP produced even greater effects on the specific activity of cortisol (7- to 9-fold increases). Similarly, higher total and also specific activities in the glucocorticoids were found in both species when ^{14}C-acetate was the radioactive precursor. By contrast, ACTH and cyclic AMP were without effect on either net hormone synthesis or the specific activity of the steroid hormones produced when either radioactive pregnenolone or progesterone were the precursors. Thus, their data identified the conversion of cholesterol to pregnenolone as a site of action of cyclic AMP on the steroidogenic pathway (Fig. 9.4). However, it still is possible that this may not be the only site at which cyclic AMP acts in steroidogenesis.

To summarize the data implicating cyclic AMP in the steroidogenic activity of ACTH: ACTH responsive adenyl cyclase activities in cell-free preparations of rat and beef adrenal tissue have been demonstrated. Positive correlations have been obtained quantitatively, qualitatively, and temporally both *in vivo* and *in vitro* between intracellular cyclic AMP levels and steroid hormone production in the rat adrenal. The methylxanthines did not act synergistically with ACTH but rather, at most concentrations, antagonized it. However, this has been partially clarified by the demonstration that theophylline interferes with the action of exogenous cyclic AMP.

Finally, the steroidogenic activity of cyclic AMP or its derivatives has been repeatedly demonstrated in a variety of species. Also, compounds which block the effect of ACTH on steroidogenesis, including puromycin and cycloheximide, which are presumed to act by inhibition of adrenal protein synthesis, and aminoglutethamide, an inhibitor of the 20α-hydroxylase (a component of the cholesterol side chain cleavage enzyme system), also inhibited the steroidogenic activity of exogenous cyclic AMP. Finally, exogenous cyclic AMP has been shown to stimulate steroidogenesis at the same site as ACTH (i.e., the conversion of cholesterol to pregnenolone).

E. Possible Mechanisms of Action of Cyclic AMP

While it is by no means clear that the only site of action of cyclic AMP on steroidogenesis is between cholesterol and pregnenolone, it is almost

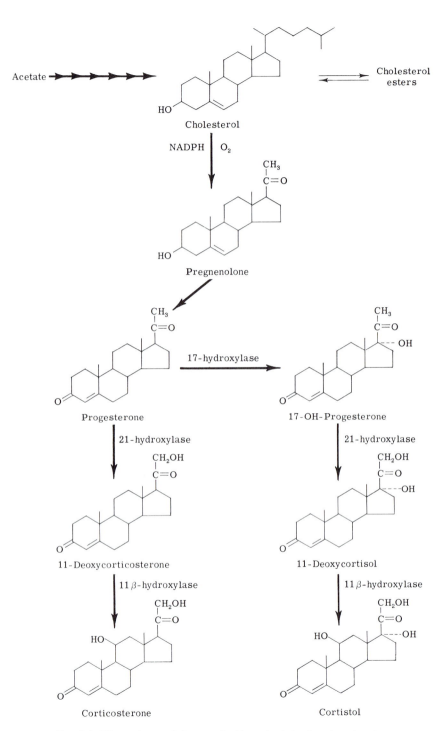

Fig. 9.4. The pathway of glucocorticoid synthesis in the adrenal cortex.

surely a site of action (cf., Karaboyas and Koritz, 1965). The enzyme system catalyzing this transformation is called the cholesterol side chain cleavage system. It is a mitochondrial system which catalyzes the insertion of O_2 into the steroid molecule with the cleavage of isocaproaldehyde from cholesterol. NADPH is required, and appears to be a mixed-function oxidase system containing cytochrome P-450 (Simpson and Estabrook, 1968). The transformations thought to be involved in the system are cholesterol → 20α-hydroxycholesterol → 20α,22ϵ-dihydroxycholesterol → pregnenolone (Satoh et al., 1966). Although side chain cleavage is a multistep process, a recent report by Hall and Young (1968) strongly suggested that the controlled step was the first one in the sequence, the 20α-hydroxylation of cholesterol. Using radioactive cholesterol and 20α-OH-cholesterol, they showed that the rate of incorporation of the former but not the latter was stimulated in response to the appropriate trophic hormone in adrenals, testes, and corpora lutea. These data confirmed an earlier report by Koritz that 20α-hydroxylation was the slow step.

Cholesterol is primarily localized in the cytosol and the cholesterol side chain cleavage system appears to be intramitochondrial (Hilf, 1965) so any proposed mechanism of action would have to take this into account. In addition, certain of the steps after pregnenolone formation are catalyzed by enzymes found in the cytosol, yet 11β-hydroxylation, which is a late step, is intramitochondrial. The same stipulation has to be made regarding NADPH, of course.

The most fundamental part of the Haynes' hypothesis was that cyclic AMP mediated the action of ACTH. However, as mentioned earlier, Haynes further postulated that cyclic AMP acted via phosphorylase activation. This was ultimately expressed by increased intracellular levels of NADPH which stimulated steroidogenesis by intact adrenal cells or homogenates. It was an extremely attractive hypothesis, because it was straightforward and very logical. However, while the participation of phosphorylase activation as a direct link in the action of cyclic AMP on steroidogenesis has not been ruled out, several lines of evidence have cast some doubt on its importance. Field et al. (1960) reported that treatment of adrenals with ACTH did not diminish the incorporation of ^{14}C from radioactive glucose to CO_2, as one would expect if ACTH were acting to increase adrenal glucose-6-phosphate concentrations. In addition, Vance et al. (1962) reported that ACTH did not cause decreased adrenal glycogen concentrations even though steroidogenesis was strongly stimulated. Also, Ferguson (1963) reported that incubation of rat adrenal quarters with either ACTH or cyclic AMP failed to produce phosphorylase activation even though steroidogenesis was strongly stimulated. In the same paper, Ferguson showed that in beef adrenocortical slices (where phosphorylase activation did occur), although puromycin blocked the steroidogenic effects of ACTH and cyclic AMP, it did not block the action of ACTH on phosphorylase activation. So far as can be determined from the reports which have thus far been published on phosphorylase activation in the rat adrenal, all measurements have been made in the presence of 5'-AMP, presumably because Haynes found that beef adrenal phosphorylase was of the liver type (i.e., inactive phosphorylase being only slightly affected by high levels of 5'-AMP). It would be most interesting to see if rat adrenal phosphorylase is indeed of the liver type or, conversely, if it resembles muscle phosphorylase. If the latter should be the case, then only by measuring the ratios of phosphorylase a to a plus b could one expect to demonstrate an effect of ACTH.

This rejoinder is offered only in the spirit of objectivity. We have no particular prejudice for the involvement of phosphorylase activation in hormone actions, and have on several occasions tried to deemphasize it. Indeed, it seems most likely to us that phosphorylase does not play a trigger but rather a supportive role in the steroidogenic mechanism. However, we did feel obliged to indicate that really complete evidence for such a dissociation has yet to be presented.

Although phosphorylase activation does not appear to directly control steroidogenesis, NADPH, as a cofactor of side chain cleavage, is still a possible site of action. However, several stipulations have to be made. First, increased NADPH levels in response to ACTH have not been demonstrated. Attempts to do so have been made, but the negative results were equivocal because of the limitations of the methodologies available. In addition, since side chain cleavage is intramitochondrial, the net change in total cellular NADPH/NADP ratios might be so small as to be undetectable except in terms of NADPH turnover. Finally, experiments showing the steroidogenic effect of exogenous NADPH may not have been relevant to the mechanism of action of cyclic AMP. It is hard to visualize so polar a molecule penetrating cells well, and evidence indicating that this action of exogenous NADPH was actually on enzymes which had leaked into the medium has been published (Tsang and Carballeira, 1966). In addition, Halkerston (1968) has shown that treatment of adrenal tissue with trypsin, which preferentially destroyed the steroidogenic activity of damaged cells, was without significant effect on the steroidogenic response of the tissues to ACTH or cyclic AMP but markedly reduced the effect of NADPH (more properly, NADP + G-6-P).

However, these objections by no means rule out the participation of increased *intracellular* NADPH in the control of steroidogenesis. Péron had suggested that cyclic AMP might be stimulating the net transport of reducing equivalents between the cytosol and mitochondrial NADPH pools via the malate shuttle. Péron and Tsang (1969) later felt this mechanism to be untenable in the rat adrenal since the shuttle did not appear to be present. However, the mechanism may still be operable in bovine adrenocortical tissue, where the shuttle is present (Simpson and Estabrook, 1968).

The involvement of protein synthesis in the control of steroidogenesis by cyclic AMP has been suggested by a number of experiments. Ferguson (1963) reported that puromycin completely abolished the effects of ACTH and cyclic AMP on steroidogenesis in rat and beef adrenals and that it also severely depressed adrenal protein synthesis. However, no effect of ACTH on increasing adrenal protein synthesis was detected. He also found that puromycin did not inhibit the steroidogenic action of exogenous NADPH. These data with puromycin suggested (but by no means proved) that the steroidogenic action of cyclic AMP involved protein synthesis. In addition, they would indicate that the action of NADPH on steroidogenesis occurred at some point along the steroidogenic pathway past that at which puromycin antagonized the action of cyclic AMP, or that perhaps NADPH acted through a different mechanism.

It should be mentioned at this point that for several years there were conflicting data about the relationship of ACTH and adrenal protein synthesis, since ACTH or exogenous cyclic AMP were reported to inhibit, stimulate, or have no effect upon the incorporation of radioactive amino acids into adrenal protein (Hilf, 1965). Ferguson and his co-workers (1967) have demonstrated that the inhibition of adrenal protein synthesis seen *in vitro* is probably the result of the high medium corticosterone levels produced in response to ACTH. It would appear then, that in the absence of the adverse effects of high concentrations of glucocorticoids adrenal protein synthesis may be increased in response to ACTH.

Garren and his co-workers (1965) investigated the action of inhibitors of protein synthesis *in vivo*. They found that the injection of cycloheximide into hypophysectomized rats which had been maximally stimulated with an i.v. dose of ACTH resulted in a rapid decline in the production of corticosterone. In the absence of cycloheximide, however, corticosterone synthesis was maintained at a high rate. They concluded from these observations that if ACTH were acting through the formation of a new protein, it must have a rapid turnover. They calculated the half-life of this protein as 8 min, so therefore ACTH (or cyclic AMP) would be continually required for its synthesis. Concentrations of actinomycin D which strongly inhibited RNA synthesis did not antagonize the stimulation of steroidogenesis by ACTH, indicating that the action of cyclic

AMP was at a level of protein synthesis past RNA synthesis (Ney *et al.*, 1966).

Cycloheximide did not antagonize the conversion of acetate to cholesterol in rats under conditions where protein synthesis and steroidogenesis were blocked (Davis and Garren, 1966), nor did it block the synthesis of corticosterone from pregnenolone, progesterone, or deoxycorticosterone. Prior injection of cycloheximide blocked the depletion of adrenal cholesterol which is normally seen during ACTH-stimulated steroidogenesis but not the hydrolysis of cholesterol esters to free cholesterol in response to ACTH (which appears to be another action of cyclic AMP) at a time when both cholesterol side chain cleavage and corticosterone production were completely antagonized. Davis and Garren (1968) confirmed that the site of action of ACTH was the cholesterol side chain cleavage system, and also showed that it was here that cycloheximide inhibited the action of ACTH.

Experiments with inhibitors of protein synthesis suffer from the nonspecificity of these agents and their actions as general cellular poisons. It is difficult if not impossible to ascribe the actions of, for example, puromycin on steroidogenesis solely to the inhibition of protein synthesis, for the compound has a variety of actions, one of which is the inhibition of the cyclic nucleotide phosphodiesterase, at least in adipose tissue and diaphragm (Appleman and Kemp, 1966). In addition, even if the decreased steroidogenesis is a manifestation of decreased protein synthesis, it is difficult to decide if protein synthesis is actually involved in the action of cyclic AMP. If one of the components of the cholesterol side chain cleavage system had a short half-life, steroid hormone formation would be dependent upon protein synthesis, yet the action of cyclic AMP to stimulate the process could be completely independent of protein synthesis.

Farese (1967) has obtained some evidence for effects of ACTH or cyclic AMP on adrenal protein synthesis. The addition of supernatant fractions of adrenal homogenates produced inhibition of mitochondrial side chain cleavage. Incubation of adrenals with either compound prior to preparing the supernatant resulted in increased amounts of a factor which stimulated (or at least relieved the inhibition of) cholesterol side chain cleavage. This factor was heat and trypsin-labile, precipitated by $(NH_4)_2SO_4$, fractionated on Sephadex like a protein, and the accumulation of this material during the incubation of the intact adrenals was inhibited by puromycin or cycloheximide. However, as Farese pointed out, the effects obtained in this system were small, and there was some nonspecificity in that certain nonadrenal materials (e.g., bovine serum albumin) stimulated cholesterol side chain cleavage.

A variety of possible mechanisms have been synthesized from the available data. Garren and his colleagues have suggested that the hypothetical rapidly turning-over protein, once synthesized in response to cyclic AMP, facilitates the translocation of cholesterol from the cytosol to the mitochondrial side chain cleavage system.

Roberts *et al.* (1967) reported that cyclic AMP directly stimulated the conversion of radioactive cholesterol to pregnenolone in sonically disrupted mitochondria. This finding, so attractive in its simplicity as compared to the schemes involving protein synthesis, NADPH, etc., has been questioned by Koritz *et al.* (1968). They found that cyclic AMP inhibited the conversion of pregnenolone to progesterone in similar preparations, and that the increase in radioactive pregnenolone was due to decreased breakdown rather than increased synthesis.

Another mechanism has been proposed by Koritz (1968) who had earlier found that mitochondrial swelling produced by several procedures resulted in increased rate of NADPH-supported steroidogenesis from endogenous precursors. This, coupled with the knowledge that pregnenolone is an inhibitor of the side chain cleavage system (Koritz *et al.*, 1968), led them to postulate that cyclic AMP acts to facilitate the escape of pregnenolone from the mitochondria, thereby releasing the side chain cleavage system from product inhibition.

In summary then, the action of cyclic AMP on the stimulation of the cholesterol side chain cleavage system is thought, but by no means proved, to involve the availability of NADPH, or protein synthesis, or both. However, it is obvious that other possibilities exist. For example, the importance of protein kinase systems in the actions of cyclic AMP (see Chapter 5) is only now becoming appreciated, and this sort of mechanism is at least as attractive a speculation as any of the others offered thus far.* Also, it should be reemphasized that cyclic AMP may very well be acting at sites besides the conversion of cholesterol to pregnenolone. For example, Roberts and his co-workers (1964) have described the stimulation of the 11β-hydroxylase system of rat adrenal homogenates by cyclic AMP. Although the fact that 11-hydroxylation is not rate-limiting somewhat diminishes the importance of the system in the control of steroidogenesis, this by no means excludes it.

F. Other Actions of ACTH Mediated by Cyclic AMP

Steroidogenesis is not the only effect of ACTH on the adrenal. Others include the trophic effect (i.e., the growth and maintenance of adreno-

---

*See Section V for a summary of recent developments.

cortical function), increased glucose oxidation, ascorbic acid depletion, the hydrolysis of cholesterol esters, increased adrenal blood flow, and glycogenolysis. Aside from the pure phenomenology associated with seeing if cyclic AMP is involved in these effects of ACTH, a rather fundamental question may also be answered. That is, is the interaction of ACTH with adenyl cyclase only one of several actions of the hormone, or is it the only action of ACTH?

The trophic effect does appear to involve cyclic AMP. Ney (1969) has shown that dibutyryl cyclic AMP injected subcutaneously in gelatin partially maintained adrenal weight and function in hypophysectomized rats. Earlier, Grahame-Smith *et al.* (1967) had attempted to dissociate cyclic AMP from the trophic effect by studying the action of ACTH on cyclic AMP levels in atrophic adrenals. Hypophysectomized rats were maintained until adrenal weight had fallen and the steroidogenic response to ACTH had disappeared. If ACTH had had no effect on cyclic AMP levels in the atrophic adrenal, it could not have been involved in the trophic action of ACTH. However, the cyclic AMP response to ACTH was undiminished in the atrophic adrenals (Fig. 9.5). Jones *et al.* (1970)

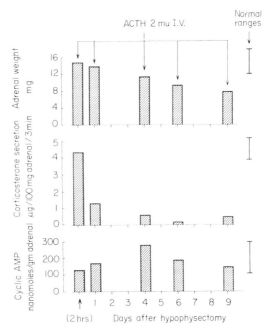

Fig. 9.5. The effects of hypophysectomy on adrenal weight and on the steroidogenic and cyclic AMP responses of rat adrenals *in vivo*. (From Grahame-Smith *et al.*, 1967.)

have reported that the oxidation of exogenous ^{14}C-glucose was directly proportional to the amount of steroid hormone synthesized, and the effects of ACTH on glucose oxidation, as on steroidogenesis, were faithfully mimicked by exogenous cyclic AMP. In addition, Earp *et al.* (1970) reported that exogenous cyclic AMP caused ascorbic acid depletion in the rat adrenal cortex, and Denicola *et al.* (1968) reported decreased ^{14}C-ascorbic acid uptake in its presence. Davis (1969) reported that dibutyryl cyclic AMP, like ACTH, stimulated the hydrolysis of adrenal cholesterol esters to cholesterol. Thus far, those actions of ACTH on the adrenal which have been tested can be explained by the activation of adenyl cyclase.

III. CYCLIC AMP AND STEROIDOGENESIS IN THE GONADS

The mechanism by which LH stimulates progesterone synthesis in the corpus luteum closely parallels the adrenal system. Marsh and Savard (1964) established an *in vitro* system using bovine corpora luteal slices which responded to LH and then performed a series of experiments designed to see if the Haynes hypothesis was applicable to the corpus luteum. They found that LH specifically increased the activity of luteal phosphorylase concomitant with increased rates of steroidogenesis, and that exogenous cyclic AMP stimulated progesterone synthesis. In addition, they found that cyclic AMP caused the same patterns of incorporation of radioactive steroid precursors as LH.

In collaborative experiments (Marsh *et al.*, 1966) the relationships between cyclic AMP and steroidogenesis in the corpus luteum were studied. LH, at quantitatively reasonable concentrations, increased cyclic AMP levels before progesterone synthesis in slices of bovine corpus luteum, while LH inactivated by exposure to hydrogen peroxide at room temperature, prolactin, ACTH, epinephrine, and glucagon failed to stimulate either cyclic AMP accumulation or steroidogenesis (Table 9.III). More recently, LH caused reproducible increases in cyclic AMP production by homogenates of bovine corpus luteum, and this was shown to be due to the activation of the adenyl cyclase system rather than to an inhibition of the phosphodiesterase (Marsh, 1970a). Puromycin, which was known to antagonize the actions of LH and exogenous cyclic AMP on luteal steroidogenesis, did not antagonize the stimulatory effects of LH on cyclic AMP levels. Hall and Koritz (1965) showed that LH and exogenous cyclic AMP were acting between cholesterol and pregnenolone. Thus, the mechanisms of action of LH on the corpus luteum and

TABLE 9.III. Effect of Various Substances on Accumulation of Cyclic AMP[a]

Experiment and addition	Cyclic AMP (nanomoles/gm tissue)
Experiment 1	
None	0.44
Luteinizing hormone (2.0 μg/ml)	14.0
Inactivated luteinizing hormone (2.0 μg/ml)	0.35
Experiment 2	
None	0.36
Luteinizing hormone (2.0 μg/ml)	6.76
Prolactin (2.0 μg/ml)	0.14
Experiment 3	
None	0.33
Luteinizing hormone (2.0 μg/ml)	2.77
Prolactin (2.0 μg/ml)	0.38
ACTH (0.3 unit/ml)	0.45
Epinephrine (0.01 μmole/ml)	0.43
Glucagon (10 μg/ml)	0.41

[a] From Marsh *et al.*, 1966.

ACTH on the adrenal cortex were virtually identical, except that the hormone-adenyl cyclase interaction had differing specificities, and the final steroid products were different because the enzymes metabolizing pregnenolone and progesterone were distributed differently in the two tissues.

The interstitial cells of the rabbit ovary synthesize progestational steroids in response to LH, and Dorrington and her co-workers have provided considerable evidence for the involvement of cyclic AMP in this process. Exogenous cyclic AMP specifically stimulated steroidogenesis, and it was potentiated by theophylline (Dorrington and Kilpatrick, 1967). In ovarian homogenates, using an assay measuring cyclic AMP formation from ^{14}C-ATP and destruction by following the disappearance of ^{3}H-cyclic AMP, Dorrington and Baggett (1969) showed that LH stimulated the adenyl cyclase system and was ineffective on the phosphodiesterase. Human chorionic gonadotropin, which is active steroidogenically also stimulated cyclic AMP accumulation, while FSH and ACTH were ineffective. Thus, although correlations between intracellular levels of cyclic AMP and steroidogenesis in the ovary have not been reported, it appears almost certain that LH is acting through cyclic AMP.

Similarly, Sandler and Hall (1966) reported that exogenous cyclic AMP mimicked the *in vitro* effect of ICSH (LH) on the conversion of cholesterol-7α-^{3}H to testosterone-^{3}H and also the conversion of endogenous precursors to testosterone.

Stimulatory effects of ICSH (LH) on cyclic AMP levels in minced rat testes have been observed (Baird and Butcher, unpublished observations). Recently, Murad *et al.* (1969a) reported that ICSH stimulated adenyl cyclase activity in cell-free fractions of rat testis. FSH also stimulated adenyl cyclase, and the effects of ICSH and FSH were not additive. Very high concentrations of epinephrine had a mild stimulatory effect which was additive with the gonadotropins, and which was blocked by propranolol.

IV. SUMMARY

In conclusion then, it would appear that the control of the formation of steroid hormones is another function of cyclic AMP. At the risk of belaboring the point, the steroidogenic tissues present a rather good example of cells doing with cyclic AMP only what they are enzymatically equipped to do. Unfortunately, there are not yet data available to decide if cyclic AMP is involved in aldosterone formation, i.e., from corticosterone to aldosterone (Fig. 9.1), although it is clear that increased levels of the nucleotide would at the very least supply more corticosterone.

Finally, it now appears that the release of adenohypophyseal hormones, including ACTH, TSH, and growth hormone in response to hypotholamic releasing factors also involves cyclic AMP (Chapter 10, Section X). Thus, cyclic AMP is involved at two points in the elegantly integrated endocrine relays leading to the ultimate arrival of steroid and thyroid hormones at their target cells.

V. ADDENDUM

Ney *et al.* (1969) reported that a transplantable adrenal carcinoma, which produces a low level of corticosterone, was not stimulated by ACTH *in vivo* or *in vitro,* nor by exogenous cyclic AMP *in vitro.* Further, ACTH failed to produce significant changes in cyclic AMP levels in the tumor under conditions where levels in normal adrenal tissues were drastically elevated. Curiously, homogenates of the tumor, fortified with ATP and Mg^{2+} were at least as responsive to ACTH as were homogenates of normal adrenals. ATP concentrations in the tumor were lower than in normal adrenals, but the importance of this in terms of the lack of effect of ACTH on cyclic AMP levels was unclear. Phosphodiesterase activity in the tumor was considerably less than that in normal tissue. In any event, no matter what the defects may ultimately turn out to be, it is perhaps important to consider the possibility that adenyl cyclase activity, as measured in cell-free systems, may not be a wholly valid index of hormone action in the very complicated milieu of the intact cell.

Gill and Garren (1969) reported that cyclic AMP is bound specifically to a protein associated with microsomes. This binding protein was purified about 100-fold, and was later shown to have protein kinase inhibitory activity (Gill and Garren, 1970). Presumably the binding protein is similar to the regulatory subunit of skeletal muscle protein kinase (Reimann *et al.*, 1971; see also addendum to Chapter 5).

Grower and Bransome (1970) reported that ACTH or cyclic AMP caused increased labeling and amounts of several protein fractions derived from cultures of mouse adrenocortical tumor cells. The changes in the labeling of cytosol proteins were evident in less than 30 minutes, and the authors felt that this provided support for the proposal that the actions of cyclic AMP might be mediated by rapid and selective effects on protein synthesis. Evidence for an exponential decay process was presented by Farese *et al.* (1969).

Schulster *et al.* (1970) provided additional strong support for the hypothesis that cyclic AMP mediates the effects of ACTH on the adrenal cortex. They found that cyclic AMP mimicked the effects of ACTH on the superfused rat adrenal gland, and that cycloheximide antagonized the effects of both ACTH and cyclic AMP on steroidogenesis but did not prevent increased steroidogenesis in response to progesterone. Their data were consistent with an action of cyclic AMP involving the induced synthesis of a rapidly turning-over protein.

Kowal and Fiedler (1969) reported that cyclic AMP and adenosine, 5'-AMP, ADP, and ATP all stimulated steroidogenesis in monolayer cultures of a functional mouse adrenal tumor. However, their experiments were rather long (2–24 hours) and because of the limited amount of tissue available in these studies, the authors were unable to measure cyclic AMP levels in the cells incubated with various nucleotides. Using a different cell line, Kowal (1969) found that cyclic AMP was only a third as effective as ACTH while other adenine nucleotides had no ACTH-like activity at all. Propranolol at concentrations greater than 10 μg/ml inhibited steroidogenesis in response to either ACTH or cyclic AMP.

Hechter *et al.* (1969) reported that the adenyl cyclase activity of bovine adrenal cortex was distributed in all particulate fractions obtained by differential centrifugation. A considerable amount of the total adenyl cyclase activity was found in association with mitochondrial fractions and the highest specific activity of the enzyme was found in this fraction.

However, as in almost all studies with adenyl cyclase involving differential centrifugation, considerable amounts were also found in the "nuclear" and "microsomal" fractions. Activity in all fractions was stimulated by ACTH and fluoride but not by TSH, glucagon, MSH, or vasopressin.

Lefkowitz *et al.* (1970) reported the preparation of pure monoiodo ACTH-^{125}I which was biologically active. The iodinated hormone was bound by adrenocortical extracts containing ACTH-sensitive adenyl cyclase but not by extracts which did not contain cyclase. Unlabeled ACTH inhibited the binding of radioactive ACTH, and derivatives of ACTH with varying biological potency inhibited binding in direct proportion to their biological activity. Other proteins including insulin and several unrelated iodinated hormones did not affect binding. The "solubilized" enzyme used in these experiments was prepared by sonication in the presence of a phospholipid and fluoride. Although electron microscopy of active extracts disclosed the presence of small membrane fragments and granules, the activity nevertheless behaved on agarose gels like a globular protein. The molecular weight was estimated at between 3 and 7×10^6 daltons (Pastan *et al.*, 1970).

The ability of exogenous cyclic GMP to stimulate adrenal steroidogenesis (Glinsmann *et al.*, 1969) is discussed in Chapter 11. The prostaglandins also stimulate steroidogenesis (Flack *et al.*, 1969), and Kuehl *et al.* (1970b) have suggested that a prostaglandin may actually mediate the effect of LH on cyclic AMP in ovarian tissue. The effects of the prostaglandins are discussed in the next chapter.

Cirillo *et al.* (1969) found that cultured granulosa cells retained their sensitivity to cyclic AMP even after sensitivity to LH had been lost. Channing and Seymour (1970) reported that exogenous cyclic AMP but not other nucleotides could mimic LH and FSH in causing luteinization of cultured granulosa cells.

CHAPTER 10

Other Hormones

The hormones in this chapter are discussed together because, when we started writing this monograph, relatively less was known about their actions in relation to cyclic AMP than was known about the hormones discussed in earlier chapters. Vasopressin was an exception even then, so that our original intent was to write about it in some detail. However, the information explosion in the case of certain other hormones has been extraordinary, and we have had to expand this chapter accordingly. Our present aim is to summarize the evidence that cyclic AMP is involved in the actions of these hormones, and to call attention to some of the features that may be helpful in understanding certain responses and processes other than the one under discussion. In the future, it may be possible to argue that once you have seen one hormonal response mediated by cyclic AMP, you've seen them all. At present this cannot be said to be the case.

I. VASOPRESSIN

A. Permeability

Vasopressin, the antidiuretic principle of the posterior pituitary gland, enhances the permeability of a number of epithelial membranes to water, sodium, and certain small molecules such as urea. The evidence that these effects are mediated by cyclic AMP is now very substantial, and will be summarized in this section. More detailed discussions may be found in reviews by Orloff and Handler (1964, 1967), and references cited therein.

The simplest structure that responds to vasopressin with a change in permeability is probably the urinary bladder of the toad and other amphibian species. It is thought to be composed of a single layer of relatively uniform epithelial cells, and is therefore a simpler system than amphibian skin, a multilayered epithelial structure that also responds to vasopressin. Both the toad bladder and frog skin have been used extensively as models for mammalian renal tubular tissue, which is more complex and more difficult to study in isolation than the amphibian systems.

Epithelial cells from these sources possess at least two permeability barriers each. The serosal membrane, basal in position, separates the cell interior from the blood-bathed extracellular fluid. The mucosal membrane, apical in position, separates the cell interior from the external environment. Vasopressin is effective only when applied to the serosal surface, while the characteristic permeability changes that it produces occur predominately in the apical membrane (Civan and Frazier, 1968; Ganote *et al.*, 1968). There is now evidence that of the four types of epithelial cells that have been distinguished (granular cells, mitochondria-rich cells, goblet cells, and basal cells), vasopressin may affect only the granular cells (DiBona *et al.*, 1969).

The suggestion that the effects of vasopressin might be mediated by cyclic AMP was based initially on the observation that exogenous cyclic AMP could mimic the hormone in all respects when applied to the serosal surface of the toad bladder. To measure water permeability, Orloff and Handler (1962) divided each bladder into two sacs, into each of which dilute Ringer's solution was introduced. Each sac was then placed in a beaker containing normal saline, and net water movement along the osmotic gradient estimated by measuring the weight loss of the sac at appropriate intervals. Under control conditions (sac A in Table 10.I) the rate of this movement was very low, but the introduction of either vasopressin or cyclic AMP (sac B) led to a striking increase. Theophylline was also effective, as indicated in Table 10.I, and was found to act synergistically with either vasopressin or exogenous cyclic AMP. Neither 5'-AMP nor ATP had any effect in these experiments.

TABLE 10.I. Effects of Vasopressin, Exogenous Cyclic AMP, and Theophylline on Water Permeability of the Isolated Toad Bladder[a]

| | Net flow (μl/min) | |
Addition to sac B	Sac A	Sac B
Vasopressin (100 milliunits/ml)	0.8	26.5
Cyclic AMP (10 mM)	1.0	26.7
Theophylline (20 mM)	0.6	18.2

[a]Outer solution (urine side) was 40 mOsm/kg of water, while inner solution (blood side) was 200 mOsm/kg. Agents were added to inner solution only. Data from Orloff and Handler (1962).

Vasopressin also stimulates sodium transport across the toad bladder (Orloff and Handler, 1967) and frog skin (Baba *et al.*, 1967), although this effect is not prominent in the mammalian kidney, and may not occur there at all. In the amphibian systems, the enhanced Na^+ flux is reflected by an increase in short-circuit current across the epithelial membranes, and this increase in current seems to result almost entirely from the increased sodium transport (Baba *et al.*, 1967; Rider and Thomas, 1969). Vasopressin also causes an increase in short-circuit current across isolated rabbit ileal mucosa (Field *et al.*, 1968), although in this case there is evidence that other ions, especially chloride and bicarbonate, may be more important than changes in sodium flux.

In all of these cases the effects of vasopressin can be mimicked by exogenous cyclic AMP and/or theophylline. The similarity of the effects of vasopressin and exogenous cyclic AMP has been demonstrated for the toad bladder (Orloff and Handler, 1967), frog skin (Baba et al., 1967; Bourguet, 1968; Bastide and Jard, 1968; Cuthbert and Painter, 1968b), isolated rabbit ileal mucosa (Field et al., 1968), and also in the isolated perfused rabbit collecting tubule (Grantham and Burg, 1966; Grantham and Orloff, 1968). In the latter system, the effect was limited to an increase in permeability to water, with no effect on urea permeability.

The effects of vasopressin on permeability are shared by other hormones to different extents depending on the system being studied. The naturally occurring antidiuretic hormone in man is arginine vasopressin, which is much more potent in the human kidney than any of the other octapeptide analogs studied. Oxytocin is almost ineffective in mammalian kidney tubules. By contrast, the amphibian skin and bladder are less discriminating, and respond to a variety of octapeptide analogs, including oxytocin (Bourguet, 1968; Bastide and Jard, 1968; Jard et al., 1968). In addition, frog skin responds to serotonin (Sayoc and Little, 1967) and β-adrenergic stimulation (Jard et al., 1968; Watlington, 1968a) with an increase in permeability to sodium, and, by analogy with other systems (see Chapter 6 and Section VII following), it seems reasonable to suppose that these effects also are mediated by cyclic AMP.

Resistance to the permeability effects of vasopressin, theophylline, and cyclic AMP may develop. Edelman et al. (1964) found that the development of spontaneous resistance to vasopressin in toad bladders did not prevent the response to exogenous cyclic AMP, just as resistance to the nucleotide did not necessarily preclude a response to vasopressin. Further study along these lines, especially involving the measurement of intracellular cyclic AMP levels, may lead to a better understanding of the phenomenon of drug resistance in general. Parisi et al. (1969) found that after a maximal response to theophylline, the frog bladder ceased to respond to all three agents.

A puzzling finding by Parisi et al. (1969) was that when the epithelial layer of the frog bladder was isolated from the peritoneal serosa and adjacent connective tissue by microdissection, the permeability response of the epithelium to oxytocin was essentially the same as in the total bladder, whereas the response to exogenous cyclic AMP or theophylline was drastically reduced. Bourguet and Morel (1967) had previously shown that oxytocin and vasotocin affected water and sodium transport differentially, and one interpretation of these data had been that the two peptides might affect the accumulation of cyclic AMP in different cells or cell compartments (Orloff and Handler, 1967). Another possibility that should be considered is that an important component of the action of oxytocin may be unrelated to changes in the level of cyclic AMP.

Again restricting our attention to vasopressin, further evidence that its action in the toad bladder is mediated by cyclic AMP was provided by Handler *et al.* (1965), who showed that the hormone could increase the intracellular level of cyclic AMP in the toad bladder *in vitro*. This is illustrated by the data in Table 10.II, which also shows the synergistic effect of arginine vasopressin and theophylline. Other active analogs were not studied in these experiments, but two polypeptides known to have no effect on permeability in the toad bladder, angiotensin and insulin, were shown to likewise have no effect on cyclic AMP accumulation.

TABLE 10.II, Effect of Vasopressin and Theophylline on the Level of Cyclic AMP in the Urinary Bladder of the Toad[a]

Treatment	Cyclic AMP (mμmoles/gm)	Increase (%)
Control	0.45	—
Vasopressin (100 milliunits/ml)	0.56	25
Theophylline (10 mM)	0.80	78
Vasopressin + theophylline	1.67	271

[a]Sections of bladder were incubated for 40 minutes in Ringer's solution at 25°C, with the indicated concentrations of vasopressin or theophylline or both for the final 20 minutes. Data from Handler *et al.* (1965).

Adenyl cyclase has not been studied in broken cell preparations from toad bladder or frog skin, but some interesting studies have been carried out with kidney preparations. Stimulation by vasopressin in particulate preparations from mammalian kidney homogenates was demonstrated some years ago (Anderson and Brown, 1963; Brown *et al.,* 1963), and it has more recently been shown (Chase and Aurbach, 1968; Melson *et al.,* 1970) that vasopressin-sensitive adenyl cyclase in the rat kidney is located almost exclusively in tubules from the renal medulla. By contrast, adenyl cyclase from cortical tissue responded primarily to parathyroid hormone (see Section IV). These observations were in line with the physiologically determined sites of action of these hormones. Vasopressin is thought to exert its antidiuretic effect in mammals by increasing the permeability to water primarily in the collecting tubules, although perhaps to some extent in the distal tubules as well. That exogenous cyclic AMP is capable of mimicking the effect of vasopressin in the isolated perfused collecting tubule (Grantham and Burg, 1966) has already been mentioned.

Senft and his colleagues (Senft, 1968; Senft *et al.*, 1968b) measured the level of cyclic AMP in rat kidney in response to hydration and dehydration, and found that during periods of antidiuresis the level in the cortex increased more than in portions from the outer medulla, which does not seem to fit with the above mentioned studies on adenyl cyclase. However, these *in situ* experiments are subject to the criticism that the kidneys were frozen by excising them and dropping them into a mixture of dry ice and acetone. Under these conditions the cortex would be frozen quite rapidly while the medullary tissue would be frozen more gradually. The importance of rapid freezing was well-demonstrated by Namm and Mayer (1968), who showed that if isolated rat hearts were frozen by clamping them between two blocks of aluminum at the temperature of liquid nitrogen, then a fourfold increase in the level of cyclic AMP could be demonstrated in response to epinephrine. If, on the other hand, the hearts were allowed to simply fall into a mixture of dry ice and acetone, then the concentration of cyclic AMP in extracts from the frozen hearts was found to be the same whether epinephrine had been given or not. Presumably this reflects an imbalance of phosphodiesterase over adenyl cyclase activity as the tissue temperature passes from $37°$ to $-70°C$. It thus seems likely, based on the *in vitro* adenyl cyclase studies (Melson *et al.,* 1970), that if a procedure could be devised for freezing all parts of the kidney at the same rapid rate, then the level of cyclic AMP in the medulla would increase more during antidiuresis than in the cortex. The difficulty of devising such a procedure that would allow subsequent anatomical dissection is a major technical challenge in studies dealing not only with the kidney but with all heterogenous tissues, notably the brain. The complexity of the kidney was, of course, one of the factors that led to the search for simpler systems with which meaningful studies on permeability could be carried out. As mentioned, the urinary bladder of the toad has been especially useful in this regard.

The mechanism by which cyclic AMP acts to increase the permeability to water and related molecules is quite unknown, as is the mechanism of the stimulatory effect on sodium transport in anuran membranes. Although cyclic AMP and 5'-AMP have both been reported to inhibit the Na^+-K^+-dependent ATPase obtained from several mammalian tissues (Mozsik, 1969, 1970), it is of interest that Morel (1970) found this enzyme from frog skin to be stimulated by cyclic AMP. Another effect reported to occur in a broken cell preparation of dog kidney was an apparent stimulation of phosphatidylinositol kinase activity (Cunningham, 1968). The relation of this effect to the effects observed in more highly organized systems is unknown at present. In line with the idea that all of the physiologically important effects of cyclic AMP are mediated by the stimulation of a protein kinase, a very active cyclic AMP-dependent kinase has been found in particulate and soluble fractions of a preparation of frog bladder epithelial cells (Jard and Bastide, 1970).

The effect of vasopressin on water permeability can be dissociated from the effect on sodium transport, even in those systems where both effects normally occur. It was known, for example, that high concentrations of calcium in the solution bathing the serosal or blood surface of the toad bladder tended to inhibit the effect of vasopressin on water and urea movement, but not on the active transport of sodium. Petersen and Edelman (1964) confirmed this observation, but found that the effect of exogenous cyclic AMP on water permeability was unaffected by calcium. They were thus led to postulate two segregated sites of cyclic AMP production in the toad bladder, one controlling water and urea movement, the other controlling sodium transport. Both were postulated to be sensitive to stimulation by vasopressin, but only one (that controlling water and urea movement) could be inhibited by calcium.

This hypothesis was tested in more detail by Argy *et al.* (1967). They found that the magnesium ion was a more effective inhibitor of the effect on osmotic water flow and urea flux than was calcium. High concentrations of Mg^{2+} (8 mM) inhibited the response to vasopressin and all of the analogs of vasopressin that were tested, and also inhibited the response to theophylline but not that to exogenous cyclic AMP. By contrast, high concentrations of Ca^{2+} (10 mM) inhibited the response to only some of the analogs of vasopressin. The response to arginine vasopressin was markedly inhibited by calcium, whereas the responses to lysine vasopressin, arginine vasotocin, and oxytocin were not significantly affected. As with Mg^{2+}, the response to theophylline was inhibited while that to cyclic AMP was not. In all cases, osmotic flow and urea flux went hand in hand, and could not be dissociated. Neither Ca^{2+} nor Mg^{2+} had any discernible effect on the stimulation of sodium transport in response to any of these agents.

Although these results are difficult to interpret at the molecular level, several hypotheses can be considered. If it is assumed that the effects on both water and sodium transport are mediated by cyclic AMP, then it seems necessary to postulate segregated sites within the toad bladder for the production and action of cyclic AMP. These could occur in separate cells, if some cells are specialized for water transport and others for sodium transport, or they could occur in the same cells, if the two compartments are segregated all the way from the serosal membrane (where adenyl cyclase presumably occurs) to the mucosal membrane (where the changes in permeability take place). An alternate possibility suggested by Orloff and Handler (1967) was that the effect of vasopressin on sodium transport might be mediated by a different cyclic nucleotide, the production of which is unaffected by Ca^{2+} or Mg^{2+}. Cyclic AMP could mimic its effect on sodium transport, but it could not mimic the effect of cyclic AMP on water permeability. Although it seemed unlikely for a time, the probability that this hypothesis might be correct was greatly increased by the demonstration (Bourgoignie *et al.*, 1969) that cyclic GMP (see Chapter 11) shares the ability of cyclic AMP to stimulate sodium transport, but has no effect on water permeability. It remains to be seen whether vasopressin can actually stimulate the formation of cyclic GMP in the toad bladder. Further speculation should probably be postponed until this important p,oint is established.

Although exogenous cyclic AMP seems to mimic the permeability effects of vasopressin quite accurately in simpler amphibian systems and also in the isolated perfused collecting tubule of the rabbit, its effects on renal function *in vivo* may be more complex. This is not surprising in view of the multiplicity of cell types that may be affected by cyclic AMP, and especially when it is realized that even vasopressin itself may at times produce diuresis (Grinnell *et al.*, 1968). Both diuresis (Abe *et al.*, 1968) and antidiuresis (Levine, 1968b) have been reported in response to cyclic AMP, whereas others could see no change in urine flow (e.g., Rasmussen *et al.*, 1968).

B. Factors Modifying the Effect of Vasopressin on Permeability

1. Theophylline

The ability of theophylline to potentiate vasopressin was mentioned previously in support of the hypothesis that the effects of vasopressin on permeability are mediated by cyclic AMP. In the case of the toad bladder, there is evidence that the effect of theophylline by itself (Orloff and Handler, 1962; Gulyassy and Edelman, 1967) is also mediated by cyclic AMP. Thus, theophylline does inhibit the rate of destruction of cyclic AMP in homogenates of toad bladder epithelial cells (Gulyassy, 1968) and increases the intracellular level of cyclic AMP when applied to the toad bladder *in vitro* (Handler *et al.*, 1965). The latter effect is presumably the result of continued adenyl cyclase activity in the face of reduced phosphodiesterase activity. The finding that theophylline loses its capacity to stimulate water flux in bladders maintained at 2°C in the absence of hormones, while retaining its capacity to potentiate these hormones (Bourguet, 1968), is compatible with this view.

In the more complex permeability barrier represented by the frog skin, effects other than phosphodiesterase inhibition may contribute to the overall response to theophylline. For example, Cuthbert and Painter (1968a) deduced that theophylline increased the permeability of the mucosal membrane to chloride ions, an effect apparently not shared by either vasopressin or exogenous cyclic AMP (Cuthbert and Painter, 1968b). The conclusion by Cuthbert and Painter that it is "not necessary to involve cyclic AMP in order to explain the effects of either ADH or theophylline" is probably correct, and raises a point in the philosophy of science that is largely beyond the scope of the present monograph. As evidence accumulates in support of any given hypothesis, it is often difficult to decide at which point it becomes "necessary" to invoke something in order to provide a scientific explanation of the phenomenon under study. At the present time, cyclic AMP stands accused of being the second messenger for vasopressin in several epithelial tissues. Although it may be many years before a final verdict can be reached, our impression is that Orloff and Handler and their colleagues and others, acting for the prosecution, have built a fairly substantial case; the arguments raised by Cuthbert and Painter, acting for the defense, seem weak by comparison. In particular, the suggestion that the effect of theophylline on chloride transport might be responsible for the effect on sodium transport was not supported by the demonstration that theophylline also increased short-circuit current in the absence of chloride (Rider and Thomas, 1969). As pointed out by others (e.g., Bernard, 1865; Schueler, 1960; Platt, 1964), hypotheses lead very precarious lives. They can be established only with difficulty, but they can often be destroyed by a single crucial experiment.

In the kidney, both vasopressin and theophylline increased the permeability to water when added individually to the isolated perfused rabbit collecting tubule (Grantham and Burg, 1966; Grantham and Orloff, 1968), but the possibility that the two agents might act synergistically in this system was apparently not tested. It could perhaps be assumed that this would occur, based on the studies with the toad bladder and frog skin, although the overall effect in the whole kidney is anything but synergistic. The diuretic effect of theophylline results at least in part from a direct effect on the kidney (Lockett and Gwynne, 1968), but whether or not this can be related to an effect on phosphodiesterase in some cells is unknown at present.

2. Other Drugs that Affect the Kidney

In view of the effects of cyclic AMP on permeability in various model systems, including the isolated collecting tubule of the rabbit, it would not be surprising if cyclic AMP were found to play an important role in the actions of many diuretic and antidiuretic agents. At the present time, however, such a role is conjectural only.

One diuretic that has been studied with the possible role of cyclic AMP in mind is *furosemide*. Ferguson (1966) found that this agent had no effect on water permeability of the isolated toad bladder under any conditions, and had no effect on resting short-circuit current. However, the drug did antagonize the stimulation of sodium transport produced by either vaso-pressin, caffeine, or exogenous cyclic AMP. It has been suggested that furosemide is more potent than the benzothiadiazide diuretics because, like ethacrynic acid but unlike the latter agents, it has the ability to inhibit sodium reabsorption in the loop of Henle and the ascending limb of the nephron (Laragh, 1968). Thus, if cyclic AMP facilitates Na$^+$ transport in the tubular epithelia of these segments as it does in the toad bladder, then inhibition of this effect by furosemide would suggest a plausible mech-anism for at least part of the diuretic activity of this compound. The com-plexity of the mammalian kidney has to date precluded further testing of this hypothesis.

The possibility that phosphodiesterase inhibition in some renal cells might lead to diuresis has long seemed worth considering in view of the known actions of theophylline and other methylxanthines. The available evidence, however, offers little or no support for such a possibility. Cer-tainly there is no correlation between diuretic activity in general and the ability to inhibit cardiac phosphodiesterase activity *in vitro* (Moore, 1968). For example, hydrochlorothiazide, a powerful diuretic, had almost no activity as a phosphodiesterase inhibitor, whereas diazoxide, which is not diuretic even though structurally related to hydrochlorothiazide, was relatively potent as a phosphodiesterase inhibitor. Theophylline and caf-feine were more potent than any of the benzothiadiazide derivatives. However, it is obvious that there are many mechanisms by which a drug or group of drugs could act to produce diuresis, and it is not clear that phosphodiesterase inhibition in cardiac preparations *in vitro* is a reliable indicator of what may or may not occur in the kidney *in situ*.

For example, Senft *et al.* (1968a) injected large doses of furosemide and hydrochlorothiazide intravenously in rats, and then measured phos-phodiesterase activity in homogenates (and fractions thereof) prepared from different parts of the excised kidneys. Following the injection of furosemide, activity was found to be reduced only in the inner medulla, whereas after hydrochlorothiazide only renal cortical activity was re-duced. It should be stressed that these findings were not in line with what is known about the sites of action of these compounds as diuretic agents (Laragh, 1968), and indeed Senft and his colleagues interpreted their data as suggesting a possible mechanism for the paradoxical *antidiuretic* ac-tivity of the compounds. The point to be noted, however, is that the results of this study could not have been predicted at all simply on the basis of *in vitro* experiments. Larger doses of these compounds had previously been found to inhibit phosphodiesterase in liver but not in skeletal muscle (Senft *et al.*, 1966).

A puzzling finding with chlorpropamide, a synthetic antidiuretic agent, was reported by Mendoza (1969). Chlorpropamide had no effect by itself on water permeability of the toad bladder, but tended to increase the response to vasopressin and very markedly increased the response to theophylline. These findings were in line with the drug's known antidiuretic activity in mammals *in vivo*. The puzzling finding, however, was that the drug actually inhibited the response of the toad bladder to exogenous cyclic AMP. The reason for this observation is not obvious at present.

In summary, there is presently no substantial evidence that the methylxanthines or any other agents produce diuresis by inhibiting phosphodiesterase in the kidney. However, we feel that this possibility should still be kept in mind. Some drugs may produce diuresis by interfering with the action of cyclic AMP, but here again more research is needed.

3. Aldosterone

Little is known about how aldosterone stimulates sodium transport across epithelial membranes, although new protein synthesis seems clearly to be involved (Sharp and Leaf, 1968). It seems obvious that both this and the resulting effect on sodium transport will require the utilization of energy, although it is doubtful that aldosterone acts by making more energy available (Handler *et al.*, 1969). The finding that pretreatment of isolated toad bladders with aldosterone enhanced the effect of either vasopressin or exogenous cyclic AMP on sodium transport (Goodman *et al.*, 1969) suggests that at least part of the effect of aldosterone is exerted on the same cells that are affected by vasopressin. It seems possible that a common mechanism may underlie the ability of steroids to enhance sensitivity to cyclic AMP, and we can hope that continued study of the effect of aldosterone in the toad bladder will contribute to a better understanding of similar effects elsewhere. These would include, for example, the effects of glucocorticoids in the liver (Friedmann *et al.*, 1967) and vitamin D in the intestine (Harrison and Harrison, 1970).

4. Adrenergic Stimulation

The catecholamines inhibit the water permeability response of the toad bladder to vasopressin or theophylline but not to exogenous cyclic AMP

(Handler *et al.*, 1968b). This is an α-receptor response involving inhibition of cyclic AMP accumulation (Turtle and Kipnis, 1967b). As discussed previously in Chapter 6, the mechanism by which α-receptors mediate this effect is unknown. The question of whether β-receptors are also present in epithelial cells at certain times of the year will require further study. Their presence in the toad bladder was indicated in the experiments of Turtle and Kipnis by a rise in the level of cyclic AMP in response to epinephrine in the presence of phentolamine, but most of these receptors could have been located in smooth muscle (Eggena *et al.*, 1968).

High concentrations of β-adrenergic blocking agents may also inhibit the permeability response to vasopressin (Strauch and Langdon, 1964). It is very doubtful, however, that this involves an interaction with adrenergic receptors. Among a variety of biogenic amines tested, only tyramine was found to produce an increase in permeability to water (Strauch and Langdon, 1969). The significance of this observation is unknown, and whether tyramine can stimulate adenyl cyclase in toad bladder epithelia has not been studied.

In the frog skin, the catecholamines produce different effects depending on the type of receptor involved. α-Receptors mediate inhibition of sodium transport in the presence of vasopressin (Watlington, 1969), presumably by inhibiting the formation or accumulation of cyclic AMP. A fall in the level of cyclic AMP in response to α-agonists was demonstrated in dorsal frog skin (Abe *et al.*, 1969a,b), but ventral skin was unfortunately not studied in these experiments (except in response to MSH, which had no effect). It could be predicted that vasopressin would increase the level of cyclic AMP in ventral skin whereas α-agonists would produce the opposite effect, but these predictions remain to be verified experimentally.

The catecholamines may also stimulate Na^+ transport and increase the permeability of the frog skin to water. Both effects are mediated by adrenergic β-receptors (Watlington, 1968a,b) and are enhanced by theophylline (Bastide and Jard, 1968), hence probably result from the stimulation of adenyl cyclase. However, the cellular locus of these receptors has not been established. It can be inferred from the study by Watlington (1969) that at least some α-receptors occur in the same cells that are affected by vasopressin, but β-receptors may occur primarily in other cells, possibly mucous cells, as suggested by Watlington (1968a,b). This would be compatible with most of the observations reported by Jard *et al.* (1968). Although not helpful in deciding the type of cell in which the change occurred, it can be noted that isoproterenol did produce a slight increase in the level of cyclic AMP in dorsal frog skin (Abe *et al.*, 1969b).

It is clear that renal function in mammals could be affected by adrenergic stimulation in a great variety of ways. However, inhibition of the antidiuretic effect of vasopressin, shown in humans (Fisher, 1968) and rats (Liberman *et al.*, 1970), seems quite analogous to the inhibition of water permeability seen in the isolated toad bladder. It seems probable on the basis of these studies that α-receptors will be found to suppress the accumulation of cyclic AMP in the same cells that are affected by vasopressin. One approach to this problem might involve the measurement of cyclic AMP in the perfusate of the isolated perfused collecting tubule. Unfortunately, although Takahashi *et al.* (1966) reported that the injection of vasopressin did increase the urinary excretion of cyclic AMP, this has not been confirmed by others (see Chapter 11, Section VII).

Finally, in this section, we can note that the catecholamines also inhibit the permeability response to vasopressin in the isolated rabbit ileal mucosa (Field and McColl, 1968). This is an α-adrenergic effect presumably resulting from a fall in the level of cyclic AMP, although cyclic AMP has so far not been measured in this preparation.*

5. Prostaglandins

Prostaglandin E_1 (PGE$_1$) inhibits the permeability response of the toad bladder to vasopressin or theophylline, but not to exogenous cyclic AMP (Orloff *et al.*, 1965; Eggena *et al.*, 1970). By analogy with the effects of PGE$_1$ in fat cells (Butcher and Baird, 1968; see also Section III,B in Chapter 8 and Section IX below), it seems probable that the effect in the toad bladder involves a fall in the level of cyclic AMP. The active principle in the toad bladder extracts studied by Overweg (1966) may have been a prostaglandin.

In the frog skin, PGE$_1$ increases sodium transport (Fassina *et al.*, 1969) possibly by a mechanism similar to that of neurohypophyseal hormones or catecholamines (Jard *et al.*, 1968). Fassina and her colleagues interpreted their data as indicating a primary action of PGE$_1$ on calcium transport (since excess Ca^{2+} inhibited the response), but stimulation of adenyl cyclase seems more likely. The prostaglandins have been shown to have this effect in a variety of tissues, and, as discussed in more detail in Section IX, are the only naturally occurring compounds, other than the catecholamines, that are known to produce some of their effects by way of an increase in the level of cyclic AMP and others by way of a decrease. However, in the absence of additional data, the mechanism of action of PGE$_1$ on the frog skin remains unknown.

The effects of PGE$_1$ on the permeability of the isolated collecting tubule

*See, however, Section XIV.

seem quite complex, not to mention their effects on renal function *in vivo* (Johnston *et al.,* 1967). In the collecting tubule, Grantham and Orloff (1968) found that PGE_1 produced a small increase in water permeability by itself, inhibited the response to vasopressin, acted synergistically with theophylline, and had no effect on the response to exogenous cyclic AMP. The interpretation offered by Grantham and Orloff was that both PGE_1 and vasopressin might be acting to stimulate the same adenyl cyclase, but that the presence of PGE_1 interferes with the greater effect of vasopressin. Although their results were compatible with this hypothesis, an interpretation that may be more consistent with observations made in other tissues is that collecting tubules contain at least two types of cells, one of which resembles toad bladder cells in that PGE_1 and vasopressin exert opposing effects on the level of cyclic AMP. The other cell type differs in that it contains adenyl cyclase that can be stimulated by PGE_1. The physiological significance of such an arrangement seems quite obscure, but should not be dismissed for that reason alone. Whether it reflects reality or not will have to be decided by future research.

6. Ions and Other Agents

Several agents not previously mentioned have been studied in relation to cyclic AMP, and the results can be summarized here.

In addition to the effects of Mg^{2+} and Ca^{2+} mentioned in Section A above, potassium ions also have an effect. Removal of K^+ from the solution bathing the serosal surface of the toad bladder leads to a rapid fall in short-circuit current with no immediate change in water permeability. However, Finn *et al.* (1966) found that after longer periods in K^+-free solutions, the water permeability response to vasopressin was also diminished. Ouabain in the presence of K^+ produced similar effects. Since the response to exogenous cyclic AMP was also diminished under these conditions, it would appear that the K^+-dependent step in the permeability response to vasopressin is subsequent to the formation of cyclic AMP.

By contrast, cysteine or thioglycollate inhibited the response of the toad bladder to vasopressin or theophylline without affecting the response to exogenous cyclic AMP (Handler and Orloff, 1964). These agents may therefore interfere with the activity of adenyl cyclase.

Anaerobiosis and a variety of metabolic inhibitors, including iodoacetate, fluoroacetate, azide, and dinitrophenol were found to prevent the response to both vasopressin and exogenous cyclic AMP (Handler *et al.,* 1966). Similar observations have been made with many other cyclic AMP-mediated responses. Other agents found to inhibit the response to both vasopressin and cyclic AMP in the toad bladder include amphotericin B

(Mendoza *et al.*, 1967) and the alkylating agents mechlorethamine and dibenamine (Gussin *et al.*, 1967).

7. Seasonal Variation

The time of the year is an important factor that may require consideration in many studies involving amphibian preparations. The response of the toad bladder to vasopressin seems to be reasonably consistent from one month to the next, although the response to norepinephrine may vary (Handler *et al.*, 1968b). A striking seasonal effect has been observed in the case of the short-circuit current response of the frog skin, with summer frogs being almost unresponsive to vasopressin (Hong *et al.*, 1968). Whether this involves an alteration in the formation or action of cyclic AMP or both remains to be determined.

C. Other Effects of Vasopressin

Vasopressin produces a variety of metabolic effects in the toad bladder. Some of these, such as phosphorylase activation (Handler and Orloff, 1963), are mediated directly by cyclic AMP, although changes in Na^+ concentration and other factors may modify the overall glycogenolytic response (Handler *et al.*, 1968a). Certain other effects, such as stimulation of palmitate oxidation (Ferguson *et al.*, 1968), seem to be entirely secondary to the effect of cyclic AMP on sodium transport.

Little is known about the mechanism of action of vasopressin and other neurohypophyseal peptides on smooth muscle. These hormones produce contraction of most smooth muscle structures, including the toad bladder (Eggena *et al.*, 1968). Oxytocin was found to stimulate contraction of the isolated rat uterus without changing the level of cyclic AMP no matter whether cyclic AMP was maintained at control levels or elevated by the addition of catecholamines (Dobbs and Robison, 1968). Vasopressin and α-adrenergic agonists act synergistically to produce vasoconstriction (Bartelstone and Nasmyth, 1965; Bartelstone *et al.*, 1967); this would be expected if α-receptors mediate a fall in the level of cyclic AMP in vascular smooth muscle and if vasopressin acts by opposing cyclic AMP. Higher concentrations of vasopressin were found to produce relaxation of the toad bladder (Eggena *et al.*, 1968). It seems possible that this could be related to stimulation of adenyl cyclase in smooth muscle, but the point has not been investigated.

Certain effects of vasopressin reported to occur *in vivo*, such as stimulation of glycogenolysis (Bergen *et al.*, 1960) and inhibition of cardiac output (Nakano, 1967), may be related to cyclic AMP but only indirectly.

II. THYROID STIMULATING HORMONE

A. Effects in the Thyroid Gland

Although certain details concerning the action of thyroid stimulating hormone (TSH, thyrotropin) remain to be elucidated, the evidence that its ability to stimulate the release of thyroid hormone depends upon cyclic AMP is reasonably complete. This evidence will be summarized in this section.

Stimulation by TSH of adenyl cyclase in thyroid preparations was noted many years ago (Klainer et al., 1962) and has been confirmed (Pastan and Katzen, 1967; Kaneko et al., 1969; Burke, 1970a). The only other agents known to be capable of stimulating thyroidal adenyl cyclase are prostaglandin E_1 (PGE_1) (Kaneko et al., 1969) and the long-acting thyroid stimulator (LATS) found in the serum of some patients with Graves' disease (Levey and Pastan, 1970); both of these agents share the ability of TSH to stimulate the thyroid. The action of LATS appears to be slower in onset and more prolonged than that of TSH (Dorrington and Munro, 1966), but in other respects the two agents seem to be similar. A bacterial protein resembling LATS has been isolated (Macchia et al., 1967), but has not been tested for its ability to stimulate adenyl cyclase.

An increase in the level of cyclic AMP in thyroid gland slices has been observed in response to TSH (Gilman and Rall, 1968a; Tonoue et al., 1970) and prostaglandin E_1 (Kaneko et al., 1969). A sample of LATS had no effect on the level of cyclic AMP in bovine thyroid slices (Gilman and Rall, 1968a), but the significance of this observation is unclear, since the activity of the LATS was not ascertained. Epinephrine had a slight effect on the level of cyclic AMP in these experiments, possibly reflecting accumulation in smooth muscle or other cells. Epinephrine has not been shown to stimulate the release of thyroid hormone by a direct effect on the thyroid (Harrison, 1964), although stimulation of ^{131}I uptake has been reported (Maayan and Ingbar, 1968). It is doubtful, however, that this effect was mediated by cyclic AMP, since it was neither inhibited by propranolol nor increased by theophylline. The possibility that it might have been the result of oxidation products (Pastan et al., 1962) was apparently not tested.* Epinephrine had no effect on the level of cyclic AMP in isolated bovine thyroid cells (Knopp et al., 1970).

Potentiation of TSH by theophylline in terms of cyclic AMP formation has been demonstrated (Gilman and Rall, 1968a), and synergism between TSH and theophylline at the functional level has been observed both *in*

---

*See, however, Maayan and Ingbar (1970).

vitro (Ensor and Munro, 1969) and *in vivo* (Bastomsky and McKenzie, 1967). Potentiation of LATS by theophylline has also been reported (Bastomsky and McKenzie, 1968).

Stimulation of thyroid hormone release by exogenous cyclic AMP or its dibutyryl derivative has been observed by many investigators (e.g., Bastomsky and McKenzie, 1967; Ensor and Munro, 1969; Rodesch *et al.*, 1969; Ahn *et al.*, 1969; Tonoue *et al.*, 1970). Several other nucleotides had a slight effect when injected *in vivo* (Bastomsky and McKenzie, 1967), but had no effect when added to thyroid glands *in vitro* (Ensor and Munro, 1969).

Thus all four of our basic criteria have been applied to the thyroid and have been met. The overall action of TSH on the thyroid gland is nevertheless complicated, and we can discuss a few of these complications now.

Among the changes that occur in the thyroid gland in response to TSH are the following: increased phospholipid turnover, increased glucose oxidation, increased endocytosis leading to the formation of intracellular colloid droplets, and increased release of thyroxine and triiodothyronine. Increased uptake of inorganic iodide and synthesis of thyroglobulin seem to be a result rather than a cause of the increased rate of release of thyroid hormones.

Of these effects, the increased rate of phospholipid turnover seems clearly *not* to be mediated by cyclic AMP. It is not shared by prostaglandin E_1 (Zor *et al.*, 1969a; Burke, 1970a) and could not be mimicked in all species by exogenous cyclic AMP (Pastan and Macchia, 1967; Schneider, 1969). The findings that this effect could be produced by LATS (J. B. Field *et al.*, 1968) and occasionally in sheep and bovine thyroid by the dibutyryl derivative of cyclic AMP (Pastan, 1966; Burke, 1968) were probably misleading, suggesting as they did that the effect might be involved in the sequence of events leading to thyroid hormone release. However, since acetylcholine also produces this effect (Altman *et al.*, 1966), its significance is quite obscure. Acetylcholine does not stimulate adenyl cyclase or increase the level of cyclic AMP in the thyroid (Pastan and Katzen, 1967; Kaneko *et al.*, 1969) and is not known to affect the release of thyroid hormones.

The effect of TSH on glucose oxidation can be separated into at least two components. Low concentrations inhibit the conversion of glucose 1-^{14}C to $^{14}CO_2$ whereas higher concentrations stimulate glucose oxidation. The inhibitory effect of low concentrations of TSH correlates well with its effect on the endogenous level of cyclic AMP (Gilman and Rall, 1968b) and can be mimicked by the application of exogenous cyclic AMP (Pastan and Macchia, 1967). There is reason to believe, therefore, that this effect is mediated by cyclic AMP, even though its significance is far from clear.

The stimulatory effect of higher concentrations of TSH is more controversial. It is shared not only by prostaglandin E_1 (Zor *et al.*, 1969a; Rodesch *et al.*, 1969; Burke, 1970a; Oraya and Solomon, 1970) and LATS (Field *et al.*, 1968; Shishiba *et al.*, 1970), but also by a variety of other agents that do not otherwise affect the thyroid, including acetylcholine, epinephrine, serotonin, and vitamin K (Altman *et al.*, 1966). Even in response to TSH, LATS, and PGE_1 it should be noted that the stimulation of glucose oxidation correlates poorly with their other effects. The dibutyryl derivative of cyclic AMP stimulates glucose oxidation in thyroid tissue of some species (Pastan, 1966; Zor *et al.*, 1968; Burke, 1968; Rodesch *et al.*, 1969), but not others, such as the rat (Pastan and Macchia, 1967). However, Macchia *et al.* (1969) found that the addition of cyclic AMP to beef thyroid homogenates did increase the rate of glucose oxidation; this effect was markedly enhanced in the presence of theophylline, and was not produced by other nucleotides. Our own interpretation of the data summarized in this paragraph is that part of the effect of TSH, LATS, and PGE_1 on glucose oxidation in the thyroid may be mediated by cyclic AMP, in at least some species, whereas another portion of the effect is probably related to some other action. Pastan *et al.* (1962) had suggested previously that the effect of epinephrine might result from its oxidation to adrenochrome and subsequent stimulation of TPNH oxidation. A similar mechanism might account for part of the effect of some of the other agents that stimulate thyroidal glucose oxidation. The significance of this effect, if any, is far from clear.

Stimulation of endocytosis seems to be a more important event in the sequence of events leading to the release of thyroid hormones. This effect is produced by all of the agents known to increase the level of cyclic AMP in the thyroid (Pastan and Wollman, 1967; Rodesch *et al.*, 1969; Shishiba *et al.*, 1970; Burke, 1970a; Onaya and Solomon, 1970) and by exogenous cyclic AMP itself, but not by a variety of other substances tested. The only exception to this generalization is the bacterial protein mentioned previously (Macchia *et al.*, 1967), and its effect on cyclic AMP formation has not been studied.

The effect of the dibutyryl derivative of cyclic AMP on colloid droplet formation in dog thyroid slices is illustrated in Fig. 10.1., from the original report by Pastan and Wollman. The mechanism of this action of cyclic AMP is unknown, but may be related to the stimulation of exocytosis seen in many other cells (Chapter 5, Section V). Subsequent fusion of the colloid droplets with lysosomes may occur independently of cyclic AMP, although direct stimulation by the nucleotide seems not unlikely. Vaes (1968a) demonstrated an effect of the dibutyryl derivative on the release of lysosomal enzymes in bone explants maintained in tissue culture.

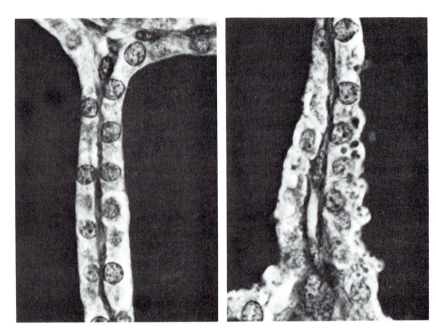

Fig. 10.1. Effect of dibutyryl cyclic AMP on histological appearance of dog thyroid cells. Left-hand panel: slices were incubated for 2 hours in a modified Krebs-Ringer bicarbonate buffer, fixed in Bouin's fluid, and stained by the periodic acid–Schiff procedure with hematoxylin counterstain. Note the smooth apical membranes of the follicular epithelial cells and the absence of pseudopods and droplets. Right-hand panel: dibutyryl cyclic AMP (250 μg/ml) present during the second hour of incubation. Note the irregularities in the apical membranes and the colloid droplets within the cells. These changes are identical to those seen in response to TSH. From Pastan and Wollman (1967).

Cyclic AMP also mimics the effect of TSH in stimulating the uptake and binding of inorganic iodide by the thyroid gland (Ahn and Rosenberg, 1968; Rodesch *et al.*, 1969). However, whether these represent separate effects of cyclic AMP or whether they follow from some other action is unknown at present. Lissitzky *et al.* (1969) found that cyclic AMP, but not 5'-AMP, could stimulate protein synthesis by thyroid polyribosomes in a cell-free system, but the significance of this is unknown at present.

Fluoride stimulates adenyl cyclase activity in broken cell preparations of the thyroid gland (Klainer *et al.*, 1962; Pastan *et al.*, 1968), as it does in similar preparations of all mammalian tissues studied, but does not cause an increase in the level of cyclic AMP when applied to intact cells (Kaneko *et al.*, 1969). In line with this, fluoride also fails to stimulate endocytosis or the release of thyroid hormones, although it does stimulate phospholipid turnover and glucose oxidation (Pastan *et al.*, 1968; Kaneko *et al.*,

1969; Rodesch *et al.*, 1969). These observations may have contributed to some of the earlier confusion in this field. As mentioned previously, how-ever, the effect of TSH on phospholipid turnover seems to be unrelated to cyclic AMP, and the effect on glucose oxidation may be related to cyclic AMP only in part. In any event, fluoride is only one of a large number of nonspecific substances capable of mimicking the effects of TSH on these particular parameters.

Most of the effects of the methylxanthines on thyroid function seem to be understandable in terms of their ability to inhibit phosphodiesterase (Gilman and Rall, 1968a; Bastomsky and McKenzie, 1968; Ensor and Munro, 1969; Rodesch *et al.*, 1969). However, their effects on the meta-bolism of phospholipids and carbohydrates seem to be at least as poorly understood as those of the other agents mentioned (Zor *et al.*, 1969a), and are most likely the result of a complex mixture of effects. The available data (Wolff and Varrone, 1969) suggest that the goitrogenic activity of the methylxanthines may be related at least in part to their ability to inhibit phosphodiesterase.

Unlike many other cyclic AMP-mediated responses, calcium does not seem to be required for any of the effects of TSH, although stimulation of glucose oxidation was reduced in the absence of Ca^{2+} (Zor *et al.*, 1968). On the other hand, removal of potassium completely prevented the effect of TSH on both adenyl cyclase activity in broken cell preparations and colloid droplet formation in slices prepared from sheep thyroid glands (Burke, 1970a). The striking dependence on K^+ observed in these ex-periments stands in contrast to broken cell preparations from many other tissues, where K^+ has been found to have relatively little effect on adenyl cyclase activity.

Even in thyroidal preparations, for that matter, the effect of K^+ may be species-dependent. In a bovine preparation incubated in a tris buffer, for example, Wolff *et al.* (1970) found the effect of added K^+ to be incon-sistant. On the other hand, they did observe a very striking inhibition by *lithium* of TSH-stimulated activity. Fluoride-stimulated activity was rela-tively more resistant to the inhibitory effect of Li^+. The effect could be overcome in either case by the addition of Mg^{2+}. It seems possible that this effect could account, at least in part, for the antithyroidal activity of lithium. For that matter, it is at least conceivable that the ability of lithium salts to inhibit adenyl cyclase, noted also in preparations of other tissues (Birnbaumer *et al.*, 1969; Dousa and Hechter, 1970), could account for many of their effects, including their known beneficial effect in the treat-ment of mania (Fann *et al.*, 1969; Gershon, 1970) and their possible bene-ficial effects in other disorders (Voors, 1969).

A number of other substances are also capable of inhibiting the effects of TSH on the thyroid gland, including lecithinase C (Pastan and Macchia, 1967), several polyene antibiotics (Butcher and Serif, 1969), adrenergic blocking agents of both types (Burke, 1969), and chlorpromazine (Onaya *et al.,* 1969; Wolff and Jones, 1970). All of these agents have been shown to inhibit the stimulation of adenyl cyclase by TSH, and some of them are not entirely nonspecific. For example, the adrenergic blocking agents were capable of inhibiting the effect of TSH at concentrations that did not prevent the effect of fluoride (Burke, 1969), and chlorpromazine actually *increased* activity in the presence of fluoride (Wolff and Jones, 1970). However, since these agents also prevent the effect of exogenous cyclic AMP on the thyroid, the mechanism by which they interfere with the action of TSH remains to be defined.

Pastan *et al.* (1966) presented evidence to show that the first step in the action of TSH on the thyroid was binding to a site or sites on the extracellular side of the cell membrane. From the more recent work by this group of investigators using adrenal cortical preparations (Lefkowitz *et al.,* 1970) and also from many other observations previously discussed, we can infer that the first step is, in fact, interaction with specific receptors for TSH, some of which are closely related to if not an integral component of the membrane adenyl cyclase system. In summary, then, the following sequence of events can be postulated. The interaction of TSH with its receptors leads to stimulation of adenyl cyclase, which in turn leads to an increase in the intracellular level of cyclic AMP. Then, by an unknown mechanism possibly involving the phosphorylation of a protein, cyclic AMP initiates endocytosis leading to the formation of colloid droplets. These droplets then fuse with lysosomes, which may or may not be directly affected by cyclic AMP, leading finally to the release of thyroid hormones. Although many details of this postulated sequence of events remain to be established, an essential role for cyclic AMP seems beyond question.

B. Effects in Other Tissues

The ability of TSH to stimulate adenyl cyclase and lipolysis in adipose tissue in some species was mentioned in Chapter 8. The physiological significance of its ability to stimulate adipose tissue is uncertain at present. Although TSH has been added as a control to preparations of many tissues other than thyroidal or adipose tissue, it has not been found to stimulate adenyl cyclase or alter the level of cyclic AMP in any of them. This would seem to correlate well with its lack of effect on other parameters in these tissues.

III. MSH AND MELATONIN

Melanocyte stimulating hormone (MSH) and melatonin can be discussed together because in dorsal frog skin, the only tissue in which their effects have been studied in relation to the level of cyclic AMP, they are physiological antagonists of one another.

Cellular aspects of the control of physiological color changes in fish, amphibians, and reptiles have been reviewed recently (Fujii and Novales, 1969; Novales and Davis, 1969; Hadley and Goldman, 1969a). The release of MSH is responsible for the ability of these animals to become dark when placed against an illuminated dark background. Although other cells may be affected to some extent, it is thought that MSH acts primarily on dermal melanophores to cause dispersion of melanin granules within these cells, leading to darkening. Evidence that this response is mediated by cyclic AMP is now substantial, and will be summarized in the following paragraphs.

First, the effect of MSH can be mimicked by exogenous cyclic AMP or its dibutyryl derivative. This has been demonstrated in amphibian preparations (Bitensky and Burstein, 1965; Novales and Davis, 1967; Abe *et al.*, 1969a; Goldman and Hadley, 1969b; Van de Veerdonk, 1969; Van de Veerdonk and Konijn, 1970), reptiles (Hadley and Goldman, 1969b; Goldman and Hadley, 1970b), and fish (Novales and Fujii, 1970). Other nucleotides have generally been found to have no effect or else the opposite effect, although 5′-AMP caused darkening of the skin of a lizard, *Anolis carolinensis* (Hadley and Goldman, 1969b).

Attempts to measure adenyl cyclase activity in broken cell preparations of frog skin have to date been unsuccessful. The reasons for this are unknown and will require further study. In the meantime, our second criterion, involving the measurement of cyclic AMP in intact cells, has been studied in some detail. As shown in Fig. 10.2, the injection of MSH in frogs caused increases in cyclic AMP and darkening of dorsal skin (Abe *et al.*, 1969a). Since ventral skin was unaffected, it seems reasonable to assume that most of the cyclic AMP produced in dorsal skins in response to MSH was produced in melanophores, although further studies would be required to establish this point.

The correlation between cyclic AMP levels and darkening in response to different hormones *in vitro* was reasonably good, as summarized in Table 10.III. Temporal studies also suggested a role for cyclic AMP in the action of MSH. Cyclic AMP levels were significantly increased in frog skin 3 minutes after the addition of MSH to the medium, and this was also the earliest time at which increased darkening was observed (Abe *et al.*, 1969a).

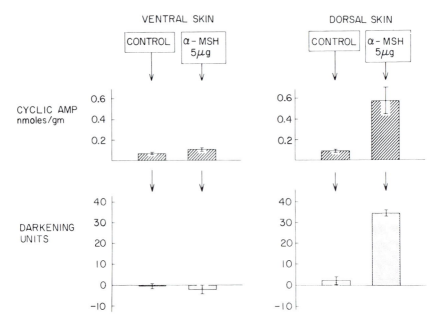

Fig. 10.2. Effect of MSH on cyclic AMP levels in ventral and dorsal frog skin *in vivo*. The hormone was injected in a volume of 0.5 ml into the dorsal lymph sac and cyclic AMP and darkening measured 30 minutes later. From Abe *et al.* (1969a).

TABLE 10.III. Relative Potencies of Three Peptides in Two Bioassay Systems[a]

Peptide studied	Cyclic AMP generating activity	Frog skin darkening activity
α-MSH (synthetic)	1000	1000
β-MSH (porcine)	334	260
ACTH (human)	40	8

[a]Cyclic AMP levels and darkening were measured in dorsal frog skins after a 30-minute incubation. Two concentrations of each hormone were used, and α-MSH was assigned an arbitrary potency of 1000 in both assays. From Abe *et al.* (1969a).

Studies of the relation between cyclic AMP levels and darkening as a function of the dose of MSH have been less satisfactory. The lowest concentration of α-MSH that produced a detectable increase in the level of cyclic AMP in frog skins *in vitro* was 0.5 ng/ml, a dose far in excess of that required to produce darkening. However, this discrepancy is probably more apparent than real. Melanophores constitute a relatively small proportion of the total mass of the frog skin, and it could be expected that biologically significant changes in the level of cyclic AMP in these cells would be difficult to detect against the background provided by other cells. The striking increase in cyclic AMP produced by higher doses of MSH (greater than tenfold in response to 50 ng/ml) suggests that melanophores, like all other cells that have been studied from this point of view, are capable of generating cyclic AMP far in excess of the amount required for maximal stimulation of cell function. Such high levels probably do not occur naturally. Using a microbiological assay, Van de Veerdonk and Konijn (1970) found that physiologically darkened skins contained approximately twice as much cyclic AMP as nondarkened skins.

Theophylline also causes increased cyclic AMP levels and skin darkening, and acts synergistically with MSH on both parameters (Abe *et al.*, 1969a; Goldman and Hadley, 1970b). Theophylline also potentiates exogenous cyclic AMP (Goldman and Hadley, 1969b). It seems likely that a major part of the effect of theophylline and other methylxanthines on skin color can be attributed to their ability to inhibit phosphodiesterase, although other mechanisms have not been ruled out. The failure of α-adenergic agonists and melatonin to inhibit skin darkening in response to theophylline, in doses that did prevent the response to MSH (Abe *et al.*, 1969b, see also below), could be interpreted as evidence for another mechanism of action. However, the mechanism by which these agents act to suppress the level of cyclic AMP is so poorly understood that it seems unwise to draw conclusions about the mechanism of action of other drugs on this basis. It would certainly not be surprising if the methylxanthines were capable of influencing melanophore function independently of their effects on phosphodiesterase, but whether they do or not remains to be seen. The recent study by Beavo *et al.* (1970b), comparing phosphodiesterase inhibition and lipolytic activity in response to a large group of analogs, suggests a paradigm that could be usefully applied in many fields, including the one under discussion.

The ability of α-adenergic agonists to oppose and of β-adrenergic agonists to mimic the response to MSH was discussed previously in Chapter 6 (Section VI, E). Since that chapter was written, an increase in the level of cyclic AMP in response to epinephrine, in the skin of a species that responds to epinephrine with skin darkening, has been reported (Van de Veerdonk and Konijn, 1970). A fall in response to α-agonists had been shown previously (Abe *et al.*, 1969b; See also Chapter 6, Fig. 6.14 and 6.15).

Melatonin also inhibits skin darkening in response to MSH, and, as shown in Fig. 10.3, likewise inhibits the rise in cyclic AMP. A difference between melatonin and norepinephrine in these experiments was that the effects of melatonin could not be blocked by α-adrenergic blocking agents. This was taken to mean that melatonin probably interacted with receptors that were separate from, but functionally similar to, adrenergic α-receptors. However, the mechanism by which these receptors suppress the accumulation of cyclic AMP is still obscure. The failure of melatonin or norepinephrine to reverse skin darkening in response to exogenous cyclic AMP, together with their lack of effect on the rate of destruction of cyclic AMP in frog skin homogenates (Abe *et al.,* 1969b), had suggested inhibition of adenyl cyclase as a likely possibility. More recently, however, Novales and Fujii (1970) observed a striking effect of epinephrine on fish melanophores in the presence of exogenous cyclic AMP. It is clear that much additional research will be required before the mechanism of α-receptor action in these and other cells is understood.

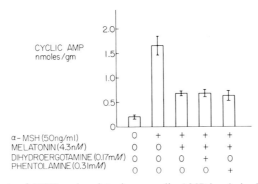

Fig. 10.3. Effects of MSH and melatonin on cyclic AMP levels in dorsal frog skin *in vitro.* Pieces of frog skin were incubated in the presence or absence of MSH or MSH + melatonin, as indicated. The difference between the second and third bars represents the cyclic AMP lowering effect of melatonin. The α-adrenergic blocking agents prevented the similar effect of norepinephrine (cf. Fig. 6.15) but did not prevent the response to melatonin. From Abe *et al.* (1969b).

An especially interesting observation by Hadley and Goldman (1969b) in lizard skins was that whereas the dibutyryl derivative of cyclic AMP produced darkening similar to that produced by MSH, exogenous cyclic AMP by itself actually produced the opposite effect. This can be added to similar observations with other systems, in which the dibutyryl derivative successfully mimicked the effect of β-adrenergic stimulation, as predicted, but where the parent compound produced the opposite effect, i.e., either an α-agonistlike or insulinlike effect (e.g., Bartelstone *et al.*, 1967; Chambaut *et al.*, 1969; Bowman and Hall, 1970). Experiments with exogenous nucleotides are very difficult to interpret, but these observations are at least compatible with the idea that one function of the system of which α-receptors and melatonin receptors (and perhaps also insulin receptors) are a part may be to stimulate the conversion of cyclic AMP to an antimetabolite of cyclic AMP. This may provide a useful working hypothesis for future research in this area.

As for the mechanism by which endogenous cyclic AMP stimulates the dispersion of melanin granules, very little is known about this. The participation of microtubular structures has been suggested (e.g., Green, 1968; Fujii and Novales, 1969; Roth *et al.*, 1970), but more research is needed. Hypertonicity of the external medium inhibits the response of cyclic AMP, and calcium seems to be required (Novales and Davis, 1969), but the significance of these observations is unclear at present.

It seems likely that many and perhaps all of the extrapigmentory effects of MSH and melatonin in other species are related to changes in the level of cyclic AMP in the cells responsible for these effects. However, except for one study in the rat, in which the ability of α-MSH to increase the level of cyclic AMP in the adrenal cortex was correlated with its effect on steroidogenesis (Grahame-Smith *et al.*, 1967), this possibility has not been investigated.

IV. HORMONES REGULATING CALCIUM HOMEOSTASIS

Nothing better illustrates the troubles that have plagued us during the writing of this book than research related to parathyroid hormone and calcitonin. When we started, there was only one report in the literature suggesting, on the basis of indirect evidence, that cyclic AMP might be involved in the actions of parathyroid hormone. Since then, all four of our basic criteria have been examined, and the evidence that cyclic AMP mediates the actions of parathyroid hormone in bone and kidney is now substantial. In addition, this information has been applied to the solution of clinical problems, and some progress has been made toward understanding the mechanism of action of calcitonin. The rapid progress in this area stands as a great credit to the investigators involved, and illustrates again (see Platt, 1964) how useful the systematic application of the scientific method can be.

A. Parathyroid Hormone

1. The Phosphaturic Effect

Parathyroid hormone (PTH) acts directly on the kidney to increase the urinary excretion of inorganic phosphate and certain other ions. The principal site of this action is thought to be the proximal part of the nephron, although the question of whether the phosphaturia results primarily from increased secretion or decreased reabsorption of phosphate appears to be still unsettled.

Chase and Aurbach (1967) were the first to observe that the injection of PTH also leads to a striking increase in the urinary excretion of cyclic AMP. As discussed in more detail in Chapter 11 (Section VII) the kidney has now been established as the major source of this increment. Under normal conditions, the influence of PTH on the kidney probably accounts for a substantial part (from 25 to 50% depending on the species) of the normal urinary level of cyclic AMP. Chase and Aurbach showed that the rise in urinary cyclic AMP in response to PTH preceded the phosphaturic effect, and others (Russell *et al.*, 1968; Rasmussen *et al.*, 1968) showed that the latter response could be mimicked by cyclic AMP or its dibutyryl derivative, but not by other nucleotides.

Infusion of theophylline was found to produce a sustained rise in the urinary level of cyclic AMP, but the effects of theophylline and parathyroid hormone were additive instead of synergistic (Chase and Aurbach, 1967). Rasmussen *et al.* (1968) found that a dose of theophylline that significantly enhanced the phosphaturic effect of the dibutyryl derivative of cyclic AMP did not increase the effect of parathyroid hormone. Since these experiments were carried out in thyroparathyroidectomized rats, the results are not easily interpreted in terms of the possible release of calcitonin (cf., Section IV, B). In fact, as pointed out previously, the effects of the methylxanthines *in vivo* would be difficult to interpret even if their only action was to inhibit phosphodiesterase, and this is not the case. However, the failure of several groups to observe potentiation of PTH by theophylline seems especially interesting in view of the large increases in urinary cyclic AMP produced by PTH. Physiological concentrations of vasopressin, by contrast, have very little effect on the urinary level of cyclic AMP (Chase *et al.*, 1969b; Kaminsky *et al.*, 1970a; see also Chapter 11, Section VII). Taken together, these observations suggest the possibility that destruction by phosphodiesterase may be a relatively unimportant means by which PTH-sensitive renal cells reduce their level of cyclic AMP; instead they may dispose of it primarily by releasing it into the extracellular space. The finding that phosphaturia in response to exogenous cyclic AMP was enhanced by theophylline could mean that much of this phosphate was derived from other cells throughout the body.

In order to test their hypothesis that parathyroid hormone might stimulate adenyl cyclase in the kidney, Chase and Aurbach (1968) measured the effects of parathyroid hormone and vasopressin on cyclase activity in washed particulate preparations from homogenates of cortical and medullary renal tissue. As illustrated in Fig. 10.4, they found that PTH had a striking effect in the cortical preparation while vasopressin produced a similar effect in the medullary preparation. Conversely, PTH had only a slight effect in the medulla, while vasopressin caused a substantial but nevertheless smaller effect in cortical tissue. In a later series of experiments (Melson *et al.*, 1970), adenyl cyclase from renal cortical tubules was found to be unresponsive to vasopressin. These results are in line with physiological evidence that the effect of PTH on phosphate transfer occurs primarily in the proximal portions of the nephron, while the effects of vasopressin on permeability are produced primarily in the collecting tubules. The stimulatory effects of PTH and vasopressin on adenyl cyclase in renal homogenates were found to be additive, providing further evidence that the two hormones affect different cells. Most of the cyclase

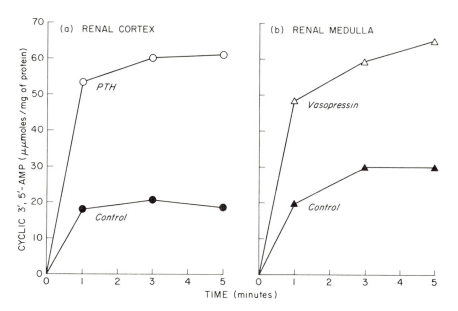

Fig. 10.4. Differential effects of PTH and vasopressin on adenyl cyclase activity of washed particulate preparations of rat renal cortex (left panel) and medulla (right panel). From Chase and Aurbach (1968).

activity in these preparations is found in the cell membrane fraction (Chase and Aurbach, 1968; Melson *et al.*, 1970).

Most of the other hormones tested have been found to have no effect on renal adenyl cyclase, although epinephrine, glucagon, and calcitonin each produced a small but statistically significant increase in the cortical tubular preparation (Melson *et al.*, 1970), and glucagon was also effective in the medulla (Marcus and Aurbach, 1969). The cellular source of the activity affected by these other hormones has not been established. The much greater effect of PTH in cortical preparations has led to the suggestion that this could be used as the basis of a bioassay for parathyroid hormone (Marcus and Aurbach, 1969; Melson *et al.*, 1970). As is commonly seen in broken cell preparations, the maximum effect of fluoride on renal adenyl cyclase is greater than that of any hormone (Dousa and Rychlik, 1968b, Melson *et al.*, 1970; Murad *et al.*, 1970a). Melson and his colleagues found that the response to PTH could reach 16% of the response to 10 mM NaF in the absence of an ATP-regenerating system, and 28% in the presence of the regenerating system.

Nagata and Rasmussen (1968) studied the formation of cyclic AMP in the rat kidney *in situ*. They found that the administration of parathyroid hormone to parathyroidectomized rats led to a fourfold increase in the renal content of cyclic AMP, and that this occurred in vitamin D-deficient rats as well as in controls. This tends to support the view that vitamin D is not permissive for the direct renal effect of the parathyroid hormone.

Cyclic AMP levels have also been measured in renal cortical tubules *in vitro* (Melson *et al.*, 1970; Nagata and Rasmussen, 1970). Parathyroid hormone produced a dose-dependent increase which seemed largely independent of the extracellular concentration of calcium.

Although the presently available evidence supports the hypothesis that an increase in the level of cyclic AMP in the proximal tubules mediates the effect of the parathyroid hormone to stimulate the secretion (or inhibit the reabsorption) of inorganic phosphate, the mechanism by which cyclic AMP might produce this effect is unknown.

2. Other Renal Effects

Parathyroid hormone and the dibutyryl derivative of cyclic AMP have been shown to stimulate gluconeogenesis from lactate in renal cortical tubules *in vitro* (Nagata and Rasmussen, 1970). Since this required the presence of Ca^{2+} in the extracellular medium, whereas cyclic AMP accumulation in response to PTH seemed independent of Ca^{2+}, Nagata and Rasmussen suggested that cyclic AMP might act in this instance by increasing the permeability of tubular membranes to Ca^{2+}. The finding by Borle (1968) that PTH could increase the uptake of Ca^{2+} by HeLa cells in tissue culture was cited as additional supporting evidence for this. Rasmussen and Tenenhouse (1968) had speculated that all of the diverse effects of cyclic AMP might be related to an alteration in calcium permeability. While this is unlikely, it does seem possible that many of the effects of cyclic AMP, including the stimulation of renal gluconeogenesis, involve an effect on the translocation of calcium ions.

Bowman (1970) found that cyclic AMP, but not 5'-AMP, stimulated gluconeogenesis from lactate and several other substrates, including glutamate, when added to the isolated perfused rat kidney. Acidosis was also an effective stimulus, but neither PTH nor the dibutyryl derivative of cyclic AMP was effective. The reasons for the differences between these results in the perfused kidney and those of Nagata and Rasmussen, using isolated tubules, are unknown at present. Although it may be largely a matter of dosage, it is nevertheless of interest that the effect of cyclic AMP itself could not be reproduced by equimolar concentrations of the derivative. Similar observations had been made in the cases of intestinal relaxation (Kim *et al.,* 1968) and inhibition of cell growth (Ryan and Heidrick, 1968). The most likely explanation for such observations is that the cells which responded to cyclic AMP lacked the capacity, at least under the conditions of these experiments, to remove the acyl group from the 2'-O position of the derivative. However, this remains to be established.

Stimulation of gluconeogenesis from glutamate, which is accompanied by the production of ammonia, had previously been demonstrated in response to exogenous cyclic AMP in renal cortical slices (Pagliara and Goodman, 1969). It has more recently been found that the addition of cyclic GMP does not stimulate gluconeogenesis in this preparation, but rather inhibits, apparently by inhibiting the conversion of glutamate to α-ketoglutarate (Pagliara and Goodman, 1970). By contrast, cyclic IMP and 5'-IMP were found to resemble cyclic AMP in that they stimulated gluconeogenesis from a variety of substrates.

3. Calcium Mobilization

After many years of controversy, it is now generally recognized that the parathyroid hormone exerts a direct effect on bone to promote the mobilization of calcium. There is now a considerable body of evidence to suggest that this action of the parathyroid hormone, like its effect in the kidney, is mediated by cyclic AMP.

Adenyl cyclase from bone is stimulated by parathyroid hormone. Chase *et al.* (1969a) prepared homogenates of rat fetal calvaria and found that cyclase activity present in the 2200 *g* fraction could be stimulated two- to threefold by addition of the hormone. This effect was initially thought to be specific for PTH, since none of a variety of other hormones tested, including glucagon, epinephrine, or ACTH, had any effect in this system.

It was later established that PTH also caused a marked rise in the level of cyclic AMP in pieces of fetal rat calvaria incubated *in vitro* (Murad *et al.*, 1970a; Chase and Aurbach, 1970). Epinephrine (acting via adrenergic β-receptors) and prostaglandins E_1 and E_2 produced smaller increments in the level of cyclic AMP (Chase and Aurbach, 1970). Although somewhat surprising, in view of the apparent lack of effect of these agents in broken cell preparations, the most surprising result of all was that *calcitonin* also increased the level of cyclic AMP in bone. Although the maximum effect of calcitonin was much smaller than that of PTH, Murad and his colleagues found that the effects of the two hormones were additive. This suggests that both hormones might act by increasing the level of cyclic AMP in bone, albeit in different cells. It had previously been thought that calcitonin, which inhibits calcium resorption, might act by reducing the level of cyclic AMP in the same cells that were affected by PTH, possibly by stimulating phosphodiesterase (Wells and Lloyd, 1968b).

Theophylline acts synergistically with PTH to increase the level of cyclic AMP in bone *in vitro* (Murad *et al.*, 1970a) and to produce hypercalcemia in thyroparathyroidectomized rats (Wells and Lloyd, 1968a). In intact animals, on the other hand, theophylline may oppose the hypercalcemic effect of PTH (Wells and Lloyd, 1967). This can now be understood partly in terms of calcitonin release, as discussed in Section IV, B below, and partly in terms of calcitonin action, if the effect of calcitonin on calcium resorption is indeed mediated by cyclic AMP. In addition, theophylline might alter blood calcium quite independently of cyclic AMP. For example, caffeine has been shown to be capable of inhibiting the uptake of calcium ions by subcellular fractions from skeletal muscle (Weber, 1968), and presumably theophylline has a similar effect. Whether this action *in vivo* would be reflected by a change in the level of blood calcium is unknown, but the possibility has not been ruled out.

The finding that both PTH and calcitonin increase the level of cyclic AMP in bone may also help to explain some puzzling observations that have been made with exogenous derivatives of cyclic AMP. Although the dibutyryl derivative may mimic PTH when injected into thyroparathyroidectomized rats (Rasmussen *et al.*, 1968), the dose-response curve is definitely biphasic (Wells and Lloyd, 1969). This is even more striking when the derivative is applied to fetal bone in tissue culture: resorption was increased in response to the derivative in concentrations up to 2×10^{-4} M, but, instead of reaching a response equal to that produced by PTH, higher concentrations produced a smaller effect, and the response was lost entirely at a concentration of 10^{-3} M (Raisz and Klein, 1969). This type of result would be expected if cyclic AMP were to mediate not only the effect of PTH but also, in other cells, the effect of calcitonin as well.

Prostaglandin E_1, which Chase and Aurbach (1970) had shown to increase the level of cyclic AMP in fetal rat bone, was shown by Klein and Raisz (1970) to stimulate bone resorption in tissue culture. This effect was seen over a wide range of concentrations, and supramaximal concentrations were not inhibitory. This might suggest that PGE_1 is capable of stimulating adenyl cyclase in·PTH-sensitive cells but not in the cells affected by calcitonin, although, as noted previously, Chase and Aurbach saw no effect of PGE_1 in their broken cell experiments. Despite its pronounced effect on cultured bone cells, PGE_1 did not raise serum calcium levels in parathyroidectomized rats (Klein and Raisz, 1970). It seems possible that exogenous PGE_1 opposes its effect in bone by stimulating the release of calcitonin. As shown by Kaneko *et al.* (1969), PGE_1 does increase the level of cyclic AMP in the thyroid gland, and it seems possible that part of this increment may occur in the parafollicular cells.

Although the cellular site of action of PTH in bone is controversial, the evidence (reviewed by Talmage, 1967; see also standard textbooks) suggests that at least two types of cells may be affected. First, PTH may act directly on osteocytes to cause the release of lysozomal enzymes, presumably by exocytosis. This effect can be mimicked by the dibutyryl derivative of cyclic AMP (Vaes, 1968a,b). Second, PTH may act on certain mesenchymal cells to stimulate their transformation to osteoclasts. This may also be mediated by cyclic AMP, although an increase in the number of osteoclasts in response to exogenous cyclic AMP had not been demonstrated at the time of this writing. Obviously a third potential site of action of PTH is represented by the osteoclasts themselves (Vaes, 1968b). Whether released primarily from osteocytes or osteoclasts, the lysozomal enzymes are thought to be responsible for the dissolution of bone matrix and the solubilization of bone mineral. Alterations in the permeability of cell membranes to calcium may also be involved (Talmage, 1967).

The cellular site of action of calcitonin is even more obscure than the site of action of PTH. Whether an increase in the level of cyclic AMP in the affected cells could account for the ability of calcitonin to inhibit bone resorption is unknown at present.

Adamson (1970) found that dibutyryl cyclic AMP stimulated amino acid uptake, the incorporation of amino acids into protein, and the incorporation of inorganic sulfate into chondroitin sulfate in embryonic chick pelvic bones *in vitro*. These effects were similar to those produced by a factor which appears in serum after injection of growth hormone.

4. Intestinal Permeability to Calcium

Harrison and Harrison (1970) found that dibutyryl cyclic AMP could stimulate the mucosal to serosal transfer of calcium across the rat small intestine. As demonstrated previously for PTH, this occurred in duodenal preparations from vitamin D-treated but not vitamin D-deficient rats. Thus the action of vitamin D in the gut may be basically similar to that of glucocorticoids in the liver (Friedmann *et al.*, 1967). To date, however, an effect of PTH on the formation of cyclic AMP in duodenal epithelia has not been demonstrated.

5. Pseudohypoparathyroidism

An interesting inquiry into the nature of pseudohypoparathyroidism has been carried out by Chase *et al.* (1969b). Pseudohypoparathyroidism is a relatively rare hereditary disease in which the chemical abnormalities of hypoparathyroidism exist in the presence of normal or even excessive levels of circulating parathyroid hormone. Chase and Aurbach (1967) had found that the characteristic phosphaturia that occurs in normal subjects in response to PTH was preceded by a remarkable rise in the urinary excretion of cyclic AMP. Patients with true (surgical or idiopathic) hypoparathyroidism responded in a similar fashion, but PTH had little or no effect on urinary cyclic AMP in patients with pseudohypoparathyroidism (Chase *et al.*, 1969b). Glucagon, by contrast, produced an increase in urinary cyclic AMP in these patients similar to that seen in normal subjects. These findings point to a defect in pseudohypoparathyroidism in the PTH-sensitive adenyl cyclase system of renal tissue (see Section IV,A,1 above), possibly involving a loss of the receptors for PTH. Although the increment in urinary cyclic AMP that occurs in response to PTH is primarily if not entirely nephrogenic (Broadus *et al.*, 1970a), and does not reflect changes occurring in bone, it seems reasonable to conclude with Chase and his colleagues that a similar defect must exist in the skeletal adenyl cyclase system in these patients. Pseudohypoparathyroidism may eventually be seen as the prototype of a number of conditions in which hormone levels seem to be normal but where the response of the target tissue to the hormone is diminished or absent.

Pseudo-pseudohypoparathyroidism refers to a syndrome in which most of the developmental abnormalities of pseudohypoparathyroidism are present but with normal serum calcium and phosphorus values. In line with this, Chase *et al.* (1969b) found that these patients responded to PTH with a rise in urinary cyclic AMP that was not strikingly different from that seen in normal patients. These two disorders seem genetically related, for they tend to occur in the same families. Perhaps the defect present in pseudohypoparathyroidism was present at one time in patients with pseudo-pseudohypoparathyroidism. We can recall here the great rapidity with which the sensitivity of rat fat cells to insulin was restored after the removal of trypsin in the experiments of Kono (1969b), and it seems possible that a similar restoration of sensitivity to other hormones could occur *in vivo*.

6. Mitogenic Activity

Parathyroid hormone is one of several hormones reported by Whitfield *et al.* (1970a) to be capable of stimulating DNA synthesis and mitosis in (and hence the proliferation of) rat thymic lymphocytes. Since this effect of PTH was enhanced by caffeine and could be mimicked by exogenous cyclic AMP, and especially in view of the known effects of PTH on renal and skeletal adenyl cyclase, it seems possible that this effect is mediated by cyclic AMP. Extracellular calcium was required for the action of PTH but not of cyclic AMP. Calcitonin and imidazole were both capable of inhibiting the response to either PTH or cyclic AMP (MacManus and Whitfield, 1970). The significance of this action of cyclic AMP has recently been discussed in an interesting paper by MacManus *et al.* (1971).

B. Calcitonin

Calcitonin is the hypocalcemic factor released from the ultimobranchial glands. These glands are distinct organs in most vertebrates, but in most mammals they eventually become associated with the thyroid gland, where they constitute the parafollicular cells. The cellular source of mammalian calcitonin was long a matter of dispute, and the alternate name "thyrocalcitonin" is now commonly used to refer to calcitonin of thyroid origin. These and other developments, including the evidence that calcitonin plays an important physiological role as an antagonist of parathyroid hormone, have been reviewed recently (Copp, 1969; Hirsch and Munson, 1969). There is now evidence that cyclic AMP may play a role in both the release and action of this hormone.

1. Release of Calcitonin

The normal physiological stimulus for calcitonin release seems to be an increase in the level of blood calcium. There is no evidence that cyclic AMP is required for this, but evidence that it may play a regulatory role came from the observation by Care and his colleagues (Care and Gitelman, 1968; Care *et al.,* 1970) that the dibutyryl derivative of cyclic AMP could stimulate calcitonin release when perfused through the thyroid *in situ.* Calcium was required for this effect, as in the release of insulin and other hormones, and the effect was enhanced by theophylline. These results have since been confirmed using thyroid slices *in vitro* (Bell, 1970).

It was later established that glucagon could also stimulate the release of calcitonin both *in vivo* (Care *et al.,* 1969; Avioli *et al.,* 1969) and *in vitro* (Bell, 1970). In view of the ability of glucagon to stimulate adenyl cyclase in other tissues, this can be taken as additional evidence of a role for cyclic AMP. Synergism between glucagon and theophylline was also demonstrated (Care *et al.,* 1970). Although the high concentrations required suggest that this is not an important physiological effect of glucagon, it can certainly explain part of the hypocalcemia that occurs in response to pharmacological doses of this hormone.

More recently it was shown that β-adrenergic catecholamines also stimulate calcitonin release (Care *et al.,* 1970). Since many other β-adrenergic effects have been shown to be mediated by cyclic AMP, this provides further evidence that cyclic AMP plays a role in the regulation of calcitonin release. Whether or not this represents an important function of the catecholamines remains to be seen.

2. The Action of Calcitonin

Calcitonin acts directly on bone to inhibit the resorption of calcium, and this seems to account satisfactorily for its well-known hypocalcemic effect *in vivo.* Since calcitonin inhibits the calcium mobilizing effect of the dibutyryl derivative of cyclic AMP (Wells and Lloyd, 1969; Klein and Raisz, 1970) as well as of PTH, and especially since the effects of calcitonin and imidazole seemed similar in this regard (Wells and Lloyd, 1968b; 1969), an appealing hypothesis was that calcitonin might act by stimulating phosphodiesterase, thereby reducing the level of cyclic AMP in the same cells affected by PTH. As discussed previously in Section IV, A, 3, this hypothesis was destroyed when it was found that calcitonin did not prevent the increase in cyclic AMP produced by PTH in bone (Chase and Aurbach, 1970). Instead, calcitonin increased the level of cyclic AMP in bone, and this effect and the effect of PTH were found to be additive (Murad *et al.,* 1970a).

These findings raised the possibility that calcitonin might oppose the action of PTH by stimulating adenyl cyclase in cells different from those affected by PTH. Although the cellular site of action of calcitonin is unknown, this general idea is appealing because of the sharply biphasic dose-response curve exhibited by the dibutyryl derivative of cyclic AMP (Raisz and Klein, 1969), suggesting that cyclic AMP does indeed inhibit its own action in bone. It is also possible that the ability of calcitonin to raise the level of cyclic AMP is not important, and that it opposes PTH independently of the level of cyclic AMP, possibly by an action similar to that of oxytocin in the uterus (see Section I,C). It is obvious that some interesting research remains to be done on the mechanism of action of calcitonin.

V. GASTROINTESTINAL HORMONES

A. Gastrin

Gastrin is a heptadecapeptide produced primarily by the pyloric glands in the antral part of the stomach; it is produced to a lesser extent in the duodenum and apparently also in the delta cells of the pancreatic islets (Lomsky *et al.,* 1969). Our impression is that acetylcholine, gastrin, histamine, and possibly other factors such as the prostaglandins may interact in a rather complex fashion in the regulation of gastric secretion. Parasympathetic stimulation or exogenous acetylcholine is capable of stimulating the release of gastrin and apparently also sensitizes parietal cells to the action of gastrin. The role of histamine as a possible mediator of the actions of gastrin or as a factor acting in addition to gastrin is controversial. For an introduction to this controversial field, the reader is referred to the proceedings of a conference edited by M. I. Grossman (1966) and to two shorter reviews by Grossman (1968) and R. J. Levine (1968). An excellent review by Thompson (1969) is also available.

Grossman (1966, 1968) has summarized the physiological actions of gastrin, i.e., those which have been seen in response to endogenously released gastrin, as follows: (1) strong stimulation of acid secretion; (2) weak stimulation of pepsin secretion; (3) weak stimulation of pancreatic flow and bicarbonate output (i.e., a weak secretinlike effect); (4) strong stimulation of pancreatic enzyme output (i.e., a strong pancreozyminlike effect); and (5) weak stimulation of hepatic biliary flow and bicarbonate onput. Whether the ability of gastrin to stimulate the release of insulin (Chapter 7, Sections II, A, 3 and 4) is physiologically important or not remains to be seen. High concentrations of gastrin may actually *inhibit* gastric secretion. The tendency of low concentrations to stimulate and of high concentrations to inhibit this function was responsible for some of the earlier confusion in this field (Grossman, 1966).

From the standpoint of the subject of this monograph, it is of interest that most of the effects of gastrin can be mimicked by exogenous cyclic AMP. More appropriately, in view of the controversial nature of this field, we should say that many of the effects of gastrin superficially *resemble* effects that can be produced by exogenous cyclic AMP. Gastric secretion will be discussed briefly in the following paragraphs and again in Section VI. Stimulation of pancreatic flow and enzyme secretion will be discussed later in connection with secretin and pancreozymin. Pepsin secretion has not to our knowledge been studied in response to cyclic AMP. Hepatic bile flow has been reported to be reduced (Homsher and Cotzias, 1965) or unchanged (Levine and Lewis, 1967) in response to cyclic AMP, in contrast to the increase that occurs in response to gastrin.

Stimulation by exogenous cyclic AMP of acid secretion by frog gastric mucosa *in vitro* was initially reported by Harris and Alonso (1965) and later confirmed by Way and Durbin (1969). However, several detailed studies of substrate and oxygen utilization (Harris and Alonso, 1965; Alonso *et al.*, 1967) have failed to clarify the mechanism by which cyclic AMP acts. Cyclic AMP increases phosphorylase activity (Nigon and Harris, 1968) and glycogenolysis (Alonso *et al.*, 1968) in this preparation, but these effects are probably not related to the stimulation of acid production. The effect of cyclic AMP is relatively specific, since equimolar or even higher concentrations of $5'$-AMP did not affect HCl secretion (Harris and Alonso, 1965; Alonso *et al.*, 1968).

Theophylline by itself may have only a slight effect on gastric secretion in mammals, but markedly increases the response to histamine (Robertson *et al.*, 1950; Mertz, 1969). In the frog preparation, theophylline was found to produce a striking response even in the absence of added histamine (Harris and Alonso, 1965). Synergism between theophylline and exogenous cyclic AMP was not demonstrated in these experiments, but careful dose-response studies were not carried out. It was later found (Harris *et al.*, 1969) that theophylline did increase the level of cyclic AMP in this preparation, and a good correlation between this effect and the effect on secretion was obtained.

Prostaglandin E_1 has been found to inhibit gastric secretion in mammals *in vivo* (Ramwell and Shaw, 1968, 1970; Robert *et al.*, 1968) and also in the frog gastric mucosa *in vitro* (Way and Durbin, 1969). In the latter preparation, PGE_1 abolished the response to gastrin and inhibited the response to histamine, but had no effect on the response to exogenous cyclic AMP. Since PGE_1 inhibits the accumulation of cyclic AMP in response to several hormones in adipose tissue (see Section II, B in Chapter 8 and Section IX in this chapter), this can be taken as evidence suggesting that gastrin may act by stimulating the formation of cyclic AMP in parietal cells.

However, whether gastrin does in fact act by this mechanism is unknown. An effect of gastrin on adenyl cyclase activity or on the level of cyclic AMP in any tissue has not to our knowledge been reported, and the predicted synergism between gastrin and theophylline has apparently not been tested.

Levine *et al.* (1966) found that cyclic AMP or any of a variety of other adenine nucleotides decreased the volume and acidity of gastric secretion when injected into human subjects. Evidence was presented to suggest that this could have been secondary to a reduction of mucosal blood flow.

B. Secretin

Secretin was the first hormone to be called a hormone, having been discovered by Bayliss and Starling in 1902, but to this day very little is known about its mechanism of action. Its most important physiological effect seems to be to stimulate the flow of secretion (primarily water and ions) from the exocrine pancreas, and there is some evidence to suggest that this effect is mediated by cyclic AMP. Thus, the response to secretin is enhanced by theophylline and can be mimicked by exogenous cyclic AMP (Case *et al.*, 1969; Scratcherd and Case, 1969). Additional indirect evidence for this is that secretin, like many other hormones that act by way of cyclic AMP in their other target tissues, stimulates adenyl cyclase in rat adipose tissue (Rodbell *et al.*, 1970).

Although the chemical structure of secretin is remarkably similar to that of glucagon, the two hormones seem to affect adenyl cyclase by way of separate receptors. Rodbell and his colleagues found that trypsin treatment of adipose tissue sufficient to prevent the effect of glucagon did not prevent the response to secretin. In liver preparations, secretin does not stimulate adenyl cyclase at all (Spiegel and Bitensky, 1969) and does not interfere with the effect of glucagon (Rodbell *et al.*, 1970). Analysis of the cardiovascular effects of glucagon and secretin (Ross, 1970) suggested that here also the two hormones probably interacted with separate receptors, at least in some vascular beds.

The ability of secretin to stimulate the release of insulin was mentioned previously (Chapter 7, Sections II, A, 3, and 4), but whether secretin acts directly on the beta cells is unknown at present. The important question of whether certain cells in pancreatic and bile ducts contain adenyl cyclase that responds specifically to secretin remains to be answered.

C. Pancreozymin

Pancreozymin stimulates the release of amylase and other enzymes from the exocrine pancreas. Evidence that this effect is mediated by cyclic AMP includes potentiation by theophylline, which effect was especially pronounced when total protein was measured (Ridderstap and Bonting, 1969). In addition, the response to pancreozymin *in vitro* was mimicked by the application of exogenous cyclic AMP to the pancreas of the mouse (Kulka and Sternlicht, 1968) and rabbit (Ridderstap and Bonting, 1969). By contrast, Case and his colleagues (1969) were unable to see an effect of either dibutyryl cyclic AMP or theophylline on enzyme secretion by the isolated cat pancreas in the presence of secretin. It should be noted that the addition of either secretin or exogenous cyclic AMP is necessary to initiate the flow of pancreatic juice in this latter preparation. It seems possible that the various gastrointestinal hormones may have slightly different roles to play in carnivores than in herbivores or omnivores. Continued study of the possible role of cyclic AMP in the actions of these hormones may provide useful information in this regard.

The ability of pancreozymin to stimulate insulin release was mentioned previously, but, as in the case of secretin, it is not certain that this represents a direct effect on the beta cells. Possible effects of pancreozymin on adenyl cyclase activity or cyclic AMP levels in the pancreas or any other organ have not been carefully studied.

D. Other Gastrointestinal Hormones

A polypeptide which seems to be chemically and biologically similar to pancreatic glucagon has been found in gastrointestinal tissue of several species (Sutherland and De Duve, 1948; Makman and Sutherland, 1964; Valverde *et al.*, 1968). Although the physiological significance of this is unknown, it is of interest that glucagon has been reported to inhibit gastric secretion in doses similar to those which produce hyperglycemia (Clarke *et al.*, 1960). Von Heimburg and Hallenbeck (1964) reported that glucagon could inhibit HCl secretion induced by either meat or exogenous gastrin, but not that induced by histamine. In fact, according to Charbon *et al.* (1963), glucagon may at times *enhance* the effect of histamine. The meaning of these various observations is not presently clear. We will only reiterate here what we said in Chapter 7, that all of the effects of glucagon that have been studied in detail, notably its effects in the liver, heart, and rat adipose tissue, have been found to be mediated by cyclic AMP. Of course there is always room for the exception that proves the rule.

A material which resembles glucagon in the sense that it cross-reacts with glucagon antibodies, but which seems to have approximately twice the molecular weight of glucagon, has also been found in extracts of gastrointestinal tissue (Unger *et al.*, 1968). This material does not affect the liver but does share with glucagon the ability to stimulate the release of insulin. As discussed in Chapter 7 (Sections II, A, 3 and 4), the biological significance of this is still unknown.

There is no evidence of which we are aware to suggest that any of the actions of cholecystokinin are mediated by cyclic AMP.

VI. HISTAMINE

A. Release of Histamine

Most of the histamine in the mammalian body is stored within granules in tissue mast cells or basophil leukocytes. Histamine is released from these cells in response to a great variety of drugs (see standard textbooks) and antigens, and there is now evidence that cyclic AMP acts to inhibit this response.

It is known from the work of Scott (see Table 6.III) that leukocytes contain adenyl cyclase that can be stimulated by catecholamines. Although not tested in Scott's experiments, it can be assumed on the basis of studies with other cells that this effect is mediated by adrenergic β-receptors, and that it would lead, in circulating leukocytes, to an increase in the level of cyclic AMP. In line with these suppositions, it has been found that β-adrenergic catecholamines do inhibit antigen-induced histamine release, that the catecholamines and theophylline act synergistically in this respect, and that the effect of the catecholamines can be reproduced by exogenous cyclic AMP (Lichtenstein and Margolis, 1968; Assem and Schild, 1969). It could be predicted on the basis of Scott's experiments that prostaglandin E_1 would also inhibit antigen-induced histamine release, although this point has not been established.

The release of histamine presumably occurs by exocytosis, and is of special interest because it stands as the only known example of this type of process being inhibited by cyclic AMP. As discussed earlier in Chapter 5 (Section V), all other exocytotic processes studied have been shown to be stimulated or else unaffected by cyclic AMP. Why the release of histamine should be different is unknown. It may be carrying teleonomic reasoning to an extreme, but we cannot resist pointing out that those release processes that are stimulated by cyclic AMP are known to be beneficial, under at least some circumstances, whereas the beneficial effects of histamine release have yet to be demonstrated.

Whether the antigen–antibody reaction on the surface of sensitized basophils leads to a fall in the intracellular level of cyclic AMP has not to our knowledge been investigated. The possible role of cyclic AMP in the immune response, in general, represents an area in which much additional research needs to be done, especially in view of its potential clinical importance. An interesting aspect of this has been discussed by Szentivanyi (1968).

B. The Actions of Histamine

Although many of the known effects of histamine may be mediated by either an increase or a decrease in the level of cyclic AMP, this has not been established in a single instance.

Kakiuchi and Rall (1968a,b) showed that histamine could produce a very striking increase in the level of cyclic AMP in rabbit brain slices, and this was later confirmed by Shimizu *et al.* (1969). This effect was presumably secondary to the stimulation of adenyl cyclase, since it was much greater in the presence of theophylline, although Sattin and Rall (1970) later reported that this depended upon the manner in which the slices were prepared. Although adenyl cyclase in broken cell preparations of most brain areas is resistant to hormonal stimulation, the activity of a rabbit cortex preparation was almost doubled by the addition of 10^{-5} M histamine (G. A. Robison and L. S. Cobb, unpublished observations).

Using slices of rabbit cerebellum, in which norepinephrine was also effective in raising the level of cyclic AMP, Kakiuchi and Rall (1968a) established that histamine and norepinephrine interacted with separate receptors. In the continued presence of either agent, the level of cyclic AMP rose and then fell, and thereafter would not respond to the addition of more of the same agent; however, addition of the other amine could still produce a response (see also Rall and Kakiuchi, 1966). It was also shown that a concentration of diphenhydramine that prevented the response to histamine had only a slight effect against norepinephrine, just as the response to norepinephrine was selectively inhibited by a β-adrenergic blocking agent. Phenoxybenzamine was also relatively specific in blocking the effect of histamine, whereas chlorpromazine inhibited the response to both agents.

It was later shown that high concentrations of potassium ions (Sattin and Rall, 1967), electrical stimulation (Kakiuchi et al., 1969), adenosine or compounds that could be converted to adenosine (Sattin and Rall, 1970), or any of several depolarizing agents (Shimizu et al., 1970c) could all increase the level of cyclic AMP when applied to brain slices in vitro, and the effect of each was found to be synergistic with that of histamine or norepinephrine. Surprisingly, the methylxanthines were found to prevent the response to either electrical stimulation or adenosine, and the reason for this is still not understood (Kakiuchi et al., 1969; Sattin and Rall, 1970).

The biological significance of these observations is obscure, at least partly because the role of histamine in the brain is so poorly understood. It can be noted that in some species, such as the rat, histamine has only a negligible effect on the level of brain cyclic AMP (Palmer et al., 1969; Shimizu et al., 1970c), whereas in most areas of the mature rabbit brain histamine is much more effective than norepinephrine (Kakiuchi and Rall, 1968b). These observations are important not for what they tell us about brain function, but primarily because they show that in at least one type of tissue histamine is capable of altering the level of cyclic AMP.

Shimizu et al. (1970a) compared histamine with a variety of analogs and metabolites with respect to their ability to stimulate cyclic AMP formation in slices of rabbit cerebral cortex. The order of potency of these agents proved to be very similar to their order of potency in stimulating gastric secretion and smooth muscle contraction, suggesting that the receptors for histamine in these tissues may be basically similar. By analogy with what is known about the adrenergic β-receptor, this finding might also point to a role for cyclic AMP in these other tissues. However, it is an old observation that the stimulatory effect of histamine on gastric secretion is poorly blocked by diphenhydramine and other antihistaminic drugs, whereas these drugs do prevent the cyclic AMP response in brain as well as the contraction of smooth muscle. The latter effect would more likely involve a decrease than an increase in the level of cyclic AMP.

Potentiation by methylxanthines, observed many years ago by Robertson et al. (1950), provides additional suggestive evidence that the stimulation of gastric secretion by histamine may be mediated by cyclic AMP. This is further supported by the demonstration that exogenous cyclic AMP stimulates HCl secretion by the isolated gastric mucosa of the frog (Harris and Alonso, 1965; Way and Durbin, 1969). However, the most important evidence, that histamine increases adenyl cyclase activity or the level of cyclic AMP in parietal cells (assuming that they represent its site of action), has yet to be presented. It is not possible at present to offer any simple interpretation of the observations linking histamine, gastrin, acetylcholine, glucagon, and the prostaglandins in the control of gastric secretion, either in terms of cyclic AMP or any other terms. Evidence that histamine may mediate the effects of gastrin has been discussed (Code, 1965; R. J. Levine, 1968) but remains controversial (see, for example, Grossman, 1966, 1968). We can note here that histamine has also been proposed as the mediator of some of the actions of several other hormones (Szego, 1965), but here again the evidence has not been universally convincing.

Another effect of histamine that is enhanced in the presence of theophylline is its positive inotropic effect, observed in isolated perfused hearts of guinea pigs (Pöch and Kukovetz, 1967) and rabbits (Dean, 1968). Histamine seems to interact in these hearts with receptors that are separate and distinct from adrenergic receptors, and the available evidence is compatible with a cyclic AMP-mediated response. As in the case of gastric secretion, however, an effect of histamine on the level of cyclic AMP in the appropriate cells has not been demonstrated. Positive inotropism was discussed previously in Chapters 6 (Section VI, B, 3) and 7 (Section I, C, 1).

Goodman (1968c) found that histamine could inhibit lipolysis in rat adipose tissue in response to either epinephrine or the dibutyryl derivative of cyclic AMP, and interpreted this finding in terms of a possible stimulation of phosphodiesterase activity. Such an effect, leading as it would to reduced levels of cyclic AMP, could also account for histamine's ability to cause smooth muscle to contract. However, as emphasized previously, an effect of histamine on the level of cyclic AMP in any peripheral tissue, including adipose tissue and smooth muscle, has yet to be demonstrated.

Thus the role of cyclic AMP in the actions of histamine remains conjectural. We can only conclude that an understanding of the function and mechanism of action of histamine seems almost as distant now as it did before cyclic AMP was discovered.

VII. SEROTONIN

Serotonin seems to play a role in invertebrates which is similar or perhaps analogous to that played by the catecholamines in higher organisms. At least one of its effects, the stimulation of glycolysis in the liver fluke, seems clearly to be mediated by cyclic AMP. The evidence for this has been reviewed by Mansour (1967), and is briefly summarized in the following paragraphs.

The liver fluke *(Fasciola hepatica)* was actually the first nonmammalian organism in which adenyl cyclase was found. Activity was highest in low-speed particulate fractions, presumably containing cell membrane fragments, and was stimulated by serotonin but not by catecholamines (Mansour *et al.,* 1960). This was in line with the lack of effect of the catecholamines on the function of these organisms.

Phosphofructokinase was later identified as the principal enzyme limiting the rate of glycolysis in these organisms. Both serotonin and cyclic AMP increased the activity of this enzyme when added to liver fluke homogenates, but only cyclic AMP was active in cell-free extracts (Mansour and Mansour, 1962). This effect was later studied in more detail by Stone and Mansour (1967a,b), and was found to involve the conversion of an inactive to an active form, as discussed previously in Chapter 5 (Section III, D). Although not established in this particular case, it seems likely, in the light of present knowledge, that the activating effect of cyclic AMP involves the stimulation of a protein kinase. As pointed out in Section III, B of Chapter 5, cyclic AMP-sensitive protein kinases are now known to be widely distributed throughout the animal kingdom.

Besides stimulating glycolysis, serotonin also stimulates the motility of the liver fluke. While it seems likely that this effect is also mediated by cyclic AMP, this has yet to be established. The possible effect of exogenous cyclic AMP on liver fluke movement has apparently not been studied.

Lysergic acid diethylamide (LSD) and amphetamine produce effects on liver fluke carbohydrate metabolism similar to those produced by serotonin, although neither drug had any effect in broken cell preparations (Mansour *et al.,* 1960; Mansour and Stone, 1970). It is conceivable that further study of the mechanism of action of these drugs in the liver fluke could throw light on the mechanism of their behavioral effects in man.

Two other effects of serotonin that can be mimicked by exogenous cyclic AMP are stimulation of sodium transport in frog skin (Sayoc and Little, 1967; Baba *et al.,* 1967) and salivary secretion in insects (Berridge and Patel, 1968). In the latter case, a striking degree of synergism between serotonin and theophylline (Berridge, 1970) stands as additional evidence that the effect is mediated by cyclic AMP. In neither case, however, has an effect of serotonin on adenyl cyclase or on the level of cyclic AMP been demonstrated.

Phosphorylase activation in response to serotonin in crustacean and molluscan muscles (Bauchan *et al.,* 1968; Rozsa, 1969) is also probably mediated by cyclic AMP, but here again an effect of serotonin on cyclic AMP formation has not been demonstrated. The mechanism by which cyclic AMP stimulates phosphorylase activation was discussed previously in Chapter 5 (see also the addendum to that chapter).

The effects of serotonin in vertebrates have not been carefully studied from the standpoint of the possible role of cyclic AMP. The stimulation of lipolysis in adipose tissue (Bieck *et al.,* 1966; see also Section II, A, 1 in Chapter 8) and of amylase secretion by the parotid gland (Babad *et al.,* 1967) are likely examples of cyclic AMP-mediated responses, but the evidence is incomplete. The only mammalian tissue in which a stimulatory effect of serotonin on cyclic AMP formation has been demonstrated is brain (Kakiuchi and Rall, 1968a; Shimizu *et al.,* 1970c). The small effect of serotonin in all of these systems, relative to the effect of the catecholamines, may reflect the decreasing importance of serotonin in higher organisms.

An interesting study involving serotonin has been reported by Feldman and Lebovitz (1970). They found that serotonin shared the ability of epinephrine to inhibit the release of insulin from the golden hamster pancreas *in vitro,* not only in response to glucose but in response to the dibutyryl derivative of cyclic AMP as well. The similarity between the excitatory effects of serotonin and the catecholamines on mammalian smooth muscle, including a tendency to interact with the same receptors, has frequently been noted (e.g., Innes and Kohli, 1969), and an interesting finding by Feldman and Lebovitz was that phentolamine blocked the inhibitory effect of both amines on insulin release. However, their finding that the effect of serotonin could be selectively antagonized by methysergide suggest that separate receptors may be involved. It seems possible that phentolamine interferes not so much with the hormone–receptor interaction per se as with some later result of this interaction. The results of Feldman and Lebovitz incidentally suggest that adrenergic α-receptors mediate an effect which inhibits not only the formation of cyclic AMP (see Chapter 6, Section III, C), but the action of cyclic AMP as well. It would appear that the same can be said of some of the receptors with which serotonin interacts.

VIII. ACETYLCHOLINE

Evidence that the release of acetylcholine at the neuromuscular junction may be under the regulatory influence of cyclic AMP was presented previously (Chapter 6, Section VI, B, 2). The possibility that cyclic AMP may also be involved in some of the actions of acetylcholine will be briefly discussed in this section.

Choline esters were shown many years ago (Murad *et al.,* 1962) to be capable of inhibiting the accumulation of cyclic AMP in broken cell preparations of dog cardiac muscle. The ability of atropine to antagonize this effect suggested that it was mediated by bona fide muscarinic receptors. A puzzling finding was that carbachol could only inhibit cyclic AMP formation (i.e., apparent adenyl cyclase activity) by 30% at most, either in the presence or absence of stimulation by catecholamines. One possible explanation for this could be that only 30% of the adenyl cyclase present in cardiac muscle is sensitive to acetylcholine, although a more complex explanation, possibly involving an indirect effect of the choline esters, may eventually come to light.

George and his colleagues (1970) reported that acetylcholine produced only an insignificant fall in the level of cyclic AMP in the isolated perfused rat heart, but this has since been found (N. D. Goldberg, personal communication) to be significant and reproducible. A more striking effect noted by these investigators was an increase in the level of cyclic GMP in response to acetylcholine. Whether the fall in cyclic AMP is secondary to the rise in cyclic GMP remains to be determined.

Some of the cardiac effects of acetylcholine could be understood, at least in principle, as the result of reduced cyclic AMP levels. These would include inhibition of phosphorylase activation (Vincent and Ellis, 1963; Haugaard and Hess, 1965) and also the negative inotropic response (Meester and Hardman, 1967; Levy and Zieske, 1969). The ability of acetylcholine to inhibit glycogenolysis without affecting the force of contraction, just as the catecholamines can increase the force of contraction without affecting phosphorylase (Chapter 6, Section VI, B, 3), is definitely compatible with this view, as is the evidence that the choline esters inhibit contractility by interfering with the production of the PIEA (Koch-Weser et al., 1964). However, the degree to which the cardiac effects of acetylcholine are in fact the result of a fall in the level of cyclic AMP (or a rise in the level of cyclic GMP) can only be determined by additional research.

It seemed possible for a time that the contraction of smooth muscle might be another cholinergic response mediated by a fall in the level of cyclic AMP. It was later found, however, that doses of acetycholine which maximally increased the tension of the longitudinal smooth muscle of the guinea pig ileum did not produce a detectable change in the level of cyclic AMP (L. Hurwitz et al., 1967, unpublished). As mentioned previously, oxytocin is also capable of stimulating smooth muscle to contract without changing the level of cyclic AMP.

Shimazu (1967) reported that stimulation of the vagus nerve in rabbits led to a rapid apparent increase in hepatic glycogen synthetase activity, but whether this can be related to a fall in the level of cyclic AMP is unknown. This possibility has not been studied in the rabbit liver, although Murad *et al.* (1962) saw no effect on cyclic AMP accumulation when they added carbachol to dog liver homogenates.

In summary, it has not been established that any of the physiological or pharmacological effects of acetylcholine occur as the result of altered cyclic nucleotide levels. However, the fact that choline esters can inhibit the formation of cyclic AMP and increase the level of cyclic GMP in cardiac tissue suggests that further investigations along these lines may be worth pursuing.

IX. THE PROSTAGLANDINS

The literature concerned with prostaglandins has grown enormously within recent years. Among reviews and monographs published since we started writing the present monograph are those by Bergström (1967), Bergström and Samuelsson (1967), Pickles (1967), Euler (1967), Bergström *et al.* (1968), Ramwell *et al.* (1968), and Ramwell and Shaw (1968, 1970). The interested reader is referred to any or all of these publications for a more comprehensive account.

The prostaglandins are a class of C_{20} fatty acids containing a cyclopentane ring. They are unusual among highly active biological compounds in that they lack nitrogen, in which respect they resemble the steroid hormones. The most widely studied has been prostaglandin E_1 (PGE$_1$), the structural formula of which is shown in Fig. 10.5. The other prostaglandins differ from PGE$_1$ in the number and location of double bonds and hydroxyl and ketone groups.

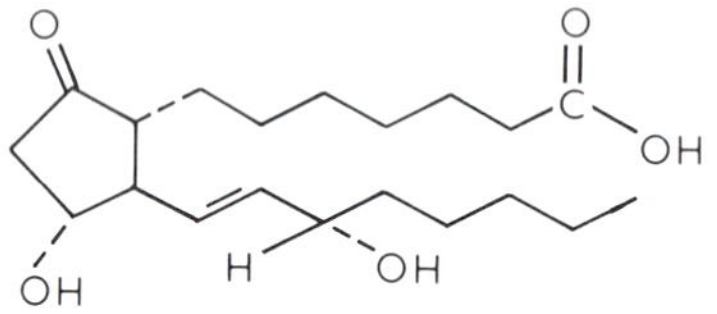

Fig. 10.5. Chemical structure of prostaglandin E_1.

The physiological role of the prostaglandins is presently unclear. Although they have been identified in many tissues and would appear to be almost ubiquitous, as yet they have been found in blood only under the most extreme conditions. In addition, many tissues have the capacity for synthesizing prostaglandins from essential fatty acids. Thus, it seems unlikely that the prostaglandins are part of a classical endocrine-target organ system. However, it is possible that they may serve as local regulators of cell function.

The prostaglandins possess a broad spectrum of pharmacological activities, including (depending upon the prostaglandin used) contraction or relaxation of smooth muscle, effects at all levels of the reproductive system, and a variety of other effects on cell function which either oppose or resemble the effects of tissue-specific hormones. In several cases these actions appear to involve changes in cyclic AMP (Table 10.IV). At present, the prostaglandins and the catecholamines are the only agents known that produce some of their effects by way of an increase in the level of cyclic AMP and others by way of a decrease. In contrast to the catecholamines, however, a relationship between the type of receptor (or the order of potency of agonists) and the direction of the change in cyclic AMP has not been noted.

The first of these changes to have been measured in response to prostaglandins was in adipose tissue. As discussed in more detail in Chapter 8 (Section III, B), low concentrations of active prostaglandins rapidly

TABLE 10.IV. Some Tissues in which Cyclic AMP Accumulation Has Been Shown to Be Affected by Prostaglandins

Tissue	Effect of Prostaglandin on cyclic AMP	References
Fat (adipocytes)	↓ versus epinephrine, ACTH, glucagon, TSH, LH	Butcher and Baird, 1968, 1970
Fat (stromovascular)	↑	Butcher and Baird, 1968
Brown fat slices	↑ alone	Butcher and Baird, 1969
	↓ versus epinephrine	
Platelets	↑ cyclase	Butcher et al., 1967; Wolfe and Shulman, 1969; Marquis et al., 1969; Scott, 1970
Platelets	↑ in intact cells	Robison et al., 1969
Thyroid	↑ cyclase	Kaneko et al., 1969
	↑ in intact cells	Zor et al., 1969b
Ovary	↑ cyclase	Marsh, 1970b
	↑ in intact cells	Kuehl et al., 1970b
Lung	↑ in intact cells	Butcher and Baird, 1968
Spleen	↑ in intact cells	Butcher and Baird, 1968
Diaphragm	↑ in intact cells	Butcher and Baird, 1968
Fetal bone	↑ in intact cells	Chase and Aurbach, 1970
Adenohypophysis	↑ cyclase	Zor et al., 1969c. 1970
	↑ in intact cells	
Heart	↑ cyclase	Sobel and Robison, 1969
Kidney medulla	↑ in slices	Beck et al., 1970

decreased cyclic AMP levels in isolated fat cells which had been stimulated by lipolytic hormones either in the presence or absence of phosphodiesterase inhibitors. Although not established experimentally, it seems likely that a similar effect occurs in other systems, such as toad bladder epithelia (Orloff *et al.*, 1965), kidney tubules (Grantham and Orloff, 1968), and cerebellar Purkinje cells (Hoffer *et al.*, 1969). Efforts to measure such an effect in slices of rat cerebellum have to date been unsuccessful (E. C. Palmer, unpublished observations), but this is not surprising in view of the fact that Purkinje cells constitute less than 1% of the mass of this tissue. In any event, the demonstrated effect in fat cells, together with the finding that prostaglandins were released from adipose tissue in response to hormonal stimulation (Shaw and Ramwell, 1968), led to the suggestion that these substances might function as part of a negative feedback system (Bergström, 1967).

Further investigation revealed that a more common effect of prostaglandins was to increase the level of cyclic AMP (Table 10.IV). This effect was noted at an early date in adipose tissue, although not in isolated fat cells, as discussed previously in Chapter 8.

Another system in which prostaglandins affect cyclic AMP levels is the platelet. Kloeze (1967) had reported that PGE_1 inhibited platelet aggregation in response to ADP and other agents, and, shortly thereafter, stimulation of human platelet adenyl cyclase activity of PGE_1 was reported (Butcher *et al.*, 1967; see also Table 6.III). It was later found (Robison *et al.*, 1969) that PGE_1 increased the levels of cyclic AMP in human platelets at low concentrations while PGE_2 was considerably less active (Fig. 10.6), in line with the order of potency of these agents as inhibitors of platelet aggregation (Weeks *et al.*, 1969). Epinephrine decreased cyclic AMP levels, and this inhibitory effect was blocked by phentolamine (Robison *et al.*, 1969; Salzman and Neri, 1969; Marquis *et al.*, 1970). The stimulatory effects of PGE_1 on platelet adenyl cyclase have been confirmed (Wolfe and Shulman, 1969; Marquis *et al.*, 1969; Vigdahl *et al.*, 1969), and synergism between PGE_1 and theophylline has been demonstrated (Cole *et al.*, 1970). It would thus appear that inhibition of platelet aggregation is one effect of the prostaglandins which is mediated by cyclic AMP.

Other tissues in which effects of the prostaglandins may be mediated by cyclic AMP include thyroid and skeletal tissue, discussed previously in Sections II, A and IV, A, 3, respectively, and the anterior pituitary, discussed in Section X following. More recently Kuehl *et al.* (1970b) have studied the effects of luteinizing hormone (LH) and prostaglandins on cyclic AMP formation and steroidogenesis in the mouse ovary, and have put forward the provocative suggestion that the effects of LH in ovarian

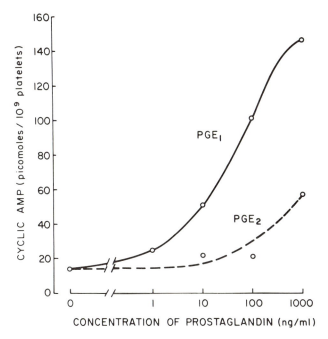

Fig. 10.6. Effect of increasing concentrations of two prostaglandins on the level of cyclic AMP in human blood platelets *in vitro*. Platelet-rich plasma was incubated for 30 minutes in the presence of 2 mM theophylline plus the indicated concentration of prostaglandin for the final 20 minutes. From Robison *et al.* (1969).

tissue may actually be mediated by a prostaglandin. This suggestion was based primarily on the finding that 7-oxa-13-prostynoic acid, which competitively antagonized the effects of PGE_1 and PGE_2 on cyclic AMP formation, appeared to antagonize the effect of LH in a similar manner. Marsh (1970a,b) had shown that both LH and PGE_1 could stimulate corpus luteal adenyl cyclase in broken cell preparations, although this does not per se argue strongly against the idea that the effect of one agent might be mediated by the other. Further studies of the mechanism of action of the prostaglandin antagonist employed by Kuehl and his colleagues should help to resolve this issue. It had previously been shown that in testicular tissue, in contrast to ovarian tissue, neither PGE_1 nor $PGF_{2\alpha}$ shared the ability of LH and FSH to stimulate adenyl cyclase (Kuehl *et al.*, 1970a), although it is conceivable that a different prostaglandin would have been effective.

In those tissues in which prostaglandins raise cyclic AMP levels, they appear to act in much the same manner as many hormones, i.e., by activating adenyl cyclase. While the precise mechanism by which this occurs is unknown, at least the site of action seems clear. By contrast, the mechanism by which prostaglandins decrease cyclic AMP levels, as in fat cells, is still obscure. As in the case of catecholamines and other agents that lower cyclic AMP levels, neither inhibition of adenyl cyclase nor stimulation of phosphodiesterase has unequivocally been demonstrated in a broken cell preparation. However, as discussed previously in Chapter 8, most of the available evidence points to inhibition of adenyl cyclase as the most likely causative factor in intact cells. The evidence presented by Stock *et al.* (1968) suggests that this inhibition may involve interference with the binding of ATP by adenyl cyclase.

One of the more intriguing aspects of the prostaglandins is that they appear to be released from tissues under hormonal stimulation. This was mentioned previously in connection with adipose tissue. Since agents which increased prostaglandin release were known to increase cyclic AMP levels, it seemed possible that cyclic AMP might be mediating the effects of the hormones on prostaglandin release. To date, however, a stimulating effect of exogenous cyclic AMP on the release of prostaglandins from any tissue has not been established (Ramwell and Shaw, 1970).

In considering the possible role of the prostaglandins in relation to cyclic AMP and the hormones that alter cyclic AMP levels, it should be borne in mind that the kinds of prostaglandins synthesized and released by tissues may be different from the exogenous prostaglandins used thus far. In addition, only in the case of fat cells and platelets have experiments with prostaglandins and cyclic AMP been carried out with anything like homogeneous cell populations. Since the prostaglandins are capable of interacting with such a variety of cell types, and since they may either increase or decrease the level of cyclic AMP, reservations must be maintained with regard to cyclic AMP measurements in mixed cell populations.

X. HYPOTHALAMIC RELEASING HORMONES

A group of hormones are released from the hypothalamus to either stimulate or inhibit the release of hormones from the anterior pituitary gland (McCann and Porter, 1969). One of these hormones, thyrotropin releasing hormone (TRH), which seems to act selectively on pituitary thyrotrophic cells to stimulate the release of TSH, has recently been identified as a tripeptide composed of proline, histidine, and pyroglutamate residues (Bowers *et al.,* 1970). Evidence that some of these hypophysiotropic hormones use cyclic AMP as a second messenger is summarized in this section.

Exogenous cyclic AMP or derivatives of cyclic AMP have been shown to stimulate the release of TSH (Bowers *et al.*, 1968b; Cehovic, 1969; Wilber *et al.*, 1969), growth hormone (Gagliardino and Martin, 1968; Wilber *et al.*, 1969; Müller *et al.*, 1969), ACTH (Fleischer *et al.*, 1969), FSH (Jutisz and de la Llosa, 1969), and LH (Ratner, 1970). In most *in vitro* experiments cyclic AMP has been specific among adenine nucleotides, although *in vivo* a variety of nucleotides may be effective when injected intravenously (Levine *et al.*, 1970).

Theophylline also stimulates the release of these hormones, and, where studied, has been shown to act synergistically with hypothalamic hormones or exogenous cyclic AMP (Schofield, 1967; Vernikos-Danellis and Harris, 1968; Bowers *et al.*, 1968b; Fleischer *et al.*, 1969; Wilber *et al.*, 1969; Jutisz and de la Llosa, 1969; Müller *et al.*, 1969; Steiner *et al.*, 1970b; Ratner, 1970). A negative result was reported by Zor *et al.* (1970), who found that while theophylline did act synergistically with a hypothalamic extract to increase the level of cyclic AMP in the rat anterior pituitary *in vitro*, it did not enhance the effect of this extract on LH release. Possibly the high concentration of theophylline used in these experiments (10 mM) was interfering with LH release at a step subsequent to cyclic AMP formation. Using a similar preparation, Ratner (1970) observed a rather striking degree of synergism between theophylline and the dibutyryl derivative of cyclic AMP on LH release.

Factors which have been shown to increase the level of cyclic AMP in the anterior pituitary, and which also stimulate the release of one or more pituitary hormones, include theophylline (Bowers *et al.*, 1968b; Fleischer *et al.*, 1969; Steiner *et al.*, 1970b; Zor *et al.*, 1970), TRH (Bowers *et al.*, 1968b), vasopressin (Fleischer *et al.*, 1969), hypothalamic extracts (Zor *et al.*, 1969c; Steiner *et al.*, 1970b), and prostaglandins (Zor *et al.*, 1970). Theophylline presumably raises the levels of cyclic AMP by inhibiting phosphodiesterase (Vernikos-Danellis and Harris, 1968) whereas the hormones act by stimulating adenyl cyclase (Zor *et al.*, 1969c; Steiner *et al.*, 1970b).

Although the effects of these various agents on cyclic AMP levels correlate qualitatively with their effects on pituitary function, quantitative correlations have been impossible to obtain. For example, theophylline acts synergistically with TRH to stimulate the release of TSH, but the effects of these agents on cyclic AMP levels in the anterior pituitary are less than additive (C. Y. Bowers and G. A. Robison, unpublished observations). This particular discrepancy would be understandable if theophylline inhibits phosphodiesterase in all pituitary cells and if TRH lowers the level of cyclic AMP in some cells while raising the level in thyrotrophs. Given the complexity of the anterior pituitary gland, containing as it does at least five and probably more than five different types of hormone-producing cells (McCann and Porter, 1969; Nakane, 1970), a high degree of correlation between the overall level of cyclic AMP and the rate of release of one or even all pituitary hormones would be surprising, to say the least.

It seems reasonable to suppose, in light of the data summarized to this point, that ACTH, GH, TSH, LH, and FSH are released primarily from different cells, each of which possesses adenyl cyclase sensitive to a different hypothalamic hormone. The level of cyclic AMP is increased in the appropriate cells in response to the appropriate hormone, and this in turn leads to hormone release, presumably by exocytosis (McCann and Porter, 1969). In line with the latter supposition, it has been shown that the removal of calcium does not prevent the rise in cyclic AMP in response to hypothalamic extracts, but does prevent the release of pituitary hormones (Steiner et al., 1970b; Zor et al., 1970; Jutisz and de la Llosa, 1970). However reasonable this general hypothesis may seem, it is not likely to become really testable until methods become available for studying the various pituitary cells in isolation. The utility of this approach is exemplified by the work of Knopp et al. (1970) on the role of cyclic AMP in thyroid cells.

Further studies of pituitary function will include, among other fascinating aspects, the possible role of cyclic AMP in feedback inhibition by other hormones. Adrenalectomy has already been shown to increase the level of cyclic AMP in the pituitary (Fleischer et al., 1969), and thyroidectomy and castration have similar effects (C. Y. Bowers and G. A. Robison, unpublished observations). Injection of triiodothyronine (T_3) reduces the level of cyclic AMP in the pituitarys of thyroidectomized rats, and prevents the response to TRH. Little is known about the mechanism of this action of T_3 except that it appears to involve protein synthesis (Bowers et al., 1968a). Dexamethasone was shown by Fleischer et al. (1969) to be capable of inhibiting the effect of dibutyryl cyclic AMP on ACTH release. It seems possible, therefore, that steroid hormones and T_3 may affect not only the formation and metabolism of cyclic AMP in pituitary cells but the action of the nucleotide as well. Studies of the control of the synthesis of cyclic AMP-binding proteins and protein kinase activity will be of great interest in this regard.

The possible role of cyclic AMP in prolactin release, which seems to be regulated not by a releasing hormone but only by an inhibitory factor (McCann and Porter, 1969), has yet to be investigated. Another question in need of study has to do with the possible role of cyclic nucleotides in regulating the synthesis, as opposed to the release, of pituitary hormones.

XI. THYROID HORMONE

Very little is presently known about how thyroxine or triiodothyronine (T_3) act to produce their often profound effects on cell function. Many of the physiologically important effects of these hormones seem clearly to involve alterations in the rate or pattern of protein synthesis (Tata, 1967), but whether these effects result from a single primary action or from many separate actions remains to be determined. At least some and possibly many of the effects of these hormones may involve cyclic AMP in one way or another, although this has not been unequivocally established in a single instance. Studies carried out with this possibility in mind are summarized briefly in this section.

As discussed previously in Chapters 6 (Section VII, A) and 8 (Section V, C), the thyroid is permissive for lipolytic hormones that act through cyclic AMP. The ability of high concentrations of T_3 to inhibit phosphodiesterase, reported initially by Mandel and Kuehl (1967), may account for the rapid enhancement of lipolysis that can be produced *in vitro* (Vaughan, 1967), but probably does not account for the physiologically important influence of the thyroid *in vivo*. Krishna *et al.* (1968b) presented evidence to suggest that thyroxine enhances lipolysis by increasing the amount of adenyl cyclase in adipose tissue. Although such an effect could account for the known influence of thyroid hormones in adipose tissue, it does not seem to occur in all tissues. For example, destruction of the thyroid gland in newborn rats prevents normal brain development but has no apparent effect on the amount of brain adenyl cyclase or on the development of sensitivity to norepinephrine (M. J. Schmidt and G. A. Robison, unpublished observations). The effect of norepinephrine on cyclic AMP in brain slices from these hypothyroid rats was the same as shown for normal rats in Fig. 6.12.

The effect of the thyroid on pituitary function seems in a sense the opposite of its effect on adipose tissue. Thus T_3 prevents the release of TSH in response to TRH, and there is evidence that this effect of T_3 involves the stimulation of protein synthesis (Bowers *et al.*, 1968a). As mentioned in the preceding section, we now know that thyroidectomy leads to an increase in the level of cyclic AMP in the rat anterior pituitary gland and that treatment with T_3 reduces this toward normal. Pretreatment with T_3 also leads to an effect which inhibits the rise in cyclic AMP normally produced by TRH (Bowers *et al.*, 1968b). The mechanism of action of T_3 in the pituitary is in need of further study, but it seems obvious that it cannot be understood in terms of an increase in the amount of adenyl cyclase.

The apparent similarities between the cardiac affects of hyperthyroidism and sympathetic stimulation have often been noted, and it was therefore of interest when Levey and Epstein (1969b) reported that T_3 and thyroxine could stimulate adenyl cyclase activity in preparations of cardiac muscle. This adenyl cyclase was apparently separate from the adenyl cyclase affected by epinephrine and glucagon. However, the effects of other analogs of thyroxine on adenyl cyclase correlated poorly with their effects on contractility, and it was later shown (McNeill *et al.*, 1969b; Frazer *et al.*, 1969) that thyroid hormones had no apparent effect on the level of cyclic AMP in the isolated perfused rat heart. Adenyl cyclase activity in cardiac preparations from hyperthyroid cats was found by Sobel *et al.* (1969b) to be the same as from normal cats, either in the presence or absence of epinephrine, whereas Levey *et al.* (1969) found reduced activity in hearts from hypothyroid cats, when measured in the presence of maximal concentrations of norepinephrine. The physiological significance of these observations, if any, is still obscure.

Casillas and Hoskins (1970) reported that T_3 and thyroxine could stimulate adenyl cyclase activity in monkey spermatozoa. This effect was biphasic, in line with the effect of these agents on spermatozoal fructose metabolism. Whether cyclic AMP affects phosphofructokinase in these cells as it is known to do in the liver fluke (Chapter 5, Section III, D) remains to be seen.

In summary, thyroid hormones are capable of influencing the formation or metabolism of cyclic AMP under certain conditions, but the role of cyclic AMP in the physiologically important effects of these hormones has yet to be established.

XII. STEROID HORMONES

Cyclic AMP may be involved in the effects of steroid hormones in several different ways. The permissive roles of glucocorticoids in hepatic carbohydrate metabolism (Friedmann *et al.*, 1967; Schaeffer *et al.*, 1969), of aldosterone in sodium transport (D. P. B. Goodman *et al.*, 1969), and of vitamin D in calcium transport (Harrison and Harrison, 1970) were mentioned previously. There is evidence in each of these cases that the steroid may enhance the sensitivity of the responding system to the action of cyclic AMP. A similar mechanism may be involved in the stimulation of thymocyte proliferation by glucocorticoids (Whitfield *et al.*, 1970b).

Although general principles concerning the mechanism of action of the steroid hormones have not been established, many of their effects are known to involve the stimulation of protein synthesis (O'Malley *et al.*, 1969). At least one of these effects, the stimulation of tyrosine transaminase induction in rat liver (Wicks, 1968, 1969) and hepatoma cells (Granner *et al.*, 1968), appears not to require cyclic AMP. This was discussed previously in Chapter 7 (Section II, B, 4).

Besides affecting the action of cyclic AMP, steroid hormones may also influence the formation of cyclic AMP. An especially interesting study in this regard has been reported by Braun and Hechter (1970). They found that adrenalectomy reduced the response of adenyl cyclase in rat fat cell membranes to ACTH, without affecting basal activity or the response to epinephrine, glucagon, or fluoride. Pretreatment with glucocorticoids, but not mineralocorticoids, restored the sensitivity to ACTH, and this restoration could be blocked by either actinomycin D or cycloheximide. It would appear that in rat fat cells the biosynthesis of a component required for the effect of ACTH — possibly an ACTH receptor or a factor coupling the receptor to adenyl cyclase — is induced by glucocorticoids at the level of gene regulation. It is of interest in this regard that the fat cells of some species (e.g., dogs and humans) have not been shown to respond to ACTH under any conditions (Carlson *et al.*, 1970).

An important study of the effects of steroid hormones on hepatic adenyl cyclase has been reported by Bitensky *et al.* (1970). In these experiments, preparations from female weanling rats were more responsive to epinephrine than were preparations from corresponding males. Pretreatment for 9 days with testosterone (or with ACTH or a synthetic glucocorticoid) reduced the response to epinephrine, and of course this was especially striking in females. By contrast, adenyl cyclase activity in the presence of glucagon was hardly affected by steroids, and progesterone and diethylstilbestrol had little or no effect even on the response to epinephrine. Bitensky and his colleagues also showed that adrenalectomy increased the response to epinephrine but not to glucagon, and that this could be reversed by glucocorticoids but not by ACTH. In line with these results, we had previously noted that while adrenalectomy did not alter the ability of glucagon to increase the level of cyclic AMP in the isolated perfused rat liver, it increased the response to epinephrine (J. H. Exton *et al.*, 1969, unpublished). Thus the effect of glucocorticoids on the hepatic response to epinephrine seems to be the result of at least two opposing effects: an increased sensitivity to the action of cyclic AMP (Friedmann *et al.*, 1967) but a decreased ability of epinephrine to stimulate the formation of cyclic AMP. The physiological significance of this is certainly far from clear. An additional observation by Bitensky and his colleagues was that the magnitude of the response to either glucagon or epinephrine decreased with age, although here again the response to epinephrine was more affected than that to glucagon.

In a related series of experiments, Weiss and Crayton (1970) showed that pineal adenyl cyclase activity in rats varied with the estrous cycle. Norepinephrine significantly increased activity at all stages except during proestrus, when estrogen levels are highest, and treatment of ovariectomized rats with estrogen reduced the response to norepinephrine. Since activity in the presence of NaF was also reduced by estrogen treatment, Weiss and Clayton suggested that the steroid might act to decrease the total amount of adenyl cyclase in the pineal. Testosterone had no effect. Whether the action of estrogen in the pineal is basically the same as that of testosterone in the liver remains to be determined. If so, and if the effect is to reduce the rate of synthesis of adenyl cyclase, it would support the earlier suggestion by Bitensky *et al.* (1968) that there are two adenyl cyclase systems in liver, one sensitive to glucagon and the other sensitive to epinephrine. On this basis, only the epinephrine-sensitive system would appear to be controlled by steroid hormones. In the absence of further data, however, other interpretations are possible.

Cyclic AMP may be involved in quite a different manner in some of the other effects of gonadotrophic steroids. Szego and Davis (1967) found that the intravenous injection of estradiol in rats led to a rapid transient increase in the uterine level of cyclic AMP. This could be prevented by pretreatment with the β-adrenergic blocking agent propranolol (Szego and Davis, 1969a) or glucocorticoids (Szego and Davis, 1969b). Rosenfeld and O'Malley (1970) later reported a transient increase in uterine adenyl cyclase activity in response to intravenous diethylstilbestrol, although this agent was ineffective when added to the homogenate *in vitro*. The effect of intravenous estrogen on uterine cyclic AMP and blockade of this effect by propranolol were confirmed by us (J. W. Dobbs and G. A. Robison, unpublished observations). We had previously shown that catecholamines could increase the level of cyclic AMP in the rat uterus *in vitro* and that this could be prevented by propranolol (Dobbs and Robison, 1968). These various observations suggest that estradiol could affect the level of cyclic AMP in the uterus by an indirect mechanism, possibly involving stimulation of adenyl cyclase by catecholamines. However, we later found (unpublished observations) that the intraperitoneal injection of a larger dose of estradiol in immature rats led to a slower and more sustained rise in the level of uterine cyclic AMP, and this was *not* prevented by propranolol even in very high doses. Thus the mechanism or mechanisms by which estradiol acts to alter the level of uterine cyclic AMP *in vivo* are unclear at present. Another unexplained observation, reported by Rosenfeld and O'Malley (1970), was that the subcutaneous injection of progesterone led to an increase in adenyl cyclase activity measured in chick oviduct homogenates. This effect took several hours to develop, reaching a peak 24 hours after injection.

A separate line of evidence implicating cyclic AMP in the actions of gonadotrophic steroids stems from studies of the effects of exogenous nucleotides. Hechter *et al.* (1967) found that the application of exogenous cyclic AMP to the rat uterus *in vitro* could mimic many of the anabolic effects characteristically produced by estrogen treatment. The results of this study were not compelling as evidence for a role of cyclic AMP because almost every other nucleotide tested produced a similar effect. However, Griffin and Szego (1968) and, later, Sharma and Talwar (1970) found that the dibutyryl derivative of cyclic AMP was much more potent and acted more rapidly than other nucleotides in stimulating uterine amino acid uptake. Singhal *et al.* (1970) reported that the injection of cyclic AMP in the presence of theophylline could reproduce some of the seminal vesicular effects of testosterone in castrate or immature rats, although in these studies nucleotide specificity was not established. The andromimetic effects of cyclic AMP on organ weight and enzyme activities could be reduced or abolished by pretreatment with either actinomycin D or cycloheximide.

Adachi and Kano (1970) have studied the effects of the gonadal hormones on cyclic AMP accumulation in washed particulate preparations of human hair follicles. The addition of testosterone had no effect, but dihydrotestosterone, which is thought to be the active androgen in at least some target tissues, markedly reduced apparent adenyl cyclase activity. Conversely, estrone (but not estradiol) was found to increase activity. Progesterone appeared to have no effect in either direction. Activity in all of these experiments was measured in the presence of NaF, and, since the incubations were carried out for 1 hour in the absence of an ATP-regenerating system, the results must be interpreted with caution. For example, the steroids may have been affecting ATPase activity rather than adenyl cyclase. Even if this were the case, however, the hormonal specificity shown in these *in vitro* experiments would still be of interest.

In summary, several observations have been reported to suggest that cyclic AMP may be involved in at least some of the actions of the steroid hormones. The role of cyclic AMP may differ considerably in different tissues and in response to different steroids. Further studies along these lines may lead to a better insight into the mechanism of action of these hormones.

XIII. GROWTH HORMONE

The effects of growth hormone seem very complex and are not well understood. Even in adipose tissue the hormone may produce a variety of effects, not all of which seem understandable in terms of a single primary action (Goodman, 1968b).

One effect which does seem to involve cyclic AMP is the stimulation of lipolysis that occurs when rat fat cells are incubated in the presence of growth hormone and a glucocorticoid (Fain *et al.*, 1966). This effect is enhanced by theophylline (Fain, 1968a,b), and Moskowitz and Fain (1970) later showed that it was associated with an increased rate of accumulation of cyclic AMP. Since the effect on either cyclic AMP or lipolysis required a lag period of at least an hour, and could be prevented by cycloheximide or puromycin (unlike the effects of the other lipolytic hormones discussed in Chapter 8), it seems possible that the effect involves an increased rate of synthesis of a protein component of the adenyl cyclase system.

Pretreatment of hypophysectomized rats with growth hormone in the absence of glucocorticoids did not alter the basal rate of glycerol production or the response to epinephrine when fat pads were incubated *in vitro*, but did significantly increase the response to theophylline (Goodman, 1969c). This apparent effect of growth hormone differs from that produced in the presence of steroids in that it was not prevented by inhibitors of protein synthesis (Goodman, 1968a). The explanation for this is not obvious at present, and is mentioned primarily to emphasize the complexity of research in this area. This complexity may decrease somewhat when more highly purified preparations of growth hormone become available.

The possibility that cyclic AMP may regulate the release of growth hormone was mentioned previously in Section X. The possibility that cyclic AMP may be involved in the physiologically important effects of this hormone on growth and development has not to our knowledge been studied.

XIV. SOME BACTERIAL AND PLANT PRODUCTS

Although they cannot be classified as hormones, some additional factors can be mentioned here because their effects resemble or alter those produced by hormones. An especially interesting example is cholera toxin. Schafer *et al.* (1970) found that the intraluminal administration of cholera toxin to dogs led, after a lag period of an hour or so, to a pronounced increase in the level of cyclic AMP in the intestinal mucosa. Sharp and Hynie (1971) showed that this effect of cholera toxin was secondary to a long-lasting increase in the activity of adenyl cyclase. Maximal activity in the presence of fluoride in epithelial homogenates from rabbit intestinal loops was not changed, but basal activity was much greater after exposure to cholera toxin. If adenyl cyclase normally exists within the cell membrane in a constrained or inhibited state, as suggested by Schramm and Naim (1970) and Perkins and Moore (1971), then perhaps the action of cholera toxin is to destroy the inhibitor. Whatever the mechanism, the effect itself seems to account satisfactorily for the excessive fluid and electrolyte loss characteristic of cholera.

Another bacterial product which affects adenyl cyclase activity, but by an obviously different mechanism, was obtained from extracts of *Clostridia perfringens*. Rosen and Rosen (1970) found that this factor, which they characterized as nondialyzable but relatively heat stable, could stimulate adenyl cyclase activity when added to broken cell preparations of frog erythrocytes, and magnified the effect of isoproterenol. By contrast, preparations of tadpole erythrocytes or rabbit cerebral cortex, which are not affected by β-adrenergic catecholamines, were likewise unaffected by the clostridial activator. Whether this factor was responsible for the TSH-like activity noted previously by Macchia *et al.* (1967; see also Section II, A of this chapter) remains to be seen.

Phosphorylase activation in response to piromin, a *Pseudomonas* polysaccharide, was mentioned previously in Chapter 5 (Section III,A,7), but whether this agent affects cyclic AMP has yet to be determined.

Smith *et al.* (1971a) have demonstrated an increase in the level of cyclic AMP in human lymphocytes in response to phytohemagglutinin (PHA), a glycoprotein present in red kidney beans. This factor has a remarkable capacity to stimulate lymphocyte transformation, an effect which is preceded by the rise in cyclic AMP and which superficially resembles the effect on rat thymocytes produced by low concentrations of exogenous cyclic AMP (MacManus and Whitfield, 1969). On the other hand, Novogrodsky and Katchalski (1970) were unable to see an effect of PHA on adenyl cyclase or cyclic AMP in rat lymph node lymphocytes, even though PGE_1 was effective, as noted also by Smith and his colleagues. It seems possible that different types of lymphocytes contain adenyl cyclase systems sensitive to different factors, but, pending a resolution of this controversy, it does not seem possible at present to define a clear-cut role for cyclic AMP in the regulation of lymphocyte function. Inhibition by cyclic AMP of PHA-induced lymphocyte transformation may be a nonspecific effect of nucleotides in general, as discussed in Chapter 5 (Section II), although the greater potency of the dibutyryl derivative (Johnson and Abell, 1970; Smith *et al.,* 1971b) raises the possibility that cyclic AMP may produce different effects depending on its intracellular level. Although isoproterenol inhibits the effect of PHA on lymphocyte transformation (Abell *et al.,* 1970), it acts synergistically with PHA to raise the level of cyclic AMP (Smith *et al.,* 1971a). It should be noted, however, that the effect of PHA on cyclic AMP levels is biphasic: the early rise was followed, several hours later, by a fall below control levels.

An interesting recent discussion of the role of cyclic AMP in lymphocytes has been presented by MacManus *et al.* (1971). Apparently the rat thymus gland contains two distinct subpopulations of lymphocytes. The larger of these (80–90% of the total cell population) is composed of small lymphocytes in a resting state; these cells do not enter a growth division cycle unless stimulated to do so by agents such as PHA. The other subpopulation consists of larger mitotically active cells which give rise to the small lymphocytes during a series of rapid cell divisions. Cyclic AMP appears to stimulate the rate at which these cells progress from the G_1 to the S phase, during which time DNA synthesis takes place. The cells then pass through the G_2 phase into mitosis, although cyclic AMP appears to have no direct effect on these latter steps. It would appear that there must be a barrier present during the G_1 phase tending to prevent the cells from entering the DNA synthetic phase. As pointed out by MacManus *et al.,* it now seems possible or even likely that this barrier is the presence of unphosphorylated histones. The phosphorylation of histones in response to cyclic AMP was discussed previously in Chapters 5 (Section III,B) and 7 (Section II,B,3).

XV. SUMMARY

Vasopressin, TSH, MSH, parathyroid hormone, serotonin, prostaglandins, and several of the hypothalamic releasing hormones produce at least some of their effects by stimulating adenyl cyclase. These agents are in this sense similar to glucagon and the catecholamines, discussed in Chapters 6 and 7, and the lipolytic and steroidogenic hormones discussed in Chapters 8 and 9.

Calcitonin, secretin, and histamine are also capable of stimulating adenyl cyclase or increasing the level of cyclic AMP in some tissues, but whether this can account for any of the physiologically important effects of these agents remains to be seen.

Melatonin and acetylcholine are capable of inhibiting the accumulation of cyclic AMP in some systems. At least one of the effects of melatonin, the antagonism of MSH in skin melanophores, seems to depend on this action. It is conceivable that some of the effects of acetylcholine are also the result of a fall in the level of cyclic AMP.

Cyclic AMP may also be involved in some of the physiologically important effects of thyroid hormone, steroid hormones, and growth hormone, although in a more complex manner than in the case of hormones that act by stimulating adenyl cyclase. The mechanism of action of these hormones is poorly understood.

The prostaglandins resemble the catecholamines in the sense that some of their effects are mediated by an increase in the level of cyclic AMP while others are mediated by a decrease.

CHAPTER 11

Other Cyclic Nucleotides *

I. INTRODUCTION

The recognition that cyclic AMP is involved in many hormone actions has stimulated interest in other cyclic 3',5'-nucleotides. Dr. Posternak has discussed their chemical preparation and properties in another chapter. Compared with the current state of knowledge about the biological role of cyclic AMP, little is known about other cyclic nucleotides; indeed as of this writing, only one, cyclic GMP, is known to occur in

*By Dr. Joel G. Hardman, Department of Physiology, Vanderbilt University School of Medicine.

400

nature and its functional significance has yet to be elucidated. Cyclic GMP was first found in a natural material by Price and his colleagues (Ashman *et al.*, 1963; Price *et al.*, 1967). These workers injected inorganic ^{32}P into rats and subsequently identified radioactive cyclic GMP in the urine by chromatographic and autoradiographic techniques. Attempts to identify other analogous compounds in nature have been limited by the lack of sensitive methods for their detection. Methods have been developed which would have permitted the detection of cyclic UMP had it been present in urine in amounts as low as 1% of the cyclic AMP present, but such levels were not detected (Hardman *et al.*, 1966; Price *et al.*, 1967). A. L. Steiner *et al.* (1970a) using sensitive radioimmunoassay techniques, were unable to detect either cyclic UMP or cyclic IMP in rat liver (the detectable level would have been 10^{-8} moles/kg).

While the emphasis in this chapter will be on cyclic GMP, potentially significant observations concerning other cyclic nucleotides will be mentioned where considered relevant. No attempt has been made here, however, to catalog all published observations concerning all other cyclic nucleotides, particularly those involving the use of these substances as pharmacological agents. Suffice it to say that at this time (Spring, 1970) there is no published evidence to strongly indicate a biological function for any cyclic nucleotide with the exception of cyclic AMP and GMP.

II. METHODS OF DETECTION

Assay methods which have been most extensively applied to the determination of cyclic GMP in biological materials involve a principle first made use of by Breckenridge (1964) in an assay for cyclic AMP. This is the conversion of the cyclic nucleotide to the corresponding 5'-nucleoside triphosphate which is measured by an enzymatic cycling system. Methods based on this principle have been used for the measurement of cyclic GMP in urine (Hardman *et al.*, 1966, 1969; Goldberg *et al.*, 1969b; Broadus *et al.*, 1970a), tissues (Ishikawa *et al.*, 1969; Goldberg *et al.*, 1969b), plasma (Broadus *et al.*, 1970a), and broken cell systems where its enzymatically catalyzed formation from GTP has been studied (Hardman and Sutherland, 1969; Ishikawa *et al.*, 1969). These methods involve (1) the conversion of cyclic GMP to 5'-GMP by cyclic nucleotide phosphodiesterase from beef heart (Butcher and Sutherland, 1962); (2) the conversion of GMP to GDP by a specific ATP:GMP phosphotransferase and ATP (Miech and Parks, 1965); and (3) the conversion of GDP to GTP by pyruvate kinase and phosphoenolpyruvate.

GTP is then determined in a cycling system consisting of a nucleoside triphosphatase (myosin), pyruvate kinase, and phosphoenolpyruvate (Hardman *et al.*, 1966, 1969; Ishikawa *et al.*, 1969) or succinic thiokinase, succinate, coenzyme A, pyruvate kinase, and phosphoenolpyruvate (Goldberg *et al.*, 1969b). The assay of Goldberg *et al.* (1969b), is more sensitive by one to two orders of magnitude than the myosin cycling method and has been successfully applied to very small amounts of tissue. The successful use of the enzymatic cycling methods requires rigorous chromatographic purification of cyclic GMP from samples of natural material because, with the exception of plasma and urine, other guanine nucleotides may be present in an excess of as much as 10,000-fold and thus cause insurmountable assay blanks unless removed.

An assay based on a different principle, the reduction of the rate of breakdown of radioactive cyclic GMP by the presence of nonradioactive cyclic GMP, has been described by Brooker *et al.* (1968), but its application to the determination of cyclic GMP in biological material has not been reported. Walaas *et al.* (1969) reported the use of thin-layer chromatographic separation followed by absorption spectrum and phosphorous content determinations to identify cyclic GMP in rat diaphragm, and Ashman *et al.* (1963) and Price *et al.* (1967) used paper chromatographic and autoradiographic techniques to identify the nucleotide in urine.

Steiner and his colleagues have devised an ingenious approach to the analysis of cyclic nucleotides. They have applied radioimmunoassay techniques to the determination of cyclic AMP (A. L. Steiner *et al.*, 1969), GMP, UMP, and IMP (Steiner *et al.*, 1970a). These techniques are highly sensitive (down to 0.03 picomole of cyclic GMP), apparently quite specific, and have been applied to studies of cyclic GMP levels in tissues.

III. CYCLIC GMP IN TISSUES

Cyclic GMP has been detected in all mammalian tissues studied as well as in several lower phyla (Ishikawa *et al.*, 1969; Goldberg *et al.*, 1969b; Walaas *et al.*, 1969; Steiner *et al.*, 1970). Levels found in various tissues of the rat have generally been between 10^{-8} and 10^{-7} moles/kg wet tissue, usually at least an order of magnitude lower than the cyclic AMP levels (Ishikawa *et al.*, 1969; Goldberg *et al.*, 1969b; Steiner *et al.*, 1970a), but Walaas *et al.* (1969) have reported 6×10^{-6} moles/kg in rat diaphragm after incubation *in vitro* without substrate. Goldberg and his colleagues (1969b) reported higher cyclic GMP levels in rat brain and liver than those found by Ishikawa *et al.* (1969), but variables such as age and sex of the rats and rapidity of tissue removal and fixation have not been thoroughly evaluated and may considerably influence levels of cyclic GMP found in tissues. Cyclic GMP levels in frog skin and muscle were in the same range as the levels in rat tissues, while slightly higher levels were found in earthworms and minnows (Ishikawa *et al.*, 1969). The cricket was a striking exception. Not only were the levels of cyclic GMP in whole crickets very high compared to tissues of the rat ($2-4 \times 10^{-6}$ moles/kg) but they were some two to three times higher than cyclic AMP levels (Ishikawa *et al.*, 1969). Steiner *et al.* (1970) reported cyclic GMP levels in mouse cerebellum which were up to tenfold higher than in other areas of mouse brain. Recently, both cyclic GMP and cyclic AMP have been detected in infertile and fertile hen and trout eggs and in trout and sea urchin sperm and bull semen (J. P. Gray, J. G. Hardman, and E. W. Sutherland, unpublished observations).

Hormonal alterations of cyclic GMP levels in tissues have not been reported, although changes in extracellular fluid levels of cyclic GMP associated with several hormones have been observed in man and rat and are discussed in Section VII of this chapter. Goldberg *et al.* (1969b) found that neither epinephrine nor glucagon changed cyclic GMP levels in rat liver *in situ* under conditions where hepatic cyclic AMP levels were elevated by both hormones. These workers also reported that alloxan diabetic rats with or without insulin treatment exhibited normal hepatic cyclic GMP levels. Walaas *et al.* (1969) saw no effect of insulin on cyclic GMP levels in rat diaphragm *in vitro;* Steiner *et al.* (1970a) saw no effect of ACTH or epinephrine on cyclic GMP levels in adipose tissue or of glucagon on cyclic GMP levels in liver.

A rapid turnover of cyclic GMP in tissue is suggested by the observations that theophylline, an inhibitor of cyclic nucleotide phosphodiesterase activity, elevated cyclic GMP levels in rat small intestine (Ishikawa *et al.*, 1969) and kidney (Goldberg *et al.*, 1969b). The observation that the rat excretes daily in the urine several times the total body content of cyclic GMP (Hardman *et al.*, 1969) further indicates a high rate of turnover of the nucleotide.

Cyclic GMP has been detected in human plasma (Broadus *et al.*, 1970a) where its level approaches 10^{-8} *M* and is usually about one-third to one-half the cyclic AMP level. Studies by Broadus *et al.* (1970a) on the disappearance of injected radioactive cyclic GMP and cyclic AMP have shown that the plasma cyclic nucleotide pool is in a dynamic steady state with a brisk rate of turnover of both nucleotides, that of cyclic GMP being somewhat more rapid. The relationship of plasma levels of the two cyclic nucleotides to their excretion in urine is discussed in Section VII of this chapter.

IV. GUANYL CYCLASE

The existence of a separate and distinct enzyme system which catalyzes the formation of cyclic GMP was suggested by observations (discussed in Section VII of this chapter) that the excretion of cyclic GMP in urine was influenced independently of the excretion of cyclic AMP. Hirata and Hayashi (1966) reported that a partially purified adenyl cyclase from *B. liquifaciens* did not use GTP as a substrate but did use dATP as a substrate for cyclic dAMP. Similar results have more recently been reported by Rosen and Rosen (1969) who used a partially purified adenyl cyclase from frog erythrocytes and by Burke (1970b) who used a thyroid homogenate. The possibility that mammalian adenyl cyclase systems use GTP as a substrate cannot be absolutely ruled out, but it is clear now that cyclic GMP is formed in large part if not entirely by an enzyme system, guanyl cyclase, which is separate and different in a number of respects from adenyl cyclase.

Guanyl cyclase uses GTP as a substrate (Hardman and Sutherland, 1969) and has been identified in all mammalian tissues so far examined. In the rat, its levels appear highest in lung (Hardman and Sutherland, 1969; White and Aurbach, 1969; Schultz *et al.*, 1969) and small intestine (Ishikawa *et al.*, 1969) and its activity is strikingly high in sea urchin sperm where levels are up to 1000-fold higher than in any mammalian tissue so far examined (Gray *et al.*, 1970). Guanyl cyclase, in contrast to adenyl cyclase, is largely soluble in most tissues of the rat (Fig. 11.1) and has been partially purified from lung and liver (Hardman and Sutherland, 1969; White and Aurbach, 1969). However, it appears to be mostly particulate in the rat small intestine (Ishikawa *et al.*, 1969) and sea urchin sperm (Gray *et al.*, 1970), and its activity in homogenates of these tissues is greatly increased by the presence of Triton.

A striking feature of guanyl cyclase is its almost absolute dependence on Mn^{2+}; barely detectable activity was seen when Mg^{2+} or Ca^{2+} was substituted for Mn^{2+} (Hardman and Sutherland, 1969; Ishikawa *et al.*, 1969; White and Aurbach, 1969; Schultz *et al.*, 1969). In preliminary experiments, the activity of guanyl cyclase was increased by Ca^{2+} in the presence of subsaturating concentrations of Mn^{2+} (J. G. Hardman and E. W. Sutherland, unpublished observations).

In contrast to adenyl cyclase, guanyl cyclase in broken cell preparations of all tissues examined is unaffected by fluoride, and neither glucagon, epinephrine, nor insulin have any detectable effect on the enzyme in cell-free systems of liver and heart (Hardman and Sutherland, 1969). The enzyme is inhibited by ATP in physiological amounts (Ishikawa

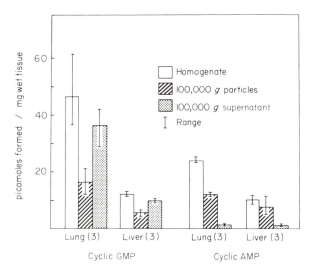

Fig. 11.1. The formation of cyclic GMP and cyclic AMP in whole homogenates and particulate and supernatant fractions of rat lung and liver. Cyclic GMP formation was determined with GTP in the incubation medium and cyclic AMP formation was determined with ATP in the incubation medium. Bars represent the mean and the vertical lines the range of three experiments. Incubation mixtures contained 2 mM ATP or GTP, 10 mM theophylline, 40 mM Tris-Cl (pH 7.4), 3 mM MnCl$_2$, and 5.5 to 11 mg of wet tissue (or the supernatant or particulate fractions from this) per milliliter. Supernatant and particulate fractions were obtained by centrifuging homogenates at 100,000 g for 1 hr. Results are expressed as amount of cyclic nucleotide formed during 10 min at 30°C. (From Hardman and Sutherland, 1969.)

et al., 1969) and by a number of other nucleotides, including all nucleoside triphosphates, oxalacetate, and by phosphoenolpyruvate (Hardman and Sutherland, 1969; White and Aurbach, 1969). Strong inhibition by Hg^{2+}, Zn^{2+}, and Cd^{2+} of guanyl cyclase is prevented by dithiothreitol or reduced glutathione (Hardman and Sutherland, 1969). The enzyme has a broad pH optimum between 7.4 and 8.0 and the apparent K_m for GTP is between 0.1 and 0.3 mM and that for Mn^{2+} is near 0.5 mM (Hardman and Sutherland, 1969; White and Aurbach, 1969; Gray *et al.,* 1970).

Questions involving the formation of pyrophosphate as the other product of the reaction and the reversibility of the guanyl cyclase reaction have not been resolved, partly because the enzyme has not yet been separated completely from pyrophosphatase and GTPase activities.

The inability of the hormones so far tested to alter the activity of guanyl cyclase *in vitro* does not rule out possible effects of the hormones on its activity *in vivo.* However, such a circumstance seems very unlikely in the case of epinephrine and glucagon in the liver, since Goldberg *et al.*

(1969b) found that neither hormone affected cyclic GMP levels in that tissue under conditions where cyclic AMP levels were elevated.

V. HYDROLYSIS OF CYCLIC GMP BY CYCLIC NUCLEOTIDE PHOSPHODIESTERASE (S)

The enzymatically catalyzed hydrolysis of cyclic GMP appears to be analogous to that of cyclic AMP. Cyclic GMP is rapidly hydrolyzed by partially purified cyclic nucleotide phosphodiesterase preparations from several sources and the reaction product has been identified chromatographically as 5'-GMP by Drummond and Perrott-Yee (1961) using rabbit brain extracts and by Beavo et al. (1970a) using the purified phosphodiesterase from beef heart described by Butcher and Sutherland (1962).

When the rates of hydrolysis of cyclic GMP and cyclic AMP have been compared at 10^{-4}–10^{-3} M substrate levels, cyclic GMP has been found to be hydrolyzed at about 70% the rate of cyclic AMP by phosphodiesterase preparations from rat brain (Cheung, 1967) and beef heart (Beavo et al., 1970a). A much slower relative rate of cyclic GMP destruction (about 33% as fast as cyclic AMP) at similar high substrate levels has been reported by Nair (1966) using a partially purified enzyme from dog heart, by Drummond and Perrott-Yee (1961) using rabbit brain extracts, and by Rosen (1970) using a partially purified preparation from frog erythrocytes. Beavo et al. (1970a) have found that the ratio of cyclic AMP hydrolysis to cyclic GMP hydrolysis at 10^{-3} M substrate levels is near 1 and varies little, if at all, among homogenates of all tissues of the rat examined as well as among subcellular fractions of several tissues despite considerable differences in absolute rates of hydrolysis among various tissues. Similar results were obtained in unpublished preliminary experiments carried out in 1966 by Dr. James E. Mitchell while a medical student at Vanderbilt. The studies of Beavo et al. and Mitchell did not confirm the report of Menahan et al. (1969) that the low-speed particulate fraction of rat liver hydrolyzed cyclic GMP about five times as fast as cyclic AMP.

At lower substrate levels, a different picture has emerged. The rate of cyclic GMP hydrolysis was found to be several times faster than cyclic AMP hydrolysis when 10^{-6} M substrate levels were used with the enzyme from beef heart as well as with homogenates or soluble fractions from a number of rat tissues (Beavo et al., 1970a). However, cyclic AMP was hydrolyzed considerably faster than cyclic GMP (at μM levels) in the presence of low-speed particulate fractions of heart and skeletal muscle. This appears to have resulted from the presence in these fractions of a phosphodiesterase differing from that in the soluble fraction in having a much lower apparent K_m for cyclic AMP.

Working with the partially purified beef heart enzyme, Beavo *et al.* (1970a) found the apparent K_m for cyclic GMP to be on the order of 10^{-6} M while that for cyclic AMP was, as previously found by Butcher and Sutherland (1962), in the 10^{-5}–10^{-4} M range (Fig. 11.2). That both nucleotides were being hydrolyzed by a single enzyme in this preparation was indicated by the finding that the nucleotides interfered with the hydrolysis of each other in a manner predictable by their apparent K_m values. As expected, cyclic GMP was a much more effective inhibitor of the hydrolysis of cyclic AMP than vice versa and the apparent K_i for cyclic AMP as an inhibitor of cyclic GMP hydrolysis was the same as its apparent K_m as a substrate. Rosen (1970) and Goren *et al.* (1970) found cyclic GMP to be a competitive inhibitor of the hydrolysis of cyclic AMP by phosphodiesterase preparations from frog erythrocytes. Here also both cyclic nucleotides seem to serve as substrates for the same enzyme. These findings *do not* lead to the presumption that cyclic GMP serves as an inhibitor of cyclic AMP hydrolysis in physiological situations, since, in most mammalian tissues, the two cyclic nucleotides appear to be present in concentrations too far below their respective apparent K_m values to effectively compete with each other as substrates for phosphodiesterase

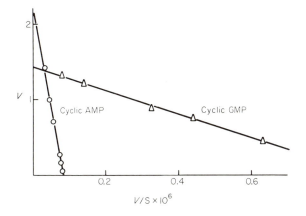

Fig. 11.2. The hydrolysis of cyclic GMP and cyclic AMP by purified cyclic nucleotide phosphodiesterase from beef heart. The slope is the apparent K_m and the intercept on the ordinate is the apparent V_{max} (Hofstee, 1952). Apparent K_m values were 28 and 1.5 μM for cyclic AMP and cyclic GMP, respectively. Units of velocity are μmoles/mg protein per 15 min at 30°C. (From Beavo *et al.*, 1970a.)

(localized higher cyclic nucleotide concentrations of course cannot be ruled out and may even be likely). However, the strong possibility is raised that high concentrations of exogenous cyclic GMP applied to some intact cells may, by competing with the naturally occurring endogenous cyclic AMP for phosphodiesterase, elevate cyclic AMP levels inside the cells (Murad *et al.*, 1970b).

The interrelationships between cyclic GMP and AMP levels and phosphodiesterase activity may, indeed, be far more complicated than one would predict on the basis of the nucleotides competing with each other as substrates for a single enzyme. Although they apparently can serve as substrates for a single enzyme, this is not an adequate explanation for a number of observations. Brooker *et al.* (1968) reported a single apparent K_m of 5×10^{-6} *M* for cyclic GMP using a brain phosphodiesterase preparation which they reported as having two K_m values for cyclic AMP. They suggested that the preparation contained distinct phosphodiesterases for each of the two nucleotides since, at substrate levels near or below their apparent K_m values, the nucleotides did not interfere with the hydrolysis of each other. The possibility that these findings could have resulted from the presence of two or more phosphodiesterases with relative rather than absolute specificities for the two nucleotides does not appear to have been ruled out. O'Dea *et al.* (1970) found cyclic AMP to be a competitive inhibitor of cyclic GMP hydrolysis by a rat brain phosphodiesterase preparation, but they saw no effect of cyclic GMP on cyclic AMP hydrolysis. Beavo *et al.* (1970a, also unpublished) studied the effects of cyclic AMP and GMP on the hydrolysis of each other by a liver phosphodiesterase preparation and found that although cyclic AMP inhibited the hydrolysis of cyclic GMP, cyclic GMP in concentrations up to $50 \mu M$ did not inhibit but, in fact, stimulated by up to two- to threefold the rate of hydrolysis of cyclic AMP at initial μM substrate levels. This effect was demonstrable with initial cyclic GMP concentrations of about 10^{-7} *M*. This observation raises the possibility that alterations in cyclic GMP levels may influence in an inverse way cyclic AMP levels in some tissues.

VI. EFFECTS OF CYCLIC GMP

A. Broken Cell Systems and Partially Purified Enzymes

Where studied in comparison with cyclic AMP, cyclic GMP has been much less potent or has had virtually no effects on enzyme activities measured in broken cell systems. Rall and Sutherland (1962) found cyclic

GMP to be only about 0.2% as potent as cyclic AMP on the activation of glycogen phosphorylase purified from dog liver. Cyclic IMP was the most potent of the other cyclic nucleotides tested, being about 2.5% as potent as cyclic AMP. Walaas *et al.* (1968) found that high concentrations of cyclic GMP and UMP produced the same maximum activation of rat skeletal muscle glycogen synthetase kinase as did cyclic AMP; however, half-maximal activation of the enzyme required, compared to cyclic AMP, some 10 and 100 times as much cyclic GMP and cyclic UMP, respectively. Schlender *et al.* (1969) reported similar findings with respect to the glycogen synthetase kinase activity of a purified protein kinase obtained from rabbit skeletal muscle. The K_a values for cyclic AMP and GMP were 0.07 and 9.9 μM, respectively. Glinsmann and Hern (1969) compared the effects of cyclic AMP with cyclic GMP, IMP, CMP, and UMP on rat liver glycogen synthetase kinase activity. Similar maximum activation was attainable with all five cyclic nucleotides but cyclic AMP produced half-maximum activation at a concentration lower by about one order of magnitude than cyclic IMP and lower by some two orders of magnitude than cyclic GMP, CMP, and UMP.

Corbin and Krebs (1969), working with a cyclic AMP-sensitive protein kinase partially purified from adipose tissue, found that cyclic GMP and UMP had some stimulatory effect on the enzyme but only in concentrations so high that a small contamination with cyclic AMP could not be ruled out. Interestingly, cyclic IMP was found to have appreciable cyclic AMP-like activity, producing half-maximal effects at 4×10^{-7} M (about eight times the cyclic AMP concentration required). Cyclic GMP has been found also to be virtually without effect on the activity of cyclic AMP-dependent protein kinases from heart and skeletal muscle (E. G. Krebs, personal communication) and on a cyclic AMP-sensitive histone phosphokinase from liver (T. A. Langan, personal communication). Cyclic GMP as well as cyclic IMP, CMP, and UMP were ineffective substitutes for cyclic AMP in reactivating ATP-inactivated pyruvate dehydrogenase from pig heart (Wieland and Siess, 1970), a process which seems to involve activation of a protein kinase.

Kuo and Greengard (1969b) compared the effects of cyclic AMP, GMP, IMP, UMP, CMP, and cyclic TMP on protein kinase preparations from several tissues. All the cyclic nucleotides except cyclic TMP could, at high concentrations (50–250 μM), stimulate protein kinase activity about as well as cyclic AMP. However, at lower concentrations only cyclic IMP approached cyclic AMP in potency; cyclic GMP and the other cyclic nucleotides were one to two orders of magnitude less potent than cyclic AMP. An exception to this was seen with a protein kinase preparation from lobster muscle. Here cyclic GMP was as effective as cyclic AMP. Kuo and Greengard (1970) were subsequently able to separate from lobster muscle two fractions of protein kinase activity. One fraction had a much higher affinity for cyclic GMP than for cyclic AMP (approx. K_m values were 0.08 μM and 4 μM, respectively) while the other fraction exhibited a stronger preference for cyclic AMP than for cyclic GMP (approx. K_m values were 0.02 μM and 1.2 μM, respectively). Kuo *et al.* (1970) also have reported the identification of protein kinases with relative specificity for cyclic GMP in some mammalian tissues. The natural substrate or substrates for these cyclic GMP-sensitive protein kinases are unknown. The elucidation of the biological role of cyclic GMP may well depend on their identification.

There is no published evidence to indicate that cyclic GMP in any way alters the action or formation of cyclic AMP in broken cell preparations. Cyclic GMP did not alter the effect of cyclic AMP on liver phosphorylase activation (Murad *et al.*, 1969b). The stimulatory effects of subsaturating levels of cyclic GMP and AMP on skeletal muscle glycogen synthetase kinase activity were additive (Schlender *et al.*, 1969). Cryer *et al.* (1969) found no effect of cyclic GMP on adenyl cyclase activity of fat cell membranes under conditions where GTP, GDP, and GMP all produced apparent inhibition.

The general finding that cyclic GMP is a very poor substitute for cyclic AMP in mammalian broken cell systems plus the generally tenfold lower mammalian tissue levels of cyclic GMP make it seem unlikely that cyclic GMP and AMP share a common regulatory role on the same processes. It may be worthwhile keeping in mind that all of the above cited effects of cyclic AMP were in systems requiring ATP. By analogy, perhaps one should not look at these systems expecting to see pronounced effects of cyclic GMP, but at systems requiring or utilizing GTP (e.g., ribosomal protein synthetic systems; phosphoenolpyruvate carboxykinase; succinic thiokinase).

B. Intact Cell or Organ Systems

In contrast to the relative ineffectiveness of cyclic GMP in mimicking cyclic AMP in cell-free systems, the two nucleotides have been observed to produce strikingly similar effects when applied in high concentrations to some intact cell systems. Cautious interpretation is called for here, since results obtained in such experiments do not of themselves indicate that cyclic GMP can produce such responses in physiological situations. Nevertheless, positive responses to exogenous cyclic GMP must be scrutinized in view of our present state of ignorance of the role of this nucleotide for they may provide leads which may subsequently be tested in other less complicated systems.

Glinsmann *et al.* (1969) studied the effects of cyclic GMP and cyclic AMP on the perfused rat liver and found that the two nucleotides produced essentially indistinguishable responses at equimolar concentrations. Glucose output, glycogenolysis, lactate uptake, urea production, and tyrosine aminotransferase induction were all stimulated, phosphorylase activity was increased, and glycogen synthetase activity (measured in the absence of glucose-6-phosphate) was decreased. Conn and Kipnis (1969) found that, at mM concentrations, cyclic GMP, IMP, and AMP all produced equal stimulation of gluconeogenesis and glycogenolysis in the perfused rat liver. Since the use of other nucleotide concentrations was not reported in this study, the estimation of relative potencies is difficult. J. H. Exton *et al.* (1970, unpublished observations) observed in addition to increased glycogenolysis and gluconeogenesis an increase in K^+ release in response to cyclic GMP by the perfused rat liver. In these experiments, cyclic GMP appeared to be about one-half as potent as cyclic AMP on all parameters examined. An interesting dissociation of the effects of the two nucleotides on the rat liver was observed by Exton *et al.* Insulin antagonized the effect of exogenous cyclic AMP on glucose release but did not alter the effect of exogenous cyclic GMP on this parameter. Levine (1969) reported, in contrast to the aforementioned findings, that cyclic GMP was much less potent than cyclic AMP in increasing glucose output of the perfused rat liver. However, it should be pointed out that a major reason for discrepancies among laboratories regarding apparent relative potencies of cyclic nucleotides in perfused organs may be the stability of cyclic nucleotides in the perfusion media. Rat blood, for example, has such high phosphodiesterase activity (A. E. Broadus, unpublished observations) that the major determinant of apparent cyclic nucleotide potency in an organ perfused with medium containing blood of this species is probably the rate of destruction of the nucleotide in the medium. Washed beef erythrocytes, on the other hand, have no detectable phosphodiesterase activity (A. E. Broadus and J. H. Exton, unpublished observations) and a medium containing them would therefore appear to be much more suitable for studies of cyclic nucleotide effects in perfused organs.

Cyclic GMP has been reported by Glinsmann *et al.* (1969) to be as potent as cyclic AMP in stimulating steroidogenesis in rat adrenal quarters *in vitro*. Mahaffee *et al.* (1970) found the steroidogenic potency of cyclic GMP on rat adrenals *in vitro* to be 20% of that of cyclic AMP while cyclic IMP was equipotent with cyclic AMP.

Another intact cell preparation in which the effects of cyclic GMP resemble at least qualitatively those of cyclic AMP is the isolated fat cell. Murad and co-workers (1970b) found that, under certain conditions (in an all Na$^+$ medium), exogenous cyclic GMP stimulated lipolysis as did cyclic AMP. Interestingly, both nucleotides inhibited hormonally stimulated lipolysis when cells were incubated in Krebs-Ringer phosphate buffer. Braun *et al.* (1969) compared the lipolytic effects of cyclic GMP and AMP on isolated fat cells incubated in Krebs-Ringer phosphate buffer with Ca^{2+}, Mg^{2+}, and K$^+$ omitted and found that, while cyclic GMP did produce lipolysis, it was much less potent in doing so than was cyclic AMP.

The qualitatively and quantitatively similar effects of cyclic AMP and cyclic GMP in perfused livers, adrenal quarters, and isolated fat cells are difficult to interpret in view of the results discussed above which indicated a relative ineffectiveness of cyclic GMP as a substitute for cyclic AMP in broken cell systems. A number of possible explanations for this apparent inconsistency may be considered. First, the relative ineffectiveness of cyclic GMP in cell-free systems may have been due to the fact that the nucleotide was not tested under optimal conditions. This is not an intuitively appealing possibility, but it is one which nevertheless cannot be dismissed. Second, cyclic GMP may have produced its effects on intact cells indirectly through elevation of intracellular cyclic AMP levels. This seems a strong possibility since, as pointed out in Section V of this chapter, cyclic GMP can, under certain conditions *in vitro*, inhibit the destruction of cyclic AMP by phosphodiesterase. Furthermore, Murad *et al.* (1970b) did find a sizeable increase in the cyclic AMP content of fat cells exposed to exogenous cyclic GMP. On the other hand, neither Glinsmann *et al.* (1969) nor J. H. Exton *et al.* (1970, unpublished observations) were able to detect any elevation in the cyclic AMP content of livers perfused with cyclic GMP. Finally, the apparent near equipotency of exogenous cyclic GMP and cyclic AMP in intact cell systems may have resulted fortuitously from a greater ability of cyclic GMP to accumulate within intact cells. This appears to have been the case in the perfused rat liver. Exton *et al.* (1970, unpublished) measured the apparent space in which exogenous cyclic AMP and GMP distributed in the perfused liver and found that the apparent cyclic AMP space was indistinguishable from the sucrose space (about 20% of liver weight) while the apparent cyclic GMP space was much larger than the sucrose space, amounting to 69% of liver weight. The apparent concentration of cyclic GMP in intracellular water then was nearly the same as in the perfusate. Thus, during perfusion with 10^{-5} to 10^{-4} M concentrations, intracellular concentrations of cyclic GMP were attained which were up to 100 times that of the intracellular cyclic AMP. Concentrations of this magnitude would be expected to affect cyclic AMP-sensitive enzymatic processes such as glycogen phosphorylase and synthetase as indicated by the observations discussed in Section VI,A. The apparent equipotency of cyclic GMP in mimicking cyclic AMP in various respects on some intact cell systems may then have no physiological relevance. Furthermore, such artifacts (if they may be called that) could obscure other more meaningful effects of cyclic GMP which might be appreciated only with the application of much lower concentrations of exogenous cyclic GMP. The means by which exogenously applied cyclic GMP is able to attain much higher concentrations than cyclic AMP in the liver is not known and deserves further exploration.

Of perhaps more importance in the search for a physiological role for cyclic GMP are effects of this agent which are unlike those of cyclic AMP. Reports of such effects are currently few in number, however. Bourgoignie *et al.* (1969) made the intriguing observation that cyclic GMP, like cyclic AMP, increased the short circuit current in the isolated toad bladder, but, unlike cyclic AMP, did not affect water permeability. Perhaps unrelated but especially interesting in light of this report is the finding of Goldberg *et al.* (1969b) that the injection of cyclic GMP into the renal artery of the dog produced an abrupt and marked increase in sodium excretion.

Pagliara and Goodman (1969, 1970) reported that cyclic GMP produced metabolic effects in kidney slices resembling those of alkalosis (a decreased production of glucose and ammonia from glutamine and glutamate but not from α-ketoglutarate and an increase in glutamate content) while cyclic AMP and cyclic IMP produced essentially opposite effects resembling those of acidosis. While the effects of cyclic GMP were not reproduced by 5'-GMP, those of cyclic IMP were largely shared by its 5' analog. These authors suggest as a result of their observation that cyclic GMP may impair the conversion of glutamate to α-ketoglutarate, perhaps by inhibiting glutamate dehydrogenase. This is a particularly intriguing suggestion since other guanine nucleotides are known to be quite potent inhibitors of this enzyme (Wolff, 1962).

VII. CYCLIC GMP AND CYCLIC AMP IN URINE

The occurrence of surprisingly large amounts of cyclic AMP in urine was established by Butcher and Sutherland (1962) who found that humans excreted from 2 to 7 μmoles of the nucleotide in 24 hr. Subsequently Price and his colleagues (Ashman *et al.*, 1963; Price *et al.*, 1967) injected inorganic ^{32}P into rats and then examined the urine for radioactive organic phosphate compounds. One substance identified was cyclic AMP. A second was cyclic GMP, and this was the first demonstration of its natural occurrence. These two substances were the only nucleotides and in fact the only identifiable organic phosphates found in rat urine. Price *et al.* (1967) detected about 1 μmole of cyclic GMP in a 24 hr collection of human urine using column and paper chromatographic procedures to isolate the nucleotide. Now it is known that the excretion of the cyclic nucleotides can be profoundly altered by hormonal factors and, furthermore, that such alteration in the excretion of one of the nucleotides occurs independently of the excretion of the other.

In man, both cyclic nucleotides are cleared from the plasma into the urine apparently only by glomerular filtration (Broadus *et al.,* 1969, 1970a). The clearances of administered tritiated cyclic GMP and cyclic AMP were the same as that of inulin and no binding of either nucleotide to plasma protein was detectable (Broadus *et al.,* 1970a). Likewise, the clearance of endogenous cyclic GMP was the same as that of inulin indicating that virtually all of this nucleotide in normal human urine is derived from plasma. In contrast, the clearance of endogenous cyclic AMP was greater than that of inulin. The ratios of the cyclic AMP clearance to inulin clearance in normal individuals ranged from about 1.2 to 2 (Broadus *et al.,* 1970a). Since the clearance of injected tritiated cyclic AMP was the same as that of inulin, the greater clearance of endogenous cyclic AMP must have been due to the direct addition to urine of cyclic AMP from kidney tissue. This was further verified by the finding that the specific activity of the tritiated nucleotide found in urine was diluted relative to that in plasma by the same proportion that the clearance of endogenous cyclic AMP exceeded that of inulin (Broadus *et al.,* 1970a).

At variance with the above findings regarding cyclic GMP excretion is the conclusion reached by Goldberg *et al.* (1969b) that a major fraction of urinary cyclic GMP in the dog comes directly from kidney tissue. This was based on the observation that there was no change in cyclic GMP excretion in dogs during an infusion of 1.8% NaCl solution which produced a marked increase in urine volume and which was presumed, although not demonstrated, to have produced an increase in glomerular filteration rate.

Several observations have been made on cyclic nucleotide excretion in rats in states of hormonal imbalance (Hardman *et al.,* 1966, 1969). Extraordinarily large amounts of cyclic GMP and cyclic AMP are excreted in rat urine. These animals excrete daily up to several times as much of the nucleotides as are contained in the whole body at a given time. A drop in cyclic GMP excretion occurred after hypophysectomy and, to a smaller extent, after adrenalectomy with little or no change in cyclic AMP excretion, and the excretion of both nucleotides was lowered by thyroparathyroidectomy (Hardman *et al.,* 1966, 1969). Hydrocortisone restored cyclic GMP excretion to normal in adrenalectomized rats as did thyroxine in thyroparathyroidectomized rats but a combination of hydrocortisone and thyroxine only partially restored cyclic GMP excretion in hypophysectomized rats. Thus a deficiency of some factor in addition to thyroid and adrenal cortical hormones appears to have been involved in the low cyclic GMP excretion after hypophysectomy. Nearly normal levels of cyclic AMP were excreted by thyroparathyroidectomized rats given thyroxine. The mechanism(s) responsible for the drop in cyclic GMP and cyclic AMP excretion in these hormone deficiencies have not been resolved. Alterations in formation and/or destruction of the nucleotides may be involved in some cases and cyclic GMP in rat blood has recently been found to be significantly lowered by hypophysectomy (W. D. Patterson, unpublished observations). Changes in hemodynamic or renal function also may be likely contributing factors, and the glomerular filteration rate is known to fall in thyroid and adrenal cortical deficiency states; furthermore, rat blood, compared to blood from man and dog, has very high phosphodiesterase activity (A. E. Broadus, unpublished observations).

Chase and Aurbach (1967) found that parathyroid hormone injection caused a striking increase in the excretion of cyclic AMP in the urine of rat and man. These authors suggested that the effect of the hormone was on the kidney since the excretion of injected radioactive cyclic AMP was unaffected. The infusion of calcium resulted in a fall in the excretion of cyclic AMP which was presumed to result from suppression of endogenous parathyroid hormone. Kaminsky *et al.* (1969, 1970a) measured both plasma and urine levels of cyclic AMP and simultaneous inulin clearance in man during parathyroid hormone infusions and demonstrated that the kidney is the principle, if not sole, source of the increased cyclic AMP appearing in urine with the hormone. Increased plasma levels did occur with injection of parathyroid hormone but these were far too small to account for the increments in urine. The studies of Kaminsky *et al.* (1970a) indicate that parathyroid hormone may be responsible for up to one-quarter to one-half of the normal urinary level of cyclic AMP (i.e., some or most of the nephrogenous component of urinary cyclic AMP).

The administration of thyrocalcitonin to humans also resulted in an elevation in the levels of cyclic AMP in plasma and urine, but, in contrast to the changes observed after parathyroid hormone, most or all of the increase in urine levels of the nucleotide could be accounted for by glomerular filtration of the elevated plasma level (Kaminsky *et al.,* 1970c).

Interestingly, Chase *et al.* (1969b) could demonstrate no change in urinary cyclic AMP in response to parathyroid hormone in pseudohypoparathyroid individuals who had abnormally high plasma levels of endogenous parathyroid hormone. The conclusion of Chase and associates that such patients suffer from a defective parathyroid hormone-sensitive renal adenyl cyclase is quite intriguing since it raises the possibility of analogous defects in other endocrinopathies. The studies by Chase *et al.* (1969b) and others by Kaminsky *et al.* (1970a) and Taylor *et al.* (1970) indicate potential diagnostic usefulness of the urine levels of cyclic AMP in patients with parathyroid disorders. Chase *et al.* (1969b) and Taylor *et al.* (1970) reported finding significantly lower than normal cyclic AMP excretion in patients with idiopathic, pseudo-, and surgical hypoparathyroidism. Kaminsky *et al.* (1970a) and Taylor *et al.* (1970) found that hyperparathyroid patients excreted abnormally high amounts of cyclic AMP and that following surgery to remove parathyroid adenoma these levels fell to within the normal range.

Kaminsky *et al.* (1970a) found that cyclic AMP levels in human urine increased many fold in response to doses of parathyroid hormone which produced no change in cyclic GMP levels. However, very large doses of parathyroid hormone did elicit modest increases in cyclic GMP excretion. The infusion of calcium, which decreased the excretion of cyclic AMP, brought about increases in cyclic GMP levels in both plasma and urine of up to two- to threefold (Kaminsky *et al.*, 1970a,b). That the effects of calcium on plasma and urine cyclic GMP levels were not mediated by thyrocalcitonin was indicated by the lack of effect of this hormone on extracellular fluid levels of cyclic GMP when it was administered in doses that did elevate cyclic AMP levels (Kaminsky *et al.*, 1970b).

Another hormone having pronounced effects on cyclic AMP excretion without changing cyclic GMP excretion is glucagon (Fig. 11.3). The subcutaneous injection of this agent into rats increased the daily excretion of cyclic AMP by some sevenfold (Hardman *et al.*, 1969). This effect was pursued in human subjects by Broadus *et al.* (1969, 1970b). Plasma levels of cyclic AMP increased in response to glucagon and the increased amount of the nucleotide appearing in urine could be accounted for entirely by renal clearance of the elevated plasma levels. The most likely source of the elevated plasma levels of cyclic AMP in response to glucagon is the liver, and Broadus *et al.* (1969, 1970b) found the nucleotide to be released from the perfused rat liver in response to glucagon. The changes in plasma and urine levels of cyclic AMP in response to glucagon are striking; up to 40-fold changes have been seen with larger doses of the hormone while detectable changes occurred with as low as 0.750 μg/kg infused intravenously over 30 min. Whether or not glucagon is

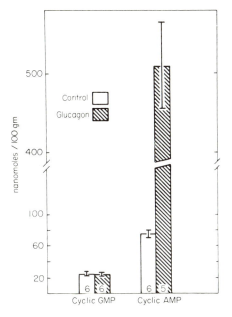

Fig. 11.3. The effect of glucagon on the 24 hr excretion of cyclic AMP and cyclic GMP in the urine of normal female rats. Glucagon (25 μg/100gm) was injected subcutaneously 4 times at 6- to 8-hr intervals during the 24-hr period of urine collection. Results are expressed as amount of nucleotide excreted in 24 hr/100 gm body weight. (From Hardman *et al.*, 1969.)

responsible for appreciable fractions of the normal plasma and urine levels of cyclic AMP cannot be ascertained at present.

The effect of antidiuretic hormone on urinary cyclic AMP has been a subject of some disagreement. Takahashi *et al.* (1966) reported that patients with diabetes insipidus excreted extremely low amounts of cyclic AMP (down to 5 % of normal) and that the administration of vasopressin to such patients (or to normal patients given a water load) caused an increase in the amount of the nucleotide excreted. Davis *et al.* (1969) and Taylor *et al.* (1970) also have reported that vasopressin administration increased cyclic AMP excretion in man. On the other hand, Chase and Aurbach (1967) saw no effect of intravenous vasopressin on cyclic AMP excretion in rats unless supraphysiological doses were used, and Chase *et al.* (1969b) stated that vasopressin injection in normal humans or patients with diabetes insipidus caused no significant increase in cyclic AMP excretion. Similarly, Kaminsky *et al.* (1970c) could detect no effect of antidiuretic doses of vasopressin on cyclic AMP excretion in humans. Furthermore, excretion of the nucleotide was unchanged in states of extreme hydration or dehydration or after the intravenous or oral administration of hypertonic sodium chloride solution in amounts that caused pronounced antidiuresis (N. I. Kaminsky and A. E. Broadus, unpublished observations). While the ability of administered vasopressin to elevate cyclic AMP excretion remains a matter of disagreement, there seems little doubt that the endogenous hormone is responsible for at most a very small fraction of urinary cyclic nucleotides.

Broadus *et al.* (1970b) found that epinephrine infusions elevated plasma and urine cyclic AMP levels. The magnitude of these changes was smaller than seen with glucagon, and amounted to about threefold in plasma with the largest dose tested. Interestingly, the increment in urine was consistently smaller than would be calculated to appear from the increment in plasma. Small but consistent increases in urine levels of cyclic GMP also were detected with epinephrine infusions. The elevation in plasma levels of cyclic AMP seen in man during epinephrine administration was prevented by the β-adrenergic receptor blocking agent propranolol but not by the α-adrenergic receptor blocking agent phentolamine (Ball *et al.*, 1970). On the other hand, the elevation in plasma cyclic GMP levels during epinephrine infusion was not blocked by propranolol but was reversed by phentolamine. The administration of norepinephrine or norepinephrine plus propranolol was more effective than epinephrine in elevating plasma cyclic GMP levels and the relatively pure β-adrenergic agent isoproterenol did not elevate cyclic GMP levels (Ball *et al.*, 1970; Kaminsky *et al.*, 1970c). An important question which is unanswered as of this writing is whether the effect of catecholamines on plasma levels of cyclic GMP is direct or indirect. An indirect effect seems the more likely possibility since catecholamines have been ineffective in increasing either guanyl cyclase activity or tissue levels of cyclic GMP (see Sections III and IV).

Although urine and plasma levels of cyclic AMP and cyclic GMP have now been shown to be changed by a number of hormonal factors, which if any of these factors are responsible for the normal extracellular fluid levels of the cyclic nucleotides is unknown. The kidney has been identified as a source of a fraction of the normal urine level of cyclic AMP and its contribution may be at least in part a result of parathyroid hormone action. The organ or organs responsible for normal plasma levels of the cyclic nucleotides have not been identified. Of additional interest is the functional significance of the presence of cyclic nucleotides in extracellular fluids. The possibility that the extrusion of cyclic nucleotides may be one means by which a cell can regulate the levels of these substances within it has been raised by the observation that avian erythrocytes rapidly extrude cyclic AMP against an apparent concentration gradient (Davoren and Sutherland, 1963a). A similar phenomenon has been suggested by experiments with *E. coli* (Makman and Sutherland, 1965).

The possibility of diagnostic usefulness of extracellular fluid levels of cyclic AMP in parathyroid disorders has been raised by the observations of Chase *et al.* (1969b), Kaminsky *et al.* (1970a), and Taylor *et al.* (1970). Paul *et al.* (1970b,c) and Abdulla and Hamadah (1970) have reported that patients with affective disorders excrete abnormal amounts of cyclic AMP (lower in depressed, higher in manic patients). A higher than normal excretion of cyclic AMP in women after the thirtieth week of pregnancy has been reported by Taylor *et al.* (1970). The relationships of the altered levels of the excreted nucleotide to the associated conditions remain to be clarified but these interesting findings further suggest that a variety of conditions involving endocrinologic or metabolic abnormalities may be reflected in alterations in extracellular fluid levels of cyclic nucleotides.

VIII. SUMMARY

Cyclic GMP is present in most tissues in concentrations generally at least tenfold lower than those of cyclic AMP. Hormones which increase cyclic AMP levels in tissues do not elevate cyclic GMP levels in tissues reported thus far but theophylline does increase levels in at least some tissues. Cyclic GMP levels in urine vary in rats in altered states of pituitary, adrenocortical, and thyroid function and its levels in plasma and urine in man are elevated by the administration of Ca^{2+} or α-adrenergic agents.

The formation of cyclic GMP from GTP is catalyzed by guanyl cyclase, an enzyme which differs from adenyl cyclase in a number of respects. It is largely soluble in most tissues, but it is virtually entirely particulate in the sperm of sea urchin where its levels are two to three orders of magnitude higher than in other sources so far examined. For maximum activity Mn^{2+} appears to be required. Fluoride and hormones tested thus far have no effect on its activity but a number of nucleotides are inhibitory.

Cyclic GMP is hydrolyzed by phosphodiesterase to 5′-GMP; in at least some cells, the same phosphodiesterase seems to hydrolyze both cyclic GMP and cyclic AMP but the existence of absolutely specific phosphodiesterases for the two nucleotides cannot be ruled out. While with some enzyme preparations the two nucleotides are competitive inhibitors of the hydrolysis of each other *in vitro*, this may have no physiological significance because of the concentrations required. With other phosphodiesterase preparations, cyclic GMP can stimulate cyclic AMP hydrolysis at quite low concentrations but the physiological significance of this is also unclear.

The biological role of cyclic GMP has not been established. Cyclic GMP does not effectively substitute for cyclic AMP in activating glycogen phosphorylase or in other systems involving cyclic AMP-dependent protein kinases when studied in cell-free systems. Cyclic GMP is generally about two orders of magnitude less potent than cyclic AMP. Cyclic GMP-sensitive protein kinases have been identified but their function is unknown.

Exogenous cyclic GMP does effectively mimic a number of effects of cyclic AMP in perfused livers and other intact cell systems but the apparent potency of cyclic GMP in these systems may result indirectly from phosphodiesterase inhibition by the high levels used or from a greater ability of exogenous cyclic GMP to accumulate within cells. A limited number of observations indicate effects of exogenous cyclic GMP unlike those of cyclic AMP in kidney tissue and these deserve further investigation.

Several hormonal factors affect levels of cyclic GMP and cyclic AMP in extracellular fluids either independently of each other or in opposite directions. The sources of extracellular cyclic nucleotides are in large part unidentified but the kidney and liver have both been shown capable of contributing cyclic AMP to the extracellular pool. A number of observations suggest that extracellular levels of the cyclic nucleotides may reflect their abnormal intracellular metabolism in states of endocrine imbalance.

IX. ADDENDUM

Following completion of this chapter, George *et al.* (1970) reported the interesting observation that cyclic GMP levels in the isolated perfused rat heart were increased by acetylcholine. Ferrendelli *et al.* (1970) reported that treatment of mice with oxytremorine led to an increase of cyclic GMP in the cerebral cortex and cerebellum. These changes in cyclic GMP in both heart and brain were prevented by atropine.

Other intriguing reports that appeared too late for inclusion were those of Emmer *et al.* (1970) and Zubay *et al.* (1970) that cyclic GMP inhibited the stimulatory effect of cyclic AMP on β-galactosidase synthesis in cell-free extracts of *E. coli*. This effect appears to result from competition with cyclic AMP for a receptor protein involved in the transcription of *lac* mRNA, as discussed in more detail in the next chapter.

The work referred to in the present chapter as the unpublished observations of J. H. Exton *et al.* will be published in 1971 in the *Journal of Biological Chemistry*.

CHAPTER 12
Cyclic AMP in Lower Organisms

I. INTRODUCTION

It now seems obvious that cyclic AMP could have been one of the earliest organic chemicals to exist on earth. Adenine has been synthesized by electron irradiation of methane, ammonia, and water, and ribose is formed from aqueous solutions of formaldehyde under even milder conditions. Adenosine and polyphosphoric acid have been shown to react under simulated prebiotic conditions to yield a mixture of phosphorylated derivatives, including 5'-AMP, ADP, and ATP. Although not measured in these experiments, it seems likely that cyclic AMP would also have been among the products of this reaction. If not, its synthesis can be easily imagined, since we know it can be produced in large yield by simple alkaline hydrolysis of ATP (see Chapter 3). For a stimulating review of this interesting field of study, chemical evolution, readers are referred to the article by Fox *et al.* (1970). The point we are trying to make here is that cyclic AMP was probably on the scene, ready to be utilized, when chemical evolution began to be supplanted by organic evolution. The question we want to ask now is in what way was it utilized?

We have seen in preceding chapters that cyclic AMP eventually came to be used by multicellular animals as a versatile regulatory agent, enabling or at least helping diverse groups of cells to function together as a coordinated unit. We are now beginning to see that the use of cyclic AMP as a regulatory agent was not an original invention on the part of multicellular animals, and it is even possible that its role as a second messenger is not unique to us.

The role of cyclic AMP in bacteria and cellular slime molds has now been studied in some detail, and we have reserved this concluding chapter for a summary of what has been learned from these studies. In addition, in Section IV, we will take the opportunity to summarize work done on the possible role of cyclic AMP in cancer.

II. BACTERIA

It was established some years ago (Makman and Sutherland, 1965) that cyclic AMP occurred in *Escherichia coli,* and that its concentration was lowered by glucose. Conversely, as shown in Fig. 12.1, the removal of glucose led to a rapid increase in the level of cyclic AMP. The full significance of this was not appreciated for several years. Our present understanding of the role of cyclic AMP in *E. coli* owes much to the work of Robert Perlman and Ira Pastan and their colleagues, just as, we might add, our present account owes much to a review of the literature by these authors (Pastan and Perlman, 1970).

A. Significance of the Glucose Effect

A major problem facing a bacterium is that it has an extremely limited amount of intracellular space in which to carry out the functions we normally associate with life. In order to survive, therefore, these organisms must operate very efficiently, and we know from the evolutionary record

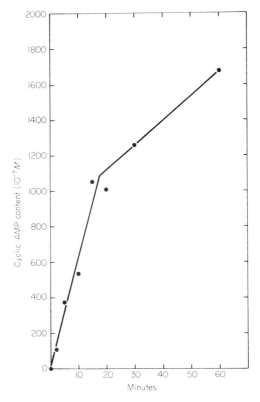

Fig. 12.1. Effect of removing glucose on the level of cyclic AMP in *E. coli*. Cells with a low initial content of cyclic AMP ($5 \times 10^{-7} M$) were grown in the presence of 1% glucose. After harvesting, the cells were resuspended in phosphate buffer, pH 7.0, and incubated for varying periods at 37°C. Incubation was stopped by rapid chilling to 10°C followed by centrifugation. Molarity was calculated on the assumption that 1 gm of the 10,000 g pellet was equivalent to 1 ml. From Makman and Sutherland (1965).

that they have done so successfully. One of the reasons for this is that, while they are capable of utilizing a great variety of energy-yielding substrates, they do not have to contain within themselves at all times all of the necessary enzymes. Instead, they carry the genetic potential for synthesizing certain groups of enzymes only when they are required. Called inducible enzymes, these proteins are synthesized in response to an inducer, which is usually a substrate for one of the enzymes involved. Now in the case of bacteria living in an environment containing many potential substrates, such as the case of *E. coli* in the mammalian gastrointestinal tract, it is clear that a decision will have to be made as to which enzymes to synthesize and which not to synthesize. Since the genetic potential exists for synthesizing so many, the synthesis of most of them will have to be suppressed. *Escherichia coli* ordinarily utilize glucose over other potential substrates such as lactose, so that in the presence of glucose the enzymes necessary for the utilization of these other substrates are either absent or else synthesized at very low rates. One of the essential roles of cyclic AMP in bacteria is to participate in this decision-making process, as outlined in more detail in the following sections.

It is interesting to note that a similar decision has to be made by mammalian cells during the early stages of embryonic differentiation (Bullough, 1967). All of these cells have the same genetic potential, and yet we know that some will give rise to contractile cells that function within the heart, others will be the ancestors of epithelial cells within the kidney, while still a third group will produce cells whose future generations will participate within the central nervous system in what we think of as thought. Just as *E. coli* in the midst of glucose do not utilize lactose, so neither do myocardial cells produce urine, and the reason seems to be basically the same, that the synthesis of some enzymes occurs while that of others is repressed. Complexity aside, the main difference between a differentiated eukaryotic cell and a bacterium seems to be the irreversible permanence with which enzyme synthesis is repressed in the eukaryotic cell. To use the example discussed in the Introduction to Chapter 5, it is difficult to imagine a mammalian liver cell suddenly acquiring the ability to produce steroid hormones in response to ACTH. And yet such a change, if it were to occur, could hardly be more remarkable than what actually does occur when *E. coli* growing in a glucose-rich medium suddenly find themselves without glucose.

Since we have emphasized the seemingly permanent nature of eukaryotic cell differentiation, as opposed to the ease with which bacteria can shift from one pattern of enzymes to another, it should be noted that even mammalian cells can change their enzyme content to some extent. A case in point is that of hepatic tyrosine transaminase, discussed in some detail in Chapter 7 (Section II,B,4). Since the role of cyclic AMP in regulating the synthesis of this enzyme superficially resembles its role in the regulation of bacterial enzyme induction, the question can be raised of whether cyclic AMP (or perhaps cyclic GMP) might also play a role in the events responsible for the more permanent type of cellular differentiation. A final answer to this question cannot be given at present, but most of the evidence suggests that cyclic AMP does not play an important role in regulating the synthesis of constitutive enzymes. For example, mutants of *E. coli* have been isolated which lack the ability to produce cyclic AMP and/or the ability to respond to the nucleotide, but which can nevertheless grow in a glucose-rich medium (Pastan and Perlman,

1970). These mutants cannot survive, however, if glucose is replaced by lactose. In broad general terms, therefore, the role of cyclic AMP in *E. coli* seems to be the same as it is in higher organisms, which is to enable the organism to survive in the face of a changing environment.

B. Mechanism of the Glucose Effect

Restricting our attention now to *E. coli,* we have already mentioned that one of the first things that happens in these cells when glucose is removed is that the level of cyclic AMP begins to rise (Fig. 12.1). Conversely, when glucose was added to cells in which the level of cyclic AMP was initially high, the level rapidly fell. The disappearance of cyclic AMP from the cells could be fully accounted for by its rapid appearance in the extracellular medium, and thereafter no further increase could be seen in the presence of glucose (Makman and Sutherland, 1965). These results suggested an effect of glucose both in suppressing the formation of cyclic AMP and in stimulating a mechanism for its rapid release.

A variety of other potential substrates were capable of reducing the level of cyclic AMP below that seen in the complete absence of an exogenous carbon source, but were much less effective than glucose (Makman and Sutherland, 1965). The possibility that their activity was related to glucose synthesis was not ruled out.

Evidence summarized by Pastan and Perlman (1970) suggests that glucose does not have to be phosphorylated or further metabolized in order to reduce the level of cyclic AMP. Glucose is phosphorylated in *E. coli* by way of a small heat-stable protein which carries phosphate on a specific histidine residue. Mutants deficient in this protein, and which therefore lack the capacity to phosphorylate glucose, are nevertheless very sensitive to the enzyme-repressing effect of glucose. On the other hand, mutants deficient in the enzyme which catalyzes the transfer of phosphate from this protein to glucose are highly resistant to the glucose effect (Pastan and Perlman, 1969b). It would therefore appear that glucose must interact with this enzyme in order to be effective, but that the actual phosphorylation of glucose is not required. This enzyme is located in the cell membrane, and may be involved in facilitating the entrance of glucose into the cell as well as its phosphorylation. It may also exist in close proximity to adenyl cyclase. Although adenyl cyclase from *E. coli* can be solubilized to some extent (Tao and Lipmann, 1969), it occurs initially in particulate fractions (Brana, 1969; Ide, 1969).

It is also possible, of course, that mutants deficient in the glucose-phosphorylating enzyme are likewise deficient in one or more other components, which may be more important in mediating the action of glucose. Since glucose suppresses the level of cyclic AMP in intact cells but has little or no effect on adenyl cyclase in cell-free extracts of *E. coli* (Tao and Lipmann, 1969), it is conceivable that the action of glucose is similar to that of insulin, prostaglandins, or α-adrenergic catecholamines in certain mammalian cells. However, this is entirely conjectural at present. The mechanism by which these various agents act to lower the level of cyclic AMP in some cells represents a major problem for future research.

The possibility that glucose might act by stimulating the activity of phosphodiesterase, which has been identified in *E. coli* and other bacterial species (Brana and Chytil, 1966; Okabayashi and Ide, 1970), can be considered. However, glucose does not affect the activity of this enzyme in cell-free extracts, and an indirect effect also seems unlikely. No evidence for cyclic AMP degradation could be obtained in the earlier experiments (Makman and Sutherland, 1965), and it has since been shown that phosphodiesterase activity may be very low (Monard *et al.,* 1969) or even undetectable (Pastan and Perlman, 1970) in certain strains of *E. coli* that nevertheless do respond to glucose. Thus the significance of bacterial phosphodiesterase is open to question.

It is possible that extrusion into the medium is a more important mechanism by which some strains of *E. coli* reduce their content of cyclic AMP than is conversion to 5'-AMP. As mentioned previously, the reduction of cyclic AMP levels in response to glucose was shown to be accompanied by release into the medium, and it seems possible that glucose may act to facilitate this process, in addition to suppressing the formation of cyclic AMP. An interesting incidental result of this released cyclic AMP is that it tends to attract certain myxamoebae which feed on *E. coli* (Konijn *et al.,* 1969a). The overall ecological significance of this phenomenon is not entirely clear, but will be discussed from the standpoint of the myxamoebae later in this chapter (Section III).

C. The Role of Cyclic AMP

As mentioned previously, a number of inducible enzymes are either not synthesized or else are synthesized at reduced rates in the presence of glucose. Evidence that the synthesis of these enzymes requires cyclic AMP, and that glucose inhibits their synthesis by reducing the level of cyclic AMP, will be presented in this and the following section. The enzyme which has been studied in the greatest detail from this point of view has been β-galactosidase, so that major emphasis will be placed on it. This enzyme, which catalyzes the hydrolysis of lactose to glucose and galactose, is one of two proteins known to be required for the utilization of lactose. The other is a galactoside permease, which permits the entry of lactose into the cell; the genetic information required for the synthesis of both of these proteins is located in the *lac* operon. This operon also codes for a third protein, thiogalactoside transacetylase, which is therefore synthesized coordinately with the other two. Its role in lactose metabolism is unknown at present.

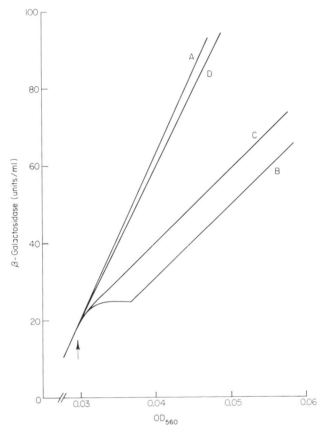

Fig. 12.2. Effect of glucose and cyclic AMP on the rate of synthesis of β-galactosidase. In the original experiment on which this figure is based, *E. coli* were grown on a medium containing glycerol as the main source of carbon. The experiment was started with the addition of IPTG (1 mM). Ten minutes later (arrow) the culture was divided into four parts with the following additions: A, none; B, 10 mM glucose; C, 10 mM glucose + 1 mM cyclic AMP; and D, 10 mM glucose + 5 mM cyclic AMP. See text for further details. From data presented by Perlman *et al* (1969).

In the absence of an inducer, these three proteins are present in *E. coli* in only very small amounts. However, the addition of lactose, or a nonmetabolizable lactose analog such as isopropylthio-β-D-galactoside (IPTG), induces the synthesis of large amounts of all three of them. This is illustrated for β-galactosidase by curve A in Fig. 12.2, where activity is plotted as a function of the turbidity of the culture, whicH, in turn, reflects the total amount of protein present. The differential rafe of β-galactosidase synthesis (the rate of enzyme synthesis relative to the rate of total protein synthesis) may vary greatly depending on the medium in which the cells are growing.

It has been common in the past to distinguish two distinct effects of glucose on the synthesis of inducible enzymes such as β-galactosidase. Both effects are illustrated in Fig. 12.2, based on an experiment in which *E. coli* were synthesizing β-galactosidase in response to IPTG in the presence of glycerol. After the addition of glucose (curve B), there is a transient period of complete or almost complete suppression of β-galactosidase synthesis. This is referred to as "transient repression." Its duration is variable, but often on the order of 20 to 30 minutes, or approximately half a generation.

After the period of transient repression, β-galactosidase synthesis resumes, but now at a lower rate than would occur in the absence of added glucose. This second effect of glucose (represented in Fig. 12.2 by the difference between curve A, on the one hand, and curve C or the second part of curve B, on the other) has been referred to as "permanent repression" to distinguish it from transient repression. It is also commonly referred to as "catabolite repression" because many carbon sources are capable of producing a similar effect. In contrast to transient repression, permanent repression seems to require glucose metabolism. Glucose-6-phosphate, for example, is more effective than glucose itself.

It was initially thought that cyclic AMP might be involved only in transient repression, and not in permanent repression. This was because the addition of 1 mM cyclic AMP (curve C in Fig. 12.2) was found to abolish transient repression in response to glucose, but did not restore the rate of β-galactosidase synthesis to its original rate (Perlman and Pastan, 1968; Ullmann and Monod, 1968; Pastan and Perlman, 1968). Instead, it restored it only to the rate characteristic of permanent or catabolite repression. Further support for this conclusion came from the finding by Goldenbaum and Dobrogosz (1968) that a concentration of cyclic AMP which did prevent the effect of glucose had only a slight effect on the repression produced by glucose-6-phosphate.

It was later shown, however, that by increasing the concentration of cyclic AMP still further (curve D of Fig. 12.2) the rate of β-galactosidase synthesis could be restored to its initial rate, either in the presence of glucose (Perlman *et al.*, 1969) or the more potent repressor *N*-acetyl-glucosamine-6-phosphate (Goldenbaum *et al.*, 1970). Broman *et al.* (1970) achieved the same result by adding either leucine or alanine to the culture medium. Under these conditions it was found that a concentration of cyclic AMP which previously prevented only transient repression (as in curve C of Fig. 12.2) was now capable of preventing permanent repression as well (as in curve D). The mechanism of this "sparing" effect of leucine or alanine is unknown at present. Other amino acids were not effective in this regard (Broman *et al.*, 1970).

It would thus appear that both transient and permanent repression by glucose are due to lowered levels of cyclic AMP. Why higher concentrations of exogenous cyclic AMP are required to reverse permanent repression is unknown at present. Perhaps some metabolite of glucose is produced which allows synthesis to proceed at a rate lower than would occur in its absence, and which simultaneously reduces sensitivity to cyclic AMP. The finding that 5 mM cyclic AMP was unable to completely overcome the repression produced by the simultaneous addition of 10 mM glucose, 10 mM glucose-6-phosphate, and 10 mM sodium gluconate (Moses and Sharp, 1970) might be taken as support for this suggestion, although other interpretations are obviously possible. Whether the levels of cyclic AMP during transient and permanent repression are the same or different is unknown at present. A detailed comparison of the effects of glucose and other carbon sources on the level of cyclic AMP in relation to their functional effects in *E. coli* has yet to be carried out.

The ability of cyclic AMP to prevent or reverse the effect of glucose seems quite specific. It was not shared by equimolar concentrations of adenine, adenosine, 2'-deoxy cyclic AMP, cyclic IMP, 2', 3'-AMP, 3'-AMP, 5'-AMP, ADP, ATP, fructose-1,6-diphosphate, or the monobutyryl or dibutyryl derivatives of cyclic AMP (Perlman and Pastan, 1968; DeCrombrugghe *et al.*, 1969). These compounds also did not interfere with the effect of cyclic AMP, although 5'-AMP was later reported to produce transient repression which could be overcome by cyclic AMP (del Campo *et al.*, 1970). Moses and Sharp (1970) reported that fructose-1,6-diphosphate and several other phosphorylated intermediates were capable of stimulating β-galactosidase synthesis in some experiments when added to EDTA-treated cells. The relevence of these experiments to catabolite repression seems doubtful, if only because glucose would be expected to increase rather than decrease the intracellular levels of these intermediates, if anything. It should also be noted that their presence in the extracellular medium would probably be an unphysiological circumstance even for bacteria.

DeCrombrugghe *et al.* (1969) showed that exogenous cyclic AMP could overcome the repression by glucose of a number of inducible enzymes in addition to β-galactosidase, including galactoside permease, arabinose permease, galactokinase, glycerokinase, serine deaminase, and thymidine phosphorylase. Stimulation of tryptophanase synthesis had been noted previously by Pastan and Perlman (1969a). It now seems clear that the synthesis of all inducible enzymes that are subject to repression by glucose can be stimulated by cyclic AMP (Pastan and Perlman, 1970). Conversely, cyclic AMP appears to have no effect on the synthesis of enzymes not subject to glucose repression, such as tryptophan synthetase and alkaline phosphatase (Pastan and Perlman, 1969a).

DeCrombrugghe *et al.* (1969) also showed that the role of cyclic AMP in the glucose effect was not restricted to *E. coli.* The ability of cyclic AMP to prevent or reverse the repression by glucose of β-galactosidase synthesis was demonstrated in a number of other gram-negative organisms, including *Salmonella, Aerobacter, Serratia,* and *Proteus.* Grampositive organisms have apparently not been studied from this point of view.

Experiments with cyclic AMP-deficient mutants have provided strong additional evidence that cyclic AMP is required for the synthesis of inducible enzymes subject to catabolite repression. Perlman and Pastan (1969) were able to isolate a mutant strain of *E. coli* which was markedly deficient in adenyl cyclase activity and which contained undetectable levels of cyclic AMP. This mutant was unable to grow on lactose, maltose, arabinose, mannitol, or glycerol, and grew relatively slowly on glucose, fructose, and galactose (Table 12.I). The addition of cyclic AMP to the medium permitted the normal utilization of all of these carbon sources. The mutant made only about 5% as much β-galactosidase as its parent strain in response to IPTG, but this was increased toward normal by the addition of exogenous cyclic AMP. As expected from earlier studies (DeCrombrugghe *et al.,* 1969), the physiological defects in this mutant were not restricted to carbohydrate metabolism. For example, the mutant was unable to produce indole in the absence of cyclic AMP because of its lack of tryptophanase. This and all of the other observed metabolic defects in this mutant were corrected by the addition of exogenous cyclic AMP.

In summary of what has been said to this point, glucose is capable of reducing the level of cyclic AMP in *E. coli,* and exogenous cyclic AMP is capable of preventing or reversing the inhibitory effect of glucose on

TABLE 12.I Ability of the Mutant Lacking Adenyl Cyclase to Ferment and Grow on Various Carbon Sources[a]

Carbon source	Fermentation		Doubling time (minutes)	
	Without cyclic AMP	With cyclic AMP	Without cyclic AMP	With cyclic AMP
Lactose	−	+	>720	85
Maltose	−	+	>720	90
Arabinose	−	+	>720	70
Glycerol	−	+	>720	90
Mannitol	−	+	>720	65
Glucose	±	+	110	65
Fructose	±	+	110	70
Galactose	±	+	110	65

[a]Fermentation was estimated using two types of indicator agars supplemented with 1% of the indicated sugars. To measure doubling times, cells were grown overnight at 37°C in the presence of 0.4% glucose, and were then diluted into fresh medium containing 0.4% of the other carbon sources, as indicated, with and without 2 mM cyclic AMP. Adapted from Perlman and Pastan (1969).

the synthesis of a number of inducible enzymes. These and other findings suggest that cyclic AMP is required for the synthesis of these enzymes, and that glucose prevents or inhibits their synthesis by reducing the level of cyclic AMP.

D. Mechanism of Action of Cyclic AMP

Current evidence suggests that cyclic AMP may regulate the synthesis of inducible enzymes in *E. coli* by at least two different mechanisms. Cyclic AMP appears to act at the level of transcription to stimulate β-galactosidase synthesis, whereas an action at the translational level may be involved in its effect on tryptophanase synthesis. These two effects are discussed separately in the following paragraphs.

1. β-Galactosidase

As reviewed by Pastan and Perlman (1970), early studies suggesting that glucose repressed the synthesis of β-galactosidase by inhibiting the formation of *lac* messenger RNA (mRNA) pointed to a transcriptional site of action for cyclic AMP. In their initial experiments, Perlman and Pastan (1968) estimated the concentration of *lac* mRNA by the capacity of *E. coli* to synthesize β-galactosidase after removal of the inducer, and showed that cyclic AMP did indeed increase the amount of *lac* mRNA in glucose-repressed cultures. Varmus *et al.* (1970) have more recently used a DNA–RNA hybridization assay to measure *lac* mRNA directly. With this assay, it was found that uninduced cultures of *E. coli* made approximately 5% as much *lac* mRNA as did cultures induced with IPTG. It was then established that the addition of glucose to induced cultures caused a marked decrease in the rate of *lac* mRNA synthesis, and that the addition of cyclic AMP could restore this rate of synthesis to normal (Table 12.II). Still more recently, DeCrombrugghe *et al.* (1970) have extended these studies to the cell-free system developed by Zubay and his colleagues. Cyclic AMP had no effect on total RNA synthesis, but stimulated *lac* mRNA synthesis markedly.

TABLE 12.II Effect of IPTG, Glucose, and Cyclic AMP on *lac* mRNA and β-Galactosidase Synthesis by *E. coli*[a]

Additions	Relative rate of synthesis of *lac* mRNA	Relative concentration of *lac* mRNA in cells	Relative rate of β-galactosidase synthesis
None	7	4	1
IPTG	100	100	100
IPTG + glucose	12	10	15
IPTG + glucose + cyclic AMP	119	82	130

[a] The rate of synthesis of *lac* mRNA (relative to the rate of total mRNA synthesis) and the amount of *lac* mRNA in the cultures were estimated by DNA–RNA hybridization techniques. Values for all parameters are expressed as percentage of the values obtained in the IPTG induced cultures. From Varmus *et al.* (1970) and Pastan and Perlman (1970).

Although our understanding of the mechanism by which cyclic AMP stimulates mRNA synthesis is still incomplete, progress in this direction has been substantial. A schematic illustration of the *lac* operon, together with the various factors which are either known or thought to be involved in its regulation, is given in Fig. 12.3. The *lac* operon, originally described by Monod and his colleagues, has now been synthesized by Shapiro *et al.* (1969). The three structural genes, *z, y,* and *a,* code for the synthesis of β-galactosidase, galactoside permease, and thiogalactoside transacetylase, respectively. The three regulatory genes of the operon are referred to as *i, p,* and *o*. The *i* gene codes for a repressor protein which binds to the *o* or operator gene, thereby preventing transcription of *lac* mRNA. Inducers such as IPTG stimulate *lac* mRNA synthesis by binding to the repressor and reducing its affinity for the *o* gene. The *p* or promoter gene

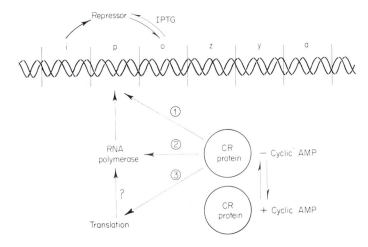

Fig. 12.3. Schematic illustration of the *lac* operon. The numbered arrows point to possible sites of action of the protein to which cyclic AMP binds in *E. coli*. See text for further description. Modified from Pastan and Perlman (1970).

controls the maximum rate of *lac* operon expression. It is thought to be the site at which RNA polymerase binds to *lac* DNA and the site at which *lac* mRNA synthesis is initiated.

Several lines of evidence support the view that the promoter is the part of the *lac* operon most directly affected by the action of cyclic AMP. Jacquet and Kepes (1969) measured the time required for the action of cyclic AMP to become resistant to actinomycin D and rifampicin. They concluded that cyclic AMP was acting to stimulate an early step in transcription, one experimentally indistinguishable from that inhibited by rifampicin. This antibiotic specifically inhibits the initiation rather than the polymerization of RNA chains.

Other evidence implicating the promoter gene in the action of cyclic AMP came from the studies of Pastan and Perlman (1968, 1970) of regulatory gene mutants. β-Galactosidase synthesis in mutants of the *i* and *o* genes was repressed by glucose and responded normally to cyclic AMP, as noted also by Ullmann and Monod (1968). By contrast, some promoter mutants were altered in their response to cyclic AMP. One mutant, carrying a deletion of most of the *p* and part of the *i* gene, was completely unresponsive to cyclic AMP. Another strain, containing a point mutation in the promoter region, was not subject to transient repression by glucose and was relatively insensitive to stimulation by cyclic AMP.

The possibility that part of the effect of cyclic AMP may be exerted elsewhere has to be kept open, especially in view of the recent studies by

Aboud and Burger (1970). They obtained evidence to suggest that cyclic AMP was acting at least in part to facilitate the translation of preformed *lac* mRNA, in line with earlier studies suggesting that glucose could inhibit this process. As discussed in the following section, Pastan and Perlman (1969a) had concluded that cyclic AMP was acting at the translational level to stimulate tryptophanase synthesis, but were unable to detect a translational effect of cyclic AMP in control of the *lac* operon. However, their studies do not seem to rule out the possibility that at least part of the effect of cyclic AMP might be secondary to an effect on translation. Aboud and Burger have suggested that the effect of cyclic AMP may occur entirely by this means.

Regardless of how the *lac* operon is ultimately affected by cyclic AMP, it is now clear that the nucleotide does not act directly. Instead, its effect appears to depend upon the presence of a protein that binds cyclic AMP with an apparent dissociation constant on the order of 10^{-6} M. This protein was discovered independently by two groups of investigators (Zubay *et al.*, 1970; Emmer *et al.*, 1970), and has been referred to by Zubay and his colleagues as CAP, for "catabolite-gene activator protein," and by the NIH group as CR protein, for "cyclic AMP receptor protein." The temptation to combine the CR and CAP designations into a third acronym is almost overwhelming, but has been resisted because it might not be acceptable to either group, and furthermore might not reflect the true importance of this protein in the action of cyclic AMP.

Chambers and Zubay (1969) had previously discovered that cyclic AMP was required in a DNA-directed cell-free system for synthesizing the enzymes of the *lac* operon. This system contains DNA (derived from ϕ80d*lac* virus containing a normal *lac* operon), a cell-free extract of *E. coli,* and a variety of substrates and cofactors found essential for RNA and protein synthesis. Zubay *et al.* (1970) were able to detect the cyclic AMP-binding protein with this system by using extracts of mutant strains of *E. coli* that were unable to grow on lactose. Some of these strains were defective because they lacked adenyl cyclase, as mentioned previously (Table 12.I), but others were unable to utilize lactose (or arabinose) even in the presence of added cyclic AMP. When one of these strains (X7901 in Table 12.III) was used as the source of the cell-free extract, only about 5% of the usual level of β-galactosidase was made (compare lines 2 and 6 in Table 12.III; strain 514 was used as the normal control in this experiment). The addition of extracts from normal strains could stimulate synthesis, but only in the presence of cyclic AMP. By using this stimulatory effect as an assay, it was possible to purify the responsible factor about 200-fold, and this was later identified as CAP. Addition of this protein with X7901 extract restored sensitivity to cyclic AMP (Table

TABLE 12.III Effect of Cyclic AMP and CAP on DNA-Directed Synthesis of
β-Galactosidase in a Cell-Free System[a]

Line	Source of bacterial extract	Other additions		Relative β-galactosidase activity
		Cyclic AMP	CAP	
1	X7901	−	−	1
2	X7901	+	−	1
3	X7901	−	+	1
4	X7901	+	+	5
5	514	−	−	1
6	514	+	−	20
7	514	−	+	1
8	514	+	+	24

[a] The conditions of this experiment are summarized in the text. Strain X7901 is a mutant strain of *E. coli* unable to grow on lactose even in the presence of cyclic AMP. Strain 514 was used as the source of CAP, which is referred to in Fig. 12.3 as "CR protein." From Zubay *et al.* (1970).

12.III, line 4), but had only a slight effect when added with strain 514 extract, presumably because the latter contained large quantities of CAP to begin with. In later experiments with X7901 extracts it was shown that stimulation of β-galactosidase synthesis was proportional to the amount of CAP added over at least a tenfold range.

A similar approach enabled Emmer *et al.* (1970) to isolate what they referred to as CR protein, and it now seems reasonably clear that it and CAP are identical. Using different techniques, both groups of investigators obtained similar values for apparent molecular weight (on the order of 42,000) and affinity for cyclic AMP. Both groups also noted that cyclic GMP could inhibit binding, and that this was proportional to its ability to inhibit the effect of cyclic AMP on β-galactosidase synthesis.

The exact function of this cyclic AMP-binding protein had not been established at the time of this writing. Travers and his colleagues (1970) have suggested that it may be one of a class of apparently similar proteins which they have termed *psi* factors. According to these investigators, the presence of an appropriate small molecule (cyclic AMP in the case of CR protein, other nucleotides in the case of other *psi* factors) might enable the protein to interact either with DNA itself (presumably at or close to the promoter region) or else with RNA polymerase, in either event enabling transcription of the appropriate section of DNA to begin.

Another idea is that the CR protein might function as part of a protein kinase system, analogous to the cyclic AMP-binding proteins found in the eukaryotic cells of higher organisms (see the addendum to Chapter 5).

Kuo and Greengard (1969a) had previously reported the existence of a cyclic AMP-sensitive protein kinase in *E. coli.* Although Emmer *et al.* (1970) could detect no protein kinase activity in association with their CR protein, it still seems possible that the latter could function as part of a protein kinase system under appropriate conditions. Current evidence suggests that in at least some mammalian cells the cyclic AMP-binding protein constitutes the regulatory subunit(s) of the complete protein kinase system (Gill and Garren, 1970; Tao *et al.,* 1970). No matter how analogous the eukaryotic and prokaryotic systems eventually prove to be, however, it seems clear that the analogy will have to break down at some point. Thus, in the mammalian systems, the evidence suggests that the regulatory subunit exerts an inhibitory influence on the catalytic subunit, such that in the complete absence of the former the latter would be fully active. By interacting with the regulatory subunit, cyclic AMP appears to interfere with its ability to inhibit the catalytic subunit. In *E. coli,* by contrast, the CR protein seems clearly to function as part of a positive control system, such that in its complete absence the affected system cannot function at all. It is possible, of course, that the discovery of an additional inhibitory factor in bacterial systems will render this difference more apparent than real.

Some additional evidence that protein phosphorylation could play a role in the action of cyclic AMP in bacteria has been provided by Martelo *et al.* (1970). These investigators added mammalian protein kinase to a system containing RNA polymerase from *E. coli* with T4 phage DNA as template, and found it to be stimulatory. The rate of RNA synthesis was further stimulated by the addition of cyclic AMP, but cyclic AMP had no effect in the absence of the kinase. The interesting finding was then made that the predicted phosphorylation of RNA polymerase occurred primarily at one or more sites on the sigma factor, with little or no phosphorylation of the core enzyme itself. The sigma factor leaves the core enzyme once transcription has been initiated. How phosphorylation of this protein might enable RNA polymerase to recognize some promoters and not others does not seem obvious, but the possibility that it might have such an effect seems worth considering.

In summary, cyclic AMP interacts with a specific cyclic AMP-binding protein in *E. coli* to stimulate the transcription of *lac* mRNA, which in turn provides the information for the synthesis of β-galactosidase and the other enzymes of the *lac* operon. Both cyclic AMP and the protein to which it binds are required for optimal synthesis of these enzymes (and presumably other inducible enzymes subject to catabolite repression). Glucose and certain other metabolites apparently inhibit the synthesis of these enzymes by reducing the level of cyclic AMP.

2. Tryptophanase

The mechanism by which cyclic AMP acts to stimulate tryptophanase synthesis seems to differ from the mechanism involved in β-galactosidase synthesis. Under the same conditions, the concentration of exogenous cyclic AMP required to restore enzyme synthesis in the presence of glucose to 50% of the unrepressed rate was much higher for tryptophanase (approximately 2 mM versus less than 1 mM for β-galactosidase) (Perlman et al., 1969; Pastan and Perlman, 1970). Also, studies on the regulation of tryptophanase synthesis in EDTA-treated cells (Pastan and Perlman, 1969a) revealed that cyclic AMP could stimulate the amount of tryptophanase made when the nucleotide was added after mRNA synthesis had been arrested either by removal of the inducer (tryptophan) or by treatment with actinomycin D or proflavine. Cyclic AMP had no apparent effect on the rate of breakdown of tryptophanase mRNA, and no evidence for an increased rate of conversion of an inactive precursor could be obtained. These results suggested that cyclic AMP was acting at the translational level to increase the rate of polypeptide chain elongation.

It will be recalled that under similar conditions cyclic AMP did not stimulate β-galactosidase synthesis, suggesting that in that case cyclic AMP was acting at the transcriptional level. On the other hand, as noted previously, Aboud and Burger (1970) found that if they allowed *lac* mRNA synthesis to proceed for longer periods before removing the inducer (or adding proflavine), then cyclic AMP would stimulate β-galactosidase synthesis. It should furthermore be noted that the experiments of Pastan and Perlman on tryptophanase synthesis did not rule out an effect at the transcriptional level, in addition to the apparent effect on translation. It seems possible, therefore, that the mechanisms by which cyclic AMP stimulates the synthesis of β-galactosidase and tryptophanase are not as different as was initially supposed, although differences obviously exist.

A possible mechanism by which cyclic AMP might act at the translational level was suggested by the work of Kuwano and Schlessinger (1970). They found that cyclic AMP was bound, in the presence of GTP, to a second protein in addition to the cyclic AMP-binding protein required for β-galactosidase synthesis. This other protein, known as the G translocation factor, or simply as "G factor," functions as or with a ribosome-dependent GTPase in the process by which successive transfer RNA molecules are shifted on the ribosome as it moves along messenger RNA. Binding of cyclic AMP to G factor was dependent on the presence of both GTP and Mg^{2+}, and was strongly inhibited by GDP. Evidence was also obtained to suggest that cyclic AMP was released from G factor during translocation, which apparently depends upon the conversion of GTP to GDP for energy. Binding seemed to be specific for cyclic AMP, since other adenine nucleotides tested were not detectably bound.

How the binding of cyclic AMP to G factor might favor the translation of some but not all mRNA, if this occurs, is unknown at present. Future studies with specific mRNA may yield results of great interest.

In summary, cyclic AMP appears to be capable of stimulating the translation of mRNA for tryptophanase and certain other inducible enzymes, but the mechanism of this effect is unknown at present.

E. Other Effects of Cyclic AMP

Effects of cyclic AMP in bacteria other than the stimulation of inducible enzyme synthesis have been noted. The most striking of these is the stimulation of flagella formation, observed in *Salmonella typhimurium* as well as in *E. coli* (Yokota and Gots, 1970). Adenyl cyclase-deficient mutants of both species were isolated and found to be incapable of forming flagella in the absence of cyclic AMP. It was known previously that flagella formation did not occur in *E. coli* in the presence of glucose. Cyclic AMP was the only one of a number of nucleotides tested capable of supporting this process. It would thus appear that in at least two species of bacteria cyclic AMP is required not only for the synthesis of inducible enzymes but also for flagella formation and, hence, motility.

An interesting additional finding mentioned by Yokota and Gots was that the gene for adenyl cyclase synthesis maps at a considerable distance from six other genes known to be required for flagella formation, and must also occur at a considerable distance from the *lac* operon. Standard text-books can be consulted for detailed chromosome maps of *S. typhimurium* and other species.

Another effect which seems to be specific for cyclic AMP, at least among adenine nucleotides, is inhibition of the activity of TPN-dependent malic enzyme of *E. coli* (Sanwal and Smando, 1969). The functional significance of this effect is unknown at present. A variety of other effects on carbohydrate metabolism, which may or may not be specific for cyclic AMP, were reported by Moses and Sharp (1970). Stimulation of the incorporation of labeled phosphate into fructose-1,6-diphosphate (and also into an unidentified intermediate) was especially striking, suggesting that cyclic AMP may stimulate phosphofructokinase activation in *E. coli* as well as in the liver fluke (see Chapter 5, Section III, D). Cyclic AMP was found to inhibit several of the light-induced reactions of isolated chromatophores from *Rhodospirillum rubrum;* some of these were also inhibited by 5′-AMP but others were not (Chaudhary and Frenkel, 1970).

Cyclic AMP has also been identified in *Brevibacterium liquifaciens* (Okabayashi *et al.,* 1963; Ide *et al.,* 1967), but its role in this organism has not been established. An intriguing feature of adenyl cyclase from this organism is its dependence on pyruvate, in the presence of which it is extremely active (Hirata and Hayaishi, 1967). Several other α-keto acids could replace pyruvate to some extent, but a variety of other agents, including alanine (which did stimulate cyclic AMP formation in cultures of *B. liquifaciens*), were inactive. None of these compounds had any significant stimulatory effect on *E. coli* adenyl cyclase (Tao and Lipmann, 1969; Ide, 1969).

It seems likely that many interesting findings will be made as cyclic AMP is studied in a greater variety of species.

III. CELLULAR SLIME MOLDS

The cellular slime molds are a fascinating group of eukaryotic organisms which are unicellular during part of their life cycle and multicellular during another. They have long been of interest to students of developmental biology because their growth and differentiation occur separately (during the unicellular and multicellular stages, respectively) instead of simultaneously as in vertebrates. They are of interest from the standpoint of this monograph because cyclic AMP is now known to play an essential role in the aggregation and hence differentiation of at least some species of slime molds. The earlier literature on these organisms has been critically reviewed by Bonner (1967), and a popular account of the role of cyclic AMP is also available (Bonner, 1969).

The species in which the role of cyclic AMP has been studied in the greatest detail has been *Dictyostelium discoideum,* and the following account applies specifically to it. A few hours after the spores have been sown in a favorable environment (which means primarily a moist environment), the capsules split and the amoebae emerge. So long as food is plentiful, these amoebae will continue to grow and multiply for long periods of time. Their favorite food appears to be living *E. coli,* although certain other organisms, such as *Aerobacter aerogenes,* seem to be almost as appetizing. Dead bacteria will also support growth, but attempts to obtain growth on chemically-defined cell-free media have met with only limited success. The amoebae are attracted by cyclic AMP (Konijn *et al.,* 1967), and it was thought for a time that the cyclic AMP released by bacteria might be the most important factor enabling the amoebae to sense and move toward food (Konijn *et al.*, 1968, 1969a). However, it is now thought that another bacterial product (unidentified at the time of this writing) is more important than cyclic AMP in this regard (Bonner *et al.*, 1970). In any event, the amoebae can be maintained in this vegetative phase indefinitely by continuously supplying fresh bacteria for the cells to eat.

As soon as the food supply is exhausted, however, two remarkable changes occur. One is that the cells in general become approximately 100 times more sensitive to the chemotactic effect of cyclic AMP (Bonner *et al.*, 1969). The other, which occurs at about the same time, is that one or a few amoebae suddenly begin to release cyclic AMP into the medium. As a result, the other amoebae begin streaming toward the cyclic AMP-producing cells. It would appear that the cells are not only attracted by cyclic AMP, but are caused by it to become "stickier" and therefore to adhere to one another (Konijn *et al.*, 1968). It would further appear that once new cells join the aggregate they are themselves stimulated to release cyclic AMP, thereby attracting still more amoebae. The whole process thus seems to work something like a telegraph relay system, with cells continuously being attracted to centers of cyclic AMP formation, and then in turn sending out a signal to attract cells from a still greater distance. The number of aggregations which arise in this way is apparently independent of the number of cells; instead, within limits, it seems to be primarily a function of territorial size (Bonner, 1967).

A few hours after aggregation has been initiated, the cells near the center of each aggregate begin to rise to form a cone-shaped mass, and the aggregate as a whole gradually turns into a slug-like pseudoplasmodium. This may now leave the initial site of aggregation, leaving a trail of slime in its wake. This sequence of events, from the vegetative stage through aggregation to the formation of the migrating pseudoplasmodia, is illustrated in the lower part of Fig. 12.4. The production of cyclic AMP by these organisms has been measured by several techniques (Barkley, 1969; Konijn *et al.*, 1969b). The amount of cyclic AMP produced is extremely low during the vegetative stage, but increases strikingly during the phase of aggregation (Fig. 12.4). Smaller but nevertheless detectable quantities are produced by the migrating pseudoplasmodia (Bonner *et al.*, 1969).

It should be noted that these phenomena involving cyclic AMP had been observed long before cyclic AMP was discovered. The chemotactic substance responsible for aggregation was initially referred to by Bonner and his colleagues as "acrasin," and was identified as cyclic AMP after high acrasin activity was detected in both human urine and extracts of

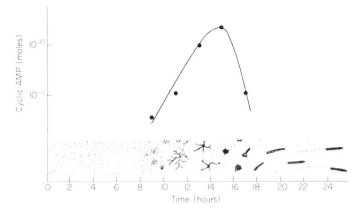

Fig. 12.4. Increased production of cyclic AMP during aggregation of *D. discoideum*. The amount of cyclic AMP released increases by two orders of magnitude during aggregation, then falls again as the slugs begin to migrate. From Bonner *et al.* (1969).

E. coli (Konijn *et al.*, 1967, 1968; Bonner, 1969). Although it now seems clear that acrasin and cyclic AMP are identical in *Dictyostelium,* it is possible that other compounds serve as the acrasin for other genera. For example, *Polysphodylium pallidum* clearly produce cyclic AMP, but the amoebae do not seem to be attracted by it (Konijn *et al.*, 1969b). Interestingly enough, these organisms also differ from *Dictyostelium* in that they apparently do not produce phosphodiesterase, which had of course been referred to at one time as "acrasinase" (see Bonner, 1967).

The production of cyclic nucleotide phosphodiesterase by *D. discoideum* was first established by Chang (1968), who also showed that the organism released large quantities of the enzyme into the extracellular medium. The activity of this phosphodiesterase is thought to play an important role in maintaining a concentration gradient of cyclic AMP, thus enabling the aggregating amoebae to orient themselves. A mathematical model based on this concept has been presented by Keller and Segel (1970). The properties of slime mold phosphodiesterase seem to be quite similar to those of mammalian phosphodiesterase, although it resembled the enzyme from *E. coli* in its lack of sensitivity to inhibition by methylxanthines (Chang, 1968). In contrast to the phosphodiesterase of *Serratia marcescens* (Okabayashi and Ide, 1970), the slime mold enzyme did require Mg^{2+} (Chang, 1968).

It will perhaps be obvious from what has been said to this point that in *Dictyostelium,* presently the most primitive multicellular organism in which cyclic AMP is known to occur, the nucleotide seems to play a role analogous to that played by hormones in higher organisms. In other words, it seems to function more as a first messenger than as a second messenger. It will be interesting in future studies to follow the role of cyclic AMP up through the phylogenetic tree, to see at which point its role as a second messenger first emerged, and for which hormone it first served as a second messenger.

The effects of cyclic AMP on slime mold amoebae have been recorded in a time-lapse motion picture by Konijn and his colleagues. Still photographs cannot begin to convey the truly remarkable nature of these effects, but Fig. 12.5 represents an attempt to illustrate two of them. In

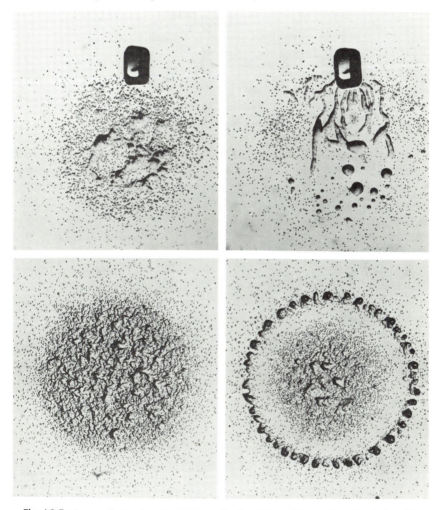

Fig. 12.5. Some effects of cyclic AMP on *D. discoideum*. See text for description. From Konijn *et al.* (1968). Copyright, University of Chicago Press.

the photograph in the upper left-hand corner, an agar block containing cyclic AMP has been placed near a drop of amoebae that are just beginning to aggregate. The photograph to the right of this was taken 57 minutes later, and it will be seen that most of the cells have now joined aggregates streaming toward the agar block. Higher concentrations of cyclic AMP will attract amoebae over longer distances, or, for the same distance, will attract them more rapidly, and this has been used as the basis of a semiquantitative assay for cyclic AMP (Konijn, 1970).

The production of cell adhesiveness is illustrated in the lower set of photographs. When a few drops of *Dictyostelium* amoebae are placed on an agar plate, the cells radiate outward as shown in the left-hand panel. When the agar contains cyclic AMP (Fig. 12.5, lower right panel), the cells spread but as they do so they give rise to a tightly clumped ring of cells. As this ring expands, it may break up into a ring of beads, as in this photograph. Konijn *et al.* (1968) have suggested, as a possible explanation of this orientation, that the cells may lower the concentration of cyclic AMP at the site of the drop by releasing phosphodiesterase, thus producing an ever-widening circular gradient which orients the ring of amoebae in their outward expansion.

Returning now to the life cycle of *D. discoideum,* the pseudoplasmodia continue their migration until the relative humidity falls below a certain point. Other factors also influence the duration of migration (Bonner, 1967), but the relative humidity seems to be of major importance. Migration may continue for many hours or even days, and during this time a certain amount of cellular differentiation will already have occurred. The cells at the tip of the advancing cell mass stop moving first; the cells of the posterior portion, which will later form the spores, continue moving until they are positioned underneath the main mass. Then the upper cells start turning into stalk cells (apparently under the influence of cyclic AMP, as discussed below). These are continuously added to from above, thus pushing the first-formed stalk cells down through the center of the cell mass toward the substratum. These stalk cells are rigid elongate cells with thick cellulose walls. When they finally reach the substratum, the continuous piling of new cells on top of what has now become a rigid stalk causes the rest of the cell mass (except for the "rear-guard cells," which were previously at the posterior tip of the migrating pseudoplasmodium, and which now surround the stalk at its base) to rise into the air. The rear-guard cells meanwhile turn into stalklike cells which together form the basal disc (Fig. 12.6).

As the culmination stage continues, the cell mass is gradually pushed higher and higher into the air, eventually becoming a mature fruiting body. Final spore differentiation begins on the upper edge of the prespore mass

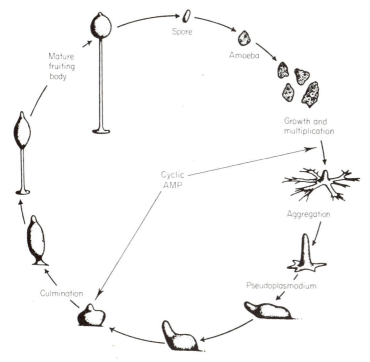

Fig. 12.6. The life cycle of *Dictyostelium discoideum,* illustrating stages at which cyclic AMP is known to be active. See text for further description.

and progresses inward and downward from that region. The spores are thus released from the lower part of the fruiting body, following which the life cycle begins anew.

Evidence that cyclic AMP plays a role in differentiation, in addition to its role in initiating aggregation, was provided recently by Bonner (1970). He found that when the agar on which the amoebae were placed contained high concentrations of cyclic AMP (as in the lower right-hand photograph in Fig. 12.5), some of the cells in and near the advancing ring would turn into stalk cells, often in clusters but sometimes separately. This was detectable 24 hours after inoculation, and by 48 hours was true of a large number of cells. By contrast, in control plates to which cyclic AMP has not been added (Fig. 12.5, lower left panel), most of the cells aggregate and form normal fruiting bodies, and the few remaining cells that fail to aggregate never differentiate into stalk cells. Stalk cell formation in response to cyclic AMP was even more striking in the presence of agents which interfere with normal development, such as actinomycin D, puromycin, or ethionine. The isolated stalk cells formed in response to exogenous cyclic AMP were indistinguishable, chemically and morphologically, from stalk cells formed in the normal course of development. It was known from earlier studies that "acrasin" is produced during culmination only at the anterior of the rising cell mass, in the region where the stalk cells are in the process of formation. Taken together, these observations provide strong evidence that cyclic AMP is required for the differentiation of at least one type of cell. Perhaps the closest analogy to this process in vertebrates would be the growth of hair and other structures which develop from epidermal cells (including scales, feathers, antlers, etc.). It is of interest that Adachi and Kano (1970) have presented evidence to suggest that cyclic AMP may indeed play a role in the growth of human hair.

Chassy *et al.* (1969) reported that other cyclic 3′,5′-nucleotides shared the ability of cyclic AMP to attract amoebae to some extent, but lacked the ability to produce the adhesiveness required for aggregate formation. Other nucleotides showed no chemotactic activity. On the other hand, a great variety of purine and pyrimidine derivatives, including both cyclic AMP and 5′-AMP at certain concentrations, were found to stimulate the rate of differentiation, i.e., more fruiting bodies were produced in a given time in the presence of these compounds than in their absence. The significance of these observations is unclear at present. Bonner (1970) found that 5′-AMP did not share the ability of cyclic AMP to induce isolated stalk cell formation.

Based on earlier studies of "acrasin," it seems possible that other roles for cyclic AMP remain to be discovered. For a more complete account of the biology of the cellular slime molds, the interested reader is referred to the monograph by Bonner (1967). As of this writing, the mechanisms by which cyclic AMP acts to stimulate the aggregation and differentiation of these organisms were unknown. It will be recalled from Chapter 6 (Section VI,D) that cyclic AMP inhibits the aggregation of blood platelets, but whether this process is really comparable to the aggregation of slime mold amoebae remains to be seen.

In summary, cyclic AMP seems to play two main roles in the life cycle of *Dictyostelium discoideum*. First, it is responsible for initiating the aggregation of the unicellular amoebae into a mass of cells which eventually coalesce to form a multicellular pseudoplasmodium. Later, after the pseudoplasmodia have migrated to their sites of culmination, cyclic AMP is apparently responsible for inducing certain cells to differentiate into stalk cells. In stimulating aggregation, cyclic AMP functions as an extracellular first messenger rather than as an intracellular second messenger. Further studies of the role of cyclic AMP in these and other lower organisms should contribute to our understanding of hormonal evolution.

IV. CANCER

A. Experiments with Tumor Cells

Cancer has been defined by Temin (1970) as a disease in which cells are genetically altered in such a way that, in a normal environment, the altered cells respond to the normal controls of cell multiplication with more multiplication than normal cells. This type of malignant transformation can occur spontaneously or after exposure to viruses, chemicals, or X-irradiation, and may involve many different properties of cells. The possibility that cyclic AMP might be involved has now been explored by several groups of investigators, as summarized in this section.

The evidence that cyclic AMP plays a role in regulating bacterial gene expression in response to environmental changes, summarized earlier in this chapter, suggests that cyclic AMP could be involved in the production of cancer quite directly. Evidence that cyclic AMP might regulate cell multiplication in at least some mammalian tissues, such as salivary gland growth in response to isoproterenol (Chapter 6, Section V,E) or thymocyte proliferation in response to exogenous cyclic AMP (Mac-Manus and Whitfield, 1969, 1970), had suggested excessively high levels of cyclic AMP as a possible factor leading to malignant cell growth. By contrast, the results of other studies point to a defect in cyclic AMP formation as a more likely concomitant of the neoplastic process.

Bürk (1968) studied the effect of exogenous cyclic AMP on two virus-transformed derivatives of a line of hamster kidney fibroblasts in tissue culture. Cyclic AMP inhibited the growth of the original as well as the transformed cells, and was more effective in this regard than 5′-AMP. Bürk also found that methylxanthines were inhibitory, but were less effective against cells transformed by a polyoma virus than against the controls or cells transformed by Rous sarcoma virus. It was then found that the polyoma cells contained lower adenyl cyclase activity than either of the other cell types.

Ryan and Heidrick (1968) found that exogenous cyclic AMP inhibited the growth of HeLa cells and a line of chick fibroblasts, whereas equimolar concentrations of 5′-AMP and the dibutyryl derivative of cyclic AMP were inactive (suggesting that these cells might lack the ability to deacylate the derivative). In line with observations mentioned in Chapter 10 (Section XII), the effect of cyclic AMP was judged similar to that of steroid hormones. In further line with these observations, Ryan and Coronel (1969) found that cyclic AMP inhibited reproduction and ovulation when injected into female mice. Here again, the dibutyryl derivative was found to be less effective. Among the effects of cyclic

AMP in these experiments was a decrease in the size of the ovary, which effect could be reversed by the injection of human chorionic gonadotropin.

Heidrick and Ryan (1970) later extended their tissue culture studies to other cell lines and other cyclic nucleotides. Cyclic AMP at a concentration of 0.3 mM was found to inhibit four tumorigenic cell lines markedly but, interestingly, had only a slight effect on the growth of a nonmalignant diploid cell line. Cyclic GMP and also 2',3'-GMP were less discriminating, producing a similar degree of inhibition in all cell lines. The pyrimidines, cyclic CMP and cyclic UMP, did not affect the growth of any of the cells.

An interesting finding by Hilz and Tarnowski (1970) has been that exogenous cyclic AMP increased the glycogen content of HeLa cells in tissue culture, acting synergistically in this respect with insulin. By contrast, the dibutyryl derivative of cyclic AMP reduced the glycogen content and prevented the opposite effect of cyclic AMP itself. One of several possibilities to explain this type of result is that HeLa cells do have the capacity to convert the dibutyryl derivative to cyclic AMP (or to the monobutyryl derivative), and that the presence of the derivative tends to prevent the further conversion of cyclic AMP to an antimetabolite of cyclic AMP.

Gericke and Chandra (1969) found that injected cyclic AMP would inhibit the growth of a subcutaneously transplanted lymphosarcoma in mice. Cyclic UMP and cyclic IMP were also tested and found to be less effective than cyclic AMP.

Following a different approach, Friedman and Pastan (1969) reported that 10 mM cyclic AMP significantly increased the antiviral activity of interferon in chick fibroblasts. Cyclic AMP by itself had no antiviral activity, and equimolar concentrations of 5'-AMP or the dibutyryl derivative of cyclic AMP did not potentiate interferon.

More recent experiments at the National Cancer Institute (I. Pastan, personal communication) have disclosed that cyclic AMP is capable of causing some transformed cells in tissue culture to revert to cells morphologically similar to untransformed cells. Cyclic AMP and theophylline have been found to act synergistically in this regard, and the N^6-monobutyryl derivative of cyclic AMP is apparently much more effective than the parent nucleotide. Although the results of these experiments have not been reported in detail, they are mentioned here because of their potential importance.*

Some interesting albeit confusing studies of adenyl cyclase activity in

* See Johnson *et al.* (1971).

broken cell preparations of tumor cells have been reported, in addition to Bürk's studies mentioned previously. Granner *et al.* (1968) reported that adenyl cyclase activity was very low and unresponsive to glucagon in preparations of cultured rat hepatoma cells. An interesting incidental finding in these experiments was that the hepatoma cells did not respond to exogenous cyclic AMP with an increase in tyrosine transaminase activity, although they did respond to glucocorticoids.

Brown *et al.* (1970) studied a series of rat hepatomas and obtained results different from those of Granner and his colleagues. They found that in the presence of NaF the activity of adenyl cyclase was *greater* than normal; a correlation was observed between cyclase activity and growth rate, with the fastest growing tumors having the highest adenyl cyclase activity. The addition of epinephrine had no effect in the presence of NaF, but it was later found (Pennington *et al.*, 1970) that epinephrine stimulated hepatoma adenyl cyclase even more than it stimulated the enzyme from normal rat liver, in the absence of NaF. This latter observation was reminiscent of an earlier finding reported by Ney *et al.* (1969), as discussed later in this section. Brown *et al.* (1969) had previously noted that the adenyl cyclase activity of a chemically-induced breast tumor was substantially greater than in preparations of the corresponding normal tissue. Here again, epinephrine was apparently more effective in stimulating the tumor-derived enzyme than the enzyme from normal cells.

Glucagon was unfortunately not tested in the experiments by Brown and Pennington and their colleagues. Makman (1970) studied adenyl cyclase in a line of cultured human liver cells, and found activity much higher than in preparations obtained from normal rat or cat liver. This activity was stimulated by NaF or epinephrine, but, unlike the enzyme from normal liver, it was not stimulated by glucagon. Makman also studied HeLa cells and a line of mouse fibroblasts, and found adenyl cyclase from both sources to be sensitive to stimulation by both fluoride and epinephrine.

Since the regenerating liver has often been likened to a growing tumor, it is of interest that Becker and Bitensky (1969) found basal adenyl cyclase activity in preparations of control and regenerating rat livers to be similar, and similarly sensitive to stimulation by either glucagon or epinephrine.

Studies comparing adenyl cyclase activity in broken cell preparations of different types of cells are difficult to interpret because of the possibility of introducing artifacts in the course of homogenization. Such artifacts might suggest differences which are not functionally important in the intact cells or they might obscure differences which are important. An especially intriguing observation in this connection was made by Ney and his colleagues (1969) studying a spontaneously occurring rat adrenocortical carcinoma. This tumor produces corticosterone but only about 1–10% as much as normal adrenal glands. It was found that ACTH did not increase the level of cyclic AMP in the tumor either *in vivo* or in slices *in vitro,* under conditions that led to 20- to 50-fold increases in normal adrenal tissue. This was in line with the inability of ACTH to stimulate steroidogenesis in the carcinoma. However, when the cells were homogenized, the surprising finding was made that ACTH actually stimulated adenyl cyclase more in the tumor preparation than it did in homogenates of normal adrenals. Conversely, the effect of NaF was smaller in the tumor preparation than in the control. The reason for the failure of ACTH to increase the level of cyclic AMP in the intact carcinoma is thus unclear. One possibility is that ACTH receptors in the tumor cells are in some manner prevented from interacting with ACTH. Alternatively, it seems possible that ATP is prevented from gaining access to the active center of adenyl cyclase. In any event, these results point to a potentially important defect in the adrenal tumor cells that would not have been revealed by adenyl cyclase studies alone. A second defect noted in these studies was that the carcinoma did not respond to exogenous cyclic AMP with an increased rate of steroidogenesis. Thus the tumor cells seemed defective not only in forming cyclic AMP but in responding to it.

It has been tacitly assumed throughout most of this monograph that cells do not possess receptors for all hormones because the genetic information required for the synthesis of these receptors has not been expressed. An alternate thought, stimulated by the work of Ney and his colleagues, is that the receptors for many hormones are ordinarily synthesized, but, at the level of the phenotype, are prevented from functioning. It could be imagined that the receptors for most hormones could be found on the surface of most cells were it not for a "covering" characteristic for each type of differentiated cell. For example, the covering characteristic of rat adipocytes would contain many "holes," such that many receptors would be exposed, whereas the covering characteristic of adrenal cortical cells would be patterned in such a way that only ACTH receptors would be allowed to function. We should emphasize that we are thinking here in functional and not necessarily anatomical terms. The hypothetical covering that we are envisioning could operate at one level to prevent a receptor from interacting with its hormone, and at a deeper level to prevent the hormone-receptor interaction from affecting the catalytic activity of adenyl cyclase. This covering could be shifted in the adrenal carcinoma cells such that the ACTH receptors are not properly exposed, even though they are properly "lined up" with adenyl cyclase (as evidenced by the ability of ACTH to stimulate in broken cells). This speculation is offered primarily to emphasize the depth of our ignorance in this area, since this may not have been emphasized strongly enough in Chapter 2. We would hope that experiments can be designed in the future to examine the basis of hormonal specificity in more detail.

In the meantime, on the basis of the observations summarized to this point, it does not seem possible to draw any conclusions about the possible role of cyclic AMP in cancer.

B. Immunological Aspects

A few studies on the possible role of cyclic AMP in the immune response have been reported. Since it is thought by some investigators that cancer as a disease may result from a failure on the part of the immune defense system, these studies can be conveniently summarized here. As in the preceding section, however, the results are to some extent contradictory and do not seem conducive to the elaboration of unifying hypotheses.

Gericke *et al.* (1970) found that antibody production by spleen cells *in vitro* could be inhibited by a variety of cyclic nucleotides, with cyclic AMP and cyclic CMP being most potent and cyclic TMP the least potent. Stimulation was not observed at any concentration. By contrast, when the nucleotides were injected into mice, cyclic AMP and cyclic GMP slightly stimulated antibody synthesis at one dose but were inhibitory at higher doses.

Ishizuka *et al.* (1970) observed only stimulation of antibody synthesis when they injected cyclic AMP into mice, and this was considerably more substantial than the effect observed by Gericke and his colleagues. The dibutyryl derivative did not seem noticeably more potent than the parent nucleotide, while equivalent doses of 5'-AMP had no effect.

The possible role of cyclic AMP in regulating lymphocyte transformation has attracted the interest of several groups of investigators. As an essential part of the immune response, lymphocytes are transformed to cells which may produce antibodies to the stimulating antigen. Phytohemagglutinin (PHA) also stimulates lymphocyte transformation, and seems to initiate a sequence of events similar to those that occur in sensitized cells in response to the second application of an antigen. Smith *et al.* (1969) reported that the dibutyryl derivative of cyclic AMP inhibited PHA-induced lymphocyte transformation, and this was confirmed by Hirschhorn *et al.* (1970). Cyclic AMP itself was much less effective, and in fact was no more effective than other nucleotides or even adenosine (Hirschhorn *et al.,* 1970; Rigby and Ryan, 1970).

The possibility that phytohemagglutinin might act by reducing the level of cyclic AMP in lymphocytes was not supported by the experiments of Novogrodsky and Katchalski (1970). Under conditions where NaF or prostaglandins E_1 or E_2 stimulated apparent adenyl cyclase activity markedly, PHA had no detectable effect. On the other hand, PHA was not tested in the presence of one of the stimulators. Using human lymphocytes, Smith *et al.* (1971a) found that PHA produced a biphasic effect: an early rise in the level of cyclic AMP was followed, several hours later, by a fall below control levels. In contrast to Novogrodsky and Katchalski, who used rat lymphocytes, Smith and his colleagues found adenyl cyclase in homogenates of human lymphocytes to be significantly increased by PHA.

MacManus and Whitfield (1969) had previously reported that low concentrations of exogenous cyclic AMP, in the range of 10^{-8}–10^{-6} $M,$ stimulated DNA synthesis and proliferation of rat thymic lymphocytes. A variety of hormones had been found to produce a similar response, and presumably some of them act by way of cyclic AMP (see also MacManus and Whitfield, 1970; Whitfield *et al.,* 1970a,b). The dibutyryl derivative of cyclic AMP was also effective, but equimolar concentrations of other nucleotides had no effect. Higher concentrations of cyclic AMP were found to be inhibitory.

Using cultured peripheral blood lymphocytes, Hirschhorn *et al.* (1970) also observed stimulation of DNA synthesis in response to cyclic AMP. Other nucleotides and adenosine had no effect or were inhibitory; cyclic AMP, but not 5′-AMP, also seemed to stimulate RNA synthesis in these experiments. Novogrodsky and Katchalski (1970), using suspensions of rat lymph node lymphocytes, observed stimulation of RNA but not DNA synthesis in response to low concentrations of the dibutyryl derivative of cyclic AMP.

Abell *et al.* (1970) studied DNA synthesis in cultured peripheral blood lymphocytes obtained from normal donors and from patients with chronic lymphocytic leukemia. The extent of DNA synthesis in response to PHA was comparable in both preparations, but only that occurring in leukemic cells was found to be sensitive to inhibition by isoproterenol. Johnson and Abell (1970) later demonstrated synergism between isoproterenol and theophylline, and showed again that the response to PHA could be prevented by the dibutyryl derivative of cyclic AMP. They also mentioned that leukemic cells were more sensitive to the derivative than were normal lymphocytes, although the results reported by Hirschhorn *et al.* (1970) seem comparable. In any event, it would appear that at least one difference between normal and leukemic lymphocytes is that the latter are endowed with adrenergic β-receptors, and thus possess an additional mechanism for increasing their content of cyclic AMP. On the other hand, Smith *et al.* (1971a) have reported that adenyl cyclase in normal lymphocytes is also sensitive to β-adrenergic stimulation, and MacManus *et al.* (1971) obtained similar data using rat lymphocytes. A puzzling finding by Smith and his colleagues was that PHA and isoproterenol acted synergistically to raise the level of cyclic AMP in human lymphocytes.

The ability of β-adrenergic agonists or exogenous cyclic AMP to inhibit antigen-induced histamine release was mentioned previously in Chapter 6 (Section VI,D). It will not be discussed further, therefore, except to note that the significance of the release of histamine and other autacoids as part of the immune response has still not been established.

Although the overall role or roles of cyclic AMP in the immune response cannot be defined at present, the observations summarized in this section may provide part of the foundation for future progress. It is at least conceivable that one or more defects related to cyclic AMP may be involved in the etiology of cancer.

V. OTHER ORGANISMS

It remains to be noted that cyclic AMP has been found in other primitive organisms in addition to those mentioned earlier in this chapter. These would include yeast (Chance *et al.*, 1965) and lower fungi (G. Meissner, G. A. Robison, and M. Delbrück, unpublished observations), although certainly the function of cyclic AMP in these organisms is presently far from clear. The presence of cyclic AMP in the cells of higher plants has still not been unequivocally established, although a recent report by Pollard (1970) suggests that barley seeds may be capable of forming cyclic AMP under some conditions.

VI. CONCLUDING REMARKS

In summary of this chapter and the entire monograph, we now know that cyclic AMP was utilized as a regulatory agent at a very early stage in the evolution of life. Long before there were metazoa, certain bacteria were using it to help them decide when to synthesize certain enzymes that were not required under all conditions. Other bacteria may have used it for other purposes. Somewhat later, the cellular slime molds picked it up and began using it in much the same fashion that our own bodies use hormones. Its most obvious and best understood function in multicellular animals is to serve as an intracellular second messenger for many hormones, which can be thought of in this context as first messengers. The greater part of this monograph represents an attempt to summarize what was known about this particular role at the time the words were written. It is possible, however, that cyclic AMP plays many other important roles in our own bodies and elsewhere that remain to be discovered.

It now seems obvious, based only on what we know about cyclic AMP, that this nucleotide was and is essential to our own survival as a species. This was discussed briefly in Chapter 7, and it was our intention at one time to continue that discussion in this concluding chapter. We have now decided not to do this, partly because the point of it seems so obvious, partly because so many of the details remain to be filled in, but mainly because we are so anxious to get the book finished. After a long dark night of writing, during which time we have witnessed the appearance of most of the papers referred to in the bibliography, the urge to return to the laboratory is almost overwhelming.

We will conclude by stating explicitly a point which may or may not seem obvious to all readers. This is that because cyclic AMP does not seem essential for many basic processes to occur (as discussed at the beginning of Chapter 2), defects in the formation, metabolism, or action of this nucleotide are likely to be of major importance in various forms of human disease. To use an example from this concluding chapter, we know that bacterial mutants lacking cyclic AMP can survive under certain well-defined environmental conditions, even though they will not be able to adapt to certain other conditions. By contrast, it is extremely difficult to imagine an organism unable to synthesize ATP or its equivalent, for such an organism could hardly survive under any conditions. Just so, in humans, we can imagine many cellular defects involving cyclic AMP and other regulatory mechanisms that would be compatible with life for long periods of time, at least long enough for the symptoms of such defects to be recorded in our textbooks of medicine and pathology. By the same token, defects at a more basic level can be imagined, but the signs and symptoms of them will be less frequently encountered.

We can therefore expect a great deal of research in the coming years to center on the role of cyclic AMP in mammalian physiology, to be concerned with elucidating the many possible defects that may lead to human disease. Our hope is that the results of this research can eventually be applied to increase the quality of human life. Although seldom acknowledged in monographs of this sort, the chief end of *all* scientific research either is or ought to be the promotion of human happiness. Research on cyclic AMP need not be clinical, of course, in order to contribute to this larger goal. For example, we expect other research with other organisms to contribute insights that will help us to understand who and what we are, and how we came to be so. Properly regarded, such insights ought to be conducive to human happiness.

Insofar as this monograph is concerned, our hope for it is that it will have contributed, in however small a way, to its own obsolescence.

Appendix on the Assay of Cyclic AMP

I. INTRODUCTION

Although a great deal of effort has gone into the development of suitable assays for cyclic AMP, as of this writing an ideal assay (one which is sensitive, accurate, and easy to establish and maintain) is not available. The first assay system for cyclic AMP to be reported was based on the ability of cyclic AMP to enhance the rate of activation liver phosphorylase (Rall and Sutherland, 1958), and with minor modifications it is still being used (Rall and Sutherland, 1960; Butcher *et al.*, 1965; Kakiuchi and Rall, 1968a). However, these modifications have been developed over a period of years, and they are scattered through several publications. Therefore, this appendix is designed to describe the phosphorylase assay in detail, and to present protocols for the preparation of the reagents, including the necessary enzymes. In addition, very brief descriptions of the other assays which have been reported are offered in Section IV.

II. THE ASSAY OF CYCLIC AMP BY THE ACTIVATION OF DOG LIVER PHOSPHORYLASE

A. Description, Characterization, and Limitations

This assay of cyclic AMP is based upon its ability to accelerate the rate of conversion of inactive phosphorylase (ILP) to the active form of the enzyme (LP). Aliquots (10 μl) of a reagent containing 1.14 μmoles of tris buffer (pH 7.4), 68 nanomoles of $MgSO_4$, 54 nanomoles of ATP, 27 nanomoles of caffeine, and 3 nanomoles of EDTA were pipetted into 7 × 50 mm culture tubes. Following this 10-μl of 10 mM tris (pH 7.4) blanks, standard cyclic AMP solutions in 10 mM tris (pH 7.4), or suitably diluted unknowns (in 10 mM tris, pH 7.4) were added to the tubes. The range of cyclic AMP standards used routinely in this system is from 0.1 to 6.0 picomoles using solutions of cyclic AMP ranging from 0.01 to 0.6 μM. The first stage of the assay was started by adding, on time, 10 μl of the enzymatic reagent. This addition consisted of 33–133 μg of purified glycogen, 0.01–0.02 units of ILP, 1–2.5 μl of an 11,000 g supernatant fraction of a 33% homogenate of dog liver containing dephosphophosphorylase kinase (PK), and 60 μmoles of 1-epinephrine in 10 mM tris (pH 7.4). The amounts of reagents vary because each dog liver supernatant fraction must be empirically balanced with optimal amounts of ILP and glycogen.

The tubes are kept in an ice water bath for exactly 30 min, and are then transferred to a 25°C water bath for 20 min. During this incubation (the first stage), ILP is transformed to LP at a rate dependent upon the concentration of cyclic AMP. The 30-min incubation in the ice water bath is not essential to the assay, but is routinely used because it provides a sensitization of the assay system to cyclic AMP. The amount of LP present at the end of the first stage is determined by adding 100 μl of a reaction mixture adjusted to pH 6.1 and containing 0.22 μmoles of 5'-AMP, 4.83 μmoles of glucose-1-phosphate, 12.7 μmoles of NaF, and 0.5 mg of glycogen, and transferring the tubes to 37°C water bath for a 30-min incubation (second stage). The reaction is terminated by the transfer of 100-μl aliquots of the reaction mixtures to tubes containing 0.5–0.75 mg of I_2, 1.0–1.5 mg of KI, and 0.02 meq of HCl in 0.5 ml. These mixtures are diluted to 7.5 ml with distilled water and the absorbance of the glycogen-iodine combination measured at 540 mμ with distilled water as the reference. The absorbance of the tris blanks is subtracted from that of the cyclic AMP standards, and the differences plotted against the known cyclic AMP concentrations. The concentrations of the cyclic nucleotide in the unknowns are determined by means of this standard curve. Four or six tris blanks are routinely incorporated with each assay, and each standard and diluted unknown is assayed in duplicate.

Typical cyclic AMP standard curves are illustrated in Fig. A.1. Curve A is the complete assay system including the 30-min cold preliminary incubation. When the cold preliminary incubation step was omitted, a decrease in the sensitivity of the assay was noted (curve B). The sensitiza-

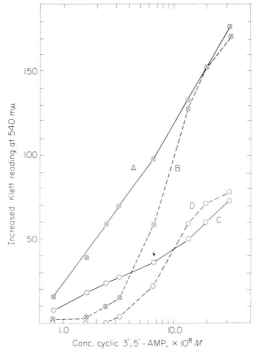

Fig. A.1. Effects of glycogen and cold preincubation on the assay of cyclic AMP. Curve A, complete system; curve B, without cold preincubation; curve C, without glycogen added in the first stage; curve D, with neither cold preincubation nor added glycogen. (From Butcher *et al.*, 1965.)

tion was most marked at the lower concentrations of cyclic AMP, and the curves intersected at 133 nM. While the effect of the cold preliminary incubation on the assay system seems of little if any physiological significance, it has been very useful because it permits accurate detection of cyclic AMP levels lower than would otherwise be possible. The mechanism by which cold preliminary incubation causes the sensitization of the assay system is not clear—the liver supernatant fraction (that is, PK) and cyclic AMP are absolutely required for the effect, while ILP is not (Butcher *et al.,* 1965).

When glycogen was omitted from the first stage of the assay, the increased optical densities produced in response to given cyclic AMP concentrations were reduced by about one-half (curve C). The addition of glycogen to the second stage (that is the stage at which active phosphorylase was assayed) did not restore the stimulation and therefore it seemed unlikely that the addition of glycogen was acting as a primer for phosphorylase. Glycogen did not protect the enzymatic components of the first stage against thermal inactivation, or did it alter the previously reported stimulatory effect of 10 mM NaF (Rall and Sutherland, 1958). The most useful property of the effect of glycogen on the standard assay system is that while it enhanced phosphorylase activation in the presence of cyclic AMP, the activation in the absence of the cyclic nucleotide was virtually unchanged. For example, while the addition of glycogen to a final concentration of 0.66% in the first stage of the assay system was without effect on the baseline, the response to 133 nM cyclic AMP was doubled (Butcher *et al.,* 1965). Krebs *et al.* (1964) reported that glycogen caused an increased activation of purified muscle phosphorylase *b* kinase in the presence of ATP, and also that the polysaccharide caused a decrease in the K_m of nonactivated phosphorylase *b* kinase for phosphorylase *b*. However, it seems unlikely that the two systems are completely analogous since Krebs *et al.* found that the effects of glycogen and cyclic AMP on the activation of phosphorylase *b* kinase were additive, which is in contrast to the response in the liver system. However, experiments with purified liver PK, complimentary to those of Krebs *et al.* with the muscle enzymes, have not been performed.

The stimulatory effect of glycogen preparations was apparently due to polysaccharides rather than to contaminating materials. Purification of glycogen with Norite or by chromatography on ion exchangers increased the activity of the preparations. Heating at 90°C in 0.1 N HCl for from 30 min to 4 hr enhanced the activity of the glycogen preparation. In addition, this treatment resulted in increased ultraviolet absorbancy, with a maximum at 260 mμ. Most or all of this ultraviolet observing material was retained on Dowex-2 at neutral pH, while the activity capable of stimulating the assay system was not. The ultraviolet absorbing substances were eluted with dilute HCl, and were inactive in the assay system. In addition, the active material was not adsorbed on Dowex-50 at either acid or neutral pH. Amylase and amylopectin were found to be more active in the assay than oyster glycogen on a weight basis, and stimulated both the baseline and the response to cyclic AMP. Rabbit liver glycogen at 0.66% was without effect on the baseline and was somewhat more active than the treated oyster glycogen, while maltose, gentiobiose, and chitin were inactive in preliminary experiments (Butcher *et al.*, 1965).

The assay of cyclic AMP by the liver phosphorylase activation method may be complicated by the presence of interfering materials. Among these are ADP, glucose-1-P and glucose-6-P, UDP-glucose, Ca^{2+}, and other unknown factors (Butcher *et al.*, 1960; Makman and Sutherland, 1964; Murad, 1965). However, anion- and cation-exchange chromatography have been used to good advantage in purifying cyclic AMP from complex mixtures such as heat-treated broken cell preparations, urine, and intact tissue preparations; the use of ^{3}H-cyclic AMP as a marker for the recovery of cyclic AMP in the experimental samples during fractionation has been validated and extensively used. Several sizes of ion-exchange columns have been utilized, and the method currently used has been adopted for convenience. In addition, the specificity of the assay system in terms of activators (that is, compounds with cyclic AMP-like activity) can be greatly enhanced by using purified cyclic nucleotide phosphodiesterase, and enzyme which is highly specific for cyclic 3′,5′-nucleotide bonds. Thus, any "cyclic AMP-like" activity found in extracts after ion-exchange chromatography which is destroyed by incubation with the phosphodiesterase is very probably cyclic AMP. In addition, the presence or absence of inhibitors of the effects of cyclic AMP on the assay system can be detected by adding known amounts of crystalline cyclic AMP along with the experimental extracts in the assay system.

The most serious problem which has been encountered with the assay system is a quantitative variability. That is, a particular set of extracts assayed under identical conditions in separate assays may give figures which are quantitatively different. The reason for this is unknown. The relative cyclic AMP activity in the extracts is the same — that is, if sample B contains twice as much cyclic AMP as sample A on day 1, it will also contain twice as much on day 2. However, the absolute values of cyclic AMP may vary. Therefore, this assay system has not been regarded as capable of measuring absolute levels of cyclic AMP but rather as a system of measuring relative cyclic AMP levels. Nonetheless, when measurements of tissue levels of cyclic AMP obtained with this method are compared with those reported by others using different methods, they have been virtually superimposable.

B. Tissue Fixation

The most critical step in the measurement of cyclic AMP levels is the fixation of the tissue. The fixation must be rapid and thorough, for the rate of turnover of cyclic AMP in tissues is high, and increased cyclic AMP levels are evanescent. Improper fixation may result in false negatives. For example, Namm and Mayer (1968) found that cyclic AMP levels in hearts exposed to epinephrine were unchanged when frozen by dropping them into freon at the temperature of liquid nitrogen, but were increased when the hearts were clamped between aluminum blocks chilled to the temperature of liquid nitrogen which is a more rapid fixation procedure. By contrast, false positives due to poor fixation have not been observed or reported.

In general, it appears that fixation by very fast freezing, for example, by clamping between aluminum blocks chilled in liquid nitrogen, followed by pulverization in impaction mortars chilled to the temperature of liquid nitrogen, and finally rapid homogenization of the frozen, powdered tissue in 0.05 or 0.1 N HCl containing about 10,000 dpm of ^{3}H-cyclic AMP is the most reliable method of fixation. Certain soft tissues, for example, the rat epididymal fat pad, or rat adrenal glands have been homogenized directly in 0.1 N HCl containing ^{3}H-cyclic AMP with good results (Butcher et al., 1965; Grahame-Smith et al., 1967). However, comparison of the more complicated methods involving freezing and the easier direct homogenization method is strongly recommended before the latter method is employed routinely.

C. Ion-Exchange Chromatography

After homogenization in HCl, the samples were placed in a boiling water bath for 12–15 min. The samples could be stored at 0°–4°C for up to 2 days or treated further at once if convenient. After cooling, the acid mixtures were centrifuged at 8000 g for 20 min and the supernatants decanted and applied to Dowex-50 columns, 15 cm high and 0.7 cm in diameter, which had been equilibrated with 0.05 N HCl. The starting materials were allowed to run over the columns, and then 0.05 N HCl sufficient to bring the volumes of each collected run through to 25 ml was added to the reservoirs. These first 25-ml fractions were discarded. Next, 40 ml of 0.05 N HCl were added to the reservoirs and allowed to run through the column. The 40-ml eluates were collected as the cyclic AMP-containing fractions. Sufficient tris buffer (pH 7.4) was added to each 40-ml fraction to bring the final concentration to 25 mM, and then the pH was adjusted to between 7.5 and 8.0 with 1.0 N NaOH, using phenol red as an indicator. These neutralized fractions were then applied to Dowex-2 columns, 5.0 cm high and 0.7 cm in diameter, which had been previously equilibrated with 0.01 M tris buffer, pH 7.4. The neutralized fractions were allowed to run through and each column was washed with 15 ml of 0.01 M tris, pH 7.4. The run-through and wash fractions were discarded. The tips of the columns were then carefully rinsed and 50-ml serum bottles were placed under the columns. Exactly 20 ml of 0.05 N HCl were added to each column and the total eluates were collected in the serum bottles. The serum bottles were tightly stoppered, and the contents were frozen in a dry ice-methanol bath and lyophilized to dryness. The contents of the serum bottles were redissolved in an appropriate amount of 0.01 M tris (pH 7.4) (usually 1.0 ml) and the pH's adjusted to between 7.2 and 7.6. Aliquots (usually 10% of the volume in which the samples were redissolved) were pipetted into a modified Bray's solution [see Section III,A,(9)] and the ^{3}H-cyclic AMP measured in a liquid scintillation counter for the recoveries of cyclic AMP, and the remainders of the extracts were available for the assay of endogenous cyclic AMP.

While there is no theoretical limit on the number of samples which can be fractionated on the columns at one time, we have in practice only rarely processed more than 36 because of the limited availability of columns and lyophilization equipment. Assays containing more than 18 experimental samples are impractically long and tedious, since each sample is assayed at two or more dilutions and each dilution in duplicate. However, errors due to the quantitative variations in the assay which occasionally occur can be detected by overlapping samples in multiple assays.

III. REAGENTS FOR THE ASSAY OF CYCLIC AMP

A. NonEnzymatic Reagents

All reagents were made in glass distilled water and stored at $-20°C$ frozen unless noted otherwise.

(1) Glucose-1-phosphate (di K$^+$), purchased from Nutritional Biochemical Corporation, Cleveland, Ohio, was recrystallized in the following manner: 2 liters of a 5% solution of G-1-P were chilled, 3 gm of Norite

were added, and the pH was adjusted to 3.5 with glacial acetic acid. The solution was brought to a concentration of 50% ethanol, filtered, and adjusted to pH 8.0 with KOH (as measured with a glass electrode), additional ethanol being added to maintain a 50% concentration. After storage at 3°C overnight, the crystals were collected and dissolved in 1.2 liters of glass distilled water and filtered at room temperature. The filtrate was again brought to a concentration of 50% ethanol. After storage at 0°C overnight, the crystals were collected by decantation and centrifugation and washed successively with 95% ethanol, absolute ethanol, absolute ethanol and absolute ether (1 : 1), and absolute ether. The yield of dry product was usually about 75% of the starting material. Glucose-1-phosphate purified in this manner may also be purchased from Nutritional Biochemical Corporation, Cleveland, Ohio.

(2) Shellfish glycogen (Nutritional Biochemical Corp.) was purified by passing 200 ml of a 10% glycogen solution in glass distilled water over a 15 cm high, 1 cm diameter column of Dowex-2 at pH 7.5. The glycogen was not retained by the ion exchanger to any appreciable extent, and the collection of the purified glycogen fraction was begun after 1 bed volume of fluid was passed through the column.

(3) Stock Iodine–2% I_2 (purchased from Merck & Co., Inc., Rahway, New Jersey, reagent grade) was dissolved in 4% KI (Merck & Co., Inc., reagent grade) and stored in the dark in a dark bottle at room temperature. Two to four weeks were allowed for the I_2 crystals to dissolve.

(4) Cyclic AMP standards–A stock solution of cyclic AMP (Schwarz Bioresearch, Inc., Orangeburg, New York) was prepared to a concentration between 10^{-4} and 10^{-3} M in 0.01 M tris and neutralized. The concentration was calculated from the absorbancy at 260 mμ and the molar extinction coefficient (14.2×10^{-3}). The stock solution was diluted to 1.0 μM in 0.01 M tris, and from this the standard solutions were prepared to contain either 0.01 M tris or 0.01 M tris and 0.01 M NaF and cyclic AMP at final concentrations of 0.01, 0.02, 0.04, 0.08, 0.1, 0.2, 0.3, 0.4, and 0.6 μM. Small aliquots (usually a 2 weeks' supply) were transferred to culture tubes for daily use.

(5) GEF–used in the assay of phosphorylase phosphatase and also in preparing phosphorylase reaction mixture for second stage of assay [see (6)]: 50 gm recrystallized glucose-1-P, 15 gm purified glycogen, 500 ml glass distilled water, and 400 ml 0.75 M NaF. Stored at $-20°C$.

(6) Phosphorylase reaction mixture (used in second stage of cyclic AMP assay): 360 ml of GEF (mixture F), 96 ml 0.025 M 5'-AMP, 38.4 ml 0.75 M NaF, and 585 ml glass distilled water. This is divided into 10-ml aliquots, stored at $-20°C$, and thawed for use.

(7) VCA (required for first stage of cyclic AMP assay): 51.6 ml glass

distilled water, 31.84 ml 1.0 M tris (pH 7.4), 6.84 ml 0.2 M MgSO$_3$, 13.68 ml 0.08 M ATP (pH 6.9), 5.44 ml 0.1 M caffeine, and 0.6 ml 0.1 M ethylenediaminetetracetic acid (pH 7.0). This is divided into 2- or 3-ml aliquots, stored at $-20°C$ and thawed for use.

(8) Calcium phosphate gel was prepared as described by Keilin and Hartree (1938), glass distilled water being used at all times.

(9) Tritiated cyclic 3',5'-AMP was measured in a liquid scintillation counter with a scintillator solvent solution consisting of 6.0 gm of naphthalene (Matheson, Coleman and Bell), 40 mg of 1,4-bis-2-(4-methyl-5-phenyloxazolyl)benzene (Packard), 400 mg of 2,5-diphenyloxazole (Packard), and 10 ml of methanol (Matheson, Coleman and Bell) in *p*-dioxane (Matheson, Coleman and Bell, spectral grade) to a total volume of 100 ml. The final tissue extract was sampled, the volume being 10% of that in which it was redissolved, and 0.01 M tris, pH 7.5, was added to a total volume of 1.0 ml. To each polyethylene vial, 20 ml of the scintillator solution were added. The starting tritiated cyclic 3',5'-AMP was counted with each series and was used as the reference point. The samples were occasionally checked for quenching by the addition of known amounts of tritiated cyclic 3',5'-AMP to the vials.

(10) Ion-exchange resins. Dowex-2 and Dowex-50 were purchased as analytical grade resins, AG2-X8 (chloride form) and AG50-X8 (hydrogen form), both 100–200 mesh, from Bio-Rad Laboratories, Richmond, California. Slurries of these resins were washed three times each with at least three volumes of distilled water followed by 2 N NaOH, distilled water, 2 N HCl, distilled water, 0.1 N NaOH, distilled water, 0.1 N HCl, and then with glass distilled water until chloride free. The slurries were then adjusted to 0.1 N HCl and stored at 2–4°C.

(11) ^{3}H-Cyclic AMP was purchased from Schwarz Bioresearch, Orangeburg, New York, at a specific activity of 1.4 C/mmole. It was purified by diluting it to contain 0.05 N HCl in a final volume of 10.0 ml and was fractionated on Dowex-50 and Dowex-2 as described in Section II, C. Tris buffer (pH 7.4) was added to the Dowex-2 eluate to a final concentration of 0.01 M and the pH adjusted to 7.4 with 1.0 N NaOH. The purified ^{3}H-cyclic AMP was stored at −20°C. Dilutions were made so as to contain about 10^5 dpm/ml, and routinely 0.1 ml of this dilution was included in the HCl used to homogenize each sample.

B. Preparation of ILP

Inactive liver phosphorylase is the key reagent in the assay of cyclic AMP. The preparation of ILP entails first the isolation of active liver phosphorylase (LP) and subsequent inactivation of the LP with purified phosphorylase phosphatase (PP). Methods for the direct preparation of ILP (that is, the purification of the inactive form of the enzyme) have also been described (Appleman *et al.,* 1966a; Kakiuchi and Rall, 1968a). This procedure is based on that described by Wosilait and Sutherland (1956), *et seq.*

1. Preparation of LP–phosphorylase

Activity of LP–phosphorylase was measured by a standard assay based on the rate of liberation of inorganic phosphate from glucose-1-P in the presence of glycogen; i.e., phosphorylase was measured in the direction of polysaccharide synthesis. A stock solution of glucose-1-P (0.036 M adjusted to pH 6.1 with HCl) containing 4.03 mg of glycogen per ml and 0.1 M NaF was stable in the frozen state and constituted the basic reagent for assay. Routinely, this reagent contained $1.4 \times 10^{-3}\,M$ 5-AMP. Enzyme dilutions were made immediately before assay in cold 0.1 M NaF. The reaction was started by the addition of 0.2 ml of enzyme to 2.8 ml of the assay reagent. Aliquots of the reaction mixture (0.5 ml) were transferred to 1.0 ml of cold 5% TCA at zero time and after incubation for 10 min at 37°C. (The remaining reaction mixture was used for an iodine-starch test, providing a rapid estimate of phosphorylase activity and also evidence that polysaccharide formation accompanied the formation of inorganic phosphate.) The reaction mixture–TCA combination was diluted to 5.0 ml by the addition of 3.5 ml of cold glass distilled water, mixed well, and centrifuged for 10 min at 2500 g at 4°C. Aliquots (usually 1 ml) of the supernatant fractions were analyzed for phosphate by the method of Fiske and SubbaRow, as adapted to the Klett-Summerson colorimeter. One unit of enzyme was defined as that amount which caused the liberation of 1.0 mg of inorganic phosphorus in 10 min when the percentage conversion of glucose-1-P was in the range of 12–22% and specific activity was expressed as units per mg of protein (Table A.I).

a. Step 1. Preparation of Filtrate. Medium sized, well-fed dogs of either sex were obtained from the pound and were injected intraperitoneally with 3 ml of a 1:1000 epinephrine solution (Parke Davis, Detroit, Michigan) 5–10 min before a lethal dose of sodium secobarbital was injected intravenously. The carotid vessels were severed and the thorax and abdomen were opened. A large cannula was inserted into the portal vein and the inferior vena cava was cut just below the heart. Several liters of cold 0.2 M NaF were perfused through the liver with intermittant pressure applied to the inferior vena cava just above the diaphragm, which

TABLE A.I. Summary of Purification and Yield[a]

Fraction	Total units	Specific activity[b]	Percentage original activity
Step 1. Filtrate	25,200	0.31	100
Step 2. 0.0–0.66 ammonium sulfate	22,786	0.44	90
Step 3. Heated 0.0–0.66 supernatant fluid	12,491	0.46	50
Step 4. 0.41–0.80 ammonium sulfate	11,400	0.57	45
Step 5. 0.41–0.80 after dialysis	10,300	0.55	41
first alcohol	9,600	4.15	38
second alcohol	8,900	5.7	35
third alcohol	7,650	6.2	30

[a] Step 1 through step 5 livers from three dogs were fractionated separately through step 3, then pooled for steps 4 and 5. The specific activity of the homogenate was about one-third that of the filtrate. No corrections have been made for sampling or for loss in precipitates. (From Sutherland and Wosilait, 1956.)

[b] See the text for unit of measurement.

increased the filling of the liver and the effectiveness of the perfusion. At the start of the perfusion, 2 ml of 1:1000 epinephrine were injected into the rubber perfusion tubing leading to the cannula. An inhibitor of phosphorylase phosphatase, NaF, and epinephrine (acting via increased cyclic AMP levels) were used in an effort to maintain the concentration of active phosphorylase at the highest level possible. The chilled perfused liver was removed and after separation of the gallbladder was immersed in chilled 0.1 M NaF. The liver was sliced into small pieces and 100-gm portions were homogenized in a Waring blendor for 2.5 min in 400 ml of cold 0.1 M NaF containing 5 mM K_2HPO_4. These and all subsequent steps were carried out at 2°–4°C unless specified otherwise. The pooled homogenates were adjusted to pH 5.7 with 1.0 N acetic acid. An amount of Hyflo Super-Cel (Fisher, Rahway, New Jersey) equal to the weight of the liver was mixed with the homogenate and the mixture was filtered under 20 pounds pressure by using a No. 0 pad in a Hormann filter press (Hormann Co., Inc., Milldale, Connecticut). The cake was washed by the addition of cold 0.1 M NaF to the top of the cake, the volume of the wash fluid equaling 1.2 times the weight of the liver.

b. Step 2. First Calcium Phosphate Gel and 0.66 Ammonium Sulfate. The turbid tan or pink filtrate was adjusted to pH 6.5 with 1.0 N KOH. Part of the phosphorylase inactivating enzyme was removed by adsorption on calcium phosphate gel. One-twelfth volume of gel was added to the filtrate and stirred occasionally for 15 min before centrifugation at

2000 g for 15 min. (Further purification of the inactivating enzyme is described in Section III, B, 2.) The supernatant fluid from the gel was adjusted to pH 7.2 with 1 N KOH and the 0.66 ammonium sulfate fraction was precipitated by the addition of 46 gm of solid ammonium sulfate (Mann Research Laboratories, New York, special enzyme grade) per 100 ml of enzyme solution. The pH was adjusted to 7.2 again with 1.0 N KOH and, after standing for 15 min, the fraction was collected by centrifugation for 20 min at 16,300 g. The supernatant fluid was discarded. Precipitates from two or three centrifugations were collected in the same bottles. The precipitates were dissolved in 0.1 M NaF (75 ml for each 100 gm of liver) and 10 ml of 0.02 M 5-AMP per 100 gm of liver were added. The 5-AMP protected LP against enzymatic inactivation and also heat denaturation in the next step. The pH was adjusted to 7.1 with 1.0 N KOH and the preparation was ready for the heat denaturation step or, alternatively, it could be stored at $-20°C$. The final volume of the solution was measured and the precipitate volume (that is, the increase in volume above the volume of added solutions) was taken as a measure of the amount of ammonium sulfate present (as 0.66 saturated solution).

c. Step 3. Heat Denaturation of Phosphorylase Phosphatase. The preparation was transferred to a large Erlenmeyer flask and washed in 30 ml of 0.1 M NaF. If the calculated concentration of ammonium sulfate was above 0.17 saturation, additional 0.1 M NaF was added until that concentration was attained. A carefully cleaned thermometer was inserted into the enzyme solution through a two-hole stopper in the Erlenmeyer flask, and the enzyme solution was heated with rapid and vigorous swirling in a water bath at between 65° and 70°C until the temperature of the enzyme solution rose to 55°C. The flask was transferred to a 55°C water bath for 5 min, removed, and chilled rapidly with swirling in an ice water bath. Heated extracts from the livers of three dogs were pooled at this stage to minimize variations in fractionation. The pooled extracts were centrifuged for 20 min at 7000 g and the precipitates discarded.

d. Step 4. Collection of 0.41 to 0.8 Ammonium Sulfate Fraction. Extracts which were rich in glycogen (as judged by simple observation of the opalescent haze characteristic of the polysaccharide) were fractionated without the addition of glycogen. However, heated extracts from some animals were clear, and to these glycogen (either powdered or dissolved) was added to bring the concentration of exogenous glycogen in the heated extract to 1%. In the absence of glycogen, much of the phosphorylase activity was associated with the 0.41 ammonium sulfate precipitate fraction, from which it was difficult to elute. The concentration of ammonium sulfate in the heated extracts was calculated, and a neutralized $(NH_4)_2SO_4$ solution saturated at room temperature was added to bring the preparation to 0.41 saturation. After 15 to 60 min of stirring in an ice water bath, the precipitates were collected by centrifugation at 7000 g for 20 min, and the precipitates discarded. Removal of this precipitate did not cause any substantial increase in specific activity, but was carried out to eliminate protein which would otherwise precipitate with ethanol in subsequent steps. The 0.41 to 0.8 ammonium sulfate fraction was obtained by adding solid $(NH_4)_2SO_4$ to the supernatant fraction of the 0.41 precipitation (27.3 gm/100 ml of supernatant solution). The pH of the enzyme preparation was maintained around neutrality after the addition of solid ammonium sulfate by adding 0.5 N KOH, using phenol red as an indicator. The enzyme solution was stirred at moderate speed on a magnetic stirrer for at least 30 min before centrifugation at 7000 g for 20 min. The LP activity was reasonably stable in 0.8 saturated $(NH_4)_2SO_4$ at 2°–4°C and was sometimes allowed to mix slowly overnight and centrifuged the next morning. Precipitates from two or three centrifugations were collected in the same bottles and the 0.8 supernatant fluids were discarded. The precipitates were dissolved in 170 ml of cold 0.1 M NaF/1000 gm of liver. The redissolved precipitates could be frozen at -20°C and kept for several days without serious loss of activity, but were usually dialyzed (Step 5) before freezing.

e. Step 5. Dialysis of 0.41–0.8 Ammonium Sulfate Fraction and Ethanol Fractionations.

The $(NH_4)_2SO_4$ concentration of the preparation was lowered by dialysis before alcohol fractionation to avoid denaturation of LP. Since traces of inactivating enzyme sometimes remained in the preparation, the dialysis was carried out in the cold versus 0.1 M NaF for periods no longer than indicated. Dialysis with continued agitation (in a precipitating jar stirred vigorously by a heavy duty magnetic stirrer) was carried out in Visking casing (size 27/32 or 27/40) versus 10 to 20 volumes of 0.1 M NaF containing 5×10^{-4} N KOH for 1½ hr, and was continued for another hour versus a fresh NaF-KOH solution. After dialysis, 0.1 volume of 0.02 M 5-AMP (pH 7.0) was added to the enzyme solution and the pH adjusted to approximately 7.0 with 1.0 N KOH. The preparation was ready for ethanol fractionation but could be stored at -20°C.

To the preparation, chilled in a salt-ice water bath and mixing vigorously on a magnetic stirrer, cold (-20°C) ethanol was added slowly until a final concentration of 25% was reached. The preparation was chilled and centrifuged at 7000 g for 20 min at -10°C in glass centrifuge bottles. The supernatant fluid was discarded and the precipitates dissolved in 0.1 M NaF (two-thirds of the volume *before* adding ethanol) and 0.02 M 5'-AMP was added (10% of the volume of the added NaF). LP was precipitated again by adding ethanol to a final concentration of 25% and the precipitate was collected and redissolved as described above. This step was then repeated a third time. The third alcohol precipitate was dissolved in cold glass distilled water (200 ml/kg of liver or, if the precipitate was large, 300 ml/kg of liver were used). The resuspended third alcohol precipitates were opalescent due to the large amounts of glycogen present. These alcohol precipitations produced a considerable degree of purification with only small losses of LP. The resuspended third alcohol precipitates were stable at $-20°C$ for extended periods of time (months or years). It should be noted that redissolved third alcohol precipitates (after dialysis overnight against water) served as the LP substrate in the assay of phosphorylase phosphatase. Therefore, aliquots of the third alcohol precipitates were occasionally set aside for this purpose.

f. Step 6. Calcium Phosphate Step. This step was designed to free phosphorylase from most of the glycogen which accompanied the preceding fractions in large amounts. The third alcohol fraction was diluted with cold glass distilled water so that it contained approximately 3 units of phosphorylase per milliliter (or so that when assayed without dilution a reading roughly equal to that of a 25 ng/ml phosphate standard was obtained, a reading in this range being desirable so that the degree of LP adsorption onto the gel could be accurately measured). Pilot experiments were routinely carried out to determine the amount of calcium phosphate gel necessary for nearly complete adsorption of the enzyme (usually 2–3% of the volume of the third diluted alcohol fractions was sufficient). The diluted resuspended alcohol precipitate was placed in an ice bath on a magnetic stirrer with vigorous stirring and calcium phosphate gel was added dropwise. The preparation was stirred for 15 min after the addition of the last of the calcium phosphate gel. The gel was collected by centrifugation at 2000 g for 10 min and the volume of the supernatant fraction measured before it was discarded. The gel was washed by resuspending it in cold 1 mM tris (pH 7.4), the volume of which was equal to the volume of the discarded supernatant fraction. The precipitate was collected by centrifugation at 2000 g for 10 min. Phosphorylase was eluted from the calcium phosphate gel by adding a volume of 10 mM potassium citrate (pH 6.5), equaling 12% of the *diluted* third alcohol volume. The potassium citrate was added at room temperature, and the elution was carried on for 10 min at 25°C with magnetic stirring. The eluted gel was packed by centrifugation at room temperature for 10 min at 7000 g. The precipitate was discarded and the supernatant fraction was taken on to the next step.

g. Step 7. Fourth Alcohol Collection and Dialysis. The calcium phosphate eluate was chilled and to each 100 ml of eluate was added 0.55 gm of NaCl and 40 ml of absolute ethanol which had been chilled to $-20°C$. The addition of ethanol was carried out in a salt-ice bath with constant and vigorous magnetic stirring. The enzyme-ethanol mixture was maintained at 3°C or lower and centrifuged at 7000 *g* for 20 min at −10°C. The supernatant fluid was discarded and the precipitate was dissolved in 0.1 *M* potassium phosphate buffer (pH 7.2) using 20 ml of buffer/1000 gm of liver. The redissolved precipitate was incubated for 30–40 min at 30°C, during which a flocculent precipitate formed. After incubation, the enzyme solution was centrifuged at 7000 *g* for 20 min. The clear yellow supernatant fluid containing the LP was decanted, and the precipitate washed with 10 ml of 0.1 *M* potassium phosphate buffer. After thorough mixing it was centrifuged at 7000 *g* for 20 min and the supernatant fraction added to the original supernatant. The precipitate was discarded. The pooled supernatant fractions were placed in Visking casing (size 27/32 or 27/40) and dialyzed against 6 volumes of cold alkalinized glass distilled water (2×10^{-5} *N* KOH) for approximately 20 hr at 2°C. The dialyzed fourth alcohol precipitate was stable when frozen at −20°C, but the preparation was usually carried through the inactivation step after the termination of the dialysis.

2. Preparation of Phosphorylase Phosphatase

The major problem encountered in preparing phosphorylase phosphatase has been that it is quite labile under the conditions which have been tested thus far. No successful method of storing the enzyme at an intermediate stage has as yet been found, and therefore it must be carried through to the final step in a single day.

a. Standard Assay. The activity of the phosphatase was determined by measuring the amount of LP inactivated during a 10 min incubation at 37°C. The reaction was started by the addition of 0.2 ml of suitably diluted enzyme in 10 m*M* tris (pH 7.4) to a reaction mixture containing 0.5 ml of 0.1 *M* tris (pH 7.2), 0.2 ml of caffeine (1.0 mg/ml), approximately 0.75 unit of LP (a dialyzed third alcohol precipitate), and water to a final volume of 1.5 ml. After 10 min of incubation at 37°C, during which LP was inactivated by the phosphatase, 0.5 ml of a solution containing 5% glucose-1-P, 1.5% glycogen, and 0.3 *M* NaF (adjusted to pH 6.1 with HCl) was added and the remaining LP activity was determined by measuring the rate of liberation of phosphate from glucose-1-P. After an additional 10 min at 37°C, a 0.5-ml aliquot of the reaction mixture was transferred to 1.0 ml cold 5% TCA and the amount of phosphate present in an equivalent of 0.1 ml of reaction mixture was determined by the Fiske and SubbaRow phosphate method. The percentage of LP inactivated was determined by the difference from a corresponding control in which the phosphatase was omitted; the values were used when the extent of inactivation was less than 65%. The activity of purified phosphatase was linear while less than 50% of the LP was inactivated. One unit of phosphatase was defined as that amount which catalyzed the inactivation of 1 unit of LP in 10 min. For further details please see Wosilait and Sutherland (1956) or Rall and Sutherland (1962).

TABLE A.II. Summary of Purification of Phosphorylase Phosphatase[a]

Fraction	Total units	Specific activity, units/mg protein	Recovery (%)
Liver homogenate[b]	129,000	2.4	100
Filtrate[b]	52,000	3.2	40
CaPO$_4$ gel supernatant[b]	24,000	2.4	–
Eluate of gel	16,500	39.0	13
First (NH$_4$)$_2$SO$_4$ (40–70%)	5,500	100	4.3
Second (NH$_4$)$_2$SO$_4$ (50%)	4,600	290	3.5

[a] From Rall and Sutherland (1960).
[b] In this example, NaCl was used in place of NaF for perfusion and homogenization. When NaF was used, aliquots of these fractions were dialyzed versus 0.01 M tris at pH 7.2 before assay.

b. Step 1. Washing of Calcium Phosphate Gel. The calcium phosphate gel from Step 2 of the LP preparation was resuspended in 0.02 M tris buffer (pH 8.0) containing 1 mM MgCl$_2$, the volume of the buffer being 1 liter per 300 to 400 gm of liver. It was centrifuged at 2000 g for 10 min at 2°–4°C and the supernatant discarded. This washing procedure was repeated and again the supernatant fluid was discarded. The gel was washed three more times under the same conditions except that the pH of the tris employed was 9.0. The volume of the final wash was reduced so that the preparation could be centrifuged in one or two bottles.

c. Step 2. Elution from the Calcium Phosphate Gel. The phosphatase was eluted from the calcium phosphate gel by resuspending the gel in 0.05 M tris (pH 10) and 0.3 M Na$_2$SO$_4$, the volume of the solution being equal to one-half the weight of the liver used. The mixture was incubated at 30°C for 10 min and then the gel collected by centrifugation at 7000 g at room temperature for 10 min. The precipitate fraction was discarded, and the clear yellow eluate was chilled in an ice water bath and adjusted to pH 8.0 with 1 N acetic acid. The repeated washings of the gel followed by elution of the enzyme resulted in consistently good purification but the yields tended to be variable and unpredictable. As a result, the phosphatase activity in the eluate was always assayed before going to further fractionation although the activity in the washes was not routinely measured.

d. Step 3. (NH$_4$)$_2$SO$_4$ Precipitation. Two-thirds volume of neutralized saturated ammonium sulfate was added to the eluate and after stirring for 15 min in an ice water bath, the solution was centrifuged at 7000 g for 10 min and the precipitate fraction discarded. Twenty-three grams of solid (NH$_4$)$_2$SO$_4$ and 1 ml of 1 N KOH/100 ml of enzyme were added to the supernatant and stirred for 15 min in an ice water bath. The (NH$_4$)$_2$SO$_4$ suspension was then centrifuged as above, the supernatant fraction was discarded, and the small precipitate dissolved in 3.0 ml of 0.1 M tris (pH 10) per 100 gm of liver. The ammonium sulfate saturation of the redissolved precipitate was calculated from its volume, assuming the saturation of the precipitate to be 0.70.

To the redissolved precipitate were added dropwise, with continuous mixing, 0.5 N KOH (2.0 ml/17 ml of redissolved precipitate) and sufficient saturated (NH$_4$)$_2$SO$_4$ solution to bring the final saturation (i.e., redissolved precipitate plus KOH) to 0.50. After 15 min of stirring in an ice water bath, the precipitate was collected by centrifugation at 7000 g for 15 min and was dissolved in 1.0 ml of 0.2 M tris (pH 9.0). Inactivation of LP was usually carried out with this redissolved precipitate immediately, or, if necessary, the phosphatase preparation could be lyophilized after the addition of 2.0 ml of 0.2 M tris (pH 9.0)/ml of enzyme solution. Although a great deal of phosphatase activity was lost during lyophilization process, the lyophilized enzyme preparation was fairly stable when stored in a desiccator at 4°–8°C.

3. Preparation of ILP (Inactivation of LP by Phosphorylase Phosphatase)

An amount of purified phosphatase calculated to inactivate 50% of the LP present in 10 min was added to the dialyzed fourth alcohol precipitate LP preparation, containing 5 mg glycogen/ml and 0.6 mM caffeine.

During the inactivation, residual LP activity was assayed as described under Section III, B, 1 with two modifications. First, 5'-AMP was omitted from the LP assay reagent and second, the incubation time was shortened to 5 min. The inactivation process was followed by periodic assays of the residual LP activity in the incubation mixture, routinely at 0, 30, and 60 min after the addition of the phosphatase and at any later times until the amount of residual LP activity was 1% or less of the original. The flask containing ILP was transferred to an ice water bath, and the ILP was distributed in small aliquots (usually 0.5 ml) in small culture tubes and frozen at −20°C. ILP under these conditions is stable for several years.

C. Preparation of Phosphorylase Kinase (PK)

The preparation of PK used in the assay of cyclic AMP is simply a crude supernatant of dog liver homogenate. Dogs which were maintained for at least 2 weeks in a state of good health and fed canned dog food *ad libitum* were sacrificed by administering a lethal dose of sodium secobarbital intravenously. The major vessels of the neck were severed and the thorax and abdomen opened. The liver was perfused with at least two volumes of cold 0.9% NaCl through the portal vein. The chilled, perfused liver was removed, and after separation of the gallbladder was chilled in 0.33 M sucrose. The liver was sliced and homogenized in 2 volumes (w/v) of 0.33 M sucrose in a Waring blendor for 30 sec, followed by centrifugation of the homogenate at 11,000 g for 20 min at 2° to 4 °C. The supernatant fraction was aspirated and, when necessary, the pH adjusted to 7.5 with 0.1 N KOH. The precipitates were discarded. After the pH was adjusted, the supernatant fraction was incubated for 20 min at 37°C to allow most of the LP present to be converted to ILP. The supernatant fraction was chilled in an ice water bath and then divided into 2-ml fractions in stoppered culture tubes and frozen rapidly in a dry ice-methyl cellosolve bath, care being taken to avoid any contamination of the preparation with methyl cellosolve. The PK activity on the supernatant fractions was stable for many months at −65°–70°C. The supernatant fractions could be stored for shorter periods of time at −20°C but such storage was undesirable because the activity in the absence of cyclic AMP tended to rise with time. In general, a tube of the supernatant fraction, once thawed for assay, was not refrozen but was discarded because the baseline activity increased after refreezing.

D. Preparation of Cyclic Nucleotide Phosphodiesterase

Myocardial tissue from fresh beef hearts were macerated in a meat grinder and then homogenized for 2 min in 2 volumes of 0.33 sucrose and 1 mM tris (pH 7.4) in a Waring commercial blendor at high speed. The homogenate was centrifuged (usually at 16,000 g for 15 min at 4°C), and the resulting supernatant fluid (which provided the starting material for the purification) was decanted through glass wool pre-washed in sucrose. The supernatant fluid was adjusted to 0.5 saturation with $(NH_4)_2SO_4$ (36 gm/100 ml extract), neutrality being maintained by the addition of 1 N KOH as required. After standing for at least 30 min with stirring (and on occasion preparations were allowed to stir overnight at 4°–8°C), the precipitate was collected by centrifugation for 20 min at 16,000 g at 4°C. The precipitate was taken up in 15% of the original extract volume in a solution containing 1 mM MgSO$_4$ and 1 mM imidazole (pH 7.5). Neutralized saturated (NH$_4$)$_2$SO$_4$ solution was added to a final concentration of 0.45 saturation and the resulting mixture stirred from 30 min to 12 hr as was convenient. The precipitate was collected by centrifugation as above and taken up in 5% of the extract volume in 1 mM MgSO$_4$ and 1 mM imidazole. The resuspended precipitate was dialyzed at 4°C against 10 to 30 volumes of 1 mM MgSO$_4$ and 1 mM imidazole (pH 7.4) with constant agitation for 12 to 24 hr, the dialysate being changed two or three times. After dialysis, the preparation was frozen at -20°C and thawed, and a heavy flocculant precipitate which appeared during dialysis and freezing was collected by centrifugation for 20 min at 20,000 g and discarded. The next step, which was the most effective but the least reproducible of the steps in this procedure was the fractionation of the PD on DEAE-cellulose. The DEAE-cellulose used was purchased from the Brown Co., type 20, reagent grade, with a capacity of 0.57 meq/gm (now offered by Carl Schleicher and Schuell of Keane, New York). It was readied for use by suspending 15 gm of the resin in 1 liter of distilled water, and pouring off the finer particles several times. The slurry was taken to a volume of 1 liter and adjusted to approximately 10 mM imidazole (pH 7.0) and 80 mM KCl. The DEAE-cellulose was put into a chromatographic tube with a sintered glass disk [0.6–0.8 gm of resin (dry wt.)/gm of protein to be applied], in a cold room at 4°C. The column was washed with several volumes of 20 mM imidazole (pH 7.0)–80 mM KCl, and the enzyme solution was also adjusted to pH 7.0 in 10 mM imidazole and 80 mM KCl. The column (1.5 cm in diameter) was packed under air pressure so that the flow rate was between 1 and 2 ml/min. The phosphodiesterase preparation was added and allowed to run through the column. As the resin became heavily loaded with protein, the flow rate tended to fall and sometimes pressure (either air or nitrogen pressure) was applied to keep the flow rates at reasonable levels. In addition, it was occasionally necessary to agitate the surface of the resin to maintain a reasonable flow rate. No adverse effects of this procedure have been noted. After the adsorption step was completed, 2 bed volumes of 10 mM imidazole (pH 7.0) and 80 mM KCl were added to the column as a washing step. The PD was eluted with 20 mM imidazole (pH 6.0) containing 0.4 M KCl. The elution was monitored by observing the appearance of a dark yellow band which formed as the eluting mixture moved down the column. PD activity invariably accompanied an intense yellow color in the eluate, and collection was continued until the color had disappeared. The fractions containing the highest specific activity were dialyzed as described above.

To the dialyzed DEAE eluate was added 1 volume of 20 mM glycine buffer (pH 10.0) and 10 mM imidazole (pH 7.0) and the pH was adjusted to 9.6 (by glass electrode). The mixture was brought to 45°C rapidly with shaking in 60°C bath, and the temperature maintained at 45°C for 16 min, and the sample was chilled rapidly. The preparation was stirred on a magnetic stirrer in ice water bath and 0.1 N acetic acid was added drop-wise. The end point of the acid step was the appearance of a heavy white haze at approximately pH 5.8 (the pH was monitored continuously with a glass electrode). The material was collected by centrifugation for 30 min at 29,000 g at 4°C and the clear supernatant solution dialyzed as above.

Calcium phosphate gel was added dropwise to the dialyzed supernatant fraction until a gel-protein ratio of 1.8 was reached. The mixture was stirred for at least 30 min, and the gel collected by centrifugation at 16,300 g for 15 min. The supernatant was discarded and the gel was washed in a volume equal to the original dialyzed supernatant volume with 100 mM imidazole, pH 6.0. It was stirred for 15 min, and then centrifuged again at 16,300 g for 15 min. The washing was repeated in the same volume of 10 mM imidazole pH 7.5, and the phosphodiesterase was eluted from the gel with a volume of 10% saturated ammonium sulfate in 100 mM glycine, pH 9.8, equal to the original supernatant volume. The preparation was stirred in the presence of this mixture for 30 min and then the gel removed by centrifugation at 7000 g for 15 min at 4°C. The supernatant solution was dialyzed as described above, and, stored at -20°C, represented the final purified preparation.

Table A.III summarizes the purification and yields of phospho-diesterase in the procedure described above. One standard unit is now defined as that amount causing the destruction of 1 μmole of cyclic AMP

TABLE A.III. Summary of Purification and Yield in Preparation of Cyclic Nucleotide Phosphodiesterase from Beef Heart[a]

Fraction	Total units	Specific activity	Original activity (%)
Extract	3150	0.2	100
First ammonium sulfate precipitate	2720	0.7	85.6
Second ammonium sulfate precipitate	2530	1.2	79.7
20,000 g Supernatant fraction	1895	1.3	59.7
Dialyzed DEAE-cellulose eluate	769	7.2	24.2
Dialyzed acid supernatant fraction	461	11.9	14.5
Dialyzed calcium phosphate gel eluate	231	25.6	7.3

[a] From Butcher and Sutherland (1962). The extract was taken as the starting material, but it should be noted that a very high (67.1) percentage of the phosphodiesterase activity in the whole homogenate was found in the particulate fraction.

in 30 min at 30°C. No problems have been encountered in reproducing this purification procedure except for the DEAE-cellulose chromatographic step. It is an unusual sort of protein chromatograph, in that the resin is heavily overloaded with protein, but this is required in order to recover any activity. Early attempts at fractionation on DEAE were unsuccessful, for no activity could be recovered as long as customary amounts of protein were applied to the resin. Despite this difficulty, it would seem that the step could be established after a few preliminary experiments. Also, caution should be taken to avoid adding too much acetic acid during the acid precipitation as pH's below 5.8 resulted in loss of supernatant PD activity which was not readily recovered from the precipitate, even when the pH was restored to neutrality.

IV. OTHER ASSAYS OF CYCLIC AMP

Krebs and his associates have developed an assay system based on the activation of skeletal muscle phosphorylase b (Posner *et al.*, 1964). It is an attractive assay because the muscle kinase has been highly purified, and has been used successfully by other groups as well. Unfortunately, the muscle system is even more tedious than liver phosphorylase because of the very high concentrations of phosphorylase b which must be used. This is necessitated by the activation of phosphorylase b by 5'-AMP, the major catabolite of cyclic AMP, whereas inactive liver phosphorylase is only sensitive to 5'-AMP.

Recently, Krebs and his co-workers have found that cyclic AMP does not activate phosphorylase b kinase directly, but rather appears to interact with a distinct protein which they have called kinase kinase. Partially purified preparations of kinase kinase catalyze the phosphorylation of phosphorylase b kinase at the expense of the terminal phosphate of ATP, and this is stimulated by cyclic AMP. In addition, kinase kinase will phosphorylate casein, and an assay for cyclic AMP based upon the rate of ^{32}P incorporation (from $ATPPP^{32}$) into casein has been reported (Walsh *et al.*, 1968).

Breckenridge (1964) reported an assay which is based on the fluorometric cycling techniques. Briefly, cyclic AMP is isolated *free* of other adenine nucleotides and converted to 5'-AMP with purified phosphodiesterase. 5'-AMP is converted to ADP by the addition of a small amount of ATP and myokinase. Next, an ATP generating system (phosphoenolpyruvate and pyruvate kinase) and an ATP utilizing system (glucose, hexokinase, glucose-6-phosphate-dehydrogenase, and $NADP^+$)

are added. NADPH is generated at a rate dependent upon the amount of ATP originally present, and this of course is the sum of the cyclic AMP in the sample plus the ATP which was added to spark the myokinase reaction.

This type of assay is extremely sensitive and direct, and can be set up with commercially available enzymes. However, the purification of cyclic AMP from the nucleotides and other interfering materials is difficult and tedious, and the problems of reagents free of fluorescent materials are well recognized by anyone who has attempted to do so. Hardman et al. (1969) have used cycling technique but generate inorganic phosphate (by hydrolyzing ATP to ADP + P_i with myosin ATPase and regenerating ATP with PEP and pyruvate kinase). Cyclic AMP is purified on Dowex-50 columns. Goldberg et al. (1969a) have reported modifications of Breckenridge's assay, utilizing thin-layer chromatography to purify cyclic AMP. Johnson et al. (1970) have used a Dupont Luminescence Biometer to measure ATP by means of the firefly luciferase-luciferin reaction. Their assay procedure for cyclic AMP was almost as sensitive as the phosphorylase method, and was linear over a much wider range.

Other methods have also appeared. Aurbach and Houston (1968) have reported a sensitive method involving the conversion of isolated cyclic AMP to 5'-AMP and then ATP, followed by measurement of ATP with a radioactive phosphate exchange reaction. Pauk and Reddy (1967) have developed a double isotope derivative dilution method which, although direct, is tedious and insensitive. More recently, Brooker et al. (1968) have reported the use of the phosphodiesterase reaction in an enzymatic radioisotope dilution.

An event of great potential importance has been the development of a radioimmunoassay for cyclic AMP and other cyclic nucleotides (Steiner et al., 1969). Cyclic AMP was succinylated at the 2'-0-position, and the free carboxyl group of this derivative was conjugated to human serum albumin. Antibodies to this conjugated protein were then obtained from rabbits. The resulting assay, based on the competition of labeled and unlabeled cyclic AMP for the antibody, was reported to be sensitive to 1 to 2 picomoles of cyclic AMP. Specificity was such that cyclic AMP could be assayed in relatively crude tissue extracts. However, small but physiologically important differences may still be difficult to detect by this procedure. Several refinements may be necessary, therefore, before this becomes the assay of choice.

Still more recently, Gilman (1970) has described an assay based on the binding of cyclic AMP to an endogenous muscle protein, presumably the regulatory subunit of a protein kinase (see Chapter 5, section VI). A similar assay using a beef adrenal protein has been described by Walton and Garren (1970). Assays based on this principle are simpler than any previously described, and seem to be at least as sensitive, but whether they will be as widely applicable remains to be seen. The existence of a cyclic GMP-sensitive protein kinase in lobster muscle (Kuo and Greengard, 1970) has raised the possibility of developing a similar assay for cyclic GMP.

REFERENCES

Abboud, F. M., Eckstein, J. W., and Zimmerman, B. G. (1965). *Amer. J. Physiol.* **209**, 383.

Abdulla, Y. H., and Hamadah, K. (1970). *Lancet* **i**, 378.

Abe, Y., Morimoto, S., Yamamoto, K., and Meda, J. (1968). *Jap. J. Pharmacol.* **18**, 271.

Abe, K., Butcher, R. W., Nicholson, W. E., Baird, C. E., Liddle, R. A., and Liddle, G. W. (1969a). *Endocrinology* **84**, 362.

Abe, K., Robison, G. A., Liddle, G. W., Butcher, R. W., Nicholson, W. E., and Baird, C. E. (1969b). *Endocrinology* **85**, 674.

Abell, C. W., Kamp, C. W., and Johnson, L. D. (1970). *Cancer Res.* **30**, 717.

Aboud, M., and Burger, M. (1970). *Biochem. Biophys. Res. Commun.* **38**, 1023.

Acheson, G. H. (1966), (ed.). "Second Symposium on Catecholamines," 803 pp. Williams & Wilkins, Baltimore, Maryland.

Adachi, K., and Kano, M. (1970). *Biochem. Biophys. Res. Commun.* **41**, 884.

Adam, P. A. J., and Haynes, R. C. (1969). *J. Biol. Chem.* **244**, 6444.

Adamson, L. F. (1970). *Biochim. Biophys. Acta* **201**, 446.

Aguilar-Parada, E., Eisentraut, A. M., and Unger, R. H. (1969). *Diabetes* **18**, 717.

Ahlquist, R. P. (1948). *Amer. J. Physiol.* **153**, 586.

Ahlquist, R. P., and Levy, B. (1959). *J. Pharmacol. Exp. Ther.* **127**, 146.

Ahn, C. S., and Rosenberg, I. N. (1968). *Proc. Nat. Acad. Sci. U. S.* **60**, 830.

Ahn, C. S., Athans, J. C., and Rosenberg, I. N. (1969). *Endocrinology* **85**, 224.

Allan, W., and Tepperman, H. C. (1969). *Life Sci.* **8** (I), 307.

Alonso, D., and Harris, J. B. (1965). *Amer. J. Physiol.* **208**, 18.

Alonso, D., Rynes, R., and Harris, J. B. (1965). *Amer. J. Physiol.* **208**, 1183.

Alonso, D., Nigon, K., Dorr, I., and Harris, J. B. (1967). *Amer. J. Physiol.* **212**, 992.

Alonso, D., Park, O. H., and Harris, J. B. (1968). *Amer. J. Physiol.* **215**, 1305.

Ali, H. I. E., Antonio, A., and Haugaard, N. (1964). *J. Pharmacol. Exp. Ther.* **145**, 142.

Altman, M., Oka, H., and Field, J. B. (1966). *Biochim. Biophys. Acta* **116**, 586.

Altszuler, N., Steele, R., Rathgeb, I., and DeBodo, R. C. (1967). *Amer. J. Physiol.* **212**, 677.

Amsterdam, A., Ohad, I., and Schramm, M. (1969). *J. Cell Biol.* **41**, 753.

Anderson, W. S., and Brown, E. (1963). *Biochim. Biophys. Acta* **67**, 674.

Andersson, R., and Mohme-Lundholm, E. (1969). *Acta Physiol. Scand.* **77**, 372.

Angelakos, E. T., and Glassman, P. M. (1965). *Arch. Int. Pharmacodyn. Ther.* **154**, 82.

Appleman, M. M., and Kemp, R. G. (1966). *Biochem. Biophys. Res. Commun.* **24**, 564.

Appleman, M. M., Krebs, E. G., and Fischer, E. H. (1966a). *Biochemistry* **5**, 2101.

Appleman, M. M., Birnbaumer, L., and Torres, H. N. (1966b). *Arch. Biochem. Biophys.* **116**, 39.

Ardlie, N. G., Glew, G., and Schwartz, C. J. (1966). *Nature (London)* **212**, 415.

Ardlie, N. G., Glew, G., Schultz, B. G., and Schwartz, C. J. (1967). *Thromb. Diath. Haemorrh.* **18**, 670.

Ardrey, R. (1967). "The Territorial Imperative," 390 pp. Atheneum, New York.

Argy, W. P., Handler, J. S., and Orloff, J. (1967). *Amer. J. Physiol.* **213**, 803.

Ariens, E. J. (1966). *Advan. Drug Res.* **3**, 235.

Ariens, E. J. (1967). *Arch. Pharmakol. Exp. Pathol.* **257**, 118.

Arnold, A., and McAuliff, J. P. (1968). *Experientia* **24**, 674.

Ashford, T. P., and Burdette, W. J. (1965). *Ann. Surg.* **162**, 191.

Ashford, T. P., and Porter, K. R. (1961). *Abstr. 1st Meeting Soc. Cell Biol.,* Chicago, 1961.

Ashford, T. P., and Porter, K. R. (1962). *J. Cell Biol.* **12**, 198.

Ashman, D. F., Lipton, R., Melicow, M. M., and Price, T. D. (1963). *Biochem. Biophys. Res. Commun.* **11**, 330.

Ashmore, J., Preston, J. E., and Love, W. C. (1962). *Proc. Soc. Exp. Biol. Med.* **109**, 291.

Assem, E. S. K., and Schild, H. O. (1969). *Nature (London)* **224**, 1028.

Aulich, A., Stock, K., and Westermann, E. (1967). *Life Sci.* **6**, 929.

Aurbach, G. D., and Houston, B. A. (1968). *J. Biol. Chem.* **243**, 5935.

Avioli, L. V., Birge, S. J., Scott, S., and Shieber, W. (1969). *Amer. J. Physiol.* **216**, 939.

Axelsson, J., and Holmberg, B. (1969). *Acta Physiol. Scand.* **75**, 149.

Baba, W. I., Smith, A. J., and Townshend, M. M. (1967). *Quart. J. Exp. Physiol.* **52**, 416.

Badad, H., Ben-Zvi, R., Bdolah, A., and Schramm, M. (1967). *Eur. J. Biochem.* **1**, 96.

Bär, H., and Hechter, O. (1969a). *Anal. Biochem.* **29**, 476.

Bär, H., and Hechter, O. (1969b). *Proc. Nat. Acad. Sci. U. S.* **63**, 350.

Bär, H., and Hechter, O. (1969c). *Biochim. Biophys. Acta* **192**, 141.

Bär, H., and Hechter, O. (1969d). *Biochem. Biophys. Res. Commun.* **35**, 681.

Ball, J. H., Kaminsky, N. I., Broadus, A. E., Hardman, J. G., Sutherland, E. W., and Liddle, G. W. (1970). *Clin. Res.* **18**, 336.

Barka, T. (1965). *Exp. Cell Res.* **39**, 355.

Barka, T., and van der Noen, H. (1969). *Lab. Invest.* **20**, 377.

Barkley, D. S. (1969). *Science* **165**, 1133.

Barnea, A. (1969). *Endocrinology* **84**, 746.

Bartelstone, H. J., and Nasmyth, P. (1965). *Amer. J. Physiol.* **208**, 754.

Bartelstone, H. J., Nasmyth, P. A., and Telford, J. M. (1967). *J. Physiol. (London)*, **188**, 159.

Baserga, R. (1966). *Life Sci.* **5**, 2033.

Bastide, F., and Jard, S. (1968). *Biochim. Biophys. Acta* **150**, 113.

Bastomsky, C. H., and McKenzie, J. M. (1967). *Amer. J. Physiol.* **213**, 753.

Bastomsky, C. H., and McKenzie, J. M. (1968). *Endocrinology* **83**, 309.

Bates, R. F. L., Bruce, J. B., and Care, A. D. (1970). *J. Endocrinol.* **46**, xi.

Bauchau, A. G., Mengest, J. C., and Olivier, M. A. (1968). *Gen. Comp. Endocrinol.* **11**, 132.

Baumann, P., and Wright, B. E. (1968). *Biochemistry* **7**, 3653.

Bdolah, A., and Schramm, M. (1965). *Biochem. Biophys. Res. Commun.* **18**, 452.

Beatty, C. H., Peterson, R. D., Bocek, R. M., Craig, N. C., and Weleber, R. (1963). *Endocrinology* **73**, 721.

Beavo, J. A., Hardman, J. G., and Sutherland, E. W. (1970a). *J. Biol. Chem.* **245**, 5649.

Beavo, J. A., Rogers, N. L., Crofford, O. B., Hardman, J. G., Sutherland, E. W., and Newman, E. V. (1970b). *Mol. Pharmacol.* **6**, 597.

Beck, N. P., Field, J. B., and Davis, B. (1970). *Clin. Res.* **18**, 494.

Becker, F. F., and Bitensky, M. W. (1969). *Proc. Soc. Exp. Biol. Med.* **130**, 983.

Belford, J., and Cunningham, M. A. (1968). *J. Pharmacol. Exp. Ther.* **162**, 134.

Bell, N. H. (1970). *J. Clin. Invest.* **49**, 1368.

Belleau, B. (1967). *Ann. N. Y. Acad. Sci.* **139**, 580.

Belocopitow, E. (1961). *Arch. Biochem. Biophys.* **93**, 457.

Belocopitow, E., Appleman, M. M., and Torres, H. N. (1965). *J. Biol. Chem.* **240**, 3473.

Belocopitow, E., Fernandez, M. D. C. G., Birnbaumer, L., and Torres, H. N. (1967). *J. Biol. Chem.* **242**, 1227.

Bergen, S. S., Sullivan, R., Hilton, J. G., Willis, S. W., and Van Itallie, T. B. (1960). *Amer. J. Physiol.* **199**, 136.

Bergen, S. S., Hilton, J. G., and Van Itallie, T. B. (1966). *Endocrinology* **79**, 1065.

Bergström, S. (1967). *Science* **157**, 382.

Bergström, S., and Samuelsson, B. (1967). (eds.). "Prostaglandins," 299 pp. Wiley (Interscience), New York.

Bergström, S., Carlson, L. A., and Weeks, J. R. (1968). *Pharmacol. Rev.* **20**, 1.

Berkowitz, B. A., Tarver, J. H., and Spector, S. (1970). *Eur. J. Pharmacol.* **10**, 64.

Berman, D. A. (1966). *J. Pharmacol. Exp. Ther.* **15**., 75.

Bernard, C. (1865). "An Introduction to the Study of Experimental Medicine" (Trans. by H. C. Greene, with Intro. by L. J. Henderson), 226 pp. Macmillan, New York (published 1927); Dover paperback published 1957.

Berne, R. M. (1964). *Physiol. Rev.* **44**, 1.

Berridge, M. J. (1970). *J. Exp. Biol.* **53**, 171.

Berridge, M. J., and Patel, N. G. (1968). *Science* **162**, 462.

Berthet, J. (1959). *Amer. J. Med.* **26**, 703.

Berthet, J. (1960). *Proc. 4th Int. Congr. Biochem* **17**, 107.

Berthet, J. (1963). *In* "Comparative Endocrinology" (U. S. von Euler and H. Heller, eds.), Vol. I, pp. 410–427. Academic Press, New York.

Berthet, J., Sutherland, E. W., and Rall, T. W. (1957). *J. Biol. Chem.* **229**, 350.

Berti, F., Sirtore, C., and Usardi, M. M. (1970). *Arch. Int. Pharmacodyn. Ther.* **184**, 328.

Beviz, A., and Mohme-Lundholm, E. (1964). *Acta Physiol. Scand.* **62**, 109.

Beviz, A., and Mohme-Lundholm, E. (1965). *Acta Physiol. Scand.* **65**, 289.

Beviz, A., and Mohme-Lundholm, E. (1967). *Acta Pharmacol. Toxicol.* **25** (Suppl. 4), 21.

Beviz, A., Mohme-Lundholm, E., and Svedmyr, N. (1967). *Acta Physiol. Scand.* **69**, 213.

Beviz, A., Lundholm, L., and Mohme-Lundholm, E. (1968). *Brit. J. Pharmacol.* **34**, 198.

Bewsher, P. D., and Ashmore, J. (1966). *Biochem. Biophys. Res. Commun.* **24**, 431.

Bhakthan, N. M., and Gilbert, L. I. (1968). *Gen. Comp. Endocrinol.* **11**, 86.

Bieck, P., Stock, K., and Westermann, E. (1966). *Life Sci.* **5**, 2157.

Bieck, P., Stock, K., and Westermann, E. (1967). *Arch. Pharmakol. Exp. Pathol.* **256**, 218.

Bieck, P., Stock, K., and Westermann, E. (1968). *Life Sci.* **7**, 1125.

Bieck, P., Stock, K., and Westermann, E. (1969). *Arch. Pharmakol. Exp. Pathol.* **263**, 387.

Biel, J. H., and Lum, B. K. B. (1966). *Prog. Drug Res.* **10**, 46.

Bierman, E. L., Bagdade, J. D., and Porte, D. (1968). *Amer. J. Clin. Nutr.* **21**, 1434.

Birmingham, M. K., Kurlento, E., Lane, R., Muhlstock, B., and Traikov, H. (1960). *Can. J. Biochem. Physiol.* **38**, 1077.

Birnbaumer, L., and Rodbell, M. (1969). *J. Biol. Chem.* **244**, 3477.

Birnbaumer, L., Pohl, S. L., and Rodbell, M. (1969). *J. Biol. Chem.* **244**, 3468.

Bishop, J. S., and Larner, J. (1967). *J. Biol. Chem.* **242**, 1354.

Bishop, J. S., and Larner, J. (1969). *Biochim. Biophys. Acta* **171**, 374.

Bitensky, M. W., and Burstein, S. R. (1965). *Nature (London)* **208**, 1282.

Bitensky, M. W., Clancy, J. W., and Gamache, E. (1967). *J. Clin. Invest.* **46**, 1037.

Bitensky, M. W., Russell, V., and Robertson, W. (1968). *Biochem. Biophys. Res. Commun.* **31**, 706.

Bitensky, M. W., Russell, V., and Blanco, M. (1970) *Endocrinology* **86**, 154.

Bjersing, L., and Carstensen, H. (1964). *Biochim. Biophys. Acta* **86**, 639.

Black, I. B., and Axelrod, J. (1969). *J. Biol. Chem.* **244**, 6124.

Blackard, W. G., and Aprill, C. N. (1967). *J. Lab. Clin. Med.* **69**, 960.

Blecher, M. (1967). *Biochem. Biophys. Res. Commun.* **27**, 560.

Blecher, M., Merlino, N. S., and Ro'Ane, J. T. (1968). *J. Biol. Chem.* **243**, 3973.

Blecher, M., Merlino, N. S., Ro'Ane, J. T., and Flynn, P. D. (1969). *J. Biol. Chem.* **244**, 3423.

Blecher, M., Ro'Ane, J. T., and Flynn, P. D. (1970). *J. Biol. Chem.* **245**, 1867.

Bloom, B. M., and Goldman, I. M. (1966). *Advan. Drug Res.* **3**, 121.

Blum, J. J. (1967). *Proc. Nat. Acad. Sci. U. S.* **58**, 81.

Blumenthal, M., and Brody, T. M. (1965). *J. Allergy* **36**, 205.

Bonner, J. T. (1967). "The Cellular Slime Molds," 2nd ed., 205 pp. Princeton Univ. Press, Princeton, New Jersey.

Bonner, J. T. (1969). *Sci. Amer.* **220** (6), 78.

Bonner, J. T. (1970). *Proc. Nat. Acad. Sci. U. S.* **65**, 110.

Bonner, J. T., Barkley, D. S., Hall, E. M., Konijn, T. M., Mason, J. W., O'Keefe, G., and Wolfe, P. B. (1969). *Develop. Biol.* **20**, 72.

Bonner, J. T., Hall, E. M., Sachsenmaier, W., and Walker, B. K. (1970). *J. Bacteriol.* **102**, 682.

Borden, R., and Smith, M. (1966a). *J. Org. Chem.* **31**, 3241.

Borden, K. R., and Smith, M. (1966b). *J. Org. Chem.* **31**, 3247.

Borle, A. B. (1968). *J. Cell. Biol.* **36**, 567.

Born, G. V. R. (1967). *Thromb. Diath. Haemorrh. Suppl.* **26**, 173.

Bourgoignie, J., Guggenheim, S., Kipnis, D. M., and Klahr, S. (1969). *Science* **165**, 1360.

Bourguet, J. (1968). *Biochim. Biophys. Acta* **150**, 104.

Bourguet, J., and Morel, F. (1967). *Biochim. Biophys. Acta* **135**, 693.

Bowers, C. Y., Lee, K. L., and Schally, A. V. (1968a). *Endocrinology* **82**, 75.

Bowers, C. Y., Robison, G. A., Lee, K. L., de Balbian Verster, F., and Schally, A. V. 1968b). *Proc. Thyroid Soc. Washington, D. C.,* p. 55.

Bowers, C. Y., Schally, A. V., Enzmann, F., Bøler, J., and Folkers, K. (1970). *Endocrinology* **86**, 1143.

Bowman, R. H. (1966). *J. Biol. Chem.* **241**, 3041.

Bowman, R. H. (1970). *J. Biol. Chem.* **245**, 1604.

Bowman, W. C., and Hall, M. T. (1970). *Brit. J. Pharmacol.* **38**, 399.

Bowman, W. C., and Nott, M. W. (1969). *Pharmacol. Rev.* **21**, 27.

Bowman, W. C., and Raper, C. (1964). *Brit. J. Pharmacol.* **23**, 184.

Bowman, W. C., and Raper, C. (1967). *Ann. N. Y. Acad. Sci.* **139**, 741–753.

Bowness, J. M. (1966). *Science* **152**, 1370.

Bradham, L. S., and Woolley, D. W. (1964). *Biochim. Biophys. Acta* **93**, 475.

Bradham, L. S., Holt, D. A., and Simms, M. (1970). *Biochim. Biophys. Acta* **201**, 250.

Brady, R. O., and Gurin, S. (1950). *J. Biol. Chem.* **187**, 589.

Brana, H. (1969). *Folia Microbiol. (Prague)* **14**, 185.

Brana, H., and Chytil, F. (1966). *Folia Microbiol. (Prague)* **11**, 43.

Braun, T., and Hechter, O. (1970). *Proc. Nat. Acad. Sci. U.S.* **66**, 995.

Braun, T., Hechter, O., and Bär, H. P. (1969). *Proc. Soc. Exp. Biol. Med.* **132**, 233.

Bray, G. A. (1967). *Biochem. Biophys. Res. Commun.* **28**, 621.

Bray, G. A., and Goodman, H. M. (1968). *J. Lipid Res.* **9**, 714.

Breckenridge, B. M. (1964). *Proc. Nat. Acad. Sci. U. S.* **52**, 1580.

Breckenridge, B. M., and Johnston, R. E. (1969). *J. Histochem. Cytochem.* **17**, 505.

Breckenridge, B. M., and Norman, J. H. (1965). *J. Neurochem.* **12**, 51.

Breckenridge, B. M., Burn, J. H., and Matschinsky, F. M. (1967). *Proc. Nat. Acad. Sci. U. S.* **57**, 1893.

Bressler, R., Cordon, M. V., and Brendel, K. (1969). *Arch. Intern. Med.* **123**, 248.

Broadus, A. E., Northcutt, R. C., Hardman, J. G., Kaminsky, N. I., Sutherland, E. W., and Liddle, G. W. (1969). *Clin. Res.* **17**, 65.

Broadus, A. E., Kaminsky, N. I., Hardman, J. G., Sutherland, E. W., and Little, G. W. (1970a). *J. Clin. Invest.* **49**, 2222.

Broadus, A. E., Kaminsky, N. I., Northcutt, R. C., Hardman, J. G., Sutherland, E. W., and Liddle, G. W. (1970b). *J. Clin. Invest.* **49**, 2237.

Brodie, B. B., Davies, J. I., Hynie, S., Krishna, G., and Weiss, B. (1966).*Pharmacol. Rev.* **18**, 273.

Brodie, B. B., Krishna, G., and Hynie, S. (1969). *Biochem. Pharmacol.* **18**, 1129.

Brody, T. M., and Diamond, J. (1967). *Ann. N. Y. Acad. Sci.* **139**, 772.

Brogan, E., Kozonis, M. C., and Overy, D. C. (1969). *Lancet* **i**, 482.

Broman, R. L., Goldenbaum, P. E., and Dobrogosz, W. J. (1970). *Biochem. Biophys. Res. Commun.* **39**, 401.

Brooker, G., Thomas, L. J., and Appleman, M. M. (1968). *Biochemistry* **7**, 4177.

Brown, D., Margrath, D., and Todd, A. (1952). *J. Chem. Soc.* p. 2708.

Brown, E., Clarke, D. L., Roux, V., and Sherman, G. H. (1963). *J. Biol. Chem.* **238**, 852.

Brown, H. D., Chattopadhyay, S. K., Spjut, H. J., Spratt, J. S., and Pennington, S. N. (1969). *Biochim. Biophys. Acta* **192**, 372.

Brown, H. D., Chattopadhyay, S. K., Morris, H. P., and Pennington, S. N. (1970). *Cancer Res.* **30**, 123.

Brown, J. D., Stone, D. B., and Steele, A. A. (1969). *Metabolism* **18**, 926.

Buchanan, K. D., Vance, J. E., Dinstl, K., and Williams, R. H. (1969a). *Diabetes* **18**, 11.

Buchanan, K. D., Vance, J. E., and Williams, R. H. (1969b). *Diabetes* **18**, 381.

Buckley, N. M. (1970). *Amer. J. Physiol.* **218**, 1399.

Buckley, N. M., Tsuboe, K. K., and Zeig, N. J. (1961) *Circulation Res.* **9**, 242.

Bueding, E., and Bulbring, E. (1967). *Ann. N. Y. Acad. Sci.* **139**, 758.

Bueding, E., Bulbring, E., Kuriyama, H., and Gercken, G. (1962). *Nature (London)* **196**, 944.

Bueding, E., Kent, N., and Fisher, J. (1964). *J. Biol. Chem.* **239**, 2099.

Bueding, E., Butcher, R. W., Hawkins, J., Timms, A. R., and Sutherland, E. W. (1966). *Biochim. Biophys. Acta* **115**, 173.

Bueding, E., Bulbring, E., Gercken, G., Hawkins, J. T., and Kuriyama, H. (1967). *J. Physiol. (London)* **193**, 187.

Bürk, R. R. (1968). *Nature (London)* **219**, 1272.

Bulbring, E., Goodford, P. J., and Setekleiv, J. (1966). *Brit. J. Pharmacol.* **28**, 296.

Bullough, W. S. (1967). "The Evolution of Differentiation," 206 pp. Academic Press, New York.

Burger, M., and Brandt, W. (1935). *Z. Gesamte Exp. Med.* **96**, 375.

Burke, G. (1968). *J. Clin. Endocrinol.* **28**, 1816.

Burke, G. (1969). *Metabolism* **18**, 961.

Burke, G. (1970a). *Endocrinology* **86**, 353.

Burke, G. (1970b). *Clin. Res.* **18**, 356.

Burns, J. J., Colville, K. I., Lindsay, L. A., and Salvador, R. A. (1964). *J. Pharmacol. Exp. Ther.* **144**, 163.

Burns, T. W., and Langley, P. E. (1968). *J. Lab. Clin. Med.* **72**, 813.

Burns, T. W., and Langley, P. E. (1970). *J. Lab. Clin. Med.* **75**, 983.

Burns, T. W., Langley, P. E., and Robison, G. A. (1970). *Clin. Res.* **18**, 86.

Burton, S. D., Mondon, C. E., and Ishida, T. (1967). *Amer. J. Physiol.* **212**, 261.

Buschiazzo, H., Exton, J. H., and Park, C. R. (1970). *Proc. Nat. Acad. Sci. U. S.* **65**, 383.

Butcher, F. R., and Serif, G. S. (1969). *Biochim. Biophys. Acta* **192**, 409.

Butcher, R. W., and Baird, C. E. (1968). *J. Biol. Chem.* **243**, 1713.

Butcher, R. W., and Baird, C. E. (1969). *In* "Drugs Affecting Lipid Metabolism" (R. Paoletti, ed.), pp. 5–23. Plenum, New York.

Butcher, R. W., and Baird, C. E. (1970). *Proc. 4th Int. Congr. Pharmacol.* **4**, 6.

Butcher, R. W., and Carlson, L. A. (1970). *Acta Physiol. Scand.* **79**, 559.

Butcher, R. W., and Sutherland, E. W. (1959). *Pharmacologist* **1**(2), 63.

Butcher, R. W., and Sutherland, E. W. (1962). *J. Biol. Chem.* **237**, 1244.

Butcher, R. W., and Sutherland, E. W. (1967). *Ann. N. Y. Acad. Sci.* **139**, 849.

Butcher, R. W., Sutherland, E. W., and Rall, T. W. (1960). *Pharmacologist* **2**(2), 66.

Butcher, R. W., Ho, R. J., Meng, H. C., and Sutherland, E. W. (1965). *J. Biol. Chem.* **240**, 4515.

Butcher, R. W., Sneyd, J. G. T., Park, C. R., and Sutherland, E. W. (1966). *J. Biol. Chem.* **241**, 1652.

Butcher, R. W., Pike, J. E., and Sutherland, E. W. (1967). *In* "Prostaglandins" (S. Bergstrom and B. Samuelsson, eds.), pp. 133–138. Wiley (Interscience), New York.

Butcher, R. W., Baird, C. E., and Sutherland, E. W. (1968). *J. Biol. Chem.* **243**, 1705.

Butterfield, W. J. H., and Van Westering, W. (eds.). (1967). "Tolbutamide . . . after Ten Years." Excepta Med. Found., New York.

Cahill, G. F., Ashmore, J., Zottu, S., and Hastings, A. B. (1957a). *J. Biol. Chem.* **224**, 237.

Cahill, G. F., Zottu, S., and Earle, A. S. (1957b). *Endocrinology* **60**, 265.

Cahill, G. F., Leboeuf, B., and Flinn, R. B. (1960). *J. Biol. Chem.* **235**, 1246.

Camerini-Davalos, R. A., and Cole, H. S. (1970). "Early Diabetes," 486 pp. Academic Press, New York.

Cannon, W. B. (1939). "The Wisdom of the Body," 333 pp. Norton, New York. (Norton Library edition published 1963.)

Cannon, W. B., and Nice, L. B. (1913). *Amer. J. Physiol.* **32**, 44.

Cannon, W. B., and Rosenblueth, A. (1937). "Autonomic Neuro-Effector Systems, 229 pp. Macmillan, New York.

Care, A. D., and Gitelman, H. J. (1968). *J. Endocrinol.* **41**, xxi.

Care, A. D., Bates, R. F. L., and Gitelman, H. J. (1969). *J. Endocrinol.* **43**, iv.

Care, A. D., Bates, R. F. L., and Gitelman, H. J. (1970). *J. Endocrinol.* **48**, 1.

Carlson, L. A. (1963). *Acta Med. Scand.* **173**, 719.

Carlson, L. A. (1968). *In* "Proteins and Polypeptide Hormones" (M. Margoulies, ed.), Vol. I, p. 248. Excerpta Med. Found., Amsterdam.

Carlson, L. A., and Bally, P. R. (1965). *In* "Handbook of Physiology, Section 5: Adipose Tissue" (A. E. Renold and G. F. Cahill, Jr., eds.), pp. 557–574. Amer. Physiol. Soc., Washington, D. C.

Carlson, L. A., and Hallberg, D. (1968). *J. Lab. Clin. Med.* **71**, 368.

Carlson, L. A., Liljedahl, S., Verdy, M., and Wirsen, C. (1964). *Metabolism* **13**, 227.

Carlson, L. A., Butcher, R. W., and Micheli, H. (1970). *Acta Med. Scand.* **187**, 525.

Case, R. M., Laundy, T. J., and Scratcherd, T. (1969). *J. Physiol.* **204**, 45.

Casillas, E. R., and Hoskins, D. D. (1970). *Biochem. Biophys. Res. Commun.* **40**, 255.

Castaneda, M., and Tyler, A. (1968). *Biochem. Biophys. Res. Commun.* **33**, 782.

Cehovic, G. (1969). *C. R. Acad. Sci. Paris Ser.* D **268**, 2929.

Cehovic, G., Lewis, U. J., and Vanderlaan, W. P. (1970a). *C. R. Acad. Sci. Ser.* D **270**, 3119.

Cehovic, G., Marcus, I., Gabbai, A., and Posternak, Th. (1970b). *C. R. Acad. Sci. Ser. D* **271**, 1399.

Cerasi, E., and Luft, R. (1967). *Diabetes* **16**, 615.

Cerasi, E., and Luft, R. (1969). *Hormone Metabolic Res.* **1**, 162.

Cerasi, E., and Luft, R. (1970). *In* "Pathogenesis of Diabetes" (E. Cerasi and R. Luft, eds.), pp. 17–40. Wiley (Interscience), New York.

Challoner, D. R., and Steinberg, D. (1966). *Amer. J. Physiol.* **211**, 897.

Chambaut, A., Eboue-Bonis, D., Hanoune, J., and Clauser, H. (1969). *Biochem. Biophys. Res. Commun.* **34**, 283.

Chambers, D. A., and Zubay, G. (1969). *Proc. Nat. Acad. Sci. U. S.* **63**, 118.

Chambers, J. W., and Bass, A. D. (1970). *Fed. Proc.* **29**, 616 Abs.

Chambers, J. W., Georg, R. H., and Bass, A. D. (1970). *Endocrinology* **87**, 366.

Chance, B., and Schoener, B. (1964). *Biochem. Biophys. Res. Commun.* **17**, 416.

Chance, B., Schoener, B., and Elsaesser, S. (1965). *J. Biol. Chem.* **240**, 3170.

Chang, Y. Y. (1968). *Science* **161**, 57.

Channing, C. P., and Seymour, J. F. (1970). *Endocrinology* **87**, 165.

Charbon, G. A., Sebus, J., Kool, D. S., and Hoekstra, M. H. (1963). *Gastroenterology* **44**, 805.

Chase, L. R., and Aurbach, G. D. (1967). *Proc. Nat. Acad. Sci. U. S.* **58**, 518.

Chase, L. R., and Aurbach, G. D. (1968). *Science* **159**, 545.

Chase, L. R., and Aurbach, G. D. (1970). *J. Biol. Chem.* **245**, 1520.

Chase, L. R., Fedak, S. A., and Aurbach, G. D. (1969a). *Endocrinology* **84**, 761.

Chase, L. R., Melson, G. L., and Aurbach, G. D. (1969b). *J. Clin. Invest.* **48**, 1832.

Chassy, B. M., Love, L. L., and Krichevsky, M. I. (1969). *Proc. Nat. Acad. Sci. U. S.* **64**, 296.

Chaudhary, A. H., and Frenkel, A. W. (1970). *Biochem. Biophys. Res. Commun.* **39**, 238.

Chen, G., Schueler, F. W., and Geiling, E. M. K. (1946). *Fed. Proc.* **5**, 170.

Chen, S., and Tchen, T. T. (1970). *Biochem. Biophys. Res. Commun.* **41**, 964.

Chenoweth, M. B., and Ellman, G. L. (1957). *Annu. Rev. Physiol.* **19**, 121.

Chesney, T. McC., and Schofield, J. G. (1969). *Diabetes* **18**, 627.

Cheung, W. Y. (1966). *Biochim. Biophys. Acta* **115**, 235.

Cheung, W. Y. (1967). *Biochemistry* **6**, 1079.

Cheung, W. Y. (1969). *Biochim. Biophys. Acta* **191**, 303.

Cheung, W. Y (1970a). *Biochem. Biophys. Res. Commun.* **38**, 533.

Cheung, W. Y (1970b). *Life Sci.* **9** (II), 861.

Cheung, W. Y., and Salganicoff, L. (1967). *Nature (London)* **214**, 90.

Cheung, W. Y., and Williamson, J. R. (1965). *Nature (London)* **207**, 979.

Chytil, F. (1968). *J. Biol. Chem.* **243**, 893.

Cirillo, V. J., Andersen, O. F., Ham, E. A., and Gwatkin, R. B. L. (1969). *Expt. Cell Res.* **57**, 139.

Civan, M. V., and Frazier, H. S. (1968). *J. Gen. Physiol.* **51**, 589.

Civen, M., Trimmer, B. M., and Brown, C. B. (1967). *Life Sci.* **6**, 1331.

Clark, A. J. (1937). *In* "Heffter's Handbuch der Experimentellen Pharmakologie" (W. Heubner and J. Schuller, eds.), Vol. 4. Springer, Berlin.

Clarke, S. D., Neill, D. W., and Welbourn, R. B. (1960). *Gut* **1**, 146.

Claycomb, W. C., and Kilsheimer, G. S. (1969). *Endocrinology* **84**, 1179.

Cockburn, F., Hull, D., and Walton, I. (1967). *Brit. J. Pharmacol.* **21**, 568.

Code, C. F. (1965). *Fed. Proc.* **24**, 1311.

Cohen, K. L., and Bitensky, M. L. (1969). *J. Pharmacol. Exp. Ther.* **169**, 80.

Cole, B., Robison, G. A., and Hartmann, R. C. (1970). *Fed. Proc.* **29**, 316 Abs.

Comsa, J. (1950). *Amer. J. Physiol.* **161**, 550.

Conn, H. O., and Kipnis, D. M. (1969). *Biochem. Biophys. Res. Commun.* **37**, 319.

Cook, W. H., Lipkin, D., and Markham, R. (1957). *J. Amer. Chem. Soc.* **79**, 3607.

Coore, H. G., and Randle, R. J. (1964). *Biochem. J.* **93**, 66.

Copp, D. H. (1969). *J. Endocrinol.* **43**, 137.

Corbin, J. D., and Krebs, E. G. (1969). *Biochem. Biophys. Res. Commun.* **36**, 328.

Corbin, J. D., and Park, C. R. (1969). *Fed. Proc.* **28**, 702.

Corbin, J. D., Reimann, E. M., Walsh, D. A., and Krebs, E. G. (1970b). *J. Biol. Chem.* **245**, 4849.

Corbin, J. D., Sneyd, J. G. T., Butcher, R. W., and Park, C. R. (1970a). *J. Biol. Chem.* submitted for publication.

Cori, C. F. (1931). *Physiol. Rev.* **11**, 143.

Cori, G. T., and Cori, C. F. (1945). *J. Biol. Chem.* **158**, 321.

Cori, C. F., and Welch, A. D. (1941). *J. Amer. Med. As.* **116**, 2590.

Cornblath, M. (1955). *Amer. J. Physiol.* **183**, 240.

Cornblath, M., Randle, P. J., Parmeggiani, A., and Morgan, H. E. (1963). *J. Biol. Chem.* **238**, 1592.

Correll, J. W. (1963). *Science* **140**, 387.

Coulter, C. L. (1968). *Science* **159**, 888.

Cowgill, R. W. (1961). *J. Biol. Chem.* **234**, 3146.

Craig, A. B., and Honig, C. R. (1963). *Amer. J. Physiol.* **205**, 1132.

Craig, J. W., and Larner, J. (1966). *Nature (London)* **202**, 971.

Craig, J. W., Rall, T. W., and Larner, J. (1969). *Biochim. Biophys. Acta* **177**, 213.

Creange, J. E., and Roberts, S. (1965). *Biochem. Biophys. Res. Commun.* **19**, 73.

Crofford, O. B. (1968). *J. Biol. Chem.* **243**, 362.

Cronholm, L. S., and Fishel, C. W. (1968). *J. Bacteriol.* **95**, 1993.

Cryer, P. E., Jarett, L., and Kipnis, D. M. (1969). *Biochim. Biophys. Acta* **177**, 586.

Cseh, G. (1970). *Acta Biochim. Biophys.* **5**, 1.

Cuatrecasas, P. (1969). *Proc. Nat. Acad. Sci. U. S.* **63**, 450.

Cunningham, E. B. (1968). *Biochim. Biophys. Acta* **165**, 574.

Cushman, P., Alter, S., and Hilton, J. G. (1966). *J. Endocrinol.* **34**, 271.

Cuthbert, A. W., and Painter, E. (1968a). *J. Pharm. Pharmacol.* **20**, 492.

Cuthbert, A. W., and Painter, E. (1968b). *J. Physiol. (London)* **199**, 593.

Dale, H. H. (1906). *J. Physiol.* **34**, 163.

Dalton, C., Quinn, J. B., Burghardt, C. R., and Sheppard, H. (1970). *J. Pharmacol. Exp. Therap.* **173**, 270.

Danforth, W. H. (1965). *J. Biol. Chem.* **240**, 588.

Danforth, W. H., and Helmreich, E. (1964). *J. Biol. Chem.* **239**, 3133.

Danforth, W. H., and Lyon, J. B. (1964). *J. Biol. Chem.* **239**, 4047.

Danforth, W. H., Helmreich, E., and Cori, C. F. (1962). *Proc. Nat. Acad. Sci. U. S.* **48**, 1191.

Davidson, I. W. F., Salter, J. M., and Best, C. H. (1960). *Amer. J. Clin. Nutr.* **8**, 540.

Davies, J. I. (1968). *Nature (London)* **128**, 349.

Davis, B., Zor, U., Kaneko, T., and Mintz, D. H. (1969). *Clin. Res.* **17**, 458.

Davis, C. H., Schliselfeld, L. H., Wolfe, D. P., Leavitt, C. A., and Krebs, E. G. (1967). *J. Biol. Chem.* **242**, 4824.

Davis, W. W. (1969). *Fed. Proc.* **28**, 701.

Davis, W. W., and Garren, L. D. (1966). *Biochem. Biophys. Res. Commun.* **24**, 805.

Davis, W. W., and Garren, L. D. (1968). *J. Biol. Chem.* **243**, 5153.

Davoren, P. R., and Sutherland, E. W. (1963a). *J. Biol. Chem.* **238**, 3009.

Davoren, P. R., and Sutherland, E. W. (1963b). *J. Biol. Chem.* **238**, 3016.

Daw, J. C., and Berne, R. M. (1967). *Amer. J. Physiol.* **213**, 1480.

Dean, P. M. (1968). *Brit. J. Pharmacol.* **32**, 65.

Debons, A. F., and Schwartz, I. L. (1961). *J. Lipid Res.* **2**, 86.

DeCrombrugghe, B., Perlman, R. L., Varmus, H. E., and Pastan, I. (1969). *J. Biol. Chem.* **244**, 5825.

DeCrombrugghe, B., Varmus, H. E., Perlman, R. L., and Pastan, I. H. (1970). *Biochem. Biophys. Res. Commun.* **38**, 894.

DeDuve, C., Pressman, B. C., Gianetto, R., Watteau, B., and Appelmans, F. (1955). *Biochem. J.* **60**, 604.

DeGubareff, T., and Sleator, W. (1965). *J. Pharmacol. Exp. Ther.* **148**, 202.

DeLange, R. J., Kemp, R. G., Riley, W. D., Cooper, R. A., and Krebs, E. G. (1968). *J. Biol. Chem.* **243**, 2200.

Del Campo, F. F., Ramirez, J. M., and Canovas, J. L. (1970). *Biochem. Biophys. Res. Commun.* **40**, 77.

Denicola, A. T., Clayman, M., and Johnston, R. M. (1968). *Endocrinology* **82**, 436.

Dennis, D. T., and Coultate, T. P. (1966). *Biochem. Biophys. Res. Commun.* **25**, 187.

Dennis, E., and Westheimer, F. (1966). *J. Amer. Chem. Soc.* **88**, 3432.

Denton, R. M., and Randle, P. J. (1966). *Biochem. J.* **100**, 420.

Denton, R. M., Yorke, R. E., and Randle, P. J. (1966). *Biochem. J.* **100**, 407.

De Plaen, J., and Galansino, G. (1966). *Proc. Soc. Exp. Biol. Med.* **121**, 501.

De Robertis, E., Arnaiz, G. R. D. L., Alberici, M., Butcher, R. W., and Sutherland, E. W. (1967). *J. Biol. Chem.* **242**, 3487.

Deter, R. L., and deDuve, C. (1967). *J. Cell Biol.* **33**, 437.

DeTorrontegui, G., and Berthet, J. (1966). *Biochim. Biophys. Acta* **116**, 467.

De Vincenzi, D. L., and Hedrick, J. L. (1967). *Biochemistry* **6**, 3489.

DeWulf, H., and Hers, H. G. (1968a). *Eur. J. Biochem.* **6**, 552.

DeWulf, H., and Hers, H. G. (1968b). *Eur. J. Biochem.* **6**, 558.

DeWulf, H., Stalmans, W., and Hers, H. G. (1968). *Eur. J. Biochem.* **6**, 545.

Dhalla, N. S. (1966). *Arch. Int. Pharmacodyn. Ther.* **164**, 472.

Dhalla, N. S., and McLain, P. L. (1967). *J. Pharmacol. Exp. Ther.* **155**, 389.

Diamond, J., and Brody, T. M. (1965). *Biochem. Pharmacol.* **14**, 7.

Diamond, J., and Brody, T. M. (1966a). *J. Pharmacol. Exp. Ther.* **152**, 202.

Diamond, J., and Brody, T. M. (1966b). *J. Pharmacol. Exp. Ther.* **152**, 212.

Diamond, J., and Brody, T. M. (1966c). *Life Sci.* **5**, 2187.

DiBona, D. R., Civan, M. V., and Leaf, A. (1969). *J. Membrane Biol.* **1**, 79.

Dickerson, R. E., and Geis, I. (1969). "The Structure and Action of Proteins," 120 pp. Harper and Row, New York.

Ditzion, B. R., Paul, M. I., and Pauk, G. L. (1970). *Pharmacology* **3**, 25.

Dobbs, J. W., and Robison, G. A. (1968). *Fed. Proc.* **27**, 352.

Dole, V. P. (1961). *J. Biol. Chem.* **236**, 3125.

Dole, V. P. (1962). *J. Biol. Chem.* **237**, 2758.

Dorigotti, L., Gaetani, M., Glasser, A. H., and Turolla, E. (1969). *J. Pharm. Pharmacol.* **21**, 188.

Dorrington, J. H., and Baggett, B. (1969). *Endocrinology* **84**, 989.

Dorrington, J. H., and Kilpatrick, R. (1967). *Biochem. J.* **104**, 725.

Dorrington, K. J., and Munroe, D. S. (1966). *Clin. Pharmacol. Ther.* **7**, 788.

Dotevall, G., and Kock, N. G. (1963). *Gastroenterology* **45**, 364.

Douglas, W. W. (1968). *Brit. J. Pharmacol.* **34**, 451.

Dousa, T., and Hechter, O. (1970). *Life Sci.* **9** (I), 765.

Dousa, T., and Rychlik, I. (1968a). *Life Sci.* **7,** 1039.

Dousa, T., and Rychlik, I. (1968b). *Biochim. Biophys. Acta* **158,** 484.

Dousa, T., and Rychlik, I. (1970a). *Biochim. Biophys. Acta* **204,** 1.

Dousa, T., and Rychlik, I. (1970b). *Biochim. Biophys. Acta* **204,** 10.

Dreisbach, R. H. (1967). *Proc. Soc. Exp. Biol. Med.* **126,** 279.

Drummond, G. I. (1967). *Fortschr. Zool.* **18,** 359.

Drummond, G. I., and Bellward, G. (1970). *J. Neurochem.* **17,** 475.

Drummond, G. I., and Duncan, L. (1966). *J. Biol. Chem.* **241,** 5893.

Drummond, G. I., and Duncan, L. (1968). *J. Biol. Chem.* **243,** 5532.

Drummond, G. I., and Duncan, L. (1970). *J. Biol. Chem.* **245,** 976.

Drummond, G. I., and Perrott-Yee, S. (1961). *J. Biol. Chem.* **236,** 1126.

Drummond, G. I., and Powell, C. A. (1970). *Mol. Pharmacol.* **6,** 24.

Drummond, G. I., Gilgan, M. W., Reiner, E. J., and Smith, M. (1964a). *J. Amer. Chem. Soc.* **86,** 1626.

Drummond, G. I., Keith, J., and Gilgan, M. W. (1964b). *Arch. Biochem. Biophys.* **105,** 156.

Drummond, G. I., Valadares, J. R. E., and Duncan, L. (1964c). *Proc. Soc. Exp. Biol. Med.* **117,** 307.

Drummond, G. I., Duncan, L., and Friesen, A. J. D. (1965). *J. Biol. Chem.* **240,** 2778.

Drummond, G. I., Duncan, L., and Hertzman, E. (1966). *J. Biol. Chem.* **241,** 5899.

Drummond, G. I., Harwood, J. P., and Powell, C. A. (1969). *J. Biol. Chem.* **244,** 4235.

Duffus, C. M., and Duffus, J. H. (1969). *Experientia* **25,** 581.

Dunn, A., Chenoweth, M., and Schaeffer, L. D. (1969). *Biochim. Biophys. Acta* **177,** 11.

Earp, H. S., Watson, B. S., and Ney, R. L. (1970). *Endocrinology* **87,** 118.

Ebadi, M. S., Weiss, B., and Costa, E. (1970). *Science* **170,** 188.

Edelman, I. S., Peterson, M. J., and Gulyassy, P. F. (1964). *J. Clin. Invest.* **43,** 2185.

Edelman, P. M., and Schwartz, I. L. (1966). *Amer. J. Med.* **40,** 695.

Edelman, P. M., Edelman, J. C., and Schwartz, I. L. (1966). *Nature (London)* **210,** 1017.

Eggena, P., Schwartz, J. L., and Walter, R. (1968). *Life Sci.* **7,** 979.

Eggena, P., Schwartz, I. L., and Walter, R. (1970). *J. Gen. Physiol.* **56,** 250.

Eik-Nes, K. B. (1969). *Amer. J. Physiol.* **217,** 1764.

Eisen, H. J., and Goodman, H. M. (1969). *Endocrinology* **84,** 414.

Eisenstein, A. B. (1967). *Amer. J. Clin. Nutr.* **20,** 282.

Elliott, F. S., Levin, F. M., and Shoemaker, W. C. (1965) *J. Pharmacol. Exp. Ther.* **150,** 61.

Elliott, K. A. C. (1955). *In* "Methods in Enzymology" (S. P. Colowick and N. O. Kaplan, eds.), Vol. I, pp. 3–9. Academic Press, New York.

Ellis, S. (1956). *Pharmacol. Rev.* **8,** 485.

Ellis, S., and Beckett, S. B. (1963). *J. Pharmacol. Exp. Ther.* **142,** 318.

Ellis, S., and Vincent, N. H. (1966). *Pharmacologist* **8** (2), 185.

Ellis, S., Kennedy, B. L., Eusebi, A. J., and Vincent, N. H. (1967). *Ann. N. Y. Acad. Sci.* **139,** 826.

Elrick, H., Whipple, N., Arai, Y., and Hlad, C. J. (1959). *J. Clin. Endocrinol.* **19,** 1274.

Emmer, M., de Crombrugghe, B., Pastan, I., and Perlman, R. (1970). *Proc. Natl. Acad. Sci. U. S.* **66,** 480.

Enser, M. (1970). *Nature (London)* **226,** 175.

Ensor, J. M., and Munro, D. S. (1969). *J. Endocrinol.* **43,** 477.

Entman, M. L., Levey, G. S., and Epstein, S. E. (1969a). *Biochem. Biophys. Res. Commun.* **35,** 728.

Entman, M. L., Levey, G. S., and Epstein, S. E. (1969b). *Circ. Res.* **25,** 429.

Espinosa, A., Driscoll, S. G., and Steinke, J. (1970). *Science* **168,** 1111.

Estler, C. J., and Ammon, H. P. T. (1966). *Biochem. Pharmacol.* **15**, 2031.

Estler, C. J., and Ammon, H. P. T. (1969). *Can. J. Physiol. Pharmacol.* **47**, 427.

Euler, U. S. von (1966). *Pharmacol. Rev.* **18**, 29.

Euler, U. S. von (1967). *Clin. Pharmacol. Ther.* **9**, 228.

Euler, U. S. von, and Lishajko, F. (1969). *Acta Physiol. Scand.* **77**, 298.

Exton, J. H., and Park, C. R. (1967). *J. Biol. Chem.* **242**, 2622.

Exton, J. H., and Park, C. R. (1968a). *J. Biol. Chem.* **243**, 4189.

Exton, J. H., and Park, C. R. (1968b). *Advan. Enzyme Regul.* **6**, 391.

Exton, J. H., and Park, C. R. (1969). *J. Biol. Chem.* **244**, 1424.

Exton, J. H., Jefferson, L. S., Butcher, R. W., and Park, C. R. (1966). *Amer. J. Med.* **40**, 704.

Exton, J. H., Corbin, J. G., and Park, C. R. (1969a). *J. Biol. Chem.* **244**, 4095.

Exton, J. H., Hardman, J. G., Williams, T. F., Sutherland, E. W., and Park, C. R. (1971). *J. Biol. Chem.*, in press.

Fain, J. N. (1968a). *Endocrinology* **82**, 825.

Fain, J. N. (1968b). *Endocrinology* **83**, 548.

Fain, J. N. (1968c). *Mol. Pharmacol.* **4**, 349.

Fain, J. N., and Loken, S. C. (1969). *J. Biol. Chem.* **244**, 3500.

Fain, J. N., Galton, D. J., and Kovacev, V. P. (1966). *Mol. Pharmacol.* **2**, 237.

Fain, J. N., Reed, N., and Saperstein, R. (1967). *J. Biol. Chem.* **242**, 1887.

Falbriard, J. G., Posternak, T., and Sutherland, E. W. (1967). *Biochim. Biophys. Acta* **148**, 99.

Fann, W. E., Asher, H., and Luton, F. H. (1969). *Dis. Nervous System* **30**, 605.

Farah, A., and Tuttle, R. (1960). *J. Pharmacol. Exp. Ther.* **129**, 49.

Farese, R. V. (1967). *Biochemistry* **6**, 2052.

Farese, R. V. (1969). *Endocrinology* **85**, 1209.

Farese, R. V., Linarelli, L. G., Glinsmann, W. H., Ditzion, B. R., Paul, M. I., and Pauk, G. L. (1969). *Endocrinology* **85**, 867.

Fassina, G. (1967). *Life Sci.* **6**, 825.

Fassina, G., Carpenedo F., and Santi, R. (1969). *Life Sci.* **8** (I), 181.

Feldman, J. M., and Lebovitz, H. E. (1970). *Diabetes* **19**, 480.

Felig, P., Pozefsky, T., Marliss, E., and Cahill, G. F. (1970). *Science* **167**, 1003.

Ferrendelli, J. A., Steiner, A. L., McDougal, D. B., and Kipnis, D. M. (1970). *Biochem. Biophys. Res. Commun.* **41**, 1061.

Ferguson, D. R. (1966). *Brit. J. Pharmacol.* **27**, 528.

Ferguson, D. R., Handler, J. S., and Orloff, J. (1968). *Biochim. Biophys. Acta* **163**, 150.

Ferguson, J. J. (1963). *J. Biol. Chem.* **238**, 2754.

Ferguson, J. J., Morita, Y., and Mendelsohn, L. (1967). *Endocrinology* **80**, 521.

Field, J. B., Pastan, I., Herring, B., and Johnson, P. (1960). *Endocrinology* **67**, 801.

Field, J. B., Remer, A., Bloom, G., and Kriss, J. P. (1968). *J. Clin. Invest.* **47**, 1553.

Field, M., and McColl, I. (1968). *Fed. Proc.* **27**, 603.

Field, M., Plotkin, G. R., and Silen, W. (1968). *Nature (London)* **217**, 469.

Finder, A. G., Boyme, T., and Shoemaker, W. C. (1964). *Amer. J. Physiol.* **206**, 738.

Finger, K. F., Page, J. G., and Feller, D. R. (1966) *Biochem. Pharmacol.* **15**, 1023.

Finn, A. L., Handler, J. S., and Orloff, J. (1966). *Amer. J. Physiol.* **210**, 1279.

Fischer, E. H , and Krebs, E. G. (1966). *Fed. Proc.* **25**, 1511.

Fisher, D. A. (1968). *J. Clin. Invest.* **47**, 540.

Fisher, J. N., and Ball, E. G. (1967). *Biochemistry* **6**, 637.

Flack, J. D., Jessup, R., and Ramwell, P. W. (1969). *Science* **163**, 691.

Fleischer, N., Donald, R. A., and Butcher, R. W. (1969). *Amer. J. Physiol.* **217**, 1287.

Foa, P. P. (1968). *Ergeb. Physiol. Biol. Chem. Exp. Pharmakol.* **60**, 141.

Foa, P. P., and Galansino, G. (1962). "Glucagon: Chemistry and Function in Health and Disease," 126 pp. Thomas, Springfield, Illinois.

Forn, J., Schonhofer, P. S., Skidmore, I. F., and Krishna, G. (1970). *Biochim. Biophys. Acta* **208**, 304.

Fox, S. W., Harada, K., Krampitz, G., and Mueller, G. (1970).*Chem. Eng. News* **48**(26), 80.

Franzini-Armstrong, C. (1964). *Fed. Proc.* **23**, 887.

Fraschini, F., Mess, B., Piva, F., and Martini, L. (1968). *Science* **159**, 1104.

Frazer, A., Hess, M. E., and Shanfeld, J. (1969). *J. Pharmacol. Exp. Ther.* **170**, 10.

Fredericq, H., and Melon, L. (1922). *C. R. Soc. Biol.* **86**, 506.

Freinkel, N. (1961). *J. Clin. Invest.* **40**, 476.

Friedman, D. L., and Larner, J. (1963). *Biochemistry* **2**, 669.

Friedman, D. L., and Larner, J. (1965). *Biochemistry* **4**, 2261.

Friedman, R. M., and Pastan, I. (1969). *Biochem. Biophys. Res. Commun.* **36**, 735.

Friedmann, N., and Park, C. R. (1968). *Proc. Nat. Acad. Sci. U. S.* **61**, 504.

Friedmann, N., Exton, J. H., and Park, C. R. (1967). *Biochem. Biophys. Res. Commun.* **29**, 113.

Friesen, A. J. D., Oliver, N., and Allen, G. (1969). *Amer. J. Physiol.* **217**, 445.

Fritz, I. B. (1967). *Perspect. Biol. Med.* **10**, 643.

Fuchs, J. (1969). *Biochim. Biophys. Acta* **172**, 566.

Fujii, R., and Novales, R. R. (1969). *Amer. Zoologist* **9**, 453.

Fukuda, M. (1968). *Jap. J. Pharmacol.* **18**, 185.

Fuller, R. W., and Snoddy, H. D. (1968). *Science* **159**, 738.

Fuller, R. W., Jones, G. T., Snoddy, H. D., and Slater, I. H. (1969). *Life Sci.* **8** (I), 685.

Furchgott, R. F. (1966). *Advan. Drug Res.* **3**, 21.

Furchgott, R. F. (1967). *Ann. N. Y. Acad. Sci.* **139**, 553.

Gagliardino, J. J., and Martin, J. M. (1968). *Acta Endocrinol.* **59**, 390.

Gagliardino, J. J., Hernandez, R. E., and Rodriguez, R. R. (1968). *Experientia* **24**, 1015.

Galansino, G., D'Amico, G., Kanamcishi, D., Berlinger, F. G., and Foa, P. P. (1960). *Amer. J. Physiol.* **198**, 1059.

Galsky, A. G., and Lippincott, J. A. (1969). *Plant Cell Physiol.* **10**, 607.

Galton, D. J., and Bray, G. A. (1967). *J. Clin. Invest.* **46**, 621.

Gangarosa, L. P., and DiStefano, V. (1966). *J. Pharmacol. Exp. Ther.* **152**, 325.

Ganote, C. E., Grantham, J. J., Moses, H. L., Burg, M. B., and Orloff, J. (1968). *J. Cell Biol.* **36**, 355.

Garland, P. B., and Randle, P. J. (1963). *Nature (London)* **199**, 381.

Garren, L. D., Ney, R. L., and Davis, W. W. (1965). *Proc. Nat. Acad. Sci. U. S.* **53**, 1443.

Garren, L. D., Davis, W. W., Crocco, R. M., and Ney, R. L. (1966). *Science* **152**, 1386.

Garrett, E., Seydel, J., and Sharpen, A. (1966) *J. Org. Chem.* **31**, 2219.

Gaut, Z. N., and Huggins, C. G. (1966). *Nature (London)* **212**, 612.

George, W. J., Polson, J. B., O'Toole, A. B., and Goldberg, N. D. (1970). *Proc. Nat. Acad. Sci. U. S.* **66**, 398.

Gergely, J., (Ed.) (1964). "Biochemistry of Muscle Contraction," 582 pp. Little, Brown, Boston, Massachusetts.

Gerhart, J. C., and Pardee, A. B. (1963). *Cold Spring Harbor Symp. Quant. Biol.* **28**, 491.

Gerhart, J. C., and Schachman, H. K. (1965). *Biochemistry* **4**, 1054.

Gericke, D., and Chandra, P. (1969). *Hoppe-Seyler's Z. Physiol. Chem.* **350**, 1469.

Gericke, D., Chandra, P., Haenzel, I., and Wacker, A. (1970). *Hoppe-Seyler's Z. Physiol. Chem.* **351**, 305.

Gerritson, G. C., and Dulin, W. E. (1965). *J. Pharmacol. Exp. Ther.* **150,** 491.

Gershon, S. (1970). *Clin. Pharmacol. Ther.* **11,** 168.

Ghouri, M. S. K., and Haley, T. J. (1969). *J. Pharm. Sci.* **58,** 511.

Gill, G. N., and Garren, L. D. (1969). *Proc. Nat. Acad. Sci. U. S.* **63,** 512.

Gill, G. N., and Garren, L. D. (1970). *Biochem. Biophys. Res. Commun.* **39,** 335.

Gilman, A. G. (1970). *Proc. Nat. Acad. Sci. U. S.* **67,** 305.

Gilman, A. G., and Rall, T. W. (1968a). *J. Biol. Chem.* **243,** 5867.

Gilman, A. G., and Rall, T. W. (1968b). *J. Biol. Chem.* **243,** 5872.

Glick, G., Parmley, W. W., Wechsler, A. S., and Sonnenblick, E. H. (1968). *Circ. Res.* **22,** 789.

Glinsmann, W. H., and Hern, E. P. (1969). *Biochem. Biophys. Res. Commun.* **36,** 931.

Glinsmann, W. H., and Mortimore, G. E. (1968). *Amer. J. Physiol.* **215,** 553.

Glinsmann, W. H., Hern, E. P., Linarelli, L. G., and Farese, R. V. (1969). *Endocrinology* **85,** 711.

Gold, H. K., Prindle, K. H., Levey, G. S., and Epstein, S. E. (1970). *J. Clin. Invest.* **49,** 999.

Goldberg, A. L., and Singer, J. J. (1969). *Proc. Nat. Acad. Sci. U. S.* **64,** 134.

Goldberg, N. D., and O'Toole, A. G. (1969). *J. Biol. Chem.* **244,** 3053.

Goldberg, N. D., Villar-Palasi, C., Sasko, H., and Larner, J. (1967). *Biochim. Biophys. Acta* **148,** 665.

Goldberg, N. D., Larner, J., Sasko, H., and O'Toole, A. G. (1969a). *Anal. Biochem.* **28,** 523.

Goldberg, N. D., Dietz, S. B., and O'Toole, A. G. (1969b). *J. Biol. Chem.* **244,** 4458.

Goldenbaum, P. E., and Dobrogosz, W. J. (1968). *Biochem. Biophys. Res. Commun.* **33,** 828.

Goldenbaum, P. E., Broman, R. L., and Dobrogosz, W. J. (1970). *J. Bacteriol.* **103,** 663.

Goldman, J. M., and Hadley, M. E. (1969a). *J. Pharmacol. Exp. Ther.* **166,** 1.

Goldman, J. M., and Hadley, M. E. (1969b). *Gen. Comp. Endocrinol.* **13,** 151.

Goldman, J. M., and Hadley, M. E. (1970a). *Arch. Int. Pharmacodyn. Ther.* **183,** 239.

Goldman, J. M., and Hadley, M. E. (1970b). *Brit. J. Pharmacol.* **39,** 160.

Goldschlager, N., Robin, E., Cowan, C. M., Leb, G., and Bing, R. J. (1969). *Circulation* **40,** 829.

Goodman, D. B. P., Allen, J. E., and Rasmussen, H. (1969). *Proc. Nat. Acad. Sci. U. S.* **64,** 330.

Goodman, D. B. P., Rasmussen, H., DiBella, F., and Guthrow, C. E. (1970). *Proc. Nat. Acad. Sci. U. S.* **67,** 652.

Goodman, H. M. (1966). *Proc. Soc. Exp. Biol. Med.* **121,** 5.

Goodman, H. M. (1968a). *Endocrinology* **82,** 1027.

Goodman, H. M. (1968b). *Endocrinology* **83,** 300.

Goodman, H. M. (1968c). *Nature (London)* **219,** 1053.

Goodman, H. M. (1969a). *Biochim. Biophys. Acta* **176,** 60.

Goodman, H. M. (1969b). *Proc. Soc. Exp. Biol. Med.* **130,** 97.

Goodman, H. M. (1969c). *Proc. Soc. Exp. Biol. Med.* **132,** 821.

Goodman, H. M. (1970). *Endocrinology* **86,** 1064.

Goodman, H. M., and Knobil, E. (1959). *Proc. Soc. Exp. Biol. Med.* **100,** 195.

Goodridge, A. G. (1964). *Comp. Biochem. Physiol.* **13,** 1.

Goodridge, A. G., and Ball, E. G. (1965). *Comp. Biochem. Physiol.* **16,** 367.

Goodwin, B. C. (1967). *Ann. N. Y. Acad. Sci.* **138,** 748.

Gordon, R. S., and Cherkes, A. (1958). *Proc. Soc. Exp. Biol. Med.* **97,** 150.

Gordon, D. B., and Hesse, D. H. (1961). *Amer. J. Physiol.* **201,** 1123.

Gordon, P., and Dowben, R. M. (1966). *Amer. J. Physiol.* **210,** 728.

Gordon, R. D., Kuchel, O., Liddle, G. W., and Island, D. P. (1967). *J. Clin. Invest.* **46**, 599.

Goren, E., Erlichman, J., Rosen, O. M., and Rosen, S. M. (1970). *Fed. Proc.* **29**, 602 Abs.

Gorman, C. K., Salter, J. M., and Penhos, J. C. (1967). *Metabolism* **16**, 1140.

Govier, W. C. (1965). *Fed. Proc.* **24**, 300.

Govier, W. C. (1968). *J. Pharmacol. Exp. Ther.* **159**, 82.

Grahame-Smith, D. G., Butcher, R. W., Ney, R. L., and Sutherland, E. W. (1967). *J. Biol. Chem.* **242**, 5535.

Grand, R. J., and Gross, P. R. (1969). *J. Biol. Chem.* **244**, 5608.

Grand, R. J., and Gross, P. R. (1970). *Proc. Nat. Acad. Sci. U. S.* **65**, 1081.

Grande, F. (1969). *Proc. Soc. Exp. Biol. Med.* **130**, 711.

Grande, F., and Prigge, W. F. (1970). *Amer. J. Physiol.* **218**, 1406.

Granner, D., Chase, L. R., Aurbach, G. D., and Tomkins, G. M. (1968). *Science* **162**, 1018.

Grantham, J. J., and Burg, M. B. (1966). *Amer. J. Physiol.* **211**, 255.

Grantham, J. J., and Orloff, J. (1968). *J. Clin. Invest.* **47**, 1154.

Gray, G. D. (1966) *Arch. Biochem. Biophys.* **113**, 502.

Gray, J. P., Hardman, J. G., Bibring, T. W., and Sutherland, E. W. (1970). *Fed. Proc.* **29**, 608 Abs.

Green, L. (1968). *Proc. Nat. Acad. Sci. U. S.* **59**, 1179.

Green, D. E., and Goldberger, R. F. (1961). *Amer. J. Med.* **30**, 666.

Green, H. N., and Stoner, H. B. (1950). "Biological Actions of the Adenine Nucleotides," 221 pp. H. K. Lewis, London.

Greengard, O. (1969a). *Science* **163**, 891.

Greengard, O. (1969b). *Biochem. J.* **115**, 19.

Greengard, P., and Costa, E. (eds.). (1970). "Role of Cyclic AMP in Cell Function," 386 pp. Raven Press, New York.

Greengard, P., Hayaishi, O., and Colowick, S. P. (1969a). *Fed. Proc.* **28**, 467.

Greengard, P., Rudolph, S. A., and Sturtevant, J. M. (1969b). *J. Biol. Chem.* **244**, 4798.

Greer, C. M., Pinkston, J. O., Baxter, J. H., and Brannon, E. S. (1937). *J. Pharmacol Exp. Ther.* **62**, 189.

Griffin, D. M., and Szego, C. M. (1968). *Life Sci.* **7**, 1017.

Grinnell, E. H., Kramar, J. L., Duff, W. M., and Lydon, T. E. (1968). *Endocrinology* **83**, 199.

Grodsky, G. M., and Bennett, L. L. (1966). *Diabetes* **15**, 910.

Grodsky, G., Landahl, H., Curry, D., and Bennett, L. (1970). *In* "Early Diabetes" (R. A. Camerini-Davalos and H. S. Cole, eds.), pp. 45–50. Academic Press, New York.

Grossman, M. I. (Ed.) (1966) "Gastrin," 351 pp. Univ. California, Berkeley, California.

Grossman, M. I. (1968). *Fed. Proc.* **27**, 1312.

Grower, M. F., and Bransome, E. D. (1970). *Science* **168**, 483.

Gulbenkian, A., Schobert, L., Nixon, C., and Tabachnick, I. I. A. (1968). *Endocrinology* **83**, 885.

Gulland, J., and Story, L. (1938). *J. Chem. Soc.* p. 692.

Gulyassy, P. F. (1968). *J. Clin. Invest.* **47**, 2458.

Gulyassy, P. F., and Edelman, I. S. (1965). *Biochim. Biophys. Acta* **102**, 185.

Gulyassy, P. F., and Edelman, I. S. (1967). *Amer. J. Physiol.* **212**, 740.

Gussin, R. Z., Miksche, U., and Farah, A. (1967). *J. Pharmacol. Exp. Ther.* **156**, 606.

Hadley, M. E., and Goldman, J. M. (1969a). *Amer. Zoologist* **9**, 489.

Hadley, M. E., and Goldman, J. M. (1969b). *Brit. J. Pharmacol.* **37**, 650.

Hagen, J. H. (1961). *J. Biol. Chem.* **236**, 1023.

Hales, C. N., and Milner, R. D. G. (1968). *J. Physiol. (London)* **194**, 725.

Hales, C. N., Chalmers, F. M., Perry, M. C., and Wade, D. R. (1968). *In* "Protein and Polypeptide Hormones" (M. Margoulies, ed.), Part II, pp. 432–443. Excerpta Medica Found. Amsterdam.

Halkerston, I. D. K. (1968), *In* "Functions of the Adrenal Cortex" (K. W. McKerns, ed.), Vol. I, pp. 399–461. Appleton-Century-Crafts, New York.

Halkerston, I. D. K., Feinstein, M., and Hechter, O. (1964). *Endocrinology* **74**, 649.

Halkerston, I. D. K., Feinstein, M., and Hechter, O. (1965). *Endocrinology* **76**, 801.

Halkerston, I. D. K., Feinstein, M., and Hechter, O. (1966). *Proc. Soc. Exp. Biol. Med.* **122**, 896.

Hall, P. F., and Koritz, S. B. (1965). *Biochemistry* **4**, 1037.

Hall, P. F., and Young, D. G. (1968). *Endocrinology* **82**, 559.

Hammermeister, K. E., Yunis, A. A., and Krebs, E. G. (1965). *J. Biol. Chem.* **240**, 986.

Hanabusa, K., and Kobayashi, H. (1967). *J. Biochem.* **61**, 662.

Handler, J. S., and Orloff, J. (1963). *Amer. J. Physiol.* **205**, 298.

Handler, J. S., and Orloff, J. (1964). *Amer. J. Physiol.* **206**, 505.

Handler, J. S., Butcher, R. W., Sutherland, E. W., and Orloff, J. (1965). *J. Biol. Chem.* **240**, 4524.

Handler, J. S., Petersen, M., and Orloff, J. (1966). *Amer. J. Physiol.* **211**, 1175.

Handler, J. S., Preston, A. S., and Rogulski, J. (1968a). *J. Biol. Chem.* **243**, 1376.

Handler, J. S., Bensinger, R., and Orloff, J. (1968b). *Amer. J. Physiol.* **215**, 1024.

Handler, J. S., Preston, A. S., and Orloff, J. (1969). *J. Biol. Chem.* **244**, 3194.

Hanze, A. R. (1968). *Biochemistry* **7**, 932.

Hardman, J. G., and Sutherland, E. W. (1965). *J. Biol. Chem* **240**, PC 3704.

Hardman, J. G., and Sutherland, E. W. (1969). *J. Biol. Chem.* **244**, 6363.

Hardman, J. G., Mayer, S. E., and Clark, B. (1965). *J. Pharmacol. Exp. Ther.* **150**, 341.

Hardman, J. G., Davis, J. W., and Sutherland, E. W. (1966). *J. Biol. Chem.* **241**, 4812.

Hardman, J. G., Davis, J. W., and Sutherland, E. W. (1969). *J. Biol. Chem.* **244**, 6354.

Harris, J. B., and Alonso, D. (1965). *Fed. Proc.* **24**, 1368.

Harris, J. B., Nigon, K., and Alonso, D. (1969). *Gastroenterology* **57**, 377.

Harrison, H. C., and Harrison, H. E. (1970). *Endocrinology* **86**, 756.

Harrison, T. S. (1964). *Physiol. Rev.* **44**, 161.

Harvey, S. C., Wang, C. Y., and Nickerson, M. (1952). *J. Pharmacol. Exp. Ther.* **140**, 363.

Hashimoto, T., Sasaki, H., and Yoshikawa, H. (1967). *Biochem. Biophys. Res. Commun.* **27**, 368.

Haugaard, E. S., and Haugaard, N. (1954). *J. Biol. Chem.* **206**, 641.

Haugaard, E. S., and Stadie, W. C. (1953). *J. Biol. Chem.* **200**, 753.

Haugaard, N., and Hess, M. E. (1965). *Pharmacol. Rev.* **17**, 27.

Haynes, R. C. (1958). *J. Biol. Chem.* **233**, 1220.

Haynes, R. C., and Berthet, L. (1957). *J. Biol. Chem.* **225**, 115.

Haynes, R. C., Koritz, S. B., and Peron, F. G. (1959). *J. Biol. Chem.* **234**, 1421.

Haynes, R. C., Sutherland, E. W., and Rall, T. W. (1960). *Recent Progr. Hormone Res.* **16**, 121.

Hechter, O., and Halkerston, I. D. K. (1964). *Perspect. Biol. Med.* **7**, 183.

Hechter, O., and Halkerston, I. D. K. (1965). *Ann. Rev. Physiol.* **27**, 133.

Hechter, O., Yoshinaga, K., Halkerston, I. D. K., Cohn, C., and Dodd, P. (1966). *In* "Molecular Basis of Some Aspects of Mental Activity," (P. Karlson, ed.), Vol. 1, pp. 291–341. Academic Press, New York.

Hechter, O., Yoshinaga, K., Halkerston, I. D. K., and Birchall, K. (1967). *Arch. Biochem. Biophys.* **122**, 449.

Hechter, O., Bär, H., Matsuba, M., and Soifer, D. (1969). *Life Sci.* **8**, 935.

Heidrick, M. L., and Ryan, W. L. (1970). *Cancer Res.* **30,** 376.

Heim, T., and Hull, D. (1966). *J. Physiol. (London)* **187,** 271.

Heimberg, M., and Fizette, N. B. (1963). *Biochem. Pharmacol.* **12,** 392.

Heimberg, M., Fizette, N. B., and Klausner, H. (1964). *J. Amer. Oil Chemists' Soc.* **41,** 774.

Heimberg, M., Weinstein, I., and Kohout, M. (1969). *J. Biol. Chem.* **244,** 5131.

Helmreich, E., and Cori, C. F. (1964). *Proc. Nat. Acad. Sci. U.S.* **51,** 131.

Helmreich, E., and Cori, C. F. (1966). *Pharmacol. Rev.* **18,** 189.

Henion, W. F., and Sutherland, E. W. (1957). *J. Biol. Chem.* **224,** 477.

Henion, W. F., Sutherland, E. W., and Posternak, T. (1967). *Biochim. Biophys. Acta* **148,** 106.

Hepp, K. D., Menahan, L. A., Wieland, O., and Williams, R. H. (1969). *Biochim. Biophys. Acta* **184,** 554.

Herman, S. H., and Ramey, E. R. (1960). *Amer. J. Physiol.* **199,** 226.

Hess, M. E., Hottenstein, D., Shanfeld, J., and Haugaard, N. (1963). *J. Pharmacol. Exp. Ther.* **141,** 274.

Hess, M. E., Aronson, C. E., Hottenstein, D. W., and Karp, J. (1969). *Endocrinology* **84,** 1107.

Hess, W. R. (1948). "Die funktionelle Organization des vegetativen Nervensystems." Schwabe, Basel.

Hilf, R. (1965). *New Engl. J. Med.* **273,** 798.

Hilton, J. G., Kruesi, O. R., Nedeljkovic, R. I., and Scian, L. F. (1961). *Endocrinology* **68,** 908.

Hilz, H., and Tarnowski, W. (1970). *Biochem. Biophys. Res. Commun.* **40,** 973.

Himms-Hagen, J. (1967). *Pharmacol. Rev.* **19,** 367.

Hirata, M., and Hayaishi, O. (1965). *Biochem. Biophys. Res. Commun.* **21,** 361.

Hirata, M., and Hayaishi, O. (1966). *Biochem. Biophys. Res. Commun.* **24,** 360.

Hirata, M., and Hayaishi, O. (1967). *Biochim. Biophys. Acta* **149,** 1.

Hirsch, P. F., and Munson, P. L. (1969). *Physiol. Rev.* **49,** 548.

Hirschhorn, R., Grossman, J., and Weissmann, G. (1970). *Proc. Soc. Exp. Biol. Med.* **131,** 1361.

Hirshfield, I. N., and Koritz, S. B. (1966). *Endocrinology* **78,** 165.

Ho, R. J., and Jeanrenaud, B. (1967). *Biochim. Biophys. Acta* **144,** 61.

Ho, R. J., Ho, S. J., and Meng, H. C. (1965). *Metabolism* **14,** 1010.

Ho, R. J., Jeanrenaud, B., and Renold, A. E. (1966). *Experientia* **22,** 86.

Ho, R. J., Jeanrenaud, B., Posternak, T., and Renold, A. E. (1967). *Biochim. Biophys. Acta* **144,** 74.

Hoffer, B. J., Siggins, G. R., and Bloom, F. E. (1969). *Science* **166,** 1418.

Hofstee, B. H. J. (1952). *Science* **116,** 329.

Hogeboom, G. H. (1955). *In* "Methods in Enzymology" (S. P. Colowick and N. O. Kaplan, eds.), Vol. 1, pp. 16–19. Academic Press, New York.

Hokin, M. R. (1969). *J. Neurochem.* **16,** 127.

Holmes, P. A., and Mansour, T. E. (1968). *Biochim. Biophys. Acta* **156,** 275.

Holt, P. G., and Oliver, I. T. (1969a). *Biochemistry* **8,** 1429.

Holt, P. G., and Oliver, I. T. (1969b). *FEBS Lett.* **5,** 89.

Homsher, E., and Cotzias, G. C. (1965). *Nature (London)* **208,** 687.

Honda, F., and Imamura, H. (1968). *Biochim. Biophys. Acta* **161,** 267.

Hong, S. K., Park, C. S., Park, Y. S., and Kim, J. K. (1968). *Amer. J. Physiol.* **215,** 439.

Honig, C. R., and Stam, A. C. (1967). *Ann. N. Y. Acad. Sci.* **139,** 724.

Honig, C. R., and Van Nierop, C. (1964). *Biochim. Biophys. Acta* **86,** 355.

Hornbrook, K. R., and Brody, T. M. (1963a). *J. Pharmacol. Exp. Ther.* **140**, 295.

Hornbrook, K. R., and Brody, T. M. (1963b). *Biochem. Pharmacol.* **12**, 1407.

Hubert-Habart, M., and Goodman, L. (1969). *Chem. Commun.* p. 740.

Huijing, F. (1966). *Fed. Proc.* **25**, 449.

Huijing. F., and Larner, J. (1966a). *Biochem. Biophys. Res. Commun.* **23**, 259.

Huijing, F., and Larner, J. (1966b). *Proc. Nat. Acad. Sci. U.S.* **56**, 647.

Huijing, F., Villar-Palasi, C., and Larner, J. (1965). *Biochem. Biophys. Res. Commun.* **20**, 380.

Hull, D., and Segall, M. M. (1965). *J. Physiol. (London)* **181**, 449.

Humes, J. L., Mandel, L. R., and Kuehl, F. A. (1968). *In* "Prostaglandin Symposium of the Worchester Foundation for Experimental Biology" (P. W. Ramwell and J. E. Shaw, eds.), pp. 79–91. Wiley (Interscience), New York.

Humes, J. L., Rounbehler, M., and Kuehl, F. A. (1969). *Anal. Biochem.* **32**, 210.

Huttunen, J. K., Steinberg, D., and Mayer, S. E. (1970). *Proc. Nat. Acad. Sci. U. S.* **67**, 290.

Huxley, J. S. (1935). *Biol. Rev.* **10**, 427.

Hynie, S., Krishna, G., and Brodie, B. B. (1966). *J. Pharmacol. Exp. Ther.* **153**, 90.

Hynie, S., Krishna, G., and Brodie, B. B. (1969). *Pharmacology* **2**, 129.

Ide, M. (1969). *Biochem. Biophys. Res. Commun.* **36**, 42.

Ide, M., Yoshimoto, A., and Okabayashi, T. (1967). *J. Bacteriol.* **94**, 317.

Imura, H., Matsukura, S., Matsuyama, H., Setsuda, T., and Miyake, T. (1965). *Endocrinology* **76**, 933.

Inchiosa, M. A. (1967). *Biochem. Pharmacol.* **16**, 329.

Ingram, G. I. C., and Vaughan Jones, R. (1966). *J. Physiol. (London)* **187**, 447.

Innes, I. R., and Kohli, J. D. (1969). *Brit. J. Pharmacol.* **35**, 383.

Innes, I. R., and Nickerson, M. (1965). *In* "The Pharmacological Basis of Therapeutics" (L. S. Goodman and A. Gilman, eds.), 3rd ed., pp. 477–520. Macmillan, New York.

Ishikawa, E., Ishikawa, S., Davis, J. W., and Sutherland, E. W. (1969). *J. Biol. Chem.* **244**, 6371.

Ishizuka, M., Gafni, M., and Braun, W. (1970). *Proc. Soc. Exp. Biol. Med.* **134**, 963.

Jacob, M. I., and Berne, R. M. (1960). *Amer. J. Physiol.* **198**, 322.

Jacquet, M., and Kepes, A. (1969). *Biochem. Biophys. Res. Commun.* **36**, 84.

Jaim-Etcheverry, G., and Zieher, L. M. (1968). *Endocrinology* **83**, 917.

James, T. N. (1965). *J. Pharmacol. Exp. Ther.* **149**, 233.

James, T. N. (1967). *Bull. N. Y. Acad. Med.* **43**, 1041.

James, T. N., Bear, E. S., Lang, K. F., and Green, E. W. (1968). *Amer. J. Physiol.* **215**, 1366.

Janakidevi, K., Dewey, V. C., and Kidder, G. W. (1966). *J. Biol. Chem.* **241**, 2576.

Jard, S., and Bastide, F. (1970). *Biochem. Biophys. Res. Commun.* **39**, 559.

Jard, S., Bastide, F., and Morel, F. (1968). *Biochim. Biophys. Acta* **150**, 125.

Jardetzky, C. (1962). *J. Amer. Chem. Soc.* **84**, 62.

Jeanrenaud, B. (1968). *Ergeb. Physiol. Biol. Chem. Exp. Pharmakol.* **60**, 57.

Jeanrenaud, B., and Hepp, D. (eds.). (1970). "Adipose Tissue: Regulation and Metabolic Functions," 426 pp. Academic Press, New York.

Jeanrenaud, B., and Renold, A. E. (1960). *J. Biol. Chem.* **235**, 2217.

Jefferson, L. S., Exton, J. H., Butcher, R. W., Sutherland, E. W., and Park, C. R. (1968). *J. Biol. Chem.* **243**, 1031.

Jergil, B., and Dixon, G. H. (1970). *J. Biol. Chem.* **245**, 425.

Joel, C. D. (1966). *J. Biol. Chem.* **241**, 814.

Johnson, G. S., Friedman, R. M., and Pastan, I. (1971). *Proc. Nat. Acad. Sci. U. S.* **68**, 425.

Johnson, L. D., and Abell, C. W. (1970). *Cancer Res.* **30**, 2718.

Johnson, R. A., Hardman, J. G., Broadus, A. E., and Sutherland, E. W. (1970). *Anal. Biochem.* **35,** 91.

Johnston, H. H., Herzog. J. P., and Lauler, D. P. (1967). *Amer. J. Physiol.* **213,** 939.

Jones, D. J., Nicholson, W. E., Liddle, G. W., and Finnegan, A. W. (1970). *Proc. Soc. Exp. Biol. Med.* **133,** 764.

Jost, J.-P., Khairallah, E. A., and Pitot, H. C. (1968). *J. Biol. Chem.* **243,** 3057.

Jost, J.-P., Hsie, A. W., and Rickenberg, H. V. (1969). *Biochem. Biophys. Res. Commun.* **34,** 748.

Jost, J.-P., Hsai, A., Hughes, S. D., and Ryan, L. (1970). *J. Biol. Chem.* **245,** 351.

Jungas, R. L. (1966). *Proc. Nat. Acad. Sci. U.S.* **56,** 757.

Jungas, R. L., and Ball, E. G. (1963). *Biochemistry* **2,** 383.

Jutisz, M., and de la Llosa, M. P. (1969). *C. R. Acad. Sci. Ser. D* **268,** 1636.

Jutisz, M., and de la Llosa, M. P. (1970). *Endocrinology* **86,** 761.

Kabat, D. (1970). *Biochemistry* **9,** 4160.

Kakiuchi, S., and Rall, T. W. (1968a). *Mol. Pharmacol.* **4,** 367.

Kakiuchi, S., and Rall, T. W. (1968b). *Mol. Pharmacol.* **4,** 379.

Kakiuchi, S., and Yamazaki, R. (1970). *Proc. Jap. Acad.* **46,** 387.

Kakiuchi, S., Rall, T. W., and McIlwain, H. (1969). *J. Neurochem.* **16,** 485.

Kaminsky, N. I., Broadus, A. E., Hardman, J. G., Ginn, H. E., Sutherland, E. W., and Liddle, G. W. (1969). *J. Clin. Invest.* **48,** 42a.

Kaminsky, N. I., Broadus, A. E., Hardman, J. G., Jones, D. J., Ball, J. H., Sutherland, E. W., and Liddle, G. W. (1970a). *J. Clin. Invest.* **49,** 2387.

Kaminsky, N. I., Ball, J. H., Broadus, A. E., Hardman, J. G., Sutherland, E. W., and Liddle, G. W. (1970b). *Clin. Res.* **18,** 528.

Kaminsky, N. I., Ball, J. H., Broadus, A. E., Hardman, J. G., Sutherland, E. W., and Liddle, G. W. (1970c). *Trans. Amer. Ass. Physicians* **83,** 235.

Kaneko, T., Zor, U., and Field, J. B. (1969). *Science* **163,** 1062.

Kaplan, N. M. (1965). *J. Clin. Invest.* **44,** 2029.

Karaboyas, G. C., and Koritz, S. B. (1965). *Biochemistry* **4,** 462.

Karam, J. H., Grasso, S. G., Wegienka, L. C., Grodsky, G. M., and Forsham, P. H. (1966). *Diabetes* **15,** 571.

Karlson, P., and Sekeris, C. E. (1966). *Acta Endocrinol.* **53,** 505.

Kasture, A. V., Shingvekar, D. S., and Dorle, A. K. (1966). *Nature (London)* **212,** 1598.

Katz, R. L., and Mandel, I. D. (1968). *Proc. Soc. Exp. Biol. Med.* **128,** 1140.

Kawashi, A., Kashimoto, T., and Yoshida, H. (1969). *Jap. J. Pharmacol.* **19,** 494.

Keilin, D., and Hartree, E. F. (1938). *Proc. Royal Soc. Ser.* **B124,** 397.

Keller, E. F., and Segel, L. A. (1970). *J. Theoret. Biol.* **26,** 399.

Kemp, R. G., and Krebs, E. G. (1967). *Biochemistry* **6,** 423.

Kennedy, B. L., and Ellis, S. (1969). *J. Pharmacol. Exp. Ther.* **168,** 137.

Ketterer, H., Eisentraut, A. M., and Unger, R. H. (1967). *Diabetes* **16,** 283.

Khairallah, E. A., and Pitot, H. C. (1967). *Biochem. Biophys. Res. Commun.* **29,** 269.

Khorana, H., Tener, G., Wright, R., and Moffatt, J. (1957). *J. Amer. Chem. Soc.* **79,** 430.

Kilo, C., Devrim, S., Bailey, R., and Recant, L. (1967). *Diabetes* **16,** 377.

Kim, J. H., and Miller, L. L. (1969). *J. Biol. Chem.* **244,** 1410.

Kim, T. S., Shulman, J., and Levine, R. A. (1968). *J. Pharmacol. Exp. Ther.* **163,** 36.

Kipnis, D. M., and Noall, M. W. (1958). *Biochim. Biophys. Acta* **28,** 226.

Kitabchi, A. E., Solomon, S. S., and Brush, J. S. (1970). *Biochem. Biophys. Res. Commun.* **39,** 1065.

Klainer, L. M., Chi, Y. M., Freidberg, S. L., Rall, T. W., Sutherland, E. W. (1962). *J. Biol. Chem.* **237,** 1239.

Klein, D. C., and Raisz, L. G. (1970). *Endocrinology* **86,** 1436.

Klein, D. C., Berg, G. R., Weller, J., and Glinsmann, W. (1970a). *Science* **167,** 1738.

Klein, D. C., Berg, G. R., and Weller, J. (1970b). *Science* **168,** 979.

Kloeze, J. (1967). *In* "Prostaglandins" (S. Bergström and B. Samuelsson, eds.), pp. 241–252. Almqvist & Wiksell, Stockholm; Wiley (Interscience), New York.

Knopp, J., Stolc, V., and Tong, W. (1970). *J. Biol. Chem.* **245,** 4403.

Kobata, A., Kida, M., and Ziro, S. (1961). *J. Biochem.* **50,** 275.

Kobayashi, S., Yago, N., Morisaki, M., Ichii, S., and Matsuba, M. (1963). *Steroids* **2,** 167.

Koch-Weser, J., and Blinks, J. R. (1963). *Pharmacol. Rev.* **15,** 601.

Koch-Weser, J., Berlin, C. M., and Blinks, J. R. (1964). *In* "Pharmacology of Cardiac Function" (O. Krayer, ed.), pp. 63–72. Macmillan (Pergamon), New York.

Kohn, R. (1932). *Arch. Exp. Pathol. Pharmakol.* **167,** 216.

Konijn, T. M. (1969). *J. Bacteriol.* **99,** 503.

Konijn, T. M. (1970). *Experientia* **26,** 367.

Konijn, T. M., van de Meene, J. G. C., Bonner, J. T., and Barkley, D. S. (1967). *Proc. Nat. Acad. Sci. U.S.* **41,** 1152.

Konijn, T. M., Barkley, D. S., Chang, Y. Y., and Bonner, J. T. (1968). *Amer. Natur.* **102,** 225.

Konijn, T. M., van de Meene, J. G. C., Chang, Y. Y., Barkley, D. S., and Bonner, J. T. (1969a). *J. Bacteriol.* **99,** 510.

Konijn, T. M., Chang, Y. Y., and Bonner, J. T. (1969b). *Nature (London)* **224,** 1211.

Kono, T. (1969a). *J. Biol. Chem.* **244,** 1772.

Kono, T. (1969b). *J. Biol. Chem.* **244,** 5777.

Koritz, S. B. (1968). *In* "Functions of the Adrenal Cortex" (K. W. McKerns, ed.), Vol. I, pp. 27–48. Appleton-Century-Crofts, New York.

Koritz, S. B., and Hall, P. F. (1965). *Biochemistry* **4,** 2740.

Koritz, S. B., Yun, J., and Ferguson, J. J. (1968). *Endocrinology* **82,** 620.

Korner, A., and Raben, M. S. (1964). *Nature (London)* **203,** 1287.

Kovacev, V., and Scow, R. O. (1966). *Amer. J. Physiol.* **210,** 1199.

Kowal, J. (1969). *Biochemistry* **8,** 1821.

Kowal, J., and Fiedler, R. P. (1969). *Endocrinology* **84,** 1113.

Krahl, M. E. (1951). *Ann. N. Y. Acad. Sci.* **54,** 649.

Krause, E.-G., Hale, W., Kallabis, E., and Wollenberger, A. (1970). *J. Mol. Cell. Cardiol.* **1,** 1.

Krebs, H. A., and Eggleston, L. V. (1938). *Biochem. J.* **32,** 913.

Krebs, E. G., and Fischer, E. H. (1956). *Biochim. Biophys. Acta* **20,** 150.

Krebs, E. G., and Fischer, E. H. (1960). *Ann. N. Y. Acad. Sci.* **88,** 378.

Krebs, E. G., and Fischer, E. H. (1962). *Advan. Enzymol.* **24,** 263.

Krebs, E. G., Love, D. S., Bratvold, G. E., Trayser, K. A., Meyer, W. L., and Fischer, E. H. (1964). *Biochemistry* **3,** 1022.

Krebs, E. G., DeLange, R. J., Kemp, R. G., and Riley, W. D. (1966). *Pharmacol. Rev.* **18,** 163.

Krebs, E. G., Huston, R. B., and Hunkeler, F. L. (1968). *Advan. Enzyme Regul.* **6,** 245.

Kreisberg, R. A., and Williamson, J. R. (1964). *Amer. J. Physiol.* **207,** 721.

Krishna, G., Weiss, B., Davies, J. I., and Hynie, S. (1966). *Fed. Proc.* **25,** 719.

Krishna, G., Weiss, B., and Brodie, B. B. (1968a). *J. Pharmacol. Exp. Ther.* **163,** 379.

Krishna, G., Hynie, S., and Brodie, B. B. (1968b). *Proc. Nat. Acad. Sci. U.S.* **59,** 884.

Krishna, G., Ditzion, B. R., and Gessa, G. L. (1968c). *Proc. Int. Union Physiol. Sci.* **7,** 247.

Krishna, G., Moskowitz, J., Dempsey, P., and Brodie, B. B. (1970). *Life Sci.* **9(I),** 1353.

Kuehl, F. A., Patanelli, D. J., Tarnoff, J., and Humes, J. L. (1970a). *Bio. Reprod.* **2**, 154.

Kruger, F. A., Leighty, E. G., and Weissler, A. M. (1967). *J. Clin. Invest.* **46**, 1080.

Kuehl, F. A., Humes, J. L., Tarnoff, J., Cirillo, V. J., and Ham, E. A. (1970b). *Science* **169**, 883.

Kukovetz, W. R. (1968). *Arch. Pharmakol. Exp. Pathol.* **260**, 163.

Kukovetz, W. R., and Pöch, G., (1967a). *Arch. Pharmakol. Exp. Pathol.* **256**, 310.

Kukovetz, W. R., and Pöch, G. (1967b). *Biochem. Pharmacol.* **16**, 1429.

Kukovetz, W. R., and Pöch, G. (1967c). *J. Pharmacol. Exp. Ther.* **156**, 514.

Kukovetz, W. R., and Pöch, G. (1970a). *Arch. Pharmakol. Exp. Pathol.* **266**, 236.

Kukovetz, W. R., and Pöch, G. (1970b). *Arch. Pharmakol. Exp. Pathol.* **267**, 189.

Kukovetz. W. R., Hess, M. E., Shanfeld, J., and Haugaard, N. (1959). *J. Pharmacol. Exp. Ther.* **127**, 122.

Kulka, R. G., and Sternlicht, E. (1968). *Proc. Nat. Acad. Sci. U.S.* **61**, 1123.

Kuo, J. F., and DeRenzo, E. C. (1969). *J. Biol. Chem.* **244**, 2252.

Kuo, J. F., and Dill, I. K. (1968). *Biochem. Biophys. Res. Commun.* **32**, 333.

Kuo, J. F., and Greengard, P. (1969a). *J. Biol. Chem.* **244**, 3417.

Kuo, J. F., and Greengard, P. (1969b). *Proc. Nat. Acad. Sci. U.S.* **64**, 1349.

Kuo, J. F., and Greengard, P. (1970). *J. Biol. Chem.* **245**, 2493.

Kuo, J. F., Holmlund, C. E., and Dill, I. K. (1966). *Life Sci.* **5**, 2257.

Kuo, J. F., Krueger, B., Sanes, J. R., and Greengard, P. (1970). *Biochim. Biophys. Acta* **208**, 509.

Kupiecki, F. P. (1969). *Life Sci.* **8** (II), 645.

Kupiecki, F. P., and Marshall, N. B. (1968). *J. Pharmacol. Exp. Ther.* **160**, 166.

Kuschke, H. J., Klusmann, H., and Schölkens, B. (1966). *Klin. Wochenschr.* **44**, 1297.

Kuwano, M., and Schlessinger, D. (1970). *Proc. Nat. Acad. Sci. U. S.* **66**, 146.

Kuzuya, T., and Kanazawa, Y. (1969). *Diabetologia* **5**, 248.

Kypson, J., Triner, L., and Nahas, G. G. (1968). *J. Pharmacol. Exp. Ther.* **159**, 8.

Laborit, H., and Weber, B. (1967). *Agressologie* **8**, 37.

Lacy, P. E. (1967). *New Engl. J. Med.* **276**, 187.

Lacy, P. E., Howell, S. L., Young, D. A., and Fink, C. J. (1968a). *Nature (London)* **219**, 1177.

Lacy, P. E., Young, D. A., and Fink, J. C. (1968b). *Endocrinology* **83**, 1155.

Lambert, A. E., Jeanrenaud, B., and Renold, A. E. (1967a). *Lancet* **i**, 819.

Lambert, A. E., Vecchio, D., Gonet, A., Jeanrenaud, B., and Renold, A. E. (1967b). *In* "Tolbutamide . . . after Ten Years" (W. J. H. Butterfield and W. V. Westering. eds.), pp. 61–82. Excerpta Medica Found., Amsterdam.

Lambert, A. E., Junod, A., Stauffacher, W., Jeanrenaud, B., and Renold, A. E. (1969a). *Biochim. Biophys. Acta* **184**, 529.

Lambert, A. E., Jeanrenaud, B., Junod, A., and Renold, A. E. (1969b). *Biochim. Biophys. Acta* **184**, 540.

Lamble, J. W. (1970). *Life Sci.* **9** (I), 463.

Landon, J., Wynn, V., Houghton, B. J., and Cooke, J. N. C. (1962). *Metabolism* **11**, 513.

Lands, A. M., Arnold, A., McAuliff, J. P., Luduena, F. P., and Brown, T. G. (1967). *Nature (London)* **214**, 597.

Lands, A. M., Luduena, F. P., and Buzzo, H. J. (1969). *Life Sci.* **8** (I), 373.

Langan, T. A. (1968). *Science* **162**, 579.

Langan, T. A. (1969a). *J. Biol. Chem.* **244**, 5763.

Langan, T. A. (1969b). *Proc. Nat. Acad. Sci. U. S.* **64**, 1276.

Laragh, J. H. (1968). *Amer. Heart J.* **75**, 564.

LaRaia, P. J., and Reddy, W. J. (1969). *Biochim. Biophys. Acta* **177**, 189.

LaRaia, P. J., Craig, R. J., and Reddy, W. J. (1968). *Amer. J. Physiol.* **215**, 968.

Larner, J. (1966). *Trans. N. Y. Acad. Sci.* **29**, 192.

Larner, J., and Sanger, F. (1965). *J. Mol. Biol.* **11**, 491.

Larner, J., Villar-Palasi, C., and Richman, D. S. (1959). *Ann. N. Y. Acad. Sci.* **82**, 345.

Larner, J., Villar-Palasi, C., Goldberg, N. D., Bishop, J. S., Huijing, F., Wenger, J. I., Sasko, H., and Brown, N. B. (1968). *Advan. Enzyme Regul.* **6**, 409.

Lawrence, A. M. (1966). *Proc. Nat. Acad. Sci. U.S.* **55**, 316.

Lebovitz, H. E., and Pooler, K. (1967). *Endocrinology* **81**, 558.

Lefkowitz, R. J., Roth, J., Pricer, W., and Pastan, I. (1970). *Proc. Nat. Acad. Sci. U.S.* **65**, 745.

Leibman, K. C., and Heidelberger, C. (1955). *J. Biol. Chem.* **216**, 823.

Leloir, L. F., and Cardini, C. E. (1957). *J. Amer. Chem. Soc.* **79**, 6340.

Levene, P., and Tipson, R. (1937). *J. Biol. Chem.* **121**, 131.

Levey, G. S. (1970). *Biochem. Biophys. Res. Commun.* **38**, 86.

Levey, G. S., and Epstein, S. E. (1969a). *Circ. Res.* **24**, 151.

Levey, G. S., and Epstein, S. E. (1969b). *J. Clin. Invest.* **48**, 1663.

Levey, G. S., and Pastan, I. (1970). *Life Sci.* **9** (I), 67.

Levey, G. S., Skelton, C. L., and Epstein, S. E. (1969). *J. Clin. Invest.* **48**, 2244.

Levine, R. A. (1965). *Amer. J. Physiol.* **208**, 317.

Levine, R. A. (1968a). *Metabolism* **17**, 34.

Levine, R. A. (1968b). *Clin. Sci.* **34**, 253.

Levine, R. A. (1969). *Fed. Proc.* **28**, 707.

Levine, R. A. (1970). *Clin. Pharmacol. Ther.* **11**, 238.

Levine, R. A., and Lewis, S. E. (1967). *Amer. J. Physiol.* **213**, 768.

Levine, R. A., and Lewis, S. E. (1969). *Biochem. Pharmacol.* **18**, 15.

Levine, R. A., and Vogel, J. A. (1966). *J. Pharmacol. Exp. Ther.* **151**, 262.

Levine, R. A., Pesch, L. A., Klatskin, G., and Giarman, N. J. (1964). *J. Clin. Invest.* **43**, 797.

Levine, R. A., Cafferata, E. P., and McNally, E. F. (1966). *Clin. Res.* **14**, 301.

Levine, R. A., Dixon, L. M., and Franklin, R. B. (1968). *Clin. Pharmacol.* **9**, 168.

Levine, R. A., Lewis, S. E., Shulman, J., and Washington, A. (1969). *J. Biol. Chem.* **244**, 4017.

Levine, R. A., Oyama, S., Kagan, A., and Glick, S. M. (1970). *J. Lab. Clin. Med.* **75**, 30.

Levine, R. J. (1968). *Fed. Proc.* **27**, 1341.

Levy, B. (1966). *J. Pharmacol. Exp. Ther.* **151**, 413.

Levy, B., and Wilkenfeld, B. E. (1968). *Brit. J. Pharmacol.* **34**, 604.

Levy, B., and Wilkenfeld, B. E. (1969). *Eur. J. Pharmacol.* **5**, 227.

Levy, M. M., and Zieske, H. (1969). *Circ. Res.* **24**, 303.

Lewis, S. B., Exton, J. H., Ho, R. J., and Park, C. R. (1970). *Fed. Proc.* **29**, 379 Abs.

Liberman, B., Klein, L. A., and Kleeman, C. R. (1970). *Proc. Soc. Exp. Biol. Med.* **133**, 131.

Lichtenstein, L. M., and Margolis, S. (1968). *Science* **161**, 902.

Linarelli, L. G., and Farese, R. V. (1970). *Pharmacology* **4**, 33.

Linarelli, L. G., Weller, J. L., and Glinsmann, W. H. (1970). *Life Sci.* **9** (II), 535.

Lipkin, D., Markham, R., and Cook, W. H. (1959a). *J. Amer. Chem. Soc.* **81**, 6075.

Lipkin, D., Cook, W. H., and Markham, R. (1959b). *J. Amer. Chem. Soc.* **81**, 6198.

Lissitzky, S., Mante, S., Attali, J., and Cartouzou, G. (1969). *Biochem. Biophys. Res. Commun.* **35**, 437.

Lockett, M. F., and Gwynne, H. L. (1968). *J. Pharm. Pharmacol.* **20**, 688.

Logan, M. E., And Cotten, M. deV. (1967). *J. Pharmacol. Exp. Ther.* **155**, 242.

Lomsky, R., Langer, F., and Vortel, V. (1969). *Nature (London)* **223**, 618.

Loten, E. G., and Sneyd, J. G. T. (1970). *Biochem. J.* **120**, 187.

Loubatieres, A., Mariani, M. M., Alric, R., and Chapal, J. (1967). *In* "Tolbutamide . . . after Ten Years" (W. J. H. Butterfield and W. V. Westering, eds.), pp. 100–113. Excerpta Medica Found., Amsterdam.

Love, W. C., Carr, L., and Ashmore, J. (1963). *J. Pharmacol. Exp. Ther.* **140**, 287.

Lowry, O., and Passonneau, J. V. (1964). *Arch. Exp. Pathol. Pharmakol.* **248**, 185.

Lowry, O. H., and Passonneau, J. V. (1966). *J. Biol. Chem.* **241**, 2268.

Lowry, O. H., Schulz, D. W., and Passonneau, J. V. (1964). *J. Biol. Chem.* **239**, 1947.

Lowry, O. H., Schulz, D. W., and Passonneau, J. V. (1967). *J. Biol. Chem.* **242**, 271.

Lu, F. C. (1952). *Brit. J. Pharmacol.* **7**, 637.

Lucchesi, B. R. (1968). *Circ. Res.* **22**, 777.

Lucchesi, B. R., Whitsitt, L. S., and Stickney, J. L. (1967). *Ann. N. Y. Acad. Sci.* **139**, 940.

Lucchesi, B. R., Stutz, D. R., and Winfield, R. A. (1969). *Circ. Res.* **25**, 183.

Lum, B. K. B., Kermani, M. H., and Heilman, R. D. (1966). *J. Pharmacol. Exp. Ther.* **154**, 463.

Lundholm, L., Mohme-Lundholm, E., and Svedmyr, N. (1966). *Pharmacol. Rev.* **18**, 255.

Lundholm, L., Rall, T., and Vamos, N. (1967). *Acta Physiol. Scand.* **70**, 127.

Lundholm, L., Mohme-Lundholm, E., and Vamos, N. (1969). *Acta Physiol. Scand.* **75**, 187.

Lutherer, L. O., Fregly, M. J., and Anton, A. H. (1969). *Fed. Proc.* **28**, 1238.

Lyon, J. B., and Mayer, S. E. (1969). *Biochem. Biophys. Res. Commun.* **34**, 459.

Lyon, J. B., and Porter, J. (1963). *J. Biol. Chem.* **238**, 1.

Lyon, J. B., Porter, J., and Robertson, M. (1967). *Science* **155**, 1550.

Lythgoe, B., Smith, H., and Todd, A. (1947). *J. Chem. Soc.* p. 355.

Maayan, M. L., and Ingbar, S. H. (1968). *Science* **162**, 124.

Maayan, M. L., and Ingbar, S. H. (1970). *Endocrinology* **87**, 588.

McCann, S. M., and Porter, J. C. (1969). *Physiol. Rev.* **49**, 240.

Macchia, V., Bates, R. W., and Pastan, I. (1967). *J. Biol. Chem.* **242**, 3726.

Macchia, V., Meldolesi, M. F., and Maselli, P. (1969). *Endocrinology* **85**, 895.

McChesney, E. W., MacAuliff, J. P., and Blumberg, H. (1949). *Proc. Soc. Exp. Biol. Med.* **71**, 220.

McGuire, J. (1970). *Arch. Dermatol.* **101**, 173.

McKeel, D. W., and Jarett, L. (1970). *J. Cell Biol.* **44**, 417.

Mackrell, D. J., and Sokal, J. E. (1969). *Diabetes* **18**, 724.

MacManus, J. P., and Whitfield, J. F. (1969). *Exp. Cell. Res.* **58**, 188.

MacManus, J. P., and Whitfield, J. F. (1970). *Endocrinology* **86**, 934.

MacManus, J. P., Whitfield, J. F., and Youdale, T. (1971). *J. Cell. Physiol.* **77**, 103.

McNeill, J. H., Nassar, M., and Brody, T. M. (1969a). *J. Pharmacol. Exp. Ther.* **165**, 234.

McNeill, J. H., Maschek, L. D., and Brody, T. M. (1969b). *Can. J. Physiol. Pharmacol.* **47**, 913.

Maddaiah, V. T., and Madsen, N. B. (1966a). *Biochim. Biophys. Acta* **121**, 261.

Maddaiah, V. T., and Madsen, N. B. (1966b). *J. Biol. Chem.* **241**, 3873.

Mahaffee, D., Watson, B., and Ney, R. L. (1970). *Clin. Res.* **18**, 73.

Makman, M. H. (1970). *Science* **170**, 1421.

Makman, M. H., and Sutherland, E. W. (1964). *Endocrinology* **75**, 127.

Makman, R. S., and Sutherland, E. W. (1965). *J. Biol. Chem.* **240**, 1309.

Malaisse, W., and Malaisse-Lagae, F. (1968). *Arch. Int. Pharmacodyn. Ther.* **171**, 235.

Malaisse, W. J., Malaisse-Lagae, F., and Mayhew, D. (1967). *J. Clin. Invest.* **46**, 1724.

Malamud, D. (1969). *Biochem. Biophys. Res. Commun.* **35**, 754.

Malamud, D., and Baserga, R. (1967). *Life Sci.* **6**, 1765.

Malamud, D., and Baserga, R. (1968a). *Exp. Cell. Res.* **50**, 581.

Malamud, D., and Baserga, R. (1968b). *Science* **162**, 373.

Mallette, L. E., Exton, J. H., and Park, C. R. (1969a). *J. Biol. Chem.* **244**, 5713.

Mallette, L. E., Exton, J. H., and Park, C. R. (1969b). *J. Biol. Chem.* **244**, 5724.

Mandel, L. R., and Kuehl, F. A. (1967). *Biochem. Biophys. Res. Commun.* **28**, 13.

Mansour, T. E. (1966). *Pharmacol. Rev.* **18**, 173.

Mansour, T. E. (1967). *Fed. Proc.* **26**, 1179.

Mansour, T. E., and Mansour, J. M. (1962). *J. Biol. Chem.* **237**, 629.

Mansour, T. E., and Stone, D. B. (1970). *Biochem. Pharmacol.* **19**, 1137.

Mansour, T. E., Sutherland, E. W., Rall, T. W., and Bueding, E. (1960). *J. Biol. Chem.* **235**, 466.

Marcus, A. J., and Zucker, M. B. (1965). "The Physiology of Blood Platelets," pp. 49–56. Grune & Stratton, New York.

Marcus, R., and Aurbach, G. D. (1969). *Endocrinology* **85**, 801.

Marinetti, G. V., Ray, T. K., and Tomasi, V. (1969). *Biochem. Biophys. Res. Commun.* **36**, 185.

Marquis, N. R., Vigdahl, R. L., and Tavormina, P. A. (1969). *Biochem. Biophys. Res. Commun.* **36**, 965.

Marquis, N. R., Becker, J. A., and Vigdahl, R. L. (1970). *Biochem. Biophys. Res. Commun.* **39**, 783.

Marsh, J. M. (1970a). *J. Biol. Chem.* **245**, 1596.

Marsh, J. M. (1970b). *FEBS Lett.* **7**, 283.

Marsh, J. M., and Savard, K. (1964). *J. Biol. Chem.* **239**, 1.

Marsh, J. M., and Savard, K. (1966). *Steroids* **8**, 133.

Marsh, J. M., Butcher, R. W., Savard, K., and Sutherland, E. W. (1966). *J. Biol. Chem.* **241**, 5436.

Martelo, O. J., Woo, S. L. C., Reimann, E. M., and Davie, E. W. (1970). *Biochemistry* **9**, 4807.

Martin, J. M., and Gagliardino, J. J. (1967). *Nature (London)* **213**, 630.

Masters, T. N., and Glaviano, V. V. (1969). *J. Pharmacol. Exp. Ther.* **167**, 187.

Mayer, S. E., and Krebs, E. G. (1970). *J. Biol. Chem.* **245**, 3153.

Mayer, S. E., Moran, N. C., and Fain, J. (1961). *J. Pharmacol. Exp. Ther.* **134**, 18.

Mayer, S. E., Cotten, M. deV., and Moran, N. C. (1963). *J. Pharmacol. Exp. Ther.* **139**, 275.

Mayer, S. E., Williams, B. J., and Smith, J. M. (1967). *Ann. N. Y. Acad. Sci.* **139**, 686.

Mayer, S. E., Namm, D. H., and Rice, L. (1970). *Circ. Res.* **26**, 225.

Mayhew, D. A., Wright, P. H., and Ashmore, J. (1969). *Pharmacol. Rev.* **21**, 183.

Meester, W. D., and Hardman, H. F. (1967). *J. Pharmacol. Exp. Ther.* **158**, 241.

Meisler, M. H., and Langan, T. A. (1969). *J. Biol. Chem.* **244**, 4961.

Melson, G. L., Chase, L. R., and Aurbach, G. D. (1970). *Endocrinology* **86**, 511.

Menahan, L. A., and Wieland, O. (1967). *Biochem. Biophys. Res. Commun.* **29**, 880.

Menahan, L. A., and Wieland, O. (1969a). *Eur. J. Biochem.* **9**, 55.

Menahan, L. A., and Wieland, O. (1969b). *Eur. J. Biochem.* **9**, 182.

Menahan, L. A., Ross, B. D., and Wieland, O. (1968). *Biochem. Biophys. Res. Commun.* **30**, 38.

Menahan, L. A., Hepp, K. D., and Wieland, O. (1969). *Eur. J. Biochem.* **8**, 435.

Mendicino, J., Beaudreau, C., and Bhattacharyya, R. N. (1966). *Arch. Biochem. Biophys.* **116**, 436.

Mendoza, S. A. (1969). *Endocrinology* **84**, 411.

Mendoza, S. A., Handler, J. S., and Orloff, J. (1967). *Amer. J. Physiol.* **213**, 1263.

Merlevede, W., and Riley, G. A. (1966). *J. Biol. Chem.* **241**, 3517.

Mersmann, H. J., and Segal, H. L. (1967). *Proc. Nat. Acad. Sci. U.S.* **58**, 1688.

Mertz, D. P. (1969). *Experientia* **25**, 269.

Metzger, B. E., Glaser, L., and Helmreich, E. (1968). *Biochemistry* **7**, 2021.

Micheel, F., and Heesing, A. (1961). *Chem. Ber.* **94**, 1814.

Michelakis, A. M., Caudle, J., and Liddle, G. W. (1969). *Proc. Soc. Exp. Biol. Med.* **130**, 748.

Michelson, A., and Todd, A. (1949). *J. Chem. Soc.* p. 2476.

Miech, R. P., and Parks, R. E. (1965). *J. Biol. Chem.* **240**, 351.

Miller, L. L. (1965). *Fed. Proc.* **24**, 737.

Miller, W. L., and Krake, J. J. (1963). *Proc. Soc. Exp. Biol. Med.* **113**, 784.

Millonig, G., and Porter, K. R. (1960). *Proc. Eur. Reg. Conf. Electron Microsc.* **2**, 655.

Milman, L. S., and Yurowitzki, Y. G. (1967). *Biochim. Biophys. Acta* **146**, 301.

Milner, R. D. G. (1969). *J. Endocrinol.* **44**, 267.

Misra, R. P., and Berman, L. B. (1969). *Amer. J. Med.* **47**, 337.

Mitznegg, P., Heim, F., and Meythaler, B. (1970). *Life Sci.* **9** (I), 121.

Miyamoto, E., Kuo, J. F., and Greengard, P. (1969). *J. Biol. Chem.* **244**, 6395.

Mohme-Lundholm, E. (1962). *Acta Physiol. Scand.* **55**, 225.

Mohme-Lundholm, E. (1963). *Acta Physiol. Scand.* **59**, 74.

Mohme-Lundholm, E., and Vamos, N. (1967). *Acta Pharmacol. Toxicol.* **25**, 87.

Mommaerts, W. F. H. M., Seraydarian, K., and Uchida, K. (1963). *Biochem. Biophys. Res. Commun.* **13**, 58, 1963.

Monard, D., Janecek, J., and Rickenberg, H. V. (1969). *Biochem. Biophys. Res. Commun.* **35**, 584.

Mondon, C. E., and Mortimore, G. E. (1967). *Amer. J. Physiol.* **212**, 173.

Montague, W., Howell, S. L., and Taylor, K. W. (1967). *Nature (London)* **215**, 1088.

Moore, P. F. (1968). *Ann. N. Y. Acad. Sci.* **150**, 256.

Moore, P. F., Ioris, L. C., and McManus, J. M. (1968). *J. Pharm. Pharmacol.* **20**, 368.

Moran, N. C. (1966). *Pharmacol. Rev.* **18**, 503.

Moran, N. C. (1967). *Ann. N. Y. Acad. Sci.* **139**, 649.

Moran, N. C., and Perkins, M. E. (1958). *J. Pharmacol. Exp. Ther.* **124**, 223.

Morel, F. (1970). *In* "The Role of Adenyl Cyclase and Cyclic Adenosine 3′,5′-Monophosphate in Biological Systems" (T. W. Rall, M. Rodbell, and P. G. Condliffe, eds.), in press. Government Printing Office, Washington, D. C.

Morgan, H. E., and Parmeggiani, A. (1964a). *J. Biol. Chem.* **239**, 2435.

Morgan, H. E., and Parmeggiani, A. (1964b). *J. Biol. Chem.* **239**, 2440.

Morgan, H. E., Neely, J. R., Wood, R. E., Liebecq, L., Liebermeister, H., and Park, C. R. (1965). *Fed. Proc.* **24**, 1040.

Mortimore, G. E. (1961). *Amer. J. Physiol.* **200**, 1315.

Mortimore, G. E., and Mondon, C. E. (1970). *J. Biol. Chem.* **245**, 2375.

Mortimore, G. E., King, E., Mondon, C. E., and Glinsmann, W. H. (1967). *Amer. J. Physiol.* **212**, 179.

Morton, R. K. (1955). *In* "Methods in Enzymology" (S. P. Colowick and N. O. Kaplan, eds.), Vol. I, pp. 25–51. Academic Press, New York.

Moses, V., and Sharp, P. B. (1970). *Biochem. J.* **118**, 481.

Mosinger, B., and Vaughan, M. (1967). *Biochim. Biophys. Acta* **144**, 569.

Moskowitz, J., and Fain, J. N. (1970). *J. Biol. Chem.* **245**, 1101.

Mozsik, G. (1969). *Eur. J. Pharmacol.* **7**, 319.

Mozsik, G. (1970). *Eur. J. Pharmacol.* **9**, 107.

Mühlbachova, E., Solyom, A., and Puglisi, K. (1967). *Eur. J. Pharmacol.* **1**, 321.

Müller, E. E., Pecile, A., Naimzada, M. K., and Ferrario, G. (1969). *Experientia* **25**, 750.

Müller-Oerlinghausen, B., Schwabe, U., Hasselblatt, A., and Schmidt, F. H. (1968). *Life Sci.* **7**, 593.

Murad, F. (1965). *Fed. Proc.* **24**, 150.

Murad, F., and Vaughan, M. (1969). *Biochem. Pharmacol.* **18**, 1053.

Murad, F., Chi, Y. M., Rall, T. W., and Sutherland, E. W. (1962). *J. Biol. Chem.* **237**, 1233.

Murad, F., Strauch, B. S., and Vaughan, M. (1969a). *Biochim. Biophys. Acta* **177**, 591.

Murad, F., Rall, T. W., and Vaughan, M. (1969b). *Biochim. Biophys. Acta* **192**, 430.

Murad, F., Brewer, H. B., and Vaughan, M. (1970a). *Proc. Nat. Acad. Sci. U. S.* **65**, 446.

Murad, F., Manganiello, V., and Vaughan, M. (1970b). *J. Biol. Chem.* **245**, 3352.

Mustard, J. F., and Packham, M. A. (1970). *Pharmacol. Rev.* **22**, 97.

Nagata, N., and Rasmussen, H. (1968). *Biochemistry* **7**, 3728.

Nagata, N., and Rasmussen, H. (1970). *Proc. Nat. Acad. Sci. U. S.* **65**, 368.

Nair, K. G. (1962). *Fed. Proc.* **21**, 84.

Nair, K. G. (1966). *Biochemistry* **5**, 150.

Nakane, P. K. (1970). *J. Histochem. Cytochem.* **18**, 9.

Nakano, J. (1967). *J. Pharmacol. Exp. Ther.* **157**, 19.

Nakano, J., Ishii, T., and Gin, A. C. (1968). *Pharmacology* **1**, 183.

Nakano, J., Oliver, R., and Ishii, T. (1970). *Pharmacology* **3**, 273.

Namm, D. H., and Mayer, S. E. (1968). *Mol. Pharmacol.* **4**, 61.

Namm, D. H., Mayer, S. E., and Maltbie, M. (1968). *Mol. Pharmacol.* **4**, 522.

Nayler, W. G. (1967). *Amer. Heart J.* **73**, 379.

Neely, J. R., Liebermeister, H., Battersby, E. J., and Morgan, H. E. (1967a). *Amer. J. Physiol.* **212**, 804.

Neely, J. R., Liebermeister, H., and Morgan, H. E. (1967b). *Amer. J. Physiol.* **212**, 815.

Neely, J. R., Bowman, R. H., and Morgan, H. E. (1969). *Amer. J. Physiol.* **216**, 804.

Neville, D. M. (1960). *J. Biophys. Biochem. Cytol.* **8**, 413.

Ney, R. L. (1969). *Endocrinology* **84**, 168.

Ney, R. L., Davis, W. W., and Garren, L. D. (1966). *Science* **153**, 896.

Ney, R. L., Dexter, R. N., Davis, W. W., and Garren, L. D. (1967). *J. Clin. Invest.* **46**, 1916.

Ney, R. L., Hochella, N. J., Grahame-Smith, D. G., Dexter, R. N., and Butcher, R. W. (1969). *J. Clin. Invest.* **48**, 1733.

Nigon, K., and Harris, J. B. (1968). *Amer. J. Physiol.* **215**, 1299.

Northrop, G. (1968). *J. Pharmacol. Exp. Ther.* **149**, 22.

Northrop, G., and Parks, R. E. (1964a). *J. Pharmacol. Exp. Ther.* **145**, 87.

Northrop, G., and Parks. R. E. (1964b). *J. Pharmacol. Exp. Ther.* **145**, 135.

Novales, R. R., and Davis, W. J. (1967). *Endocrinology* **81**, 283.

Novales, R. R., and Davis, W. J. (1969). *Amer. Zoologist* **9**, 479.

Novales, R. R., and Fujii, R. (1970). *J. Cell. Physiol.* **75**, 133.

Novogrodsky, A., and Katchalski, E. (1970). *Biochim. Biophys. Acta* **215**, 291.

O'Dea, R. F., Haddox, M. K., and Goldberg, N. D. (1970). *Fed. Proc.* **29**, 473 Abs.

Ohneda, A., Aguilar-Parada, E., Eisentraut, A. M., and Unger, R. H. (1969). *Diabetes* **18**, 1.

Okabayashi, T., and Ide, M. (1970). *Biochim. Biophys. Acta* **220**, 116.

Okabayashi, T., Yoshimoto, A., and Ide, M. (1963). *J. Bacteriol.* **86**, 930.

Olson, C. B., Braveny, P., and Blinks, J. R. (1967). *Pharmacologist* **9**, 223.

O'Malley, B. W., McGuire, W. L., Kohler, P. O., and Korenman, S. G. (1969). *Recent Progr. Hormone Res.* **25**, 105.

Onaya, T., and Solomon, D. H. (1970). *Endocrinology* **86**, 423.

Onaya, T., Solomon, D. H., and Davidson, W. D. (1969). *Endocrinology* **85**, 150.

Orci, L., Pictet, R., Forssmann, W. G., Renold, A. E., and Rouiller, C. (1968). *Diabetologia* **4**, 56.

Orci, L., Lambert, A. E., Rouiller, C., Renold, A. E., and Samols, E. (1969). *Hormone Metabolic Res.* **1**, 108.

Orloff, J., and Handler, J. S. (1962). *J. Clin. Invest.* **41**, 702.

Orloff, J., and Handler, J. S. (1964). *Amer. J. Med.* **36**, 686.

Orloff, J., and Handler, J. (1967). *Amer. J. Med.* **42**, 757.

Orloff, J., Handler, J. S., and Bergström, S. (1965). *Nature (London)* **205**, 397.

Overweg, N. I. A. (1966). *J. Pharmacol. Exp. Ther.* **153**, 314.

Øye, I. (1965). *Acta Physiol. Scand.* **65**, 251.

Øye, I. (1967). "The Mode of Action of Adrenaline in the Isolated Rat Heart." Universitetsforlaget, Oslo.

Øye, I., and Sutherland, E. W. (1966). *Biochim. Biophys. Acta* **127**, 347.

Ozand, P., and Narahara, H. T. (1964). *J. Biol. Chem.* **239**, 1964.

Ozawa, E., Hosoi, K., and Ebashi, S. (1967). *J. Biochem.* **61**, 531.

Pagliara, A. S., and Goodman, A. D. (1969). *J. Clin. Invest.* **48**, 1408.

Pagliara, A. S., and Goodman, A. D. (1970). *Amer. J. Physiol.* **218**, 1301.

Palmer, E. C., Sulser, F., and Robison, G. A. (1969). *Pharmacologist* **11**, 258.

Palmer, E. C., Robison, G. A, and Sulser, F. (1971). *Biochem. Pharmacol.* **20**, 236.

Paloyan, E., Paloyan, D., and Harper, P. V. (1967). *Metabolism* **16**, 35.

Papavasiliou, P. S., Miller, S. T., and Cotzias, G. C. (1968). *Nature (London)* **220**, 74.

Parisi, M., Ripoche, P., and Bourguet, J. (1969). *Pflügers Arch.* **309**, 59.

Park, C. R., Crofford, O. B., and Kono, T. (1968). *J. Gen. Physiol.* **52**, 296s.

Park, J. H., Meriwether, B. P., Park, C. R., Mudd, S. H., and Lipmann, F. (1956). *Biochim. Biophys. Acta* **22**, 403.

Parmley, W. W., and Braunwald, E. (1967). *J. Pharmacol. Exp. Ther.* **158**, 11.

Parmley, W. W., Glick, G., and Sonnenblick, E. H. (1968). *New Eng. J. Med.* **279**, 12.

Pastan, I. (1966). *Biochem. Biophys. Res. Commun.* **25**, 14.

Pastan, I., and Katzen, R. (1967). *Biochem. Biophys. Res. Commun.* **29**, 792.

Pastan, I., and Macchia, V. (1967). *J. Biol. Chem.* **242**, 5757.

Pastan, I., and Perlman, R. L. (1968). *Proc. Nat. Acad. Sci. U. S.* **61**, 1336.

Pastan, I., and Perlman, R. L. (1969a). *J. Biol. Chem.* **244**, 2226.

Pastan, I., and Perlman, R. L. (1969b). *J. Biol. Chem.* **244**, 5836.

Pastan, I., and Perlman, R. L. (1970). *Science* **169**, 339.

Pastan, I., and Wollman, S. H. (1967). *J. Cell Biol.* **35**, 262.

Pastan, I., Herring, B., Johnson, P., and Field, J. B. (1962). *J. Biol. Chem.* **237**, 287.

Pastan, I., Roth, J., and Macchia, V. (1966). *Proc. Nat. Acad. Sci. U. S.* **56**, 1802.

Pastan, I., Macchia, V., and Katzen, R. (1968). *Endocrinology* **83**, 157.

Pastan, I., Pricer, W., and Blanchette-Mackie, J. (1970). *Metabolism* **19**, 809.

Pauk, G. L., and Reddy, W. J. (1967). *Anal. Biochem.* **21**, 298.

Paul, M. I., Pauk, G. L., and Ditzion, B. R. (1970a). *Pharmacology* **3**, 148.

Paul, M. I., Ditzion, B. R., and Janowsky, D. S. (1970b). *Lancet* **i**, 88.

Paul, M. I., Ditzion, B. R., Pauk, G. L., and Janowsky, D. S. (1970c). *Amer. J. Psychiat.* **126**, 1493.

Paul, M. I., Cramer, H., and Goodwin, F. K. (1970d). *Lancet* **i**, 996.

Penhos, J. C., and Krahl, M. E. (1963). *Amer. J. Physiol.* **204**, 140.

Penhos, J. C., Wu, C. H., Daunas, J. A., and Levine, R. (1967). *Biochim. Biophys. Acta* **144**, 678.

Pennington, S. N., Brown, H. D., Chattopadhyay, S., Conaway, C., and Morris, H. P. (1970). *Experientia* **26**, 139.

Perkins, J. P., and Moore, M. M. (1971). *J. Biol. Chem.* **246**, 62.

Perlman, R. L., and Pastan, I. (1968). *J. Biol. Chem.* **243**, 5420.

Perlman, R. L., and Pastan, I. (1969). *Biochem. Biophys. Res. Commun.* **37**, 151.

Perlman, R. L., de Crombrugghe, B., and Pastan, I. (1969). *Nature (London)* **223**, 810.

Peron, F. G., and Tsang, C. P. W. (1969). *Biochim. Biophys. Acta* **180**, 445.

Petersen, M. J., and Edelman, I. S. (1964). *J. Clin. Invest.* **43**, 583.

Peterson, M. J., Patterson, C., and Ashmore, J. (1968). *Life Sci.* **7**, 551.

Pickles, V. R. (1967). *Biol. Rev.* **42**, 614.

Platt, J. R. (1964). *Science* **146**, 347.

Pöch, G., and Kukovetz, W. R. (1967). *J. Pharmacol. Exp. Ther.* **156**, 522.

Pohl, S. L., Birnbaumer, L., and Rodbell, M. (1969). *Science* **164**, 566.

Pohto, P. (1968). *J. Oral Ther. Pharmacol.* **4**, 467.

Pollard, C. J. (1970). *Biochim. Biophys. Acta* **201**, 511.

Pollard, C. J., and Venere, R. J. (1970). *Fed. Proc.* **29**, 670 Abs.

Porte, D. (1967a). *J. Clin. Invest.* **46**, 86.

Porte, D. (1967b). *Diabetes* **16**, 150.

Porte, D. (1968). *Ann. N. Y. Acad. Sci.* **150**, 281.

Porte, D. (1969). *Arch. Intern. Med.* **123**, 252.

Posner, J. B., Hammermeister, K. E., Bratvold, G. E., and Krebs, E. G. (1964). *Biochemistry* **3**, 1040.

Posner, J. B., Stern, R., and Krebs, E. G. (1965). *J. Biol. Chem.* **240**, 982.

Posternak, T., Sutherland, E. W., and Henion, W. F. (1962). *Biochim. Biophys. Acta* **65**, 558.

Posternak, T., Marcus, I., Gabbai, A., and Cehovic, G. (1969). *C. R. Acad. Sci. Ser.* **D269**, 2409.

Potter, V. R. (1955). *In* "Methods in Enzymology" (S. P. Colowick and N. O. Kaplan, eds.), Vol. I, pp. 10–15. Academic Press, New York.

Powell, C. E., and Slater, I. H. (1958). *J. Pharmacol. Exp. Ther.* **122**, 480.

Poyart, C. F., and Nahas, G. G. (1968). *Mol. Pharmacol.* **4**, 389.

Price, J. M., and Nayler, W. G. (1967). *Arch. Int. Pharmacodyn. Ther.* **166**, 390.

Price, T. D., Ashman, D. F., and Melicow, M. M. (1967). *Biochim. Biophys. Acta* **138**, 452.

Pryor, J., and Berthet, J. (1960). *Biochim. Biophys. Acta* **43**, 556.

Pullman, T. N., Lavender, A. R., and Aho, I. (1967). *Metabolism* **16**, 358.

Raben, M. S., and Hollenberg, C. H. (1959). *J. Clin. Invest.* **38**, 484.

Rabinowitz, M., DeSalles, L., Meisler, J., and Lorand, L. (1965). *Biochim. Biophys. Acta* **97**, 29.

Raisz, L. G., and Klein, D. C. (1969). *Fed. Proc.* **28**, 320.

Rall, T. W., and Kakiuchi, S. (1966). *In* "Molecular Basis of Some Aspects of Mental Activity," (P. Karlson, ed.), Vol. 1, pp. 417–427. Academic Press, New York.

Rall, T. W., and Sutherland, E. W. (1958). *J. Biol. Chem.* **232**, 1065.

Rall, T. W., and Sutherland, E. W. (1960). *In* "Methods in Enzymology" (S. P. Colowick and N. O. Kaplan, eds.), Vol. V, pp. 377–391. Academic Press, New York.

Rall, T. W., and Sutherland, E. W. (1961). *Cold Spring Harbor Symp. Quant. Biol.* **26**, 347.

Rall, T. W., and Sutherland, E. W. (1962). *J. Biol. Chem.* **237**, 1228.

Rall, T. W., and West, T. C. (1963). *J. Pharmacol. Exp. Ther.* **139**, 269.

Rall, T. W., Sutherland, E. W., and Wosilait, W. D. (1956a). *J. Biol. Chem.* **218**, 483.

Rall, T. W., Wosilait, W. D., and Sutherland. (1956b). *Biochim. Biophys. Acta* **20**, 69.

Rall, T. W., Sutherland, E. W., and Berthet, J. (1957). *J. Biol. Chem.* **224**, 463.

Ramachandran, J., and Lee, V. (1970). *Biochem. Biophys. Res. Commun.* **41**, 358.

Ramey, E. R., and Goldstein, M. S. (1957). *Physiol. Rev.* **37**, 155.

Ramwell, P. W., and Shaw, J. E. (eds.). (1968). "Prostaglandin Symposium of the Worcester Foundation for Experimental Biology," 402 pp. Wiley (Interscience), New York.

Ramwell, P. W., and Shaw, J. E. (1970). *Recent Progr. Hormone Res.* **26**, 139.

Ramwell, P. W., Shaw, J. E., Clarke, G. B., Grostic, M. F., Kaiser, D. G., and Pike, J. E. (1968). *Progr. Chem. Fats Other Lipids* **9**, 231.

Raper, C. (1967). *Circ. Res.* **21** (Suppl. III), 147.

Rasmussen, H. (1970). *Science* **170**, 404.

Rasmussen, H., and Tenenhouse, A. (1968). *Proc. Nat. Acad. Sci. U. S.* **59**, 1364.

Rasmussen, H., Pechet, M., and Fast, D. (1968). *J. Clin. Invest.* **47**, 1843.

Ratner, A. (1970). *Life Sci.* **9** (I), 1221.

Ray, T. K., Tomasi, V., and Marinetti, G. V. (1970). *Biochim. Biophys. Acta* **211**, 20.

Reed, N., and Fain, J. N. (1968a). *J. Biol. Chem.* **243**, 2843.

Reed, N., and Fain, J. N. (1968b). *J. Biol. Chem.* **243**, 6077.

Reed, B. L., Finnin, B. C., and Ruffin, N. E. (1969). *Life Sci.* **8** (II), 113.

Regen, D. M., and Terrell, E. B. (1968). *Biochim. Biophys. Acta* **170**, 95.

Reik, L., Petzold, G. L., Higgins, J. A., Greengard, P., and Barrnett, R. J. (1970). *Science* **168**, 382.

Reimann, E. M., and Walsh, D. A. (1970). *Fed. Proc.* **29**, 601 Abs.

Reimann, E. M., Brostrom, C. O., Corbin, J. D., King, C. A., and Krebs, E. G. (1971). *Biochem. Biophys. Res. Commun.* **42**, 187.

Reiter, M. (1964). *In* "Pharmacology of Cardiac Function," (O. Krayer, ed.), pp. 25–42. Macmillan (Pergamon), New York.

Reiter, M. (1965). *Experientia* **21**, 87.

Renold, A. E. (1970). *New Engl. J. Med.* **282**, 173.

Renold, A. E., Crofford, O. B., Stauffacher, W., and Jeanrenaud, B. (1965). *Diabetologia* **1**, 4.

Reynolds, R. C., and Haugaard, N. (1967). *J. Pharmacol. Exp. Ther.* **156**, 417.

Richardson, H. B., Shorr, E., and Loebel, R. O. (1930). *J. Biol. Chem.* **86**, 551.

Ridderstap, A. S., and Bonting, S. L. (1969). *Pflugers Arch.* **313**, 62.

Riddick, D. H., Kregenow, F., and Orloff, J. (1969). *Fed. Proc.* **28**, 339.

Rider, J., and Thomas, S. (1969). *Brit. J. Pharmacol.* **37**, 539P.

Rieser, P. (1966). *Amer. J. Med.* **40**, 759.

Rigby, P. G., and Ryan, W. L. (1970). *European J. Clin. Biol. Res.* **15**, 774.

Riley, G. A. (1963). *Fed. Proc.* **22**, 258.

Riley, G. A., and Haynes, R. C. (1963). *J. Biol. Chem.* **238**, 1563.

Riley, G. A., and Wahba, W. W. (1969). *Pharmacologist* **11**, 253.

Riley, W. D., DeLange, R. J., Bratvold, G. E., and Krebs, E. G. (1968). *J. Biol. Chem.* **243**, 2209.

Rinard, G. A., Okuno, G., and Haynes, R. C. (1969). *Endocrinology* **84**, 622.

Rizack, M. A. (1964). *J. Biol. Chem.* **239**, 392.

Rizack, M. A. (1965). *Ann. N. Y. Acad. Sci.* **131**, 250.

Rizack, M. A. (1967). *Anal. Biochem.* **20**, 192.

Robert, A., Nezamis, J. E., and Phillips, J. P. (1968). *Gastroenterology* **55**, 481.

Roberts, S., Creange, J. E., and Fowler, D. D. (1964). *Nature (London)* **203**, 759.

Roberts, S., Creange, J. E., and Young, P. L. (1965). *Biochem. Biophys. Res. Commun.* **20**, 446.

Roberts, S., McCune, R. W., Creange, J. E., and Young, P. L. (1967). *Science* **158**, 372.

Robertson, C. R., Rosiere, C. E., Blickenstaff, D., and Grossman, M. I. (1950). *J. Pharmacol. Exp. Ther.* **99**, 362.

Robison, G. A., Butcher, R. W., Øye, I., Morgan, H. E., and Sutherland, E. W. (1965). *Mol. Pharmacol.* **1**, 168.

Robison, G. A., Butcher, R. W., and Sutherland, E. W. (1967a). *Ann. N. Y. Acad. Sci.* **139,** 703.

Robison, G. A., Exton, J. H., Park, C. R., and Sutherland, E. W. (1967b). *Fed. Proc.* **26,** 257.

Robison, G. A., Arnold, A., and Hartmann, R. C. (1969). *Pharmacol. Res. Commun.* **1,** 325.

Robison, G. A., Coppen, A. J., Whybrow, P. C., and Prange, A. J. (1970). *Lancet* **ii,** 1028.

Rodbell, M. (1964). *J. Biol. Chem.* **239,** 375.

Rodbell, M. (1967a). *J. Biol. Chem.* **242,** 5744.

Rodbell, M. (1967b). *J. Biol. Chem.* **242,** 5751.

Rodbell, M., and Jones, A. B. (1966). *J. Biol. Chem.* **241,** 140.

Rodbell, M., Jones, A. B., Chiappe de Cingolani, G. E., and Birnbaumer, L. (1968). *Recent Progr. Hormone Res.* **24,** 215.

Rodbell, M., Birnbaumer, L., and Pohl, S. L. (1970). *J. Biol. Chem.* **245,** 718.

Rodesch, F., Neve, P., Willems, C., and Dumont, J. E. (1969). *Eur. J. Biochem.* **8,** 26.

Roll, P. M., Weinfeld, H., Carroll, E., and Brown, G. B. (1956). *J. Biol. Chem.* **220,** 439.

Rosell-Perez, M., and Larner, J. (1964). *Biochemistry* **3,** 81.

Rosell-Perez, M., Villar-Palasi, C., and Larner, J. (1962). *Biochemistry* **1,** 763.

Rosen, O. M. (1970). *Arch. Biochem. Biophys.* **137,** 435.

Rosen, O. M., and Erlichman, J. (1969). *Arch. Biochem. Biophys.* **133,** 171.

Rosen, O. M., and Rosen, S. M. (1968). *Biochem. Biophys. Res. Commun.* **31,** 82.

Rosen, O. M., and Rosen, S. M. (1969). *Arch. Biochem. Biophys.* **131,** 449.

Rosen, O. M., and Rosen, S. M. (1970). *Arch. Biochem. Biophys.* **141,** 346.

Rosen, O. M., Erlichman, J., and Rosen, S. M. (1970). *Mol. Pharmacol.* **6,** 524.

Rosenfeld, M. G., and O'Malley, B. W. (1970). *Science* **168,** 253.

Ross, G. (1970). *Amer. J. Physiol.* **218,** 1166.

Ross, B. D., Hems, R., Freedland, R. A., and Krebs, H. A. (1967). *Biochem. J.* **105,** 869.

Rossi, E. C. (1968). *Thromb. Diath. Haemorrh.* **19,** 53.

Roth, L. E., Pihlaja, D. J., and Shigenaka, Y. (1970). *J. Ultrastr. Res.* **30,** 7.

Rowe, G. G., Afonso, S., Gurtner, H. P., Chelius, C. J., Lowe, W. C., Castillo, C. A., and Crumpton, C. W. (1962). *Amer. Heart J.* **64,** 228.

Rozenberg, M. C., and Holmsen, H. (1968). *Biochem. Biophys. Acta* **155,** 342.

Rozsa, K. S. (1969). *Life Sci.* **8** (II), 229.

Rudman, D., Di Girolamo, M., Malkin, M. F., and Garcia, L. A. (1965). *In* "Handbook of Physiology, Section 5: Adipose Tissue" (A. E. Renold and G. F. Cahill, Jr., eds.), pp. 533–540. Amer. Physiol. Soc., Washington, D. C.

Russell, R. G. G., Casey, P. A., and Fleisch, H. (1968). *Proc. 6th Eur. Symp. Calcified Tissues,* Lund, Sweden, 1968.

Ryan, W. L., and Coronel, D. M. (1969). *Amer. J. Obstet. Gynecol.* **105,** 121.

Ryan, W. L., and Heidrick, D. M. (1968). *Science* **162,** 1484.

Salomon, Y., and Schramm, M. (1970). *Biochem. Biophys. Res. Commun.* **38,** 106.

Salzman, E. W., and Neri, L. L. (1969). *Nature (London)* **224,** 609.

Samols, E., Marri, G., and Marks, V. (1966). *Diabetes* **15,** 855.

Sanbar, S. S. (1968). *Metabolism* **17,** 631.

Sandhu, R. S., and Hokin, M. R. (1967). *Fed. Proc.* **26,** 391.

Sandhu, R. S., and Hokin, M. R. (1968). *J. Chromatogr.* **35,** 136.

Sandler, R., and Hall, P. F. (1966). *Endocrinology* **79,** 647.

Sandow, A. (1965). *Pharmacol. Rev.* **17,** 265.

Sanwal, B. D., and Smando, R. (1969). *Biochem. Biophys. Res. Commun.* **35,** 486.

Sarkar, N. K. (1969). *Life Sci.* **8** (II), 435.

Sasaki, T., Litwack, G., and Baserga, R. (1969). *J. Biol. Chem.* **244,** 4831.

Satchell, D. G., Freeman, S. E., and Edwards, S. V. (1968). *Biochem. Pharmacol.* **17,** 45.

Satoh, P., Constantopoulos, G., and Tchen, T. T. (1966). *Biochemistry* **5,** 1646.

Sattin, A., and Rall, T. W. (1967). *Fed. Proc.* **26,** 707.

Sattin, A., and Rall, T. W. (1970). *Mol. Pharmacol.* **6,** 13.

Sayoc, E. F., and Little, J. M. (1967). *J. Pharmacol. Exp. Ther.* **155,** 352.

Schaeffer, L. D., Chenoweth, M., and Dunn, A. (1969). *Biochim. Biophys. Acta* **192,**292.

Schafer, D. E., Lust, W. D., Sircar, B., and Goldberg, N. D. (1970). *Proc. Nat. Acad. Sci. U. S.* **67,**851.

Schild, H. O. (1967). *Brit. J. Pharmacol.* **31,** 578.

Schimmer, B. P., Ueda, K., and Sato, G. H. (1968). *Biochem. Biophys. Res. Commun.* **32,** 806.

Schlender, K. K., Wei, S. H., and Villar-Palasi, C. (1969). *Biochim. Biophys. Acta* **191,**272.

Schmidt, M. J., Palmer, E. C., Dettbarn, W. D., and Robison, G. A. (1970). *Develop. Psychobiol.* **3,**53.

Schneider, P. B. (1969). *J. Biol. Chem.* **244,** 4490.

Schneyer, C. A. (1969). *Proc. Soc. Exp. Biol. Med.* **131,** 71.

Schofield, J. G. (1967). *Nature (London)* **215,** 1382.

Schotz, M. C., and Page, I. H. (1960). *J. Lipid Res.* **1,** 466.

Schramm, M., and Naim, E. (1970). *J. Biol. Chem.* **245,** 3225.

Schueler, F. W. (1960). "Chemobiodynamics and Drug Design." 638 pp. Blakiston, New York.

Schulster, D., Tait, A. S., Tait, J. F., and Mrotek, J. (1970). *Endocrinology* **86,** 487.

Schultz, G., Senft, G., Losert, W., and Sitt, R. (1966). *Arch. Exp. Pathol. Pharmakol.* **253,** 372

Schultz, G., Bohme, E., and Munske, K. (1969). *Life Sci.* **8 (II),** 1323.

Scott, J. B., Daugherty, R. M., Dabney, J. M., and Haddy, F. J. (1965). *Amer. J. Physiol.* **208,** 813.

Scott, N. S., and Falconer, I. R. (1965). *Anal. Biochem.* **13,** 71.

Scott, R. E. (1970). *Blood* **35,** 514.

Scratcherd, T., and Case, R. M. (1969). *Gut* **10,** 957.

Scriabine, A., and Bellet, S. (1967). *Experientia* **23,** 648.

Seery, V. L., Fischer, E. H., and Teller, D. C. (1967). *Biochemistry* **6,** 3315.

Selye, H., Veilleux, and Cantin, M. (1961). *Science* **133,** 44.

Senft, G. (1968). *Arch. Pharmakol. Exp. Pathol.* **259,**117.

Senft, G., Losert, W., Schultz, G., Sitt, R., and Bartelheimer, H. K. (1966). *Arch. Pharmakol. Exp. Pathol.* **255,** 369.

Senft, G., Munske, K., Schultz, G., and Hoffmann, M. (1968a). *Arch. Pharmakol. Exp. Pathol.* **259,** 344.

Senft, G., Hoffmann, M., Munske, K., and Schultz, G. (1968b). *Pflugers Arch.* **298,**348.

Senft, G., Sitt, R., Losert, W., Schultz, G., and Hoffmann, M. (1968c). *Arch. Pharmakol. Exp. Pathol.* **260,** 309.

Senft, G., Schultz, G., Munske, K., and Hoffmann, M. (1968d). *Diabetologia* **4,** 322.

Seraydarian, K., and Mommaerts, W. F. H. M. (1965). *J. Cell Biol.* **26,** 641.

Shafrir, E., Sussman, K. E., and Steinberg, D. (1959). *J. Lipid Res.* **1,** 109.

Shanfeld, J., Frazer, A., and Hess, M. E. (1969). *J. Pharmacol. Exp. Ther.* **169,** 315.

Shanta, T. R., Woods, W. D., Waitzman, M. B., and Bourne, G. H. (1966). *Histochemie* **7,** 177.

Shapiro, J., Machattie, L., Eron, L., Ihler, G., Ippen, K., and Beckwith, J. (1969). *Nature (London)* **224,** 768.

Sharma, G. P., and Eiseman, B. (1966). *Surgery* **59,** 66.

Sharma, S. K., and Talwar, G. P. (1970). *J. Biol. Chem.* **245**, 1513.

Sharp, G. W. G., and Leaf, A. (1968). *J. Gen. Physiol.* **51**, 271s.

Sharp, G. W. G., and Hynie, S. (1971). *Nature (London)* **229**, 266.

Shaw, J. E., and Ramwell, P. W. (1968). *J. Biol. Chem.* **243**, 1498.

Shein, H. M., and Wurtman, R. J. (1969). *Science* **166**, 519.

Shen, L. C., Fall, L., Walton, G. M., and Atkinson, D. E. (1968). *Biochemistry* **7**, 4041.

Sheppard, H., and Burghardt, C. (1969). *Biochem. Pharmacol.* **18**, 2576.

Sheppard, H., and Burghardt, C. (1970). *Mol. Pharmacol.* **6**, 425.

Shima, K., and Foa, P. P. (1968). *Clin. Chim. Acta* **22**, 511.

Shimazu, T. (1967). *Science* **156**, 1256.

Shimazu, T., and Amakawa, A. (1968a). *Biochim. Biophys. Acta* **165**, 335.

Shimazu, T., and Amakawa, A. (1968b). *Biochim. Biophys. Acta* **165**, 349.

Shimazu, T., and Fukuda, A. (1965). *Science* **150**, 1607.

Shimizu, H., Daly, J. W., and Creveling, C. R. (1969). *J. Neurochem.* **16**, 1609.

Shimizu, H., Daly, J. W., and Creveling, C. R. (1970a). *J. Neurochem.* **17**, 441.

Shimizu, H., Daly, J. W., and Creveling, C. R. (1970b). *Mol. Pharmacol.* **6**, 184.

Shimizu, H., Creveling, C. R., and Daly, J. (1970c). *Proc. Nat. Acad. Sci. U. S.* **65**, 1033.

Shinebourne, E. A., and White, R. J. (1970). *Cardiovas. Res.* **4**, 194.

Shinebourne, E. A., Hess, M. L., White, R. J., and Hamer, J. (1969). *Cardiovas. Res.* **3**, 113.

Shishiba, Y., Solomon, D. H., and Davidson, W. D. (1970). *Endocrinology* **86**, 183.

Shoemaker, W. C., Van Itallie, T. B., and Walker, W. F. (1959). *Amer. J. Physiol.* **196**, 315.

Shrago, E., Lardy, H. A., Nordlie, R. C., and Foster, D. O. (1963). *J. Biol. Chem.* **238**, 3188.

Siggins, G. R., Hoffer, B. J., and Bloom, F. E. (1969). *Science* **165**, 1018.

Simons, J. A. (1970). *Science* **167**, 1378.

Simpson, L. L. (1968). *J. Pharm. Pharmacol.* **20**, 889.

Simpson, E. R., and Estabrook, R. W. (1968). *Arch. Biochem. Biophys.* **126**, 977.

Simpson, F. O., and Oertelis, S. J. (1962). *J. Cell Biol.* **12**, 91.

Singhal, R. L., and Lafreniere, R. (1970). *Endocrinology* **87**, 1099.

Singhal, R. L., Vijayvargiya, R., and Ling, G. M. (1970). *Science* **168**, 261.

Sinsheimer, R. L. (1969). *Amer. Sci.* **57**, 134.

Siperstein, M. D., Unger, R. H., and Madison, L. L. (1970). *In* "Early Diabetes," (R. A. Camerini-Davalos and H. S. Cole, eds.), pp. 261–271. Academic Press, New York.

Skelton, C. L., Levey, G. S., and Epstein, S. E. (1970). *Circ. Res.* **26**, 35.

Skosey, J. L. (1970). *J. Biol. Chem.* **245**, 510.

Smith, M., Drummond, G. I., and Khorana, H. G. (1961). *J. Amer. Chem. Soc.* **83**, 698.

Smith, S. S., Barboriak, J. S., and Hardman, H. F. (1967). *J. Pharmacol. Exp. Ther.* **155**, 397.

Smith, J. W., Steiner, A., Newberry, W. M., and Parker, C. W. (1969). *Fed. Proc.* **28**, 566.

Smith, J. W., Steiner, A. L., Newberry, W. M., and Parker, C. W. (1971a). *J. Clin. Invest.* **50**, 432.

Smith, J. W., Steiner, A. L., Newberry, W. M., and Parker, C. W. (1971b). *J. Clin. Invest.* **50**, 442.

Sneyd, J. G. T., Corbin, J. D., and Park, C. R. (1968). *In* "Pharmacology of Hormonal Polypeptides and Proteins," (N. Bach, L. Martini, and R. Paoletti, eds.), pp. 367–376. Plenum, New York.

Sobel, B. E., and Robison, A. (1969). *Circulation* **40**, III-189.

Sobel, B. E., Dempsey, P. J., and Cooper, T. (1968). *Biochem. Biophys. Res. Commun.* **33**, 758.

Sobel, B. E., Henry, P. D., Robison, A., Bloor, C., and Ross, J. (1969a). *Circ. Res.* **24**, 507.

Sobel, B. E., Dempsey, P. J., and Cooper, T. (1969b). *Proc. Soc. Exp. Biol. Med.* **132**, 6.

Soderling, T. R., and Hickenbottom, J. P. (1970). *Fed. Proc.* **29**, 601 Abs.

Sokal, J. E. (1966). *Amer. J. Med.* **41**, 331.

Sokal, J. E. (1968). *Diabetes* **17**, 256.

Sokal, J. E., and Ezdinli, E. Z. (1967). *J. Clin. Invest.* **46**, 778.

Sokal, J. E., Sarcione, E. J., and Henderson, A. M. (1964). *Endocrinology* **74**, 930.

Sokal, J. E., Aydin, A., and Kraus, G. (1966). *Amer. J. Physiol.* **211**, 1334.

Sollman, T., and Pilcher, J. D. (1911). *J. Pharmacol. Exp. Ther.* **3**, 63.

Solomon, S. S., Brush, J. S., and Kitabchi, A. E. (1970). *Science* **169**, 387.

Solyom, A., Puglisi, L., and Muhlbachova, E. (1967). *Biochem. Pharmacol.* **16**, 521.

Somlyo, A. V., and Somlyo, A. P. (1966). *Amer. Heart J.* **71**, 568.

Somlyo, A. V., and Somlyo, A. P. (1968). *J. Pharmacol. Exp. Ther.* **159**, 129.

Somlyo, A. V., Haeusler, G., and Somlyo, A. P. (1970). *Science* **169**, 490.

Sonnenblick, E. H. (1962). *Fed. Proc.* **21**, 975.

Søvik, O., Øye, I., and Rosell-Perez, M. (1966). *Biochim. Biophys. Acta* **124**, 26.

Spiegel, A. M., and Bitensky, M. W. (1969). *Endocrinology* **85**, 638.

Staehelin, M., Barth, P., and Desaulles, P. A. (1965). *Acta Endocrinol.* **50**, 55.

Stam, A. C., and Honig, C. R. (1965). *Amer. J. Physiol.* **209**, 8.

Stanton, H. C., Brenner, G., and Mayfield, E. D. (1969). *Amer. Heart J.* **77**, 72.

Starcich, R., Barbaresi, F., Volta, G., and Manfredi, M. (1967). *Minerva Cardioangiol.* **15**, 844.

Steele, R. (1966). *Ergeb. Physiol. Biol. Chem. Exp. Pharmacol.* **57**, 91.

Steinberg, D. (1966). *Pharmacol. Rev.* **18**, 217.

Steinberg, D., and Vaughan, M. (1967). *In* "Prostaglandins," (S. Bergstrom and B. Samuelsson, eds.), pp. 109–121. Almquist & Wiksell, Stockholm.

Steinberg, D., Vaughan, M., Nestel, P. J., Strand, O., and Bergstrom, S. (1964). *J. Clin. Invest.* **43**, 1533.

Steiner, A. L., Kipnis, D. M., Utiger, R., and Parker, C. (1969). *Proc. Nat. Acad. Sci. U. S.* **64**, 367.

Steiner, A. L., Parker, C. W., and Kipnis, D. M. (1970a). *J. Clin. Invest.* **49**, 43a.

Steiner, A. L., Peake, G. T., Utiger, R. D., Karl, I. E., and Kipnis, D. M. (1970b). *Endocrinology* **86**, 1354.

Steiner, C., Wit, A. L., and Damato, A. N. (1969). *Circ. Res.* **24**, 167.

Steiner, D. F. (1966). *Vitam. Hormo. (N. Y.)* **24**, 1.

Steiner, D. F., Younger, L., and King, J. (1965). *Biochemistry* **4**, 740.

Steiner, D. F., Clark, J., Nolan, C., Rubenstein, A. H., Margoliash, E., Aten, B., and Oyer, P. (1969). *Recent Progr. Hormone Res.* **25**, 207.

Steinmetz, J., Lowry, L., and Yen, T. T. T. (1969). *Diabetologia* **5**, 373.

Steitz, T., and Lipscomb, W. (1965). *J. Amer. Chem. Soc.* **87**, 2488.

Stetten, D., and Stetten, M. R. (1960). *Physiol. Rev.* **40**, 505.

Stock, K., and Westermann, E. (1966). *Life Sci.* **5**, 1667.

Stock, K., Aulich, A., and Westermann, E. (1968). *Life Sci.* **7**, 113.

Stone, D., and Hechter, O. (1954). *Arch. Biochem. Biophys.* **51**, 457.

Stone, D. B., and Mansour, T. E. (1967a). *Mol. Pharmacol.* **3**, 161.

Stone, D. B., and Mansour, T. E. (1967b). *Mol. Pharmacol.* **3**, 177.

Strauch, B. S., and Langdon, R. G. (1964). *Biochem. Biophys. Res. Commun.* **16**, 27.

Strauch, B. S., and Langdon, R. G. (1969). *Arch. Biochem. Biophys.* **129**, 277.

Strauss, D., and Fresco, J. (1965). *J. Amer. Chem. Soc.* **87**, 1374.

Streeto, J. M. (1969). *Metabolism* **18**, 968.

Streeto, J. M., and Reddy, W. J. (1967). *Anal. Biochem.* **21**, 416.

Strubelt, O. (1968). *Biochem. Pharmacol.* **17**, 156.

Strubelt, O. (1969). *Arch. Int. Pharmacodyn. Ther.* **179**, 215.

Studzinski, G. P., and Grant, J. K. (1962). *Nature (London)* **193**, 1075.

Sussman, K. E., and Vaughan, G. D. (1967). *Diabetes* **16**, 449.

Sutherland, E. W. (1949). *J. Biol. Chem.* **180**, 1279.

Sutherland, E. W. (1950). *Recent Progr. Hormone Res.* **5**, 441.

Sutherland, E. W. (1951a). *Ann. N. Y. Acad. Sci.* **54**, 693.

Sutherland, E. W. (1951b). *In* "Phosphorus Metabolism," (W. D. McElroy and B. Glass, eds.), Vol. I, pp. 53–66. Johns Hopkins Press, Baltimore, Maryland.

Sutherland, E. W. (1952). *In* "Phosphorus Metabolism," (W. D. McElroy and B. Glass, eds.), Vol. II, pp. 577–596. Johns Hopkins Press, Baltimore, Maryland.

Sutherland, E. W. (1956). *Proc. 3rd Int. Congr. Biochem,* pp. 318–327. Academic Press, New York.

Sutherland, E. W. (1962). *Harvey Lect. Ser.* **57**, 17–33.

Sutherland, E. W. (1965). *In* "Pharmacology of Cholinergic and Adrenergic Transmission," (G. B. Koelle, W. W. Douglas, and A. Carlsson, eds.), pp. 317–318. Macmillan (Pergamon), New York.

Sutherland, E. W., and Cori, C. F. (1951). *J. Biol. Chem.* **188**, 531.

Sutherland, E. W., and de Duve, C. (1948). *J. Biol. Chem.* **175**, 663.

Sutherland, E. W., and Rall, T. W. (1957). *J. Amer. Chem. Soc.* **79**, 3608.

Sutherland, E. W., and Rall, T. W. (1958). *J. Biol. Chem.* **232**, 1077.

Sutherland, E. W., and Rall, T. W. (1960). *Pharmacol. Rev.* **12**, 265.

Sutherland, E. W., and Robison, G. A. (1966). *Pharmacol. Rev.* **18**, 145.

Sutherland, E. W., and Wosilait, W. D. (1955). *Nature (London)* **175**, 169.

Sutherland, E. W., and Wosilait, W. D. (1956). *J. Biol. Chem.* **218**, 459.

Sutherland, E. W., Rall, T. W., and Menon, T. (1962). *J. Biol. Chem.* **237**, 1220.

Sutherland, E. W., Øye, I., and Butcher, R. W. (1965). *Recent Progr. Hormone Res.* **21**, 623.

Svedmyr, N. (1965). *Acta Pharmacol. Toxicol.* **23**, 103.

Svedmyr, N. (1966). *Acta Physiol. Scand.* **66**, 257.

Swislocki, N. I. (1970a). *Biochim. Biophys. Acta* **201**, 242.

Swislocki, N. I. (1970b). *Analyt. Biochem.* **38**, 260.

Symons, R. H. (1970). *Biochem. Biophys. Res. Commun.* **38**, 807.

Szego, C. M. (1965). *Fed. Proc.* **24**, 1343.

Szego, C. M., and Davis, J. S. (1967). *Proc. Nat. Acad. Sci. U. S.* **58**, 1711.

Szego, C. M., and Davis, J. S. (1969a). *Mol. Pharmacol.* **5**, 470.

Szego, C. M., and Davis, J. S. (1969b). *Life Sci.* **8** (I), 1109.

Szentivanyi, A. (1968). *J. Allergy* **42**, 203.

Tabachnick, I. I. A., and Gulbenkian, A. (1968). *Ann. N. Y. Acad. Sci.* **150**, 204.

Tabachnick, I. I. A., Gulbenkian, A., and Seidman, F. (1965). *J. Pharmacol. Exp. Ther.* **150**, 455.

Tagawa, H., and Vander, A. J. (1970). *Circ. Res.* **26**, 327.

Takahashi, K., Kamimura, M., Shinko, T., and Tsuji, S. (1966). *Lancet* **ii**, 967.

Talmage, R. V. (1967). *Clin. Orthop.* **54**, 163.

Tao, M., and Lipmann, F. (1969). *Proc. Nat. Acad. Sci. U. S.* **63**, 86.

Tao, M., and Schweiger, M. (1970). *J. Bacteriol.* **102**, 138.

Tao, M., Salas, M. L., and Lipmann, F. (1970). *Proc. Nat. Acad. Sci. U.S.* **67**, 408.

Tarui, S., Nonaka, K., Ikura, Y., and Shima, K. (1963). *Biochem. Biophys. Res. Commun.* **13**, 329.

Tata, J. R. (1967). *Nature (London)* **213**, 566.

Taunton, O. D., Roth, J., and Pastan, I. (1969). *J. Biol. Chem.* **244**, 247.

Taussky, H. H., Washington, A., Zubillaga, E., and Milhorat, A. T. (1962). *Nature (London)* **126,** 1100.

Taylor, A. L., Davis, B. B., Pawlson, G., Josimovich, J. B., and Mintz, D. H. (1970). *J. Clin. Endocrinol.* **30,** 316.

Temin, H. M. (1970). *Perspect. Biol. Med.* **14,** 11.

Therriault, D. G., and Winters, V. G. (1970). *Life Sci.* **9** (I), 421.

Therriault, D. G., Morningstar, J. F., and Winters, V. G. (1969). *Life Sci.* **8** (II), 1353.

Thomas, H. J., and Montgomery, J. A. (1968). *J. Med. Chem.* **11,** 44.

Thomas, D. P., Gurewich, V., and Stuart, R. K. (1968). *J. Lab. Clin. Med.* **71,** 955.

Thompson, J. C. (1969). *Ann. Rev. Med.* **20,** 291.

Tomasi, V., Koretz, S., Ray, T. K., Dunnick, J., and Marinetti, G. V. (1970). *Biochim. Biophys. Acta* **211,** 31.

Tomasz, A. (1965). *Nature (London)* **208,** 155.

Tonoue, T., Tong, W., and Stolc, V. (1970). *Endocrinology* **86,** 271.

Torres, H. N., Marechal, L. R., Bernard, E., and Belocopitow, E. (1968). *Biochim. Biophys. Acta* **156,** 206.

Travers, A. A., Kamen, R. I., and Schleif, R. F. (1970). *Nature (London)* **228,** 748.

Triner, L., and Nahas, G. G. (1966). *J. Pharmacol. Exp. Ther.* **153,** 569.

Triner, L., Overweg, N. I. A., and Nahas, G. G. (1970). *Nature (London)* **225,** 282.

Tsai, S., Belfrage, P., and Vaughan, M. (1970). *J. Lipid Res.* **11,** 466.

Tsang, C. P. W., and Carballeira, A. (1966). *Proc. Soc. Exp. Biol. Med.* **122,** 1031.

Tsujimoto, A., Tanino, S., Nishiue, T., and Kurogochi, Y. (1965). *Jap. J. Pharmacol.* **15,** 441.

Turtle, J. R., and Kipnis, D. M. (1967a). *Biochemistry* **6,** 3970.

Turtle, J. R., and Kipnis, D. M. (1967b). *Biochem. Biophys. Res. Commun.* **28,** 797.

Turtle, J. R., Littleton, G. K., and Kipnis, D. M. (1967). *Nature (London)* **213,** 727.

Tuttle, R. R. (1970). *Fed. Proc.* **29,** 611 Abs.

Ueda, I., Fukushima, K., and Loehning, R. W. (1965). *Arch. Int. Pharmacodyn. Ther.* **158,** 234.

Ui, M. (1965). *Amer. J. Physiol.* **209,** 359.

Ullmann, A., and Monod, J. (1968). *FEBS Lett.* **2,** 57.

Unger, R. H. (1966). *Diabetes* **15,** 500.

Unger, R. H., Ketterer, Dupre, H. J., and Eisentraut, A. M. (1967). *J. Clin. Invest.* **46,** 630.

Unger, R. H., Ohneda, A., Valverde, I., Eisentraut, A. M., and Exton, J. H. (1968). *J. Clin. Invest.* **47,** 48.

Unger, R. H., Aguilar-Parada, E., Müller, W. A., and Eisentraut, A. M. (1970). *J. Clin. Invest.* **49,** 837.

Usher, D., Dennis, E., and Westheimer, F. (1965). *J. Amer. Chem. Soc.* **87,** 2320.

Vaes, G. (1968a). *Nature (London)* **219,** 939.

Vaes, G. (1968b). *J. Cell Biol.* **39,** 676.

Valverde, I., Rigopoulou, D., Exton, J. H., Ohneda, A., Eisentraut, A., and Unger, R. H. (1968). *Amer. J. Med. Sci.* **255,** 415.

Vance, V. K., Girard, F., and Cahill, G. F., Jr. (1962). *Endocrinology* **71,** 113.

Vance, J. E., Buchanan, K. D., Challoner, D. R., and Williams, R. H. (1968). *Diabetes* **17,** 187.

Van de Veerdonk, F. C. G. (1969). *Gen. Comp. Endocrinol.* **12,** 658.

Van de Veerdonk, F. C. G., and Konijn, T. M. (1970). *Acta Endocrinol.* **64,** 364.

Van Rossum, J. M. (1965). *J. Pharm. Pharmacol.* **17,** 202.

Varmus, H. E., Perlman, R. L., and Pastan, I. (1970). *J. Biol. Chem.* **245,** 2259.

Vaughan, M. (1960). *J. Biol. Chem.* **235,** 3049.

Vaughan, M. (1964). *Amer. J. Physiol.* **207,** 1166.

Vaughan, M. (1967). *J. Clin. Invest.* **46,** 1482.

Vaughan, M., and Murad, F. (1969). *Biochemistry* **8,** 3092.

Vaughan, M., and Steinberg, D. (1963). *J. Lipid Res.* **4,** 193.

Vaughan, M., and Steinberg, D. (1965). *In* "Handbook of Physiology, Section 5: Adipose Tissue" (A. S. Renold and G. F. Cahill, Jr., eds.), pp. 239–252. Amer. Physiol. Soc., Washington, D. C.

Vecchio, D., Luyckx, A., Zahnd, G. R., and Renold, A. E. (1966). *Metabolism* **15,** 577.

Venner, H. (1966). *Hoppe-Seyler's Z. Physiol. Chem.* **344,** 189.

Vernikos-Danellis, J., and Harris, C. G. (1968). *Proc. Soc. Exp. Biol. Med.* **128,** 1016.

Vigdahl, R. L., Marquis, N. R., and Tavormina, P. A. (1969). *Biochem. Biophys. Res. Commun.* **37,** 409.

Villar-Palasi, C., and Wenger, J. I. (1967). *Fed. Proc.* **26,** 563.

Villar-Trevino, S., Shull, K. H., and Farber, E. (1963). *J. Biol. Chem.* **238,** 1757.

Villee, C. A., Gordon, E. E., Loring, J. M., and Wellington, F. M. (1955). *Fed. Proc.* **14,** 297.

Vincent, N. H., and Ellis, S. (1963). *J. Pharmacol. Exp. Ther.* **139,** 60.

Vinuela, E., Salas, M. L., Salas, M., and Sols, A. (1964). *Biochem. Biophys. Res. Commun.* **15,** 243.

Von Bruchhausen, F. (1968). *Hoppe-Seyler's Z. Physiol. Chem.* **349,** 1437.

Von Heimburg, R. L., and Hallenbeck, G. A. (1964). *Gastroenterology* **47,** 531.

Voors, A. W. (1969). *Lancet* **ii,** 1337.

Vulliemoz, Y., Triner, L., and Nahas, G. (1968). *Life Sci.* **7,** 1063.

Walaas, E., and Walaas, O. (1970). *Acta Physiol. Scand.* **78,** 393.

Walaas, O., Walaas, E., and Osaki, S. (1968). *In* "Control of Glycogen Metabolism" (W. J. Whelan, ed.), pp. 139–152. Academic Press, New York.

Walaas, O., Walaas, E., and Wick, A. N. (1969). *Diabetologia* **5,** 79.

Waldstein, S. S. (1966). *Ann. Rev. Med.* **17,** 123.

Walsh, D. A., Perkins, J. P., and Krebs, E. G. (1968). *J. Biol. Chem.* **243,** 3763.

Walton, G. M., and Garren, L. D. (1970). *Biochemistry* **9,** 4223.

Watenpaugh, K., Dow, J., Jensen, L. H., and Furberg, S. (1968). *Science* **159,** 206.

Watlington, C. O. (1968a). *Amer. J. Physiol.* **214,** 1001.

Watlington, C. O. (1968b). *Biochem. Physiol.* **24,** 965.

Watlington, C. O. (1969). *Biochim. Biophys. Acta* **193,** 394.

Way, L., and Durbin, R. P. (1969). *Nature (London)* **221,** 874.

Weber, A. (1968). *J. Gen. Physiol.* **52,** 760.

Weeks, J. R., Sekhar, N. C., and Ducharme, D. W. (1969). *J. Pharm. Pharmacol.* **21,** 103.

Weintraub, B., Sarcione, E. J., and Sokal, J. E. (1969). *Amer. J. Physiol.* **216,** 521.

Weis, L. S., and Narahara, H. T. (1969). *J. Biol. Chem.* **244,** 3084.

Weiss, B. (1969a). *J. Pharmacol. Exp. Ther.* **166,** 330.

Weiss, B. (1969b). *J. Pharmacol. Exp. Ther.* **168,** 146.

Weiss, B., and Costa, E. (1967). *Science* **156,** 1750.

Weiss, B., and Costa, E. (1968a). *J. Pharmacol. Exp. Ther.* **161,** 310.

Weiss, B., and Costa, E. (1968b). *Biochem. Pharmacol.* **17,** 2107.

Weiss, B., and Crayton, J. (1970). *Endocrinology* **87,** 527.

Weiss, B., and Maickel, R. P. (1968). *Int. J. Neuropharmacol.* **7,** 393.

Weiss, B., Davies, J. I., and Brodie, B. B. (1966). *Biochem. Pharmacol.* **15,** 1553.

Wekstein, D. R., and Zolman, J. F. (1969). *Fed. Proc.* **28,** 1023.

Weller, M., and Rodnight, R. (1970). *Nature (London)* **225,** 187.

Wells, H. (1967). *Amer. J. Physiol.* **212,** 1293.

Wells, H., and Lloyd, W. (1967).*Endocrinology* **81**, 139.

Wells, H., and Lloyd, W. (1968a). *Endocrinology* **82**, 468.

Wells, H., and Lloyd, W. (1968b).*Endocrinology* **83**, 521.

Wells, H., and Lloyd, W. (1969). *Endocrinology* **84**, 861.

White, A. A., and Aurbach, G. D. (1969). *Biochim. Biophys. Acta* **191**, 686.

White, J. E., and Engel, F. L. (1958). *J. Clin. Invest.* **37**, 1556.

Whitehouse, F. W., and James, T. N. (1966). *Proc. Soc. Exp. Biol. Med.* **122**, 823.

Whitfield, J. F., MacManus, J. P., and Rixon, R. H. (1970a). *J. Cell. Physiol.* **75**, 213.

Whitfield, J. F., MacManus, J. P., and Rixon, R. H. (1970b). *Proc. Soc. Exp. Biol. Med.* **134**, 1170.

Wicks, W. D. (1968).*Science* **160**, 997.

Wicks, W. D. (1969). *J. Biol. Chem.* **244**, 3941.

Wicks, W. D., Kenney, F. T., and Lee, K. (1969). *J. Biol. Chem.* **244**, 6008.

Wieland, O., and Siess, E. (1970). *Proc. Nat. Acad. Sci. U. S.* **65**, 947.

Wilber, J. F., Peake, G. T., and Utiger, R. D. (1969). *Endocrinology* **84**, 758.

Wilkenfeld, B. E., and Levy, B. (1968). *Arch. Int. Pharmacodyn. Ther.* **176**, 218.

Wilkenfeld, B. E., and Levy, B. (1969). *J. Pharmacol. Exp. Ther.* **169**, 61.

Williams, B. J., and Mayer, S. E. (1966). *Mol. Pharmacol.* **2**, 454.

Williams, G. A., Bowser, E. N., and Henderson, W. J. (1969). *Endocrinology* **85**, 537.

Williams, H. E., and Field, J. B. (1961). *J. Clin. Invest.* **40**, 1841.

Williams, R. H., Walsh, S. A., Hepp, D. K., and Ensinck, J. W. (1968a). *Metabolism* **17**, 653.

Williams, R. H., Walsh, S. A., and Ensinck, J. W. (1968b). *Proc. Soc. Exp. Biol. Med.* **128**, 279.

Williams, R. H., Little, S. A., and Ensinck, J. W. (1969). *Amer. J. Med. Sci.* **258**, 190.

Williams, T. F., Exton, J. H., Park, C. R., and Regen, D. M. (1968). *Amer. J. Physiol.* **215**, 1200.

Williamson, J. R. (1965). *Nature (London)* **206**, 473.

Williamson, J. R. (1966). *Pharmacol. Rev.* **18**, 205.

Williamson, John R., Kreisberg, R. A., and Felts, P. W. (1966). *Proc. Nat. Acad. Sci. U. S.* **56**, 247.

Williamson, Joseph R., Garcia, A., Renold, A. E., and Cahill, G. F. (1966). *Diabetes* **15**, 183.

Williamson, J. R., Cheung, W. Y., Coles, H. S., and Herczeg, B. E. (1967). *J. Biol. Chem.* **242**, 5112.

Wing, D. R., and Robinson, D. S. (1968). *Biochem. J.* **109**, 841.

Wolfe, S. M., and Shulman, N. R. (1969). *Biochem. Biophys. Res. Commun.* **35**, 265.

Wolff, J. (1962). *J. Biol. Chem.* **237**, 236.

Wolff, J., and Jones, A. B. (1970). *Proc. Nat. Acad. Sci. U. S.* **65**, 454.

Wolff, J., and Varrone, S. (1969). *Endocrinology* **85**, 410.

Wolff, J., Berens, S. C., and Jones, A. B. (1970). *Biochem. Biophys. Res. Commun.* **39**, 77.

Wollenberger, A., and Krause, E. (1968). *Amer. J. Cardiol.* **22**, 349.

Wollenberger, A., Krause, E., and Heier, G. (1969). *Biochem. Biophys. Res. Commun.* **36**, 664.

Wong, K. K., Symchowicz, S., Staub, M. S., and Tabachnick, I. I. A. (1967). *Life Sci.* **6**, 2285.

Wool, I. G., Stirewalt, W. S., Kurilhara, K., Low, R. B., Bailey, P., and Oyer, D. (1968). *Recent Prog. Hormone Res.* **24**, 139.

Wosilait, W. D., and Sutherland, E. W. (1956).*J. Biol. Chem.* **218**, 469.

Wurtman, R. J., Axelrod, J., and Kelly, D. E. (1968). "The Pineal," 199 pp. Academic Press, New York.

Yago, N., Kobayashi, S., Morisaki, M., Ichii, S., and Matsuba, M. (1963). *Steroids* **2**, 175.

Yamamoto, I., Inoki, R., and Kojima, S. (1968). *Eur. J. Pharmacol.* **3**, 123.

Yamamoto, M., and Massey, K. L. (1969). *Comp. Biochem. Physiol.* **30**, 941.

Yen, T. T. T., Steinmetz, J., and Wolff, G. L. (1970). *Hormone Metab. Res.* **2**, 200.

Yeung, D., and Oliver, I. T. (1968a). *Biochem. J.* **108**, 325.

Yeung, D., and Oliver, I. T. (1968b). *Biochemistry* **7**, 3231.

Yip, A. T., and Larner, J. (1969). *Physiol. Chem. Phys.* **1**, 383.

Yokota, T., and Gots, J. S. (1970). *J. Bacteriol.* **103**, 513.

Yunis, A. A., and Arimura, G. K. (1966a). *Biochim. Biophys. Acta* **118**, 325.

Yunis, A. A., and Arimura, G. K. (1966b). *Biochim. Biophys. Acta* **118**, 335.

Yunis, A. A., and Krebs, E. G. (1962). *J. Biol. Chem.* **237**, 34.

Zierler, K. L. (1966). *Amer. J. Med.* **40**, 735.

Zollinger, R. M., and Ellison, E. H. (1955). *Ann. Surg.* **142**, 709.

Zollinger, R. M., Elliott, D. W., Endahl, G. E., Grant, G. N., Goswitz, J. W., and Taft, D. A. (1962). *Ann. Surg.* **156**, 570.

Zor, U., Lowe, I. P., Bloom, G., and Field, J. B. (1968). *Biochim. Biophys. Res. Commun.* **33**, 649.

Zor, U., Bloom, G., Lowe, I. P., and Field, J. B. (1969a). *Endocrinology* **84**, 1082.

Zor, U., Kaneko, T., Lowe, I. P., Bloom, G., and Field, J. B. (1969b). *J. Biol. Chem.* **244**, 5189.

Zor, U., Kaneko, T., Schneider, H. P. G., McCann, S. M., Lowe, I. P., Bloom, G., Borland, B., and Field, J. B. (1969c). *Proc. Nat. Acad. Sci. U. S.* **63**, 918.

Zor, U., Kaneko, T., Schneider, H. P. G., McCann, S. M., and Field, J. B. (1970). *J. Biol. Chem.* **245**, 2883.

Zubay, G., Schwartz, D., and Beckwith, J. (1970). *Proc. Nat. Acad. Sci. U. S.* **66**, 104.

SUBJECT INDEX

A

Acetate, 255, 274
  incorporation into fatty acids, 183
Acetylcholine
  effects of, 382–384
    cardiac, 197, 267, 383
    on cyclic AMP (lowering), 383
    on cyclic GMP
      in heart, 383, 420
    inotropic, 197
N-Acetyltransferase, 230
Acidosis, 222, 282
Acrasin, 441
ACTH, see adrenocorticotropic hormone
Acyl derivatives of cyclic AMP, 101, see
    also Derivatives of cyclic AMP
Adenosine, 83, 214, 231
  effects of, 102–104
    on brain cyclic AMP, 212, 379
    cardiovascular, 201
    on cyclic AMP accumulation, 104
    pharmacological, 98–104
  inhibition of lymphocyte transformation
    by, 452
Adenosine triphosphate (ATP)
  importance of, 10, 18, 454
Adenyl cyclase, 73–84
  activity in tumor cells, 449
  in adipose tissue, 221, 287, 288, 293–
    294, 313
  adrenal, see Adrenal adenyl cyclase
  in adrenocortical carcinoma, 450
  assay of, 82
    contamination by enzymes, 83
    importance of phosphodiesterase in-
      hibitors for, 83
  in B. liquefaciens, 79, 440
  in bone, 367–368
  in brain, see Brain, adenyl cyclase in
  cellular distribution of, 24, 25, 75

density gradient centrifugation of, 76
deficiency of, in mutants, 431–432
discovery of, 11–13
distribution
  in cardiac muscle, 168
  inside membranes, 77–78
effects of catecholamines on, 152
of cations on, 81, 357
of cholera toxin on, 397
of clostridial extracts on, 397
of enzymes on, 78
of fluoride on, 80, 397
of guanine nucleotides in activation of,
  90
in E. coli, 426, 440
in erythrocytes, 217
fractionation from, 75
in fat, 288
in heart, see Cardiac adenyl cyclase
in hepatoma cells, 449
histochemical method for, 239
hormones and
  development of hormone sensitivity to,
    81, 211, 217, 258
  fast-acting lipolytic, 293–294, 313
  relation to hormone receptors, 22, 47
K_m for ATP, 90
in kidney, 365
lability of, 37
  thermal, 79
in leukocytes, 216
in liver, see Hepatic adenyl cyclase
in liver fluke, 381
in lymphocytes, 452–453
in pineal gland, 191, 394
in platelets, 216, 386
purification of, 79
in rat liver, 229, 449
reaction, reversal of, 70, 82
solubilization of, 90, 337
species distribution, 73–75